The Pathophysiology of Preeclampsia and Eclampsia

The Pathophysiology of Preeclampsia and Eclampsia

Editors

Junie P. Warrington
Ana T. Palei
Mark W. Cunningham
Lorena M. Amaral

MDPI • Basel • Beijing • Wuhan • Barcelona • Belgrade • Manchester • Tokyo • Cluj • Tianjin

Editors
Junie P. Warrington
Neurology
University of Mississippi
Medical Center
Jackson
United States

Ana T. Palei
Department of Surgery
University of Mississippi
Medical Center
Jackson
United States

Mark W. Cunningham
Physiology & Anatomy
University of North Texas
Health Science Center
Fort Worth
United States

Lorena M. Amaral
Pharmacology & Toxicology
University of Mississippi
Medical Center
Jackson
United States

Editorial Office
MDPI
St. Alban-Anlage 66
4052 Basel, Switzerland

This is a reprint of articles from the Special Issue published online in the open access journal *Cells* (ISSN 2073-4409) (available at: www.mdpi.com/journal/cells/special_issues/Cells_Preeclampsia).

For citation purposes, cite each article independently as indicated on the article page online and as indicated below:

LastName, A.A.; LastName, B.B.; LastName, C.C. Article Title. *Journal Name* **Year**, *Volume Number*, Page Range.

ISBN 978-3-0365-6411-1 (Hbk)
ISBN 978-3-0365-6410-4 (PDF)

Cover image courtesy of Jonna Smith

Contents

About the Editors

Junie P. Warrington

Junie PAULA Warrington, PhD, FAHA, is an associate professor in the department of Neurology and associate director of the PhD program in neuroscience at the University of Mississippi Medical Center in Jackson, MS.

She received a PhD in neuroscience from the University of Oklahoma Health Sciences Center in 2011 and has published numerous peer-reviewed manuscripts on topics covering whole-brain irradiation-induced cognitive impairment, cerebrovascular physiology, preeclampsia, eclampsia, obesity, and hypoxia. Dr. Warrington's research has been continuously funded through agencies such as the American Heart Association and the National Institutes of Health.

Dr. Warrington is actively involved in several professional organizations and is active in the peer review process. She reviews manuscripts for several journals in the fields of neuroscience, physiology, hypertension, and pregnancy and serves as ad hoc reviewer for multiple grant agencies.

Using preclinical animal models, Dr. Warrington works on identifying pathophysiological mechanisms contributing to neurovascular abnormalities and seizures in hypertensive disorders of pregnancy during antepartum, peripartum, and postpartum periods. Her work employs knockout and reporter mouse models, pharmacological, molecular, and behavioral approaches to elucidate underlying mechanisms, with the goal of identifying therapeutic targets for prevention and treatment of pregnancy-associated neurological sequelae. Her ultimate mission is to change the landscape of academia by making it a more welcoming and supportive environment for future scientists.

Ana T. Palei

Ana C. Palei, PharmD, PhD is an Assistant Professor in the Department of Surgery at the University of Mississippi Medical Center (UMMC) in Jackson, MS, USA. Dr. Palei has authored over 70 peer-reviewed papers on diverse topics of cardiovascular research. She has received several awards including a National Institutes of Health (NIH)-K01 Award, an American Physiological Society (APS)-Research Career Enhancement Award, an American Heart Association (AHA)-Postdoctoral Fellowship, and the APS-Caroline tum Suden Professional Opportunity Award.

Dr. Palei has made research contributions in the fields of acute pulmonary embolism (APE) and preeclampsia (PE). Using a rodent model, she provided new insights into the pathophysiology and treatment of APE-induced cardiovascular derangements through inhibition of matrix metalloproteinases (MMPs). Her clinical studies demonstrated that polymorphisms in MMP genes may affect 1) circulating MMP levels in PE; 2) susceptibility for the development of PE; and 3) responsiveness to anti-hypertensive drugs commonly used to treat PE. Later, her studies in experimental animals showed that excessive metabolic disturbances are necessary to cause hypertension during pregnancy. Currently, Dr. Palei's research is focused on in vitro, animal, and clinical studies investigating mechanisms of and potential therapies for PE-induced hypercoagulability and heart dysfunction during gestation and postpartum. Her research has been funded by agencies from Brazil and USA.

Dr. Palei has been engaged in many outreach activities and served in multiple leadership capacities locally, nationally, and internationally. She currently serves as chair of the IASH Communications Committee and as Councilor-at-Large for Translational Research of the APS Water and Electrolytes Section. In addition, she has served as ad-hoc member on grant review panels for the NIH as well as peer-reviewer and guest editor for many international scientific journals.

Mark W. Cunningham

Dr. Mark W. Cunningham Jr., Ph.D., MBA, is an Assistant Professor in the Department of Physiology and Anatomy at the University of North Texas Health Science Center (UNTHSC). He is active in several professional societies (AHA and APS), serves as editor on numerous scientific journals, and has over 45 peer-review publications. He enjoys teaching, mentoring, learning, exploring, traveling, and community service. Dr. Cunningham is a native of the Atlanta, Georgia metropolitan area. He is a proud graduate of Morehouse College, where he received his bachelor's degree of biology in 2008. He attended the University of Florida for graduate school and obtained his doctoral degree in the medical sciences in 2014. From 2014 - 2021, he completed his Postdoctoral training and started his faculty positions as Instructor and Assistant Professor at the University of Mississippi Medical Center. Currently, he is an Assistant Professor at the UNTHSC with a research focus on maternal health and fetal programing.

Specifically, the Cunningham laboratory goal is to examine the mechanisms for dysfunction of blood vessels in the brain, cardiovascular diseases (CVD), and high blood pressure in females during and after pregnancy, and in their offspring. Serval studies, including studies by Dr. Cunningham, have shown that inflammatory molecules and mitochondrial dysfunction are upregulated in hypertensive pregnancies and may contribute to this pathophysiology. Thus, the mission of the Cunningham lab will be to determine if the blockade or interference of these factors associated with hypertensive pregnancies will improve maternal and fetal outcomes during pregnancy and later in life. The clinical implications of this work are to study sex differences in the offspring from hypertensive moms; and to importantly provide insights for therapies and approaches to improve the lives of both the mother and her fetus during pregnancy and beyond.

Lorena M. Amaral

Lorena M. Amaral, PharmD, PhD is an Assistant Professor in the Department of Pharmacology and Toxicology at the University of Mississippi Medical Center in Jackson, MS. Dr. Amaral is member of AHA, APS, and American Society for Pharmacology and Experimental Therapeutics, Society for Reproductive Investigation and Society for Maternal Fetal Medicine. She serves as ad-hoc member on grant review panels of the American Heart Association (AHA) and has served as peer-reviewer and guest editor for international scientific journal.

Dr. Amaral has authored over 57 peer-reviewed papers on diverse topics of hypertensive pregnancy disorder research. Her research has been funded by agencies from Brazil and United States, including AHA and National Institutes of Health (NIH). She has received several awards and over the past few years, she has mentored graduate students, MD fellows and served as PI or Co-investigator on a number of clinical and basic science research grants that have investigated preeclampsia and gestational hypertension.

Dr. Amaral's research is focused on translational studies investigating novel therapeutic targets for hypertensive disorders of pregnancy. Her laboratory focuses on examining the effects of progesterone and progesterone induced blocking factor-PIBF on the pathophysiology of preeclampsia. Her clinical data have shown that a large percentage of our patient population delivering at the UMMC are hypertensive and have significantly lower circulating progesterone compared to healthy pregnant women. She has shown that progesterone supplementation lowers proinflammatory cells and cytokines and stimulates placental and circulating CD4+ T helper 2 cells.

This increased TH2 population was associated with lower blood pressure and improved pup weight in the RUPP- rat mimic model of PE. Importantly, her recent data indicate that PIBF supplementation attenuates the hypertension and improves fetal IUGR in the RUPP model.

cells

MDPI

Review

The Pivotal Role of the Placenta in Normal and Pathological Pregnancies: A Focus on Preeclampsia, Fetal Growth Restriction, and Maternal Chronic Venous Disease

Miguel A. Ortega [1,2,3,*,†], Oscar Fraile-Martínez [1,2,†], Cielo García-Montero [1,2,†], Miguel A. Sáez [1,2,4], Miguel Angel Álvarez-Mon [1,2], Diego Torres-Carranza [1], Melchor Álvarez-Mon [1,2,5], Julia Bujan [1,2], Natalio García-Honduvilla [1,2], Coral Bravo [6,7,8,‡], Luis G. Guijarro [2,9,‡] and Juan A. De León-Luis [6,7,8,‡]

1 Department of Medicine and Medical Specialties, Faculty of Medicine and Health Sciences, University of Alcalá, 28801 Alcalá de Henares, Madrid, Spain; oscarfra.7@hotmail.com (O.F.-M.); cielo.gmontero@gmail.com (C.G.-M.); msaega1@oc.mde.es (M.A.S.); maalvarezdemon@icloud.com (M.A.Á.-M.); diegotc90@gmail.com (D.T.-C.); mademons@gmail.com (M.Á.-M.); mjulia.bujan@uah.es (J.B.); natalio.garcia@uah.es (N.G.-H.)
2 Ramón y Cajal Institute of Healthcare Research (IRYCIS), 28034 Madrid, Spain; luis.gonzalez@uah.es
3 Cancer Registry and Pathology Department, Hospital Universitario Principe de Asturias, 28801 Alcalá de Henares, Madrid, Spain
4 Pathological Anatomy Service, Central University Hospital of Defence-UAH, 28047 Madrid, Spain
5 Immune System Diseases-Rheumatology and Oncology Service, University Hospital Príncipe de Asturias, CIBEREHD, 28801 Alcalá de Henares, Madrid, Spain
6 Department of Public and Maternal and Child Health, School of Medicine, Complutense University of Madrid, 28040 Madrid, Spain; cbravoarribas@gmail.com (C.B.); jaleon@ucm.es (J.A.D.L.-L.)
7 Department of Obstetrics and Gynecology, University Hospital Gregorio Marañón, 28009 Madrid, Spain
8 Health Research Institute Gregorio Marañón, 28009 Madrid, Spain
9 Unit of Biochemistry and Molecular Biology (CIBEREHD), Department of System Biology, University of Alcalá, 28801 Alcalá de Henares, Madrid, Spain
* Correspondence: miguel.angel.ortega92@gmail.com; Tel.: +34-91-885-4540; Fax: +34-91-885-4885
† These authors contributed equally to this work.
‡ These authors shared senior authorship in this work.

Citation: Ortega, M.A.; Fraile-Martínez, O.; García-Montero, C.; Sáez, M.A.; Álvarez-Mon, M.A.; Torres-Carranza, D.; Álvarez-Mon, M.; Bujan, J.; García-Honduvilla, N.; Bravo, C.; et al. The Pivotal Role of the Placenta in Normal and Pathological Pregnancies: A Focus on Preeclampsia, Fetal Growth Restriction, and Maternal Chronic Venous Disease. *Cells* **2022**, *11*, 568. https://doi.org/10.3390/cells11030568

Academic Editors: Junie P Warrington, Ana T. Palei, Mark W. Cunningham and Lorena M. Amaral

Received: 21 December 2021
Accepted: 5 February 2022
Published: 6 February 2022

Abstract: The placenta is a central structure in pregnancy and has pleiotropic functions. This organ grows incredibly rapidly during this period, acting as a mastermind behind different fetal and maternal processes. The relevance of the placenta extends far beyond the pregnancy, being crucial for fetal programming before birth. Having integrative knowledge of this maternofetal structure helps significantly in understanding the development of pregnancy either in a proper or pathophysiological context. Thus, the aim of this review is to summarize the main features of the placenta, with a special focus on its early development, cytoarchitecture, immunology, and functions in non-pathological conditions. In contraposition, the role of the placenta is examined in preeclampsia, a worrisome hypertensive disorder of pregnancy, in order to describe the pathophysiological implications of the placenta in this disease. Likewise, dysfunction of the placenta in fetal growth restriction, a major consequence of preeclampsia, is also discussed, emphasizing the potential clinical strategies derived. Finally, the emerging role of the placenta in maternal chronic venous disease either as a causative agent or as a consequence of the disease is equally treated.

Keywords: placenta; preeclampsia; fetal growth restriction; maternal chronic venous disease (CVeD)

1. Introduction

The placenta is an intricate and vital organ during pregnancy, coordinating a wide variety of functions in this period. Synchronically with the fetus, the placenta is an organ that experiences incredible transformation and growth from its early development to the end of pregnancy [1]. The placenta could be considered the mastermind behind maternal physiology, orchestrating an entire organism to create a proper milieu for fetal development [2]. When this structure does not work properly, it could lead to the onset of different pregnancy complications, with serious consequences for maternofetal well-being [3]. Despite this organ being only with us during gestation, its relevance extends far beyond this period. Indeed, the fetus perceives the placenta as a reflection of the outer environment, and the different signals received through this organ may have important consequences in a newborn and even in adulthood [4]. Thus, a study of the placenta prominently progresses our understanding of human health and disease [5]. On the other hand, due to the transitory and dynamic nature of the placenta, little information could be obtained from the placenta in real time, hampering examinations of this organ. To shed light on this issue, some projects, such as The Human Placenta Project, aim to deeply explain the development, structure, and function of this complex organ [6]; however, the road to gaining further insights into the complex and critical role of the placenta during pregnancy is still very long.

In this context, the purpose of this review is to collect updated knowledge about the placenta in uncomplicated pregnancies, with the aim to understand the steps of placentation and placental growth, the cytoarchitecture of this structure, its immunology, its multiple activities, and the modulatory role of this organ in fetal development and maternal physiology, hence creating a global picture of this organ. In the second part, we focus on the role of the placenta in the pathophysiology of two hypertensive disorders: preeclampsia (PE) and chronic venous disease (CVeD). We also analyze the relevance of this organ with fetal growth restriction.

2. Placental Development, Cytoarchitecture, and Immunology

2.1. Early Development of the Placenta

The placenta is an organ that develops during pregnancy in a gradual and poorly understood process. Currently, after fecundation and successive cell divisions, the embryo undergoes complex interactions with a receptive uterus in the form of blastocysts. Consequently, if the implantation process is successful, the embryo attaches to the endometrium, invading the epithelium and maternal circulation, initiating the process of placentation [7]. During the implantation process, the blastocyst differentiates into an inner cell mass (embryo) and trophectoderm (placenta). The trophectoderm houses trophoblasts—the main drivers of the placentation process and different placental functions [8]. Simultaneously, the stromal cells in the maternal endometrium surrounding the implanting embryo develop a plethora of changes in a process designated as decidualization, which is an imperative prerequisite for implantation success [9]. Trophoblasts are semi-allogeneic cells, as they are derived from the embryo. In this sense, a possible rejection of these cells by the maternal immune system would likely be fatal. However, trophoblasts exert multiple immunoregulatory actions, leading to a maternofetal tolerance response, hence ensuring placentation and pregnancy success [10].

Just after implantation, a syncytial fusion of mononucleated trophoblasts forms the oligonucleated syncytiotrophoblasts (STBs). The remaining mononucleated trophoblasts are referred to as cytotrophoblasts (CTBs) [11]. The sequence of processes involved in the formation of the placenta is as follows: (1) In the prelacunar stage, the fusion of mononucleated cells leads to the formation of the first STBs or primary syncytium. This syncytium is the outer layer of the placenta and is in direct contact with the maternal blood, and as described later, these cells represent a major structural and functional unit of the placenta. (2) In the lacunar stage, fluid-filled spaces—named lacunae—appear within the central mass of the primary syncytium. STBs surrounding the lacunae are named trabecula. This phase occurs from day 8 to day 13 after conception. The system of trabecula and lacunae is coated with two layers free of lacunae: the basal layer, facing the endometrium, known as the cytotrophoblast shell, and a superficial layer in contact with the blastocyte, called the primary chorionic plate [12]. In these lacunae, some evidence of maternal circulation can be found, coming from the erosion of spiral arterioles and small veins from the endometrium after the invasion of this structure [13]. However, maternal circulation in the placenta is not well-established until the end of the first trimester, when hypoxia is essential for the growth and physiological development of the embryo, and the placenta is beneficial [14]. On day 12 after conception, a bilayer structure composed of the CTBs, and extra-embryonic mesodermal cells (somatopleure) is formed. This structure is the chorion, and it represents the fetal part of the placenta [15]. CTBs coming from the chorionic plate invade the syncytial mass of the trabecula, and on day 15, they reach the maternal side of the placenta, leading to their transformation in a special subtype of extravillous trophoblasts (EVTs), named endovascular trophoblasts (eEVTs) and interstitial trophoblasts (iEVTs). iEVTs remain in the endometrial decidua, whereas eEVTs start remodeling the spiral arteries, concluding with the replacement of the endothelium and the smooth muscle cells from the tunica media by trophoblasts. This leads to a set of changes in the properties of these vessels, including loss of elasticity or of vasomotor control [16]. (3) The villous stage starts between days 12 and 18 after conception. The trophoblastic trabecula starts to proliferate, forming protrusions into the maternal blood surrounding the trabecula (primary villi), which is composed of a CTB core with an outer layer of STBs [17]. Then, the extra-embryonic mesodermal cells of the chorionic plate invade the trabecula, although they stop in the distal part exclusively filled with CTBs. These are referred to as the trophoblastic cell column, being a source of EVTs [11]. In addition, the extra-embryonic mesodermal cells invade the primary villi, giving them a mesenchymal core and transforming them into secondary villi. On days 18–20, fetal capillaries appear in the core of the secondary villi, leading to the development of tertiary villi. During the first trimester of pregnancy, a system of villous trees is formed by further proliferation and branching, and the lacunae become the intervillous space. Then, the blueprint of the placenta is established [18].

2.2. Placental Anatomy and Cytoarchitecture

During the second and third trimesters of pregnancy, the placenta is an organ with rapid development and multiple changes occurring in this structure as well as in the fetus. During weeks 10–12 after fecundation, the average weight of the placenta is about 51 g, whereas the delivered or mature placenta is essentially a discoid organ with a weight of 500–600 g, a diameter of 22 cm, and a thickness of 2–4 cm. These values may vary under abnormal or pathological pregnancies [19,20]. The human placenta is composed of a fetal surface or chorionic plate, covered by the amnion, to which the umbilical cord attaches, and a maternal surface or decidual basal plate in contact with the endometrium. Between those plates is the intervillous space, in which a set of fetal villous trees (or chorionic villi) project. Overall, at least five types of villous trees have been described according to their developmental stage, structure, vessel-cell type components, vessel branches, and histologic features. Herein, the main types of villi and their features are summarized, although more detail can be found in specialized literature [21–24]. (1) Mesenchymal villi: initially, all tertiary villi are of this type. They are where the villi proliferate and perform virtually all of the endocrine activities of the placenta. At term, these villi represent less than 1% of the villous volume, as they differentiate into immature intermediate villi in the first and second trimesters and into stem villi in the third trimester. (2) Immature intermediate villi represent an advanced but immature continuation of mesenchymal villi. They may be considered the growth centers of villous trees, working as the main site of exchange during the first and second trimesters, where terminal villi have not yet differentiated. (3) Stem villi are characterized by a condensed fibrous stroma, large vessels, and microvessels and are responsible for supporting the structures of villous trees, having no impact on the endocrine activity and maternofetal exchange in the placenta. (4) Mature intermediate villi present a higher degree of fetal vascularization, important for maternofetal exchange as well as terminal villi formation. (5) Terminal villi are linked to stem villi by intermediate structures and present a high degree of capillarization and dilated sinusoids, making them a proper location for diffusive exchange. Indeed, terminal villi are critical for the transfer of oxygen/carbon dioxide, electrolytes, and nutrients between the mother and fetus. After delivery, the mature placenta consists of 15 to 28 subunits designated as "cotyledons", which are perfusion chambers partly or completely separated from others by connective tissue and irrigated by one or more maternal spiral arteries. Each cotyledon contains one or more fetal villous tree(s), a fetal artery, and a vein. What determines the greatest or lowest number of cotyledons remains elusive [25]. Stem villi are the major structural units of cotyledons. Each stem villous branches into 3–5 intermediate villi, which in turn divide into 10 to 12 terminal villi. These terminal villi represent 40% of the total villous volume, and most of them float in the intervillous space, although others are attached to the decidua, favoring structural stability for the placenta [23].

Notwithstanding that the cytoarchitectures of villi are slightly different according to type, the following integrators may be distinguished: (A) STBs are the continuous, specialized layer of epithelial cells that are in contact with the maternal blood, orchestrating maternofetal exchange. (B) CTBs are highly proliferative cells that may lead to two trophoblastic phenotypes: (1) the villous phenotype, which leads to the development of multinucleated STBs, and (2) the extravillous phenotype in CTBs that detach from placental villi and could be differentiated into iEVTs, invading the endometrial decidua and eEVTs responsible for remodeling of maternal spiral arteries [8]. (C) Fixed and free connective tissue cells are derived from the differentiation of mesenchymal cells and include fibroblasts, endothelial cells, smooth muscle cells, myofibroblasts, or macrophages that can be found at different proportions in the stroma. In the case of macrophages, they are named Hofbauer cells. Despite them being firstly differentiated from mesenchymal cells (even with prior placental circulation established), the latter recruitment of circulating monocytes also enhances the population of these cells [22]. These cells play key roles in angiogenesis, defense, immunomodulation, and villi remodeling [24]. (D) Fetal vessels comprise capillaries and sinusoids in the terminal villi, surrounded by a basement membrane, arteries and arterioles in the stem, and intermediate villi with tunica media, without elastic laminae, but with a more mature endothelial layer in comparison with venules and veins. Importantly, the lumen of these vessels is controlled by autocrine and paracrine factors, as the placenta lacks any innervation [11]. (E) Fibrinoid consists of two different kinds of extracellular-deposited materials known as fibrin-type fibrinoid and matrix-type fibrinoid. The former is composed of fibrin and maternal blood-clot products regulating the growth of villous trees and adapting the intervillous space to blood flow. The latter is a secretory product of EVTs, containing laminins, collagen IV, and heparan sulfate. These components orchestrate trophoblast invasion by interacting with cell surface integrins [26]. On the whole, the placenta is a dynamic and complex organ, and multiple cells, products, and formed structures are essential for a successful pregnancy. However, another central component, the immune system, is equally important.

2.3. *Immunology of the Placenta*

Pregnancy is notably an immune-mediated process, involving complex interactions between the semi-allogeneic fetal cells and maternal immune cells [10]. The interplay between these cells not only prevents immune rejection but also favors and creates an appropriate environment for pregnancy [27]. The placenta is a site at which the majority of immunomodulatory actions occur. The communication between maternal immune cells and fetal trophoblasts is bidirectional, and they are essentially mediated by direct contact and through the release of a plethora of autocrine, paracrine, and endocrine signals, including cytokines, growth factors, and adhesion molecules expressed on the surface of cells, such as integrins, cadherins, selectins, and immunoglobulins [7]. Through these mechanisms, placental trophoblasts orchestrate the actions of resident decidual cells, also influencing the recruitment of circulating leukocytes to the maternofetal interface [28]. In this section, we summarize the main immunobiology of the placenta.

Many maternal immune cells inhabit the endometrial decidua, including decidual natural killers (dNKs), macrophages, T cells, and dendritic cells (DCs) [29]. dNKs are a major population of leukocytes at the maternofetal interface (70% of the total) [30].

Through several pathways, dNKs mainly interact with the EVT located in the decidua, being centrally involved in fetal tolerance, EVT invasion, and spiral artery remodeling [31]. Trophoblast release of interleukin-15 (IL-15) seems to promote dNK maturation, and these cells promote decidual remodeling through the production of several cytokines, such as interferon-gamma (IFN-γ), vascular endothelial growth factor (VEGF), tumor necrosis factor–α (TNF-α), IL-8, and the chemokine (C-X-C motif) ligand 10 (CXCL10) [32]. Moreover, dNKs are also major mediators of the immune response against several pathogens such as toxoplasmosis or human cytomegalovirus (HCMV) [33]. Recently, three different subsets of NKs have been identified (dNK1, dNK2, and dNK3) [34]. These different subsets are characterized by expressing distinctive receptors and cytokine profiles and exerting different immunomodulatory actions. Decidual macrophages are the second population most commonly found in the endometrial decidua, representing around 20% of the total [30]. Similar to dNKs, Jiang and Wang [35] identified three subpopulations of macrophages according to the expression of C-C chemokine receptor type 2 (CCR2) and the glycoprotein CD11c, distinguishing between CCR2 negative CD11c low, the most abundant (~80%); CCR2 positive CD11c high (10–15%); and CCR2 negative CD11c high, which was the lowest (~5%). Through a transcriptomic analysis, they suggested that CCR2 positive CD11c high subsets were pro-inflammatory M1-like macrophages in vivo, whereas the remaining CCR2 negative subsets were more likely anti-inflammatory M2 macrophages. Thus, under non-pathological conditions, decidual macrophages are an M2 anti-inflammatory phenotype. The main functions of decidual macrophages are tissue remodeling and repair, debris clearance, angiogenesis, and immune tolerance. These macrophages produce the enzyme indoleamine 2,3-dioxygenase (IDO), which catabolizes tryptophan and hinders inflammatory T cell activation [32]. However, due to infections or an aberrant inflammatory environment, these macrophages may switch to an M1 phenotype, associated with the development of multiple pregnancy complications [36]. T cells represent up to 10 to 15% of the total decidual cells, although its presence progressively increases during the later phase of pregnancy [37]. About 45–75% are CD8 T cells or cytotoxic T lymphocytes (CTLs), and ~30–45% of the cells are CD4 T cells or T helper (Th) cells [38]. CTLs are relevant cells implicated in fetal tolerance while providing an immune defense against pathogens and viral infections [39]. Th cells are prominent regulators of pro-inflammatory/anti-inflammatory status in the maternofetal structures [37]. Effector Th mainly includes three central polarizations: (A) Th1 with pro-inflammatory actions and potentially associated with allograft rejection and pregnancy pathologies; (B) Th2, less harmful for the embryo and inversely associated with Th1 polarization; and (C) Th17, probably involved in acute inflammatory events such as infections [38]. In addition, another crucial type of T CD4+ cells named regulatory T cells (Treg) are crucial for the induction and maintenance of tolerance, especially for implantation and during the first stages of pregnancy [40]. Interestingly, the proportion and populations of each T cell subtype appear to vary during pregnancy. Thus, during the first trimester, Treg and Th1 seem to be the predominant cells, with little implication of both Th17 and Th2, creating a pro-inflammatory but controlled environment that is critical during the early stages [41]. The second trimester of pregnancy is more favorable for the mother and is mainly anti-inflammatory, with predominant Th2 responses. The third trimester and, especially, parturition is again a pro-inflammatory status, aiding the expulsion of the fetus and the placenta [42].

Globally, the role of immune systems in the placenta, especially the maternal part (decidua basal plate), is crucial for developmental and gestational success. In Figure 1, the main structures and cells present in the human placenta are summarized to create an integrative perspective of this organ.

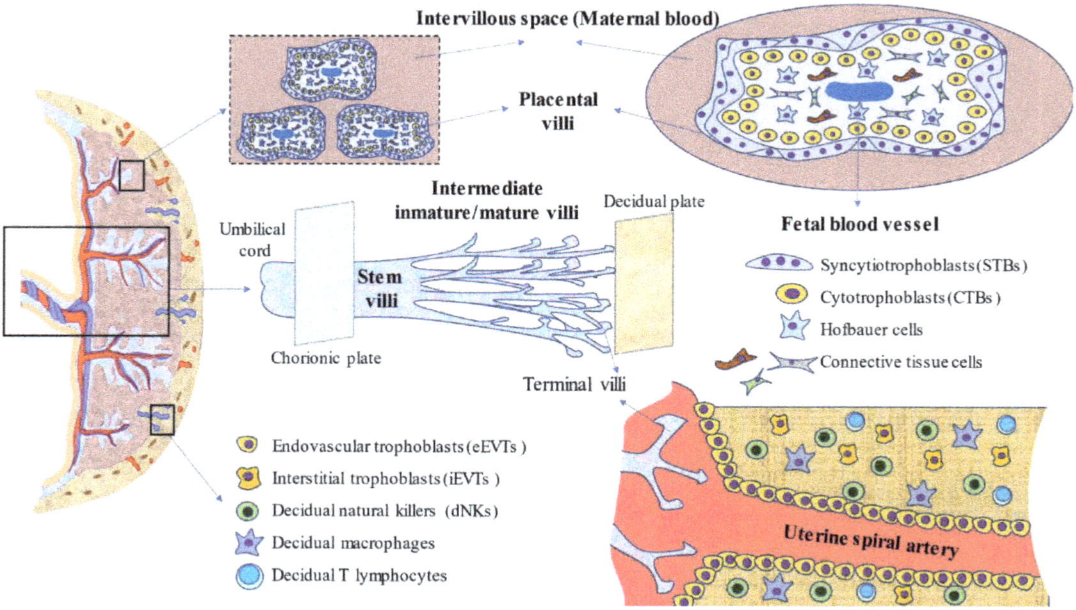

Figure 1. An integrative picture of the components of the placenta. Herein, the main cell types and structures formed are summarized. At the top, placental villi and their cells, as well as the location of the maternal and fetal blood, in the intervillous space and inside the villi, respectively, are represented. In the center of the image, the villous tree, as well as the different types of villi, are represented. The chorionic plate, or fetal surface, is covered by the amnion, where the umbilical cord is attached. The decidual basal plate or maternal surface is in contact with the endometrium. As represented at the bottom, profound remodeling of the uterine spiral arteries is mainly due to the coordinated efforts of a set of cells, mainly extravillous trophoblasts (EVTs) and immune cells, prominently represented by decidual natural killers (dNKs). Having complete knowledge of the placenta in non-pathologic pregnancies is crucial for the study of different pregnancy complications, as the placenta is responsible for a wide variety of functions, as is subsequently discussed.

2.4. Function and Activity of the Placenta during Pregnancy

Despite the relatively short time that the placenta is kept in the women's body, it should be considered the most important organ in pregnancy. Adequate functioning of this organ is critical for fetal well-being, and substantial alterations of this structure are related to the future development of chronic maladies in the offspring [43]. The essential functions of the placenta that are collected in this section include the following: maternofetal exchange, endocrine activity, barrier and defense activity, and fetal programming.

2.4.1. Placenta in Maternofetal Exchange

Maternofetal exchange is achieved through a wide variety of mechanisms, with STBs and fetal endothelial cells being two of the major mediators of this process [44]. As mentioned above, maternofetal exchange occurs in successive placental villi, with the terminal villi being those in which this process is conducted in the mature placenta. The mechanisms described in the maternofetal exchange are: (1) bulk flow/solvent drag, which implies that the movement of water and solutes are in favor of the pressure gradient; (2) net solute diffusion, depending on the concentration gradient (e.g., respiratory gases); (3) transport mediated by proteins, where different molecules are implicated in the transport (e.g., nutrients such as glucose, amino acids, ions, or fatty acids); and (4) transcytosis, a combined process of endocytosis and exocytosis (e.g., immunoglobulin G). Two potential mechanisms of transport also occurring in the placenta, such as paracellular transport and the transplacental electrical potential difference, still require much more studies to be understood [45]. Regarding nutrient exchange between the placenta to the fetus, three different mechanisms have been reported: direct transfer from the maternal blood to the fetus, placental intake of nutrients, or placental metabolism of nutrients to alternate substrate forms [46]. Indeed, the placenta is a notable metabolic organ with high oxygen and glucose consumption rates and exerts other reactions related to glucose (glycolysis, gluconeogenesis, and glycogenesis), lipids (lengthening or shortening of fatty acids and triglyceride synthesis), and protein metabolism (protein synthesis and amino acid interconversion) [47]. Inappropriate nutrient supply from the placenta to the fetus is associated with different pregnancy complications [48]. In this sense, previous studies have identified a critical role for the nutrient sensor mammalian target of rapamycin (mTOR) as a link between maternal nutrient availability and fetal growth, thereby representing an interesting marker of fetal well-being [49]. Another critical nutrient, oxygen supply to the fetus, depends on blood oxygen content and flow rate in the uterine and umbilical arteries, as well as the diffusing capacity of the placenta and oxygen use by the placenta, which may occasionally represent a relevant limitation on oxygen availability [50]. Under hypoxic conditions, the placenta may adapt to this change and enhance anaerobic glycolysis in order to ensure an adequate oxygen flow to the fetus. However, this situation may lead to decreased fetal growth as the nutrient supply to the fetus is diminished [51]. The application of various mathematical models is being investigated in order to predict placental exchange efficiency, which may be of great aid as a diagnostic or prognostic tool in terms of mother and fetus health risk [52]. Thus, the fact that the placenta nutrient exchange not only depends on the direct transfer from maternal to fetal blood but also on the metabolic status, morphology, and nutrient utilization of the placenta must be considered. Interestingly, previous studies have found that the placental exchange function may be different across male and female fetuses. Barapatre et al. [53] showed that there was a noteworthy variation in the number of female cell nuclei in STBs in comparison with those detected in males. These changes may be due to the differential environment created by sex chromosomes and may affect the nutrient supply of the placenta, and significant differences may be observed between both sexes [54]. Moreover, some authors extend the function of STBs beyond the villi at the maternofetal interface. This is the case of trophoblast debris, which ranges from multinucleated syncytial nuclear aggregates to subcellular micro and nanovesicles. The former consists of fragments from the STBs with two possible origins: (A) They derive from newly formed villi that start to sprout from existing villi, more prominently during early stages [55]. (B) They come from aged and late apoptotic STBs nuclei, which are packed into apical protrusions, forming what is known as syncytial knots. These structures are released by the STBs in maternal circulation, reaching the lungs where the local macrophages phagocyte this structure, without the activation of any inflammatory response [56]. The micro and nanovesicles (exosomes) are molecules released by the STBs that are crucial for cell-to-cell communication in the maternofetal interface [57]. Overall, trophoblast debris is a critical component that regulates immune and vascular responses in the placenta, with multiple consequences in maternofetal well-being.

2.4.2. Endocrine Activity of the Placenta

The endocrine function of the placenta is a major hallmark of the placenta. STBs are the most important source of hormones in the placenta, although it seems that other cell types in the placenta, such as the placental bed giant cells derived from EVTs, may also participate in the endocrine milieu in the placenta [58,59]. The production of several hormones in the placenta is crucial for pregnancy onset and maintenance, with multiple local and systemic effects. Costa [60] summarized some of the most important pregnancy hormones, emphasizing the role of human chorionic gonadotropin (HCG), estrogens, progesterone, placental growth hormone (PGH), placental lactogen, adiponectin, and other adipokines, leptin, resistin, pregnancy-associated plasma protein-A (PAPP-A), activin A, inhibin A, placental protein-13 (PP13), and kisspeptin. All of these hormones are crucial in regulating placentation, immune tolerance, and fetal growth and development in healthy pregnancies. Moreover, altered blood levels of most of these hormones are important biomarkers to study in pregnancy-related complications [61].

2.4.3. Placental Barrier

The placenta also represents an important mechanical, chemical, and immunological barrier, exerting diverse actions to protect the fetus. As previously defined, the immunology of the placenta is quite diverse, playing a crucial role in pregnancy homeostasis. Moreover, the immune system and trophoblast also collaborate to reach immune defense against pathogens, including bacteria, viruses, or protozoa, which could be associated with several adverse events [62]. In these times, the severe acute respiratory syndrome coronavirus 2 (SARS-CoV-2) pandemic responsible for the coronavirus disease 19 (COVID-19) has importantly hit the global population, with an important demographic and socioeconomic burden [63]. Pregnant women are considered a group vulnerable to COVID-19 [64]. Prior research has described the central role of the placenta against SARS-CoV-2 infection. The antiviral response of the placenta comprises the presence of Toll-like receptors (TLRs) and RIG-I-like receptors (RLRs) in the trophoblasts, leading to the activation of the nuclear factor-kB (NF-kb) and type III interferon (IRF3) to orchestrate the antiviral response. Moreover, microRNAs from the chromosome19 miRNA cluster (C19MC) are also major mediators of this defense [65]. Immune responses occurring in the placenta are especially effective, as cases with vertical transmission are rare, though they exist, also aiding in softening the cytokine storm in severely ill patients and in mitigating an exacerbated immune response [66]. Moreover, the presence of an own microbiome in the placenta should also be considered in the relationship between the immune system and microorganisms. Unfortunately, evidence confirming that the presence of an own microbiome in the placenta [67] are not consistent, and further studies are required in this field to unravel the presence and/or functions of the microbiota in the placenta.

The defensive role of the placenta extends far beyond the immune system and infections. Xenobiotics are essentially drugs or outer components that are not naturally produced in the body. Previous studies have found that these xenobiotics may cross the placenta to some extent via passive or active mechanisms [68]. Hopefully, the placenta and, in particular, STB counts with a range of enzymes involved not only in the synthesis of different hormones or metabolic reactions but also in detoxification and efflux of xenobiotics, acting in a similar manner to the hepatocytes in the adult [69]. These enzymes are known as xenobiotic-metabolizing enzymes (XMEs), participating in the biotransformation and elimination of maternal and fetal hormones, dietary compounds, drugs, and environmental chemicals [70]. Among these enzymes, of note is the role of cytochrome P450 (CYP), also found in hepatic cells. For instance, at the early stages, the placenta expresses CYP2C, CYP2D6, and CYP3A7, whereas, at term, CYP4B1 and CYP19 (steroid aromatase) are expressed more often. CYP1A1 is prominently induced by exposure to cigarette smoking [71]. Other enzymes of note are the multidrug resistance protein (MRP) family and P glycoprotein. This family, composed of MRP1, MRP2, and MRP3, is differentially expressed in the apical membrane of STBs and the fetal endothelium and is critical for protecting the fetus from the entry of organic anions [72]. The P glycoprotein encoded in the gene MDR1 limits the exposure of fetal hydrophobic and cationic xenobiotics acting as an active pump that leads xenobiotics back into maternal circulation [73]. If any pathological condition is established, the enzyme 11β-hydroxysteroid dehydrogenase-2 (11β-HSD2) is highly expressed in the placenta in order to transform maternal cortisol into cortisone, preventing damage associated with excessive maternal stress [74]. Of note, similar to the nutrient transfer function, 11β-HSD2 is differentially expressed in male and female fetuses. Stark et al. [75] found that this enzyme exerts reduced activity in newborn males relative to females, which may have important implications in the different male morbidities and mortalities following preterm birth. Likewise, the placenta also defends the fetus from excessive oxidative stress, which could be related to several damages, both for the mother and fetus [76]. Oxidative stress is the result of excessive oxidants or free radical production, mainly represented by reactive oxygen species (ROS) and reactive nitrogen species (RNS). Antioxidants are molecules (vitamins or enzymes) implicated in the defense of oxidative damage derived from ROS and RNS. In non-pathological pregnancies, the placenta expresses a wide variety of antioxidants, therefore preventing the development of oxidative stress, which is associated with adverse pregnancy outcomes [77,78].

2.4.4. Placenta and Maternofetal Programming

As shown, the placenta is an essential link between the mother and fetus. Any changes in the maternal organism may affect the placenta, and this could have noteworthy implications in the fetus. Alterations in the growth, vascularization, nutrient, and waste product exchange, hormone production, or metabolism may have long-term effects on the offspring's life [79]. Thus, any insult received in utero at a critical developmental stage may lead to fetal programming, which can determine the development of different diseases in adulthood [80]. From an evolutionary perspective, the placenta is the first way by which the fetus makes contact with the outer environment, receiving endogenous and exogenous signals from the mother. Past and recent environments are major determinators of maternal phenotype and, together with fetal genotype, are related to placental functionality. The fetus responds by increasing its Darwinian fitness, ensuring the probability of reproduction, and preparing the fetus to face the real world after birth [81]. Fortunately, environmental exposure throughout the life of the offspring also affects phenotype, although fetal programming should also be considered. For instance, maternal undernutrition in pregnancy negatively impacts placental development and function, also limiting fetal growth and development [82]. This fact may be understood for the fetus as a signal of food scarcity, which may lead to the activation of the "thrifty phenotype", fetal programming that is associated with increased risk from suffering from type 2 diabetes, metabolic syndrome, and cardiovascular disease in adulthood [83]. Maternal overnutrition may also lead to adverse developmental and long-term outcomes for the offspring due to the activation of multiple epigenetic mechanisms [84]. Recently, Connor et al. [85] compared the effects of undernutrition versus a high-fat diet in placental morphology and functionality. Interestingly, they obtained different adaptative responses of the placenta to both situations, concluding that, compared with undergoing less maturity and inefficient placental transport, overnutrition was associated with variations in multiple placental markers, adapting to the excessive nutrient supply. However, the relevance of ensuring an adequate diet and environment during pregnancy to influence proper development of the placenta and fetus is undeniable. In this sense, different approaches must be considered here, including the introduction of a proper dietary context and an active lifestyle, with the implementation of different and adapted training [86,87]. Limiting exposure to cortisol and stress as well as adequate sleep hygiene will favorably influence the placenta development and functionality, also having positive short and long-term outcomes for the fetus [88–90].

On the other hand, despite the relevance of placental structure, development and functioning have been extensively researched for fetal programming, compelling evidence also evaluates the pivotal role of this organ for maternal health. During pregnancy, virtually all systems and organs of the women undergo different physiological adaptations, according to the fetal necessities [91]. The cardiovascular system is prominently altered during pregnancy, showing a set of changes that are critical for ensuring an adequate blood supply to the placenta and the fetus [92]. Emerging evidence supports that if the uteroplacental circulation is not well-established, there is an increased risk of short-term and long-term cardiovascular disease for the mother [93]. Besides, it seems that changes in the inflammatory response in the placenta and on its genome and epigenome may drive substantial consequences not only for the fetus but also for maternal health [94,95]. As it will be subsequently discussed, there are different vascular disorders of pregnancy whose pathophysiological basis resides on altered placental perfusion, hence highlighting the importance of this organ for the maternofetal well-being. Moreover, a very recent longitudinal study of 33,336 women followed for 50 years showed that the placental weight to birthweight ratio was associated with long-term maternal mortality [96]. In other words, the status and development of the placenta are also crucial for maternal health during and after gestation, as well as to prevent long-term morbidity and mortality.

Collectively, the placenta exerts multiple functions that are crucial for pregnancy success. Moreover, these activities are crucial for maternal and fetal well-being both in the short and long terms, therefore supporting the capital importance of this organ in a human's life. In the next section, we summarize some of the most relevant changes occurring in the placenta and the consequences under certain pathological conditions.

3. Describing the Placenta in Pathological Conditions

3.1. The Role of Placenta in Preeclampsia

3.1.1. Introduction

PE belongs to a set of diseases defined as hypertensive disorders of pregnancy, which also include chronic hypertension, gestational hypertension, and chronic hypertension with superimposed PE. Epidemiological data indicate that PE has a prevalence of 3–5% among all pregnancies, whereas the presence of any hypertensive disorder in pregnancy is estimated to be 10% [97]. PE is a disorder of pregnancy, associated with new-onset hypertension, often accompanied by new-onset proteinuria, although this condition may be presented in the absence of this clinical sign [98]. The main diagnostic criteria of PE consist of the presence of >140 mmHg systolic blood pressure and >90 mmHg diastolic pressure manifested after 20 weeks of gestation. Traditionally, patients with increased blood pressure over 160 mmHg systolic blood pressure and 110 mmHg diastolic blood pressure were diagnosed with severe PE. However, following the current recommendations of the International Society for the Study of Hypertension in Pregnancy (ISSHP), this distinction should not be made anymore, as independent of the blood pressure, PE can deteriorate rapidly and without warning [99]. In the absence of proteinuria, new-onset hypertension may be manifested with thrombocytopenia, renal insufficiency, impaired liver function, pulmonary edema, and new-onset headache unresponsive to medication and not accounted for by alternative diagnoses or visual symptoms [100].

Multiple risk factors for suffering from PE have been identified, including preexisting medical conditions, such as antiphospholipid syndrome, hypertension, or insulin-dependent diabetes; family history or having suffered a prior event of PE; obesity; age (≥40 years old), assisted reproductive techniques; as well as nulliparity or multiple pregnancies [101,102]. From a clinical perspective, two main types of PE exist: (1) early-onset PE (EO-PE), also defined as placental PE, and (2) late-onset preeclampsia (LO-PE). This classification depends on the time of initiation of clinical symptoms, with EO-PE occurring before 34 weeks and LO-PE occurring after 34 weeks [100,103]. Moreover, the impact of PE for the fetus and the mother, serum markers, heritability, and clinical features are quite different for each presentation [104].

The consequences of PE, and especially EO-PE for both the mother and fetus, are numerous. For the mother, the most severe and life-threatening issue could be the development of a cerebrovascular hemorrhage, and as previously discussed, an increased risk of suffering from cardiovascular diseases later in life [105]. HELLP syndrome (hemolysis, elevated liver enzymes, and low platelet count) is another presentation occurring in 10–20% of women with PE, sharing many pathophysiological mechanisms [106], although HELLP syndrome has also been reported without occurring PE [107]. Eclampsia can also occur as a result of PE, consisting of the development of generalized tonic-clonic seizures more often occurring antepartum, 20 weeks after gestation, intrapartum, and postpartum, although some exceptional cases have reported the onset of eclampsia before 20 gestational weeks as well [108]. For the fetus, one of the most worrisome consequences is fetal growth restriction (FGR), as will be subsequently discussed [109]. Other possible adverse outcomes include oligohydramnios, increased risk of stillbirth, and in many cases, PE could be related to iatrogenic preterm birth, which may drive to the development of infant respiratory distress syndrome, intraventricular hemorrhage, sepsis, bronchopulmonary dysplasia, and neurodevelopmental disability [105]. Besides, there is also a profound remodeling of the vascular system of the infants affected with EO-PE, especially for preterm-born offspring [110].

3.1.2. Preventive and Therapeutic Approaches

The only definite cure for PE and its possible complications is delivery. On the other hand, prior studies have reported some benefits from using low-dose aspirin (LDA) as a prophylactic method in high-risk populations [111]. In more detail, the ASPRE (Combined Multimarker Screening and Randomized Patient Treatment with Aspirin for Evidence-Based Preeclampsia Prevention) trial show that 150 mg per day from 11–14 until 36 weeks of gestation reduced the risk of suffering from PE by 62%, hence supporting the relevance of aspirin as a unique prophylactic agent currently available [112]. The most significant effects occurred when LDA was administered before 16 weeks of pregnancy. Indeed, the initiation of this treatment after this time did not diminish the risk of suffering from PE, according to a meta-analysis conducted by Bujold et al. [113]. Besides, the prophylactic benefits from using LDA were observed to prevent preterm PE but not full-term PE, and only if starting before 16 weeks [114]. Compelling evidence has also found that calcium supplementation before and early in pregnancy may reduce the risk of women experiencing the composite outcome PE and pregnancy loss at any gestational age [115]. This could be especially useful in low-income countries, where calcium deficiency is more common [116]. Other strategies such as physical exercise, rest, reduced salt intake, and other nutritional interventions have shown insufficient evidence to be recommended as preventive measurements for PE [117].

Likewise, an early prediction of PE allows for timely initiation of preventive therapy. Moreover, conducting a rapid and early diagnosis of PE is equally important in order to perform continuous observation of the affected patient. In this sense, as it will be subsequently discussed, there are a set of different biophysical and biochemical markers being explored as major clinical features to predict the onset of PE well as to perform an early diagnosis [118,119].

Nowadays, the fact that further knowledge on the biological mechanisms of the disease would be of great aid in the clinical management of such a harmful condition cannot be denied. The placenta has a central role in the pathogenesis of PE and, more prominently, in EO-PE. Currently, an accepted hypothesis explains PE as being a two-stage disease involving the following steps: stage 1 (preclinical), characterized by defective spiral artery remodeling and trophoblast invasion, leading to cell ischemia in the placenta, with an imbalance between anti-angiogenic and angiogenic factors in favor of the former. This anti-angiogenic status may be widespread to the endothelium of the different organs, leading to stage 2 (clinical), in which the maternal syndrome is manifested and defined by systemic endothelial dysfunction accompanied by vascular inflammation, oxidative stress, and the disruption of several serum markers [120–122].

3.1.3. Pathophysiology of Early Onset/Placental Preeclampsia

Defective Spiral Artery Remodeling and Trophoblast Invasion

Failures in uterine spiral artery remodeling and trophoblast invasion are the first pathophysiological events involved in the EO-PE and FGR [123]. However, deficient spiral artery remodeling and trophoblastic invasion are difficult to study in humans, as these processes occur in the earliest stages of pregnancy and the samples are available after the first trimester, before presentation of maternal disease, or at term, after disease presentation [124]. Despite having some limitations, animal studies have shed light on the early pathogenesis of PE, especially mice and rat models. Contrary to human disease, PE must be induced in animals surgically, pharmacologically, or genetically. Moreover, the translation of these models may have some important limitations due to the different placental structures. In this context, the ASB4 deletion murine model and the Dahl S rat have been widely studied to unravel the mechanisms involved in impaired spiral artery remodeling and trophoblast invasion [125]. The precise cause of the defective spiral artery remodeling and trophoblast behavior remains elusive. The hypothesis is that failures in the trophoblast lineage at any stage of early development affect the differentiation of villous trophoblast, which ultimately is responsible for the pathogenesis of PE [56]. The origin of this aberrant functioning is unknown, although different factors have been proposed here, including genetic causes, intrinsic placental alterations affecting the trophoblast and immune system behavior, as well as extrinsic or maternal determinants [121]. All of these factors may lead to the development of acute atherosis in the spiral arteries, which appears in between 20 and 40% of women with PE [126]. In turn, this could be one cause of placental infarction that can be documented in 70% of patients with severe PE and 40% of mild PE [127].

The Antiangiogenic Status

Persistent placental ischemia/hypoxia derived from aberrant spiral artery remodeling and trophoblast invasion is another critical event involved in the pathogenesis of PE [128,129]. Sustained and augmented levels of hypoxia-inducible factor (HIF-1α) are associated with the enhanced production of anti-angiogenic components, such as soluble fms-like tyrosine kinase-1 (sFlt-1) and soluble endoglin (sEng), accompanied by reductions in pro-angiogenic markers such as VEGF and placental growth factor (PlGF) [130]. sFlt-1 is the soluble form of the Flt-1 receptor, also designated as VEGF receptor-1. Likewise, sEng is the soluble form of the Eng receptor, and in broad terms, both Flt-1 and Eng receptors are essential mediators of the angiogenesis process in tissues mainly due to their interaction with VEGF and PlGF [131,132]. Under hypoxic conditions, the preeclamptic placenta shows a decrease in PlGF production together with an increase in the expression of both Eng and Flt-1 receptors as well as their soluble forms released into maternal blood [133]. sFlt-1 binds to free VEGF and PlGF, thereby limiting the bioavailability of these angiogenic components [134]. Chronic hypoxic trophoblasts are responsible for a sustained sFlt-1 release that indeed is sufficient to cause endothelial dysfunction in vitro [135]. This effect is amplified when sFlt-1 acts with sEng, which impairs the binding of TGF-β1 to its receptors and its downstream signaling in the vasculature [136]. The binding of sFlt-1 to free circulating VEGF and sEng to TGF-β1, together with a decreased PlGF production by the trophoblasts, are at some extent responsible for the impaired local angiogenesis in the placenta and the systemic endothelial dysfunction related to PE. Additional evidence supporting the central role of these factors in the pathogenesis of PE is their relevant use in clinical practice as central biomarkers. For instance, measuring the sFlt-1/PlGF ratio allows us to define the clinical diagnosis of PE [137,138]. A PROGNOSIS (Prediction of Short-Term Outcome in Pregnant Women with Suspected Preeclampsia Study) study showed that an sFlt-1/PlGF ratio of 38 or lower had a negative predictive value of 99.3%, whereas the positive predictive value of an sFlt-1/PlGF ratio of 38 or above was 36.7% within 4 weeks. This means that a low ratio of sFlt-1/PlGF (38 or lower) can be used to predict the short-term absence of PE in women in whom the syndrome is suspected clinically [138], although a higher ratio did not have to be related to the presence of PE. Hence, analyzing this ratio could also serve to prevent inappropriate hospitalization, which has a significant economic impact [139]. Therefore, cumulative evidence supports the clinical use of measuring both sFlt-1 and PlGF in women with PE.

Vascular Inflammation

Previous studies have identified a central role of innate and adaptative immune cells in PE. dNKs and decidual macrophages are hyperactivated in PE in response to sustained hypoxia and excessive trophoblast cell debris or trophoblast necrosis [140,141]. For instance, enhanced production of angiotensin II type 1 receptor autoantibodies (AT1AA) is observed in patients with EO-PE, which is a major consequence of a renin–angiotensin–aldosterone system (RAAS) dysfunction [142]. Some authors have hypothesized that the abnormal placental development causes an exacerbated release of RAAS components and other molecules implicated in the regulation of this system, such as micro RNAs or AT1AA, probably in exosomes [143]. AT1AA is an agonist antibody secreted by B cells responsible for the activation of the angiotensin II type 1 receptor, similar to angiotensin II, with multiple consequences. Among the main targets of AT1AA, the activation of the MAP kinase (MAPK) pathway tissue factor, plasminogen activator inhibitor-1 (PAI-1), an impaired trophoblast invasion, oxidative stress markers, and augmented production of sEng through TNFα induction should be highlighted [144]. AT1AA overproduction is not the only mechanism by which the immune system mediates its pathophysiological role in PE. Simultaneously, a differentiation of Th cells into Th1 and Th17 with diminished Treg activity also influences the abnormal inflammatory response found in PE [145]. These changes are accompanied by substantial changes in the cytokine profiling, with an overproduction of type I (pro-inflammatory) cytokines IL-6 and TNF-α and a decrease in type II (anti-inflammatory) IL-4 and IL-10 [146]. In this context, some authors claimed that endometrial mesenchymal stem cells are relevant in the immune system, mediating immunosuppressive effects in the local tissue and allowing for trophoblast invasion in normal pregnancies. The precise role of these cells in the etiopathogenesis of PE remains unknown [147], although some in vitro studies have reported some promising results from using this type of cell to reduce the Th1 pro-inflammatory differentiation [148]. Further research in this field may aid in the immunoregulation occurring in the placental environment at the early stages. Some authors hypothesized that an altered placental microbiome might shed inflammatory molecules, such as lipopolysaccharides (LPS), with a resultant inflammatory switch that accompanies abnormal placentation and maternal endothelial cell activation [149]. The excessive immune activation contributes to endothelial dysfunction via several mechanisms, including the release of endothelin 1 (ET-1), ROS, and an increase in vasoconstrictors such as the proper AT1AA and angiotensin II along with a decrease in vasodilators such as nitric oxide (NO) and prostacyclin [150]. ET-1 is a direct consequence of endothelial dysfunction, related to a powerful vasoconstrictor response. ET-1 is produced due to VEGF inactivation or inhibition. Then, ET-1 triggers the production of ROS and oxidative stress in the placenta, which responds with enhanced production and release of sFlt-1, creating a vicious cycle [151].

Oxidative Stress

Oxidative stress is widely accepted as a major determinant of PE and placental diseases. Contrary to normal pregnancies, PE is characterized by increased ROS and oxidative molecules, with a decrease in antioxidant systems [152]. Virtually all placental cell types are sources of oxidative stress, including trophoblasts, endothelial cells in the placenta, Hofbauer macrophages, or stromal cells in the villi [153]. The sustained hypoxia leads to excessive inflammation and oxidative stress, two closely related conditions [154,155]. Furthermore, excessive oxidative stress may lead to a disruption in different placental cells, affecting some critical events, such as apoptosis or autophagy, participating in the pathogenesis of PE [156]. Among the most important consequences of oxidative stress, a decrease in the placental endothelial nitric oxide synthase (eNOS) strongly diminishes NO production via several mechanisms [157]. In this sense, some studies have found an inverse correlation between NO levels and levels of sFlt-1 and sEng in women with PE, thereby suggesting that somehow anti-angiogenic factors may inhibit the production of NO [158]. The diminished NO production is also potentially related to the reduced bioavailability of its precursor, L-arginine, both being observed in women with PE [159]. These alterations may promote changes in the renal and cardiac microvasculature as well as a reduction in the number of fetal nephrons in vivo, which can possibly also occur in humans [160]. Notwithstanding, a clear implication of oxidative stress is seen in the etiopathogenesis of PE: the use of antioxidants such as vitamin C, E, or n-acetylcysteine has failed to show any clinical benefit [161].

Endocrine Disruption

Many of the hormones produced by the placenta may be partly responsible for the onset and development of PE. For instance, reduced levels of hCG are secreted by STBs in the first trimester, but an increased detection in the second trimester appears to be associated with a later diagnosis of PE [162]. These differences were due to the pro-angiogenic activity of hCG, and the higher presence of this protein seems to be a compensatory mechanism of the placenta to the sustained ischemia. Moreover, decreased levels of hyperglycosylated Hcg (HhCG) secreted by EVT in the second trimester was also associated with a higher risk of PE. Importantly, HhCG, but not hCG, is involved in deficient trophoblast invasion, which may support its possible pathophysiological role in PE [163]. Other altered hormones associated with the pathophysiology of PE include androgens (testosterone and androstenedione) and leptin, which in turn are correlated with estrogen levels [164]. Other hormones not produced by the placenta, such as cortisol, arginine vasopressin (AVP), epinephrine, norepinephrine, natriuretic atrial peptide, brain natriuretic peptide, and melatonin also appear to be implicated in the pathogenesis of PE [165], hence concluding the complex role of multiple hormonal systems in these pregnancies.

Placental Aging and Damage

As mentioned above, increased cell death is observed in the placental tissue of women with EO-PE. Apart from oxidative stress, the enhanced apoptosis observed in the placental tissue related to PE is also attributable to the triggered inflammation and chronic hypoxia. Whereas in physiological pregnancies, the apoptosis process is exclusively regulated by the immune system through the extrinsic pathway in an FAS ligand (FASL) manner, the pathologic environment in PE leads to the activation of the intrinsic route and a decrease in anti-apoptotic proteins. This concludes with enhanced apoptosis of EVTs and STBs, with increased syncytial knots, a marker of accelerated aging in the placental tissue and in EO-PE [166,167]. The premature aging of placental tissue is also supported by the dysregulation of several hallmarks of aging, including telomere shortening, cell senescence, loss of proteostasis, epigenetic variations, and mitochondrial dysfunction, among others [168–171]. The acceleration in the aging process of the placenta may be caused by the pathological environment in this organ while contributing to an exacerbation of the disease.

In this section, we summarized the multiple alterations occurring in the placenta tissue both in the preclinical and clinical stages of PE, with substantial consequences not only for this tissue but also for the mother and fetus. The main results summarized in this section are represented in Figure 2. Further insights into the role of the placenta in EO-PE should significantly help deepen our understanding of the origins and development of such an intricate disease, also identifying novel biomarkers with prognostic or predictive value, or perhaps prophylactic approaches.

3.1.4. Pathophysiology of Late-Onset Preeclampsia

Contrary to EO-PE, designated as placental PE, LO-PE pathogenesis is often related to maternal pathophysiology, and neither the placenta nor fetus suffer the same significant effects as in EO-PE [172]. LO-PE presents higher rates of incidence than EO-PE, and different risk factors have been identified for both conditions. Furthermore, EO-PE is associated with a higher incidence of fetal death and maternal risk of many cardiovascular, respiratory, nervous, hepatic renal, and other morbidities in comparison with LO-PE [173,174]. These differences may be attributed to the fact that, in LO-PE, the spiral arteries seem to maintain normal behavior, with no evidence of an aberrant trophoblast invasion. The etiology of this condition is speculated to probably be more ligated not to placental but to maternal extrinsic factors [56,174]. In this sense, some authors proposed that both EO-PE and LO-PE are consequences of STB stress [175]. However, in contraposition to EO-PE, LO-PE may be secondary to intraplacental (intervillous) malperfusions due to mechanical restrictions as no evidence of altered spiral artery remodeling exists. In this line, a growing amount of evidence endorses the implication of the cardiovascular system in the origin of LO-PE, resulting from the inability of the maternal heart to meet the increased metabolic demands of an overgrown fetoplacental unit [176]. Indeed, an abnormal Doppler assessment may be reported not only in uterine arteries but also in other unrelated blood vessels [177]. Thus, the main differences between both conditions in pathophysiological terms may be the duration and causes of dysfunctional uteroplacental perfusion, although the response and behavior of the placenta and stage 2 are quite similar [175]. Moreover, notwithstanding, more studies are required to compare EO-PE with LO-PE biomarkers, and some differences have been found between them, including oxidative stress (superoxide production is higher in EO-PE in comparison with LO-PE), angiogenesis (further decrease in tyrosine kinase endothelial receptor (Tie-2) and increase in Flt-1 receptor in EO-PE), inflammatory markers (upregulation of the intercellular adhesion molecule 1 (ICAM) is more marked in EO-PE), and hormone functionality (PP13 is more decreased in EO-PE) [104]. Additionally, the placenta of women with EO-PE shows further hypomethylation of different circadian clock elements in comparison with LO-PE and non-pathological placentas, indicating a different epigenetic pattern in these genes [178]. The possible differential role of other pathological markers was previously described in EO-PE versus LO-PE, although the results obtained are not consistent [104]. Collectively, these studies revealed noteworthy variations in the placental behavior in EO-PE versus LO-PE. The fact that EO-PE is established at the early stages of placentation, leading to prolonged hypoxia, inflammation, or oxidative damage, may explain these differences. Nevertheless, LO-PE may be taken cautiously as some of the pathophysiological mechanisms reported in EO-PE also exist in this condition.

3.2. *Placenta in Fetal Growth Restriction*

3.2.1. Introduction

FGR describes a condition characterized by an abnormal umbilical artery Doppler and is frequently related to a fetus small for gestational age (SGA) (less than the tenth percentile of weight at birth) [179–181]. However, it is important to make a distinction between FGR and SGA, as recent works recognize that both SGA and FGR may be considered separately. For instance, infants with SGA could not be growth restricted, and newborns with birth weights > 10th percentile can be growth restricted. Thus, the agreed-upon definition of FGR includes birth weight less than the third percentile or three out of the following criteria: birth weight < 10th percentile, head circumference < 10th percentile, length < 10th percentile, prenatal diagnosis of FGR, and the presence of any pregnancy complication [182]. Other variables such as placental, brain, or liver weight lower than 10th percentile; brain-weight-to-liver-weight ratio higher than 4; placental-weight-to-birth-weight ratio higher than 90th percentile; and histologic or gross features of placental insufficiency/malperfusion may also be used to identify fatal cases of FGR [183]. In the case of SGA infants, it could be categorized into two major groups: (A) Constitutionally normal infants with birth weight less than the 10th percentile due to inherent factors like maternal height, weight, ethnicity, and parity, and (B) SGA infants secondary to FGR [181].

FGR affects 10–15% of pregnant women and is the second most frequent cause of perinatal mortality and morbidity [184]. The percentage of FGR births is different across regions, with the highest-burden being in Asia, accounting for nearly 75% of all global cases of FGR [185]. It is of great importance to understand that this does not mean that in these regions, 3/4 births are affected with FGR. Conversely, due to the limitation of available data in other regions, the total fertility rates, the number of inhabitants in Asian countries like India or China, and more prominently because of the relationship of some of these countries with the different risk factors that will be subsequently discussed, more cases have been reported in Asia in comparison to other continents.

The origin of the disease remains elusive. However, it is thought to be caused by an interaction of environmental and genetic factors with either a fetal, placental, or maternal origin [186].

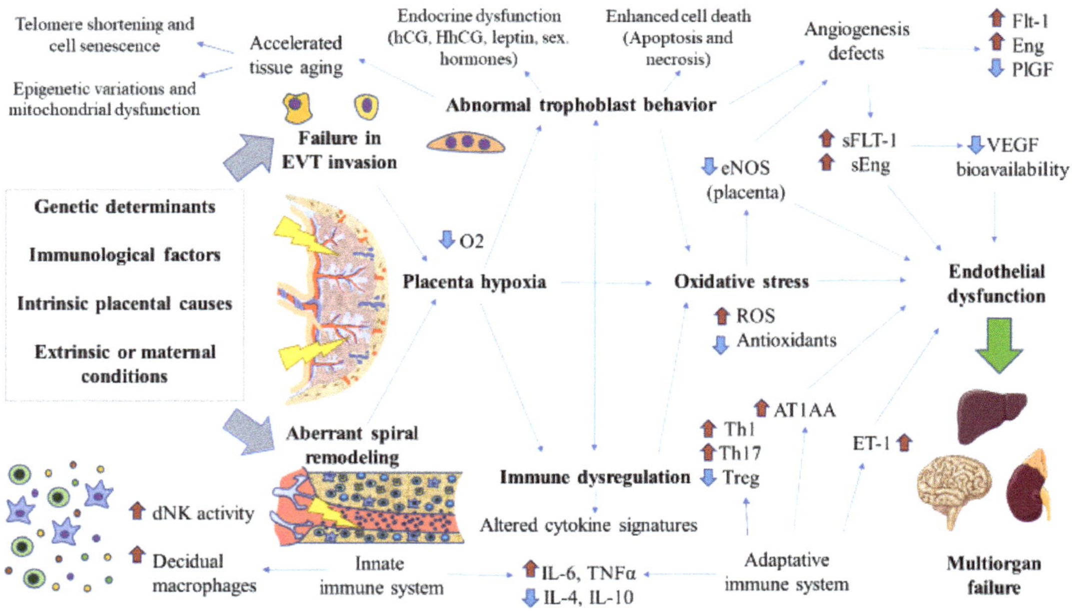

Figure 2. An overview of the main mechanisms involved in the pathogenesis of early-onset preeclampsia. As described before, aberrant spiral artery remodeling and failures in trophoblast invasion trigger placental hypoxia, leading to an exacerbated immune response affecting the innate and adaptative immune systems and cytokine production. In turn, an abnormal trophoblast behavior is also observed in the placenta, including accelerated aging, endocrine dysfunction, enhanced cell death, and angiogenesis defects, emphasizing the relevance of an imbalance between anti-angiogenic (sflt-1 and sEng) and pro-angiogenic factors (VEGF and PlGF). Almost every cell type in the placenta increases ROS production and decreases antioxidants, which are associated with oxidative stress, affecting some crucial components in the placenta, such as endothelial nitric oxide synthase, with detrimental effects on angiogenesis. Overall, oxidative stress, trophoblast alterations, and the inflammatory environment lead to systemic and profound endothelial dysfunction, eventually affecting different organs in the body and leading to the onset of clinical manifestations.

The maternal factors involved in FGR are various. One of them is, as previously stated, a PE diagnosis. Prevalent undernutrition in women from developing countries is key to the origins of FGR and consequent stunting, wasting, and other predisposing factors for child mortality [187]. The opposite case, overnutrition due to abnormal maternal diet and pregnancies associated with obesity or diabetes, is also relevant due to over-insulinisation and derived vasculopathies [188,189], as well as advanced maternal age [190]. Other factors that increase the risk of FGR are toxic substances consumed by smoking, alcohol consumption, and poly drug use, which imply lower food consumption and, thus, lack of fetal weight gain [191–193]. Not less important is the exposure to environmental chemicals, such as bisphenol A and phthalates, which impair the functions of placental hormones such as gonadotropin, estrogens, progesterone, prolactin, or growth hormone, especially in the first half of pregnancy [194,195]. Avoiding contact with polycyclic aromatic hydrocarbons is also important, especially in the first trimester of pregnancy, when they cause more severe adverse effects on fetal growth [196]. Some cases in which diffuse chronic villitis of unknown etiology can also be associated with FGR [179]. Additional maternal factors include maternal hypertension, exposure to stressful conditions, genetics, or high altitudes [197]. Fetal causes encompass some congenital malformations like congenital heart disease, congenital diaphragmatic hernia, or trisomies 13, 18, and 21 [198]. Infections caused by HCMV, rubella, or *Plasmodium falciparum*, and multiple pregnancies are also common causes of FGR [179,199]. The placental causes of FGR are mainly related to altered nutrient transport in the placenta and, more specifically, with amino acids, glucose, fatty acids, and oxygen intake [200]. Other placental factors include an incorrect umbilical cord insertion, placental tumor, a single umbilical artery, and circumvallate placenta [197].

Despite the different origins that this condition may entail, it must be remarked that FGR is, in fact, a consequence of placental vascular pathology, concretely a chronic insufficiency of oxygen and nutrients for the fetus due to impairments in maternofetal circulation [201]. The accompanying problems that newborns may have include perinatal asphyxia, hypothermia, hypoglycemia, and polycythemia, and later, they may also face growth retardation and neurodevelopmental problems, with all of these echoed in their health in adulthood [202]. Consequences in adulthood include, above all, non-communicable diseases such as metabolic syndrome, renal and cardiovascular disease, and neuro-inflammation [195]. Adaptative changes of FGR persist after birth. Although neonates show smaller left ventricular dimensions, either in symmetric or asymmetric FGR, systolic function seems less altered when compared with non-FGR neonates [203,204]. Other studies find an apparent dysfunction of systolic and diastolic signs, and coronary flow is significantly increased in neonates with FGR, but continuous monitoring of extrauterine life shows a similar heart rate in comparison to infants without FGR [205,206]. Despite these results, further research is required to denote how extra effort by a premature neonate for remodeling cardiac dimensions and functions—as it could not take place in placenta—may have its short and long-term sequelae echoed in the development of cardiovascular diseases in adulthood [207]. In contrast, embryological alveolar formation is key; uncompleted lung function configurations have shown more difficulties; and epidemiological data show correlations between FGR, low birth weight (LBW), and respiratory morbidity from infancy progress with age to adulthood [208]. High concentrations of surfactant protein D in the umbilical cord blood of premature fetuses with FGR are determinants of asthma and chronic obstructive pulmonary disease in later life [209]. Abnormalities in the RAAS promote fetal renal tubular maldevelopment [210].

3.2.2. Pathophysiological Role of the Placenta

As mentioned above, FGR may appear as a consequence of PE, especially EO-PE, sharing common pathophysiological mechanisms in the placenta. For instance, a subjacent deficient remodeling of uterine spiral arteries can be presented in FGR, just as in the case of EO-PE [211]. Malperfusion causes the privation of nutrients and oxygen to support normal aerobic growth of the fetus and then, cell stress, selective suppression of protein synthesis, and hence poorer cell proliferation with a consequent reduced villous volume—lower surface area for maternofetal traffic of substances—and villous damage [212]. The subjacent alterations in intrauterine hydrostatic pressure gradients denote that these disturbances in maternofetal circulation alter metabolism and placental immune function. Morphological and metabolic changes are more severe in cases of FGR associated with PE [211]. New magnetic resonance imaging technologies today allow for watching the redistribution of circulation and incompetent oxygen and substrate transport in response to hypoxia in FGR, translating these into slower fetal brain growth and fetal growth in general [213].

Most severe cases determine infarction and fibrin deposition. Massive intervillous fibrin deposition, so-called maternal floor infarction, compromises villous tree. It constitutes a placental disorder in which fibrinoid material is deposited within perivilli and intravilli, correlated with hypovascular/avascular villi and elevated plasma cell deciduitis, with inflammatory infiltration being determinant for villous architecture lesions [214,215].

The placenta is a pivotal structure responsible for the pathophysiology of FGR, and exposure to multiple environments during pregnancy may lead to detrimental effects in this organ. For instance, alcohol consumption and its metabolic products entail multiple changes in the placenta, including oxidation of lipids and proteins, DNA damage, mitochondrial dysfunction, apoptosis, and cellular injury [216]. This leads to a prominent placental dysfunction, decreased placental size, endocrine changes, impaired blood flow, and nutrient transport, increased rates of stillbirth and abruption, umbilical cord vasoconstriction, and LBW [217]. In this context, recent studies have demonstrated the association between alcohol consumption with impaired insulin-like growth factor 1 (IGF-1) signaling. IGF-1 has a pivotal role in mediating cell proliferation, migration, and differentiation, as well as in placentation, thereby showing that the inhibition of this molecular pathway by ethanol may be partly responsible for FGR [216]. Some questions to solve in future studies include what molecular mechanisms underlie ethanol-mediated placental dysfunction and what are the possible adaptative responses of this organ according to the fetal sex [218].

3.2.3. Screening, Predictive, and Diagnostic Biomarkers

Due to the huge impact of FGR, effective screening of this condition is of significance in limiting the detrimental effects of this condition. In this sense, some studies have proven that a combined analysis of fetal biometry and fetal growth velocity is essential for identifying a subset of SGA fetuses that are at increased risk of neonatal morbidity, roughly tripling the detection of SGA infants [219].

The simultaneous assessment of maternal and fetal biomolecular markers has been determined to study this complication, with most of them having been mentioned above for PE, sharing the upregulation of anti-angiogenic factors. The measurement of maternal serum PAPP-A is downregulated in thick placenta (>4 cm or >50% of placental length) (evaluated with ultrasound technique) [220] and is associated with poor pregnancy outcomes [221]. Altered serum levels of PAPP-A have been observed to be associated not only with FGR but also gestational diabetes and PE [222,223]. Long-term studies have found that this measurement is also associated with short stature in offspring and maternal diabetes mellitus in later life [224]. For this reason, this biomarker is key in the first trimester for a sooner prediction of placental health and prenatal outcomes to classify patients with increased risk of FGR [225]. Evidence shows that it is also relevant for monitoring the mother and her offspring in later life. In fact, PAPP-A has proven to be reliable as a single predictor for prenatal screening, and when combined with other maternal serum markers, the sensitivity of diagnosis rates increases [226].

Other biomarkers have been used lately to study the impact in newborns with FGR. The role of the extracellular matrix glycoprotein reelin in perinatal neurodevelopment has shown elevated levels significantly associated with cerebral blood redistribution [227]. Angiogenic molecules such as PlGF, which was implicated in PE, are also applicable to FGR. Very low levels of this marker correlate with significantly lower gestational age at delivery and is an indicator of urgent delivery in pregnancies at risk of adverse outcomes [228]. Additionally, according to some prospective studies, calculating the sFlt-1/PlGF ratio can predict PE and FGR after the 34th week [229]. Several prospective studies have emphasized the importance of accomplishing repeated measurements of serum values in pregnant women, such as the sFlt-1/PlGF ratio, as PE and FGR can account for the later stages [230]. Conversely, the plasma levels of sFlt1 and sEng do not show significantly different results for patients with PE compared with patients with FGR [231]. Additionally, the abovementioned anabolic hormones IGF-1 and its binding proteins (IGFBPs) present low circulating levels in FGR, endangering oxygen and nutrient transport across the placenta and, hence, its dysfunction. As IGF-1 plays a key role in fetal growth, several animal models have proposed that the administration of low doses of this hormone restores its function in FGR pathophysiology [184]. Likewise, certain studies found that, from the first trimester, low maternal-free β-hCG concentrations are associated with adverse outcomes, such as FGR, preterm birth, and LBW. On the other hand, abnormally high concentrations in the second trimester were associated with spontaneous abortion and FGR as well [232]. Additionally, recent studies have shown that circulating Epidermal Growth Factor-Like domain 7 (EGFL7) allows us to discriminate between isolated FGR and PE at different gestational ages. Levels are high in both pathological conditions, but the cutoff is significantly different, being much higher in PE [233]. In any case, standardizing screening methods employing maternal serum biomarkers to predict the window of vulnerability and sooner episodes of FGR development are thus necessary.

Finally, the diagnosis and management of FGR rely on the use of ultrasound and Doppler technologies—especially the following parameter: uterine artery Doppler velocimetry (UADV) pulsatility index (PI). Optimized imaging studies are needed for earlier detection and for preventing the late onset of FGR [197] and suppose a more cost-effective multi-parametric testing [234]. Moreover, each case should be addressed individually to determine the appropriate timing of delivery. PlGF from maternal blood can reveal urgent delivery before the 35th gestational week and has been proven to have more sensitivity when fetal flow Doppler ultrasonography cannot identify late adverse events [235]. More efforts are being put into improving monitoring tools with the aim to predict adverse outcomes earlier and to evaluate time to delivery [236]. In the neonatal period, intensive care therapies are regarded as critical research areas [201].

3.2.4. Preventive and Therapeutic Approaches

Once the FGR problem is established, it seems irreversible. Preventing undernourishment in developing countries or sedentary and chronic illnesses in developed countries constitutes a public health concern.

Nutritional intervention programs are also vitally important in children who were born with FGR and, hence, born with more susceptibility to developing metabolic disorders such as obesity. An example of fetal programming is the IGF-1 gene: from birth, high IGF-1 DNA methylation can be observed to screen infants with higher risks of developing cardiovascular disease or diabetes in adulthood [237]. The growing interest in interventions addressing lifestyle factors has shown evidence for targeting the epigenome, which could prevent morbidities in adulthood. The focus on monitoring individuals who suffered prenatal adversities such as FGR still requires further deepening. Currently, research works about this concern are few, and the guidelines maintain most of the recommendations related to micronutrient supplementation. Supplemented calcium is known to protect against LBW, and magnesium was associated with a reduced risk of SGA [238]. The controlled energy achieved by a balanced diet should always be considered. It would also be interesting to accomplish a study on the effects of zinc, group B vitamins, and other micronutrients in pregnancies with FGR.

In the same context, physical exercise, as one of the pillars of health, should always be individually advised—with adapted intensity for each woman—but the truth is that, for a long time, perceptions about physical activity in pregnancy have been inaccurate [239]. Some rodent models showed a reduced risk of metabolic diseases following FGR thanks to exercise-induced metabolic reprogramming, with improved β-cell functions [240]. Evidence in human studies suggests that some pro-angiogenic factors such as VEGF can be upregulated by exercise during pregnancy and do not correlate with preterm labor or FGR [241], with the gestational weight gain control being promoted instead [242]. Safe physical exercise for pregnant women improves insulin sensitization and glucose tolerance [243]. Likewise, regular workouts before pregnancy ensure better reprogramming of the vascular network, which may aid in preventing FGR.

Moreover, the increased risk of preterm delivery associated with alcohol consumption also supports abstinence and smoking cessation during pregnancy.

Following the actual recommendations of the FIGO (International Federation of Gynecology and Obstetrics) [244], no medical interventions to prevent FGR have been clearly established. Indeed, cumulative evidence only supports a small benefit from the use of aspirin at low doses to reduce SGA in women at risk, whereas low-molecular-weight heparin shows some promising but still inconsistent results in the prevention of FGR. The only management option in cases with high risks of hypoxia, acidosis, and intrauterine death is iatrogenic preterm birth, with the use of peripartum maternal administration of magnesium sulfate for neuroprotection and corticosteroids for fetal lung maturity in order to prevent adverse neonatal outcomes [245]. Congenital infections, such as those caused by HCMV, may be related to impaired placental development with a high probability of fetal transmission (30–40%), which has permanent hostile consequences in newborns, including not only FGR but also neurological, hearing, and vision defects [246]. Avoiding fetal infections is difficult, although the search for a vaccine to prevent these adversities is still being developed [247].

Another factor to bear in mind for pregnancy is advanced maternal age, with advanced age being more susceptible to developing FGR and to experiencing stillbirth. In any case, family planning and social progress may help women find a balance between conceiving children at an optimal biological age and their career/working life in order to avoid disease for themselves and their offspring. The main ideas reviewed in this section are collected in Figure 3.

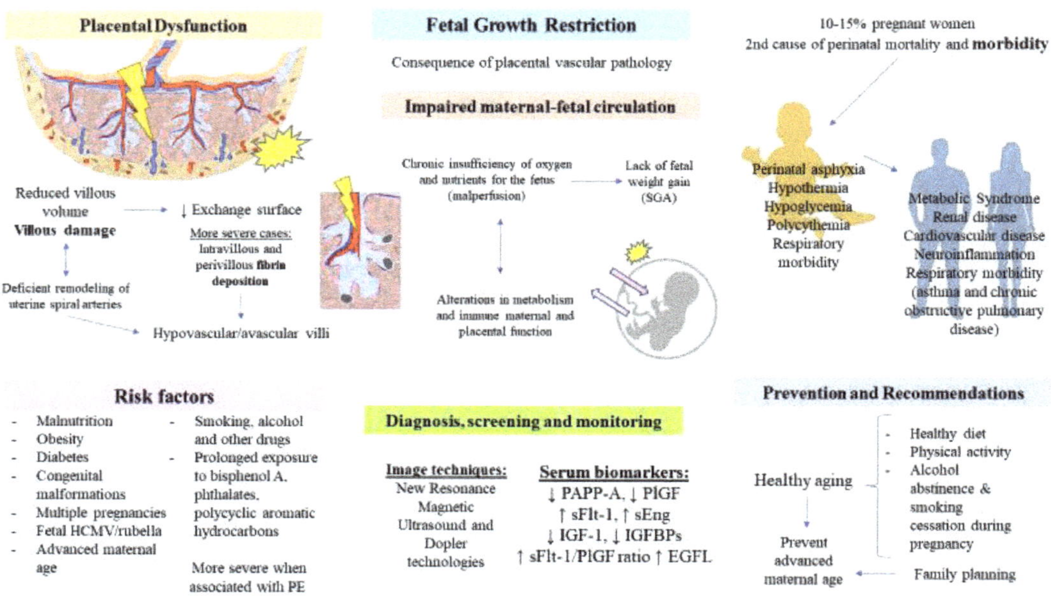

Figure 3. The vital role of the placenta in fetal growth restriction, risk factors, and consequences of the impaired maternofetal circulation. Main diagnosis, screening, and monitoring methods, as well as some preventive measures and recommendations, are also described.

3.3. Chronic Venous Disease, Clinical Manifestations, and Repercussions

CVeD is a multifactorial vascular disorder characterized by an impairment in the venous return, mainly from the lower limbs. Varicose veins (VVs) are the main presentation of CVeD [248], although a broad spectrum of clinical manifestations is classified according to the Clinical-Etiology-Anatomy-Pathophysiology (CEAP) criteria [249]. The term chronic venous insufficiency (CVI) refers to the most advanced stages of CVeD and may include dermatologic alterations and active ulcerations [250]. Notably, CVeD and, more prominently, CVI entail important socioeconomic burdens, with important consequences on the patient´s quality of life [251]. The signature of CVeD is an increase in venous pressure, referred to as venous hypertension. Venous hypertension appears as a consequence of both genetic and environmental factors. The pathophysiology of CVeD involves a profound remodeling and injury of the venous wall, valve incompetence, microcirculatory changes, hypoxia, and a significant inflammatory response secondary to venous hypertension [252]. Women, especially those who are pregnant, represent a vulnerable population that suffers from CVeD [253]. Due to the differential hemodynamic activities occurring in the cardiovascular system, exposure to different hormonal factors, and mechanical action due to proper growth of the fetus, approximately 40% of women undergo CVeD during pregnancy [252]. Moreover, the number of pregnancies also seems to be a major risk factor for suffering from CVeD [254]. Considering that CVeD has not only local but also global consequences is important. Along this line, multiple studies have confirmed the dysregulation of several systemic components, including pro-inflammatory cytokines, epigenetic markers, oxidative stress products, and some circulating parameters (estradiol, homocysteine, VEGF) [255–258].

The relationship between CVeD and the placenta is notably complex. Prior research suggests that the placenta could represent an important point of study either as a causative agent or as a consequence of CVeD. Indeed, we have demonstrated several changes in the placenta of women who develop CVeD for the first time during the third trimester of pregnancy, with no evidence or diagnosis of any health concern. For instance, as detailed above, the placenta is a central source of hormones in a pregnant woman, and previous studies have noticed the association of some of these hormones and their receptors with the development of CVeD [259–262]. Moreover, women with CVeD seem to present a higher presence of oxidative stress markers. Particularly, increased levels of NADPH oxidase 1 (NOX-1), 2 (NOX-2), inducible nitric oxide synthase (iNOS), poly (ADP-ribose) polymerase (PARP), and ERK were observed in the placenta of affected women [263]. Moreover, we evidenced augmented malondialdehyde (MDA) levels in the serum of mothers with CVeD and a decrease in the fetus serum pH. Thus, the placenta could be a major source of oxidative stress, although establishing if this may favor the onset and progression of CVeD or whether it is a response is difficult. Furthermore, we also observed evidence of hypoxic damage in the placental villi related to CVeD [264]. The hypoxic environment observed in these placentas was related to an increased number of villi together with enhanced apoptosis and syncytial knot detection in comparison with the women in the control group. Moreover, similar results were also found in the umbilical cord of these women [265], therefore evidencing the association between CVeD with hypoxia and oxidative stress in the maternofetal structures during pregnancy. The various alterations related to CVeD promote an abnormal environment in the placental tissue, leading to the development of different adaptative responses. The sustained hypoxia leads to dysregulation of the local angiogenesis and lymphangiogenesis observed by marked alterations of Flt-1, VEGF, PlGF, cluster differentiation 31 (CD31), and podoplanin (D2–40), which in turn were related to an augmented villous calcification [266,267]. Moreover, a pro-inflammatory environment was also described in the placenta of women with CVeD, accompanied by substantial changes related to the phosphatidylinositol (PI3K) pathway, Wnt/βcatenin, MAPK, the PAPP-A/IGF-1/stanniocalcin-2 axis, and insuline receptor substrate 4 (IRS-4) [263,268]. All of these molecular variations may also be related to a switch in lipidomic profiling in the placenta of women with CVeD [269]. Finally, we observed that profound remodeling affects the placental tissue, with dysregulation of collagen and elastic fibers, including an increase in type III collagen, matrix metalloprotease 9 (MMP-9), and elastogenic processes with a decrease in EGFL7 expression [270,271]. Currently, the impact of CVeD in maternal and fetal well-being is starting to be elucidated. In this sense, we found an important association between VVs in the lower extremity and intrapartum fetal compromise [272].

Despite our understanding of the effects and relationship between CVeD, placental composition and function are still in their infancy, and compelling evidence is starting to unravel the harmful effects of this condition in maternofetal structures. This disease likely presents a unique pathophysiological signature in the placenta of pregnant women. Future studies are needed to shed some light on the changes that occur in this organ and the possible implications of this dysregulation.

4. Conclusions

The placenta acts as a pivotal director in fetal and maternal processes during pregnancy and is also critical for understanding healthy and pathological pregnancies. PE, especially EO-PE, is a disease initiated and triggered in the placenta, entailing a set of consequences for both the mother and the fetus, such as FGR. Different factors may be associated with the development of PE and FGR, and some of the different dysregulated products of the placenta may be used in the screening, diagnosis, and monitoring of this condition. Different strategies that focus on correcting placental function are being investigated. Finally, although less established, pregnancy-related CVeD is a condition commonly found in pregnant women, and prior studies have started to demonstrate noteworthy placental alterations in these patients. Further studies are needed to clarify the precise relationship between CVeD and the placenta.

Author Contributions: All authors have read and agreed to the published version of this manuscript.

Funding: This study (FIS-PI18/00912, PI21/01244) was supported by the Instituto de Salud Carlos III (Plan Estatal de I+D+i 2013–2016) and co-financed by the European Development Regional Fund "A way to achieve Europe" (ERDF) and by B2017/BMD-3804 MITIC-CM and B2020/MITICAD-CM and HALEKULANI S.L.

Institutional Review Board Statement: Not applicable.

Informed Consent Statement: Not applicable.

Data Availability Statement: Not applicable.

Conflicts of Interest: The authors declare no conflict of interest.

References

1. Gude, N.M.; Roberts, C.T.; Kalionis, B.; King, R.G. Growth and function of the normal human placenta. *Thromb. Res.* **2004**, *114*, 397–407. [CrossRef]
2. Burton, G.J.; Jauniaux, E. What is the placenta? *Am. J. Obstet. Gynecol.* **2015**, *213*, S6.e1–S6.e4. [CrossRef]
3. Huppertz, B. Placental pathology in pregnancy complications. *Thromb. Res.* **2011**, *127*, S96–S99. [CrossRef]
4. Konkel, L. Lasting Impact of an Ephemeral Organ: The Role of the Placenta in Fetal Programming. *Environ. Health Perspect.* **2016**, *124*, A124–A129. [CrossRef]
5. Thornburg, K.L.; Marshall, N. The placenta is the center of the chronic disease universe. *Am. J. Obstet. Gynecol.* **2015**, *213*, S14–S20. [CrossRef] [PubMed]
6. Guttmacher, A.; Maddox, Y.; Spong, C. The Human Placenta Project: Placental structure, development, and function in real time. *Placenta* **2014**, *35*, 303–304. [CrossRef] [PubMed]
7. Kim, S.-M.; Kim, J.-S. A Review of Mechanisms of Implantation. *Dev. Reprod.* **2017**, *21*, 351–359. [CrossRef]
8. Staud, F.; Karahoda, R. Trophoblast: The central unit of fetal growth, protection and programming. *Int. J. Biochem. Cell Biol.* **2018**, *105*, 35–40. [CrossRef]
9. Ramathal, C.Y.; Bagchi, I.C.; Taylor, R.N.; Bagchi, M.K. Endometrial Decidualization: Of Mice and Men. *Semin. Reprod. Med.* **2010**, *28*, 017–026. [CrossRef]
10. Riley, J.K. Trophoblast Immune Receptors in Maternal-Fetal Tolerance. *Immunol. Investig.* **2008**, *37*, 395–426. [CrossRef]
11. Huppertz, B. The anatomy of the normal placenta. *J. Clin. Pathol.* **2008**, *61*, 1296–1302. [CrossRef] [PubMed]
12. McMaster, M.T.; Fisher, S.J. Placental Development. In *Encyclopedia of Hormones*; Academic Press: Cambridge, MA, USA, 2003; pp. 213–219. [CrossRef]
13. Carter, A.M. When is the maternal placental circulation established in man? *Placenta* **1997**, *18*, 83–87. [CrossRef]
14. Jaffe, R.; Jauniaux, E.; Hustin, J. Maternal circulation in the first-trimester human placenta—Myth or reality? *Am. J. Obstet. Gynecol.* **1997**, *176*, 695–705. [CrossRef]
15. Menon, R.; Moore, J.J. Fetal Membranes, Not a Mere Appendage of the Placenta, but a Critical Part of the Fetal-Maternal Interface Controlling Parturition. *Obstet. Gynecol. Clin. N. Am.* **2020**, *47*, 147–162. [CrossRef] [PubMed]
16. Kaufmann, P.; Black, S.; Huppertz, B. Endovascular Trophoblast Invasion: Implications for the Pathogenesis of Intrauterine Growth Retardation and Preeclampsia. *Biol. Reprod.* **2003**, *69*, 1–7. [CrossRef]
17. Castellucci, M.; Kosanke, G.; Verdenelli, F.; Huppertz, B.; Kaufmann, P. Villous sprouting: Fundamental mechanisms of human placental development. *Hum. Reprod. Updat.* **2000**, *6*, 485–494. [CrossRef]
18. Turco, M.Y.; Moffett, A. Development of the human placenta. *Development* **2019**, *146*, dev163428. [CrossRef]
19. Rasmussen, K.M.; Yaktine, A.L. *Institute of Medicine (US) and National Research Council (US) Committee to Reexamine IOM Pregnancy Weight Guidelines*; National Academies Press (US): Washington, DC, USA, 2009.
20. Fadl, S.; Moshiri, M.; Fligner, C.L.; Katz, D.S.; Dighe, M. Placental Imaging: Normal Appearance with Review of Pathologic Findings. *RadioGraphics* **2017**, *37*, 979–998. [CrossRef]
21. Architecture of Normal Villous Trees. In *Pathology of the Human Placenta*; Springer: Berlin/Heidelberg, Germany, 2006; pp. 121–173. [CrossRef]

22. Huppertz, B. Human Placentation. In *Encyclopedia of Reproduction*; Academic Press: Cambridge, MA, USA, 2018; pp. 431–439. [CrossRef]
23. Wang, Y.; Zhao, S. Structure of the Placenta. In *Vascular Biology of the Placenta*; Morgan & Claypool Life Sciences: San Rafael, CA, USA, 2010.
24. Aplin, J.D.; Lewis, R.M.; Jones, C.J. Development of the Human Placental Villus. In *Reference Module in Biomedical Sciences*; Elsevier: Amsterdam, The Netherlands, 2018. [CrossRef]
25. Barker, D.; Osmond, C.; Grant, S.; Thornburg, K.; Cooper, C.; Ring, S.; Davey-Smith, G. Maternal cotyledons at birth predict blood pressure in childhood. *Placenta* **2013**, *34*, 672–675. [CrossRef]
26. Kaufmann, P.; Huppertz, B.; Frank, H.-G. The fibrinoids of the human placenta: Origin, composition and functional relevance. *Ann. Anat. Anat. Anz.* **1996**, *178*, 485–501. [CrossRef]
27. Mor, G.; Cardenas, I. The Immune System in Pregnancy: A Unique Complexity. *Am. J. Reprod. Immunol.* **2010**, *63*, 425–433. [CrossRef] [PubMed]
28. Tong, M.; Abrahams, V.M. Immunology of the Placenta. *Obstet. Gynecol. Clin. N. Am.* **2020**, *47*, 49–63. [CrossRef] [PubMed]
29. Sanguansermsri, D.; Pongcharoen, S. Pregnancy Immunology: Decidual Immune Cells. *Asian Pac. J. Allergy Immunol.* **2008**, *26*, 171–181. [PubMed]
30. Hoo, R.; Nakimuli, A.; Vento-Tormo, R. Innate Immune Mechanisms to Protect Against Infection at the Human Decidual-Placental Interface. *Front. Immunol.* **2020**, *11*, 2070. [CrossRef]
31. Liu, Y.; Gao, S.; Zhao, Y.; Wang, H.; Pan, Q.; Shao, Q. Decidual Natural Killer Cells: A Good Nanny at the Maternal-Fetal Interface During Early Pregnancy. *Front. Immunol.* **2021**, *12*, 1684. [CrossRef]
32. Ander, S.E.; Diamond, M.S.; Coyne, C.B. Immune responses at the maternal-fetal interface. *Sci. Immunol.* **2019**, *4*, eaat6114. [CrossRef]
33. Shmeleva, E.V.; Colucci, F. Maternal natural killer cells at the intersection between reproduction and mucosal immunity. *Mucosal Immunol.* **2021**, *14*, 991–1005. [CrossRef]
34. Vento-Tormo, R.; Efremova, M.; Botting, R.A.; Turco, M.Y.; Vento-Tormo, M.; Meyer, K.B.; Park, J.-E.; Stephenson, E.; Polański, K.; Goncalves, A.; et al. Single-cell reconstruction of the early maternal–fetal interface in humans. *Nature* **2018**, *563*, 347–353. [CrossRef]
35. Jiang, X.; Wang, H. Macrophage subsets at the maternal-fetal interface. *Cell. Mol. Immunol.* **2020**, *17*, 889–891. [CrossRef]
36. Ning, F.; Liu, H.; Lash, G.E. The Role of Decidual Macrophages During Normal and Pathological Pregnancy. *Am. J. Reprod. Immunol.* **2016**, *75*, 298–309. [CrossRef]
37. Lissauer, D.; Kilby, M.; Moss, P. Maternal effector T cells within decidua: The adaptive immune response to pregnancy? *Placenta* **2017**, *60*, 140–144. [CrossRef] [PubMed]
38. Nancy, P.; Erlebacher, A. T cell behavior at the maternal-fetal interface. *Int. J. Dev. Biol.* **2014**, *58*, 189–198. [CrossRef] [PubMed]
39. Tilburgs, T.; Strominger, J.L. CD8+ Effector T Cells at the Fetal-Maternal Interface, Balancing Fetal Tolerance and Antiviral Immunity. *Am. J. Reprod. Immunol.* **2013**, *69*, 395–407. [CrossRef]
40. Shima, T.; Sasaki, Y.; Itoh, M.; Nakashima, A.; Ishii, N.; Sugamura, K.; Saito, S. Regulatory T cells are necessary for implantation and maintenance of early pregnancy but not late pregnancy in allogeneic mice. *J. Reprod. Immunol.* **2010**, *85*, 121–129. [CrossRef] [PubMed]
41. Mjösberg, J.; Berg, G.; Jenmalm, M.; Ernerudh, J. FOXP3+ Regulatory T Cells and T Helper 1, T Helper 2, and T Helper 17 Cells in Human Early Pregnancy Decidua1. *Biol. Reprod.* **2010**, *82*, 698–705. [CrossRef] [PubMed]
42. Mor, G.; Cardenas, I.; Abrahams, V.; Guller, S. Inflammation and pregnancy: The role of the immune system at the implantation site. *Ann. N. Y. Acad. Sci.* **2011**, *1221*, 80–87. [CrossRef]
43. Burton, G.J.; Fowden, A.L.; Thornburg, K.L. Placental Origins of Chronic Disease. *Physiol. Rev.* **2016**, *96*, 1509–1565. [CrossRef] [PubMed]
44. Sibley, C.P.; Birdsey, T.J.; Brownbill, P.; Clarson, L.H.; Doughty, I.; Glazier, J.D.; Greenwood, S.L.; Hughes, J.; Janssont, T.; Mylona, P.; et al. Mechanisms of maternofetal exchange across the human placenta. *Biochem. Soc. Trans.* **1998**, *26*, 86–91. [CrossRef]
45. Sibley, C.P.; Brownbill, P.; Glazier, J.D.; Greenwood, S.L. Knowledge needed about the exchange physiology of the placenta. *Placenta* **2018**, *64*, S9–S15. [CrossRef]
46. Hay, J.W.W. Placental Transport of Nutrients to the Fetus. *Horm. Res.* **1994**, *42*, 215–222. [CrossRef]
47. Hay, W.W. The Placenta: Not Just a Conduit for Maternal Fuels. *Diabetes* **1991**, *40*, 44–50. [CrossRef] [PubMed]
48. Larqué, E.; Ruiz-Palacios, M.; Koletzko, B. Placental regulation of fetal nutrient supply. *Curr. Opin. Clin. Nutr. Metab. Care* **2013**, *16*, 292–297. [CrossRef] [PubMed]
49. Jansson, T.; Aye, I.; Goberdhan, D. The emerging role of mTORC1 signaling in placental nutrient-sensing. *Placenta* **2012**, *33*, e23–e29. [CrossRef]
50. Carter, A.M. Placental Gas Exchange and the Oxygen Supply to the Fetus. *Compr. Physiol.* **2015**, *5*, 1381–1403. [CrossRef] [PubMed]
51. Murray, A.J. Oxygen delivery and fetal-placental growth: Beyond a question of supply and demand? *Placenta* **2012**, *33*, e16–e22. [CrossRef]
52. Serov, A.S.; Salafia, C.; Grebenkov, D.S.; Filoche, M. The role of morphology in mathematical models of placental gas exchange. *J. Appl. Physiol.* **2016**, *120*, 17–28. [CrossRef]

53. Barapatre, N.; Haeussner, E.; Grynspan, D.; Schmitz, C.; Von Koch, F.E.; Frank, H.-G. The Density of Cell Nuclei at the Materno-Fetal Exchange Barrier is Sexually Dimorphic in Normal Placentas, but not in IUGR. *Sci. Rep.* **2019**, *9*, 2359. [CrossRef]
54. Christians, J.K. The Placenta's Role in Sexually Dimorphic Fetal Growth Strategies. *Reprod. Sci.* **2021**, 1–13. [CrossRef]
55. Burton, G. Deportation of syncytial sprouts from the term human placenta. *Placenta* **2011**, *32*, 96–98. [CrossRef]
56. Huppertz, B. Placental Origins of Preeclampsia: Challenging the Current Hypothesis. *Hypertension* **2008**, *51*, 970–975. [CrossRef]
57. Chamley, L.; Holland, O.; Chen, Q.; Viall, C.; Stone, P.; Abumaree, M. Review: Where is the maternofetal interface? *Placenta* **2013**, *35*, S74–S80. [CrossRef] [PubMed]
58. Ji, L.; Brkić, J.; Liu, M.; Fu, G.; Peng, C.; Wang, Y.-L. Placental trophoblast cell differentiation: Physiological regulation and pathological relevance to preeclampsia. *Mol. Asp. Med.* **2013**, *34*, 981–1023. [CrossRef] [PubMed]
59. Jones, C.J.; Aplin, J.D. A re-examination of the origins of placental bed giant cells. *Placenta* **2021**, *114*, 39–41. [CrossRef] [PubMed]
60. Costa, M.A. The endocrine function of human placenta: An overview. *Reprod. Biomed. Online* **2016**, *32*, 14–43. [CrossRef] [PubMed]
61. Gomes, J.; Au, F.; Basak, A.; Cakmak, S.; Vincent, R.; Kumarathasan, P. Maternal blood biomarkers and adverse pregnancy outcomes: A systematic review and meta-analysis. *Crit. Rev. Toxicol.* **2019**, *49*, 461–478. [CrossRef]
62. Heerema-McKenney, A. Defense and infection of the human placenta. *APMIS* **2018**, *126*, 570–588. [CrossRef]
63. Ortega, M.A.; Fraile-Martínez, O.; García-Montero, C.; García-Gallego, S.; Sánchez-Trujillo, L.; Torres-Carranza, D.; Álvarez-Mon, M.; Pekarek, L.; García-Honduvilla, N.; Bujan, J.; et al. An integrative look at SARS-CoV-2 (Review). *Int. J. Mol. Med.* **2020**, *47*, 415–434. [CrossRef]
64. Cuñarro-López, Y.; Cano-Valderrama, Ó.; Pintado-Recarte, P.; Cueto-Hernández, I.; González-Garzón, B.; García-Tizón, S.; Bujan, J.; Asúnsolo, Á.; Ortega, M.A.; De León-Luis, J.A. Maternal and Perinatal Outcomes in Patients with Suspected COVID-19 and Their Relationship with a Negative RT-PCR Result. *J. Clin. Med.* **2020**, *9*, 3552. [CrossRef]
65. Zaga-Clavellina, V.; Diaz, L.; Olmos-Ortiz, A.; Godínez-Rubí, M.; Rojas-Mayorquín, A.E.; Ortuño-Sahagún, D. Central role of the placenta during viral infection: Immuno-competences and miRNA defensive responses. *Biochim. Biophys. Acta (BBA) Mol. Basis Dis.* **2021**, *1867*, 166182. [CrossRef]
66. Kreis, N.-N.; Ritter, A.; Louwen, F.; Yuan, J. A Message from the Human Placenta: Structural and Immunomodulatory Defense against SARS-CoV-2. *Cells* **2020**, *9*, 1777. [CrossRef]
67. Pelzer, E.; Gomez-Arango, L.F.; Barrett, H.L.; Nitert, M.D. Review: Maternal health and the placental microbiome. *Placenta* **2016**, *54*, 30–37. [CrossRef] [PubMed]
68. Myllynen, P.; Pasanen, M.; Vähäkangas, K. The fate and effects of xenobiotics in human placenta. *Expert Opin. Drug Metab. Toxicol.* **2007**, *3*, 331–346. [CrossRef] [PubMed]
69. Burton, G.J.; Fowden, A.L. The placenta: A multifaceted, transient organ. *Philos. Trans. R. Soc. B Biol. Sci.* **2015**, *370*, 20140066. [CrossRef]
70. Blanco-Castañeda, R.; Galaviz-Hernández, C.; Souto, P.C.D.S.; Lima, V.V.; Giachini, F.R.; Escudero, C.; Damiano, A.E.; Barragán-Zúñiga, L.J.; Martínez-Aguilar, G.; Sosa-Macías, M. The role of xenobiotic-metabolizing enzymes in the placenta: A growing research field. *Expert Rev. Clin. Pharmacol.* **2020**, *13*, 247–263. [CrossRef] [PubMed]
71. Hakkola, J.; Pelkonen, O.; Pasanen, M.; Raunio, H. Xenobiotic-Metabolizing Cytochrome P450 Enzymes in the Human Feto-Placental Unit: Role in Intrauterine Toxicity. *Crit. Rev. Toxicol.* **1998**, *28*, 35–72. [CrossRef]
72. St-Pierre, M.V.; Serrano, M.A.; Macias, R.I.R.; Dubs, U.; Hoechli, M.; Lauper, U.; Meier, P.J.; Marin, J.J.G. Expression of members of the multidrug resistance protein family in human term placenta. *Am. J. Physiol. Integr. Comp. Physiol.* **2000**, *279*, R1495–R1503. [CrossRef]
73. Ceckova-Novotna, M.; Pavek, P.; Staud, F. P-glycoprotein in the placenta: Expression, localization, regulation and function. *Reprod. Toxicol.* **2006**, *22*, 400–410. [CrossRef]
74. Martineau, M.; Papacleovoulou, G.; Abu-Hayyeh, S.; Dixon, P.; Ji, H.; Powrie, R.; Larson, L.; Chien, E.; Williamson, C. Cholestatic pregnancy is associated with reduced placental 11βHSD2 expression. *Placenta* **2013**, *35*, 37–43. [CrossRef]
75. Stark, M.J.; Wright, I.M.R.; Clifton, V.L. Sex-specific alterations in placental 11β-hydroxysteroid dehydrogenase 2 activity and early postnatal clinical course following antenatal betamethasone. *Am. J. Physiol. Integr. Comp. Physiol.* **2009**, *297*, R510–R514. [CrossRef]
76. Duhig, K.; Chappell, L.C.; Shennan, A.H. Oxidative stress in pregnancy and reproduction. *Obstet. Med.* **2016**, *9*, 113–116. [CrossRef]
77. Poston, L.; Raijmakers, M. Trophoblast Oxidative Stress, Antioxidants and Pregnancy Outcome—A Review. *Placenta* **2004**, *25*, S72–S78. [CrossRef]
78. Jones, M.L.; Mark, P.J.; Lewis, J.L.; Mori, T.A.; Keelan, J.A.; Waddell, B.J. Antioxidant Defenses in the Rat Placenta in Late Gestation: Increased Labyrinthine Expression of Superoxide Dismutases, Glutathione Peroxidase 3, and Uncoupling Protein 21. *Biol. Reprod.* **2010**, *83*, 254–260. [CrossRef] [PubMed]
79. Godfrey, K.M. The Role of the Placenta in Fetal Programming—A Review. *Placenta* **2002**, *23*, S20–S27. [CrossRef] [PubMed]
80. Myatt, L. Placental adaptive responses and fetal programming. *J. Physiol.* **2006**, *572*, 25–30. [CrossRef] [PubMed]
81. Lewis, R.; Cleal, J.; Hanson, M. Review: Placenta, evolution and lifelong health. *Placenta* **2012**, *33*, S28–S32. [CrossRef]
82. Belkacemi, L.; Nelson, D.M.; Desai, M.; Ross, M.G. Maternal Undernutrition Influences Placental-Fetal Development1. *Biol. Reprod.* **2010**, *83*, 325–331. [CrossRef]

83. Vaag, A.A.; Grunnet, L.G.; Arora, G.P.; Brøns, C. The thrifty phenotype hypothesis revisited. *Diabetologia* **2012**, *55*, 2085–2088. [CrossRef]
84. Şanlı, E.; Kabaran, S. Maternal Obesity, Maternal Overnutrition and Fetal Programming: Effects of Epigenetic Mechanisms on the Development of Metabolic Disorders. *Curr. Genom.* **2019**, *20*, 419–427. [CrossRef]
85. Connor, K.L.; Kibschull, M.; Matysiak-Zablocki, E.; Nguyen, T.T.-T.N.; Matthews, S.G.; Lye, S.J.; Bloise, E. Maternal malnutrition impacts placental morphology and transporter expression: An origin for poor offspring growth. *J. Nutr. Biochem.* **2020**, *78*, 108329. [CrossRef]
86. Ramírez-Vélez, R.; Bustamante, J.; Czerniczyniec, A.; De Plata, A.C.A.; Lores-Arnaiz, S. Effect of Exercise Training on Enos Expression, NO Production and Oxygen Metabolism in Human Placenta. *PLoS ONE* **2013**, *8*, e80225. [CrossRef]
87. Clapp, J. Influence of Endurance Exercise and Diet on Human Placental Development and Fetal Growth. *Placenta* **2006**, *27*, 527–534. [CrossRef]
88. Dahlerup, B.R.; Egsmose, E.L.; Siersma, V.; Mortensen, E.L.; Hedegaard, M.; Knudsen, L.E.; Mathiesen, L. Maternal stress and placental function, a study using questionnaires and biomarkers at birth. *PLoS ONE* **2018**, *13*, e0207184. [CrossRef] [PubMed]
89. Bronson, S.L.; Bale, T.L. The Placenta as a Mediator of Stress Effects on Neurodevelopmental Reprogramming. *Neuropsychopharmacology* **2015**, *41*, 207–218. [CrossRef] [PubMed]
90. Qiu, C.; Sanchez, S.E.; Gelaye, B.; Enquobahrie, D.A.; Ananth, C.V.; Williams, M.A. Maternal sleep duration and complaints of vital exhaustion during pregnancy is associated with placental abruption. *J. Matern. Neonatal Med.* **2014**, *28*, 350–355. [CrossRef] [PubMed]
91. Soma-Pillay, P.; Nelson-Piercy, C.; Tolppanen, H.; Mebazaa, A. Physiological changes in pregnancy. *Cardiovasc. J. Afr.* **2016**, *27*, 89–94. [CrossRef]
92. Sanghavi, M.; Rutherford, J.D. Cardiovascular Physiology of Pregnancy. *Circulation* **2014**, *130*, 1003–1008. [CrossRef] [PubMed]
93. Parks, W.T.; Catov, J.M. The Placenta as a Window to Maternal Vascular Health. *Obstet. Gynecol. Clin. N. Am.* **2019**, *47*, 17–28. [CrossRef]
94. Del Gobbo, G.; Konwar, C.; Robinson, W.P. The significance of the placental genome and methylome in fetal and maternal health. *Qual. Life Res.* **2019**, *139*, 1183–1196. [CrossRef] [PubMed]
95. Goldstein, J.A.; Gallagher, K.; Beck, C.; Kumar, R.; Gernand, A.D. Maternal-Fetal Inflammation in the Placenta and the Developmental Origins of Health and Disease. *Front. Immunol.* **2020**, *11*, 531543. [CrossRef]
96. Yeung, E.; Saha, A.; Zhu, C.; Trinh, M.; Hinkle, S.; Pollack, A.; Grantz, K.; Mills, J.; Mumford, S.; Zhang, C.; et al. Placental characteristics and risks of maternal mortality 50 years after delivery. *Placenta* **2021**, *117*, 194–199. [CrossRef]
97. Hutcheon, J.A.; Lisonkova, S.; Joseph, K. Epidemiology of pre-eclampsia and the other hypertensive disorders of pregnancy. *Best Pr. Res. Clin. Obstet. Gynaecol.* **2011**, *25*, 391–403. [CrossRef] [PubMed]
98. Filipek, A.; Jurewicz, E. Preeclampsia—A Disease of Pregnant Women. *Postepy Biochem.* **2018**, *64*, 229–232. [CrossRef] [PubMed]
99. Brown, M.A.; Magee, L.A.; Kenny, L.C.; Karumanchi, S.A.; McCarthy, F.; Saito, S.; Hall, D.R.; Warren, C.E.; Adoyi, G.; Ishaku, S. Hypertensive Disorders of Pregnancy. *Hypertension* **2018**, *72*, 24–43. [CrossRef] [PubMed]
100. Gestational Hypertension and Preeclampsia: ACOG Practice Bulletin, Number 222. *Obstet. Gynecol.* **2020**, *135*, e237–e260. [CrossRef] [PubMed]
101. Kenny, L.; English, F.; McCarthy, F. Risk factors and effective management of preeclampsia. *Integr. Blood Press. Control* **2015**, *8*, 7–12. [CrossRef]
102. Paré, E.; Parry, S.; McElrath, T.F.; Pucci, D.; Newton, A.; Lim, K.-H. Clinical Risk Factors for Preeclampsia in the 21st Century. *Obstet. Gynecol.* **2014**, *124*, 763–770. [CrossRef] [PubMed]
103. Tranquilli, A.L. Introduction to ISSHP new classification of preeclampsia. *Pregnancy Hypertens.* **2013**, *3*, 58–59. [CrossRef]
104. Raymond, D.; Peterson, E. A Critical Review of Early-Onset and Late-Onset Preeclampsia. *Obstet. Gynecol. Surv.* **2011**, *66*, 497–506. [CrossRef]
105. Bokslag, A.; van Weissenbruch, M.; Mol, B.W.; de Groot, C.J.M. Preeclampsia; short and long-term consequences for mother and neonate. *Early Hum. Dev.* **2016**, *102*, 47–50. [CrossRef]
106. Wallace, K.; Harris, S.; Addison, A.; Bean, C. HELLP Syndrome: Pathophysiology and Current Therapies. *Curr. Pharm. Biotechnol.* **2018**, *19*, 816–826. [CrossRef]
107. Berhan, Y. No Hypertensive Disorder of Pregnancy; No Preeclampsia-eclampsia; No Gestational Hypertension; No Hellp Syndrome. Vascular Disorder of Pregnancy Speaks for All. *Ethiop. J. Health Sci.* **2016**, *26*, 177–186. [CrossRef] [PubMed]
108. Magley, M.; Hinson, M.R. Eclampsia. In *StatPearls*; StatPearls Publishing: Treasure Island, FL, USA, 2021.
109. Marasciulo, F.; Orabona, R.; Fratelli, N.; Fichera, A.; Valcamonico, A.; Ferrari, F.; Odicino, F.E.; Sartori, E.; Prefumo, F. Pre-eclampsia and late fetal growth restriction. *Minerva Obstet. Gynecol.* **2021**, *73*, 435–441. [CrossRef] [PubMed]
110. Frost, A.L.; Suriano, K.; Aye, C.Y.L.; Leeson, P.; Lewandowski, A.J. The Immediate and Long-Term Impact of Preeclampsia on Offspring Vascular and Cardiac Physiology in the Preterm Infant. *Front. Pediatr.* **2021**, *9*, 625726. [CrossRef] [PubMed]
111. Ives, C.W.; Sinkey, R.; Rajapreyar, I.; Tita, A.T.; Oparil, S. Preeclampsia—Pathophysiology and Clinical Presentations. *J. Am. Coll. Cardiol.* **2020**, *76*, 1690–1702. [CrossRef]
112. Rolnik, D.L.; Wright, D.; Poon, L.C.Y.; Syngelaki, A.; O'Gorman, N.; de Paco Matallana, C.; Akolekar, R.; Cicero, S.; Janga, D.; Singh, M.; et al. ASPRE trial: Performance of screening for preterm pre-eclampsia. *Ultrasound Obstet. Gynecol.* **2017**, *50*, 492–495. [CrossRef]

113. Bujold, E.; Roberge, S.; Lacasse, Y.; Bureau, M.; Audibert, F.; Marcoux, S.; Forest, J.-C.; Giguère, Y. Prevention of Preeclampsia and Intrauterine Growth Restriction With Aspirin Started in Early Pregnancy. *Obstet. Gynecol.* **2010**, *116*, 402–414. [CrossRef]
114. Roberge, S.; Bujold, E.; Nicolaides, K. Aspirin for the prevention of preterm and term preeclampsia: Systematic review and metaanalysis. *Am. J. Obstet. Gynecol.* **2018**, *218*, 287–293.e1. [CrossRef]
115. Hofmeyr, G.J.; Manyame, S.; Medley, N.; Williams, M.J. Calcium supplementation commencing before or early in pregnancy, for preventing hypertensive disorders of pregnancy. *Cochrane Database Syst. Rev.* **2019**, *2019*, CD011192. [CrossRef]
116. Omotayo, M.O.; Dickin, K.; O'Brien, K.O.; Neufeld, L.M.; De Regil, L.M.; Stoltzfus, R.J. Calcium Supplementation to Prevent Preeclampsia: Translating Guidelines into Practice in Low-Income Countries. *Adv. Nutr. Int. Rev. J.* **2016**, *7*, 275–278. [CrossRef]
117. Maia e Holanda Moura, S.B.; Marques Lopes, L.; Murthi, P.; da Silva Costa, F. Prevention of Preeclampsia. *J. Pregnancy* **2012**, *2012*, 1–9. [CrossRef]
118. Sunjaya, A.F.; Med, B.; Sunjaya, A.P. Evaluation of Serum Biomarkers and Other Diagnostic Modalities for Early Diagnosis of Preeclampsia. *J. Fam. Reprod. Health* **2019**, *13*, 56.
119. Park, H.J.; Shim, S.S.; Cha, D.H. Combined Screening for Early Detection of Pre-Eclampsia. *Int. J. Mol. Sci.* **2015**, *16*, 17952–17974. [CrossRef] [PubMed]
120. Chaiworapongsa, T.; Chaemsaithong, P.; Yeo, L.; Romero, R. Pre-eclampsia part 1: Current understanding of its pathophysiology. *Nat. Rev. Nephrol.* **2014**, *10*, 466–480. [CrossRef] [PubMed]
121. Gathiram, P.; Moodley, J. Pre-eclampsia: Its pathogenesis and pathophysiolgy. *Cardiovasc. J. Afr.* **2016**, *27*, 71–78. [CrossRef]
122. Roberts, J.; Hubel, C. The Two Stage Model of Preeclampsia: Variations on the Theme. *Placenta* **2009**, *30*, 32–37. [CrossRef]
123. Lyall, F.; Bulmer, J.N.; Duffie, E.; Cousins, F.; Theriault, A.; Robson, S.C. Human Trophoblast Invasion and Spiral Artery Transformation: The Role of PECAM-1 in Normal Pregnancy, Preeclampsia, and Fetal Growth Restriction. *Am. J. Pathol.* **2001**, *158*, 1713–1721. [CrossRef]
124. Burke, S.D.; Karumanchi, S.A. Spiral Artery Remodeling in Preeclampsia Revisited. *Hypertension* **2013**, *62*, 1013–1014. [CrossRef]
125. Bakrania, B.A.; George, E.M.; Granger, J.P. Animal models of preeclampsia: Investigating pathophysiology and therapeutic targets. *Am. J. Obstet. Gynecol.* **2021**. Online ahead of print. [CrossRef]
126. Staff, A.C.; Johnsen, G.M.; Dechend, R.; Redman, C.W. Preeclampsia and uteroplacental acute atherosis: Immune and inflammatory factors. *J. Reprod. Immunol.* **2013**, *101–102*, 120–126. [CrossRef]
127. Predoi, C.; Grigoriu, C.; Vladescu, R.; Mihart, A. Placental damages in preeclampsia—From ultrasound images to histopathological findings. *J. Med. Life* **2015**, *8*, 62–65.
128. Soleymanlou, N.; Jurisica, I.; Nevo, O.; Ietta, F.; Zhang, X.; Zamudio, S.; Post, M.; Caniggia, I. Molecular Evidence of Placental Hypoxia in Preeclampsia. *J. Clin. Endocrinol. Metab.* **2005**, *90*, 4299–4308. [CrossRef]
129. Redman, C.W.; Sargent, I.L. Latest Advances in Understanding Preeclampsia. *Science* **2005**, *308*, 1592–1594. [CrossRef] [PubMed]
130. Gilbert, J.S.; Nijland, M.J.; Knoblich, P. Placental ischemia and cardiovascular dysfunction in preeclampsia and beyond: Making the connections. *Expert Rev. Cardiovasc. Ther.* **2008**, *6*, 1367–1377. [CrossRef] [PubMed]
131. Shibuya, M. Vascular endothelial growth factor receptor-1 (VEGFR-1/Flt-1): A dual regulator for angiogenesis. *Angiogenesis* **2006**, *9*, 225–230. [CrossRef] [PubMed]
132. Dijke, P.T.; Goumans, M.-J.; Pardali, E. Endoglin in angiogenesis and vascular diseases. *Angiogenesis* **2008**, *11*, 79–89. [CrossRef] [PubMed]
133. Gu, Y.; Lewis, D.F.; Wang, Y. Placental Productions and Expressions of Soluble Endoglin, Soluble fms-Like Tyrosine Kinase Receptor-1, and Placental Growth Factor in Normal and Preeclamptic Pregnancies. *J. Clin. Endocrinol. Metab.* **2008**, *93*, 260–266. [CrossRef]
134. Lecarpentier, E.; Tsatsaris, V. Angiogenic balance (sFlt-1/PlGF) and preeclampsia. *Ann. Endocrinol.* **2016**, *77*, 97–100. [CrossRef]
135. Zhou, Q.; Qiao, F.-Y.; Zhao, C.; Liu, H.-Y. Hypoxic trophoblast-derived sFlt-1 may contribute to endothelial dysfunction: An implication for the mechanism of trophoblast-endothelial dysfunction in preeclampsia. *Cell Biol. Int.* **2010**, *35*, 61–66. [CrossRef]
136. Venkatesha, S.; Toporsian, M.; Lam, C.; Hanai, J.-I.; Mammoto, T.; Kim, Y.M.; Bdolah, Y.; Lim, K.-H.; Yuan, H.-T.; Libermann, T.A.; et al. Soluble endoglin contributes to the pathogenesis of preeclampsia. *Nat. Med.* **2006**, *12*, 642–649. [CrossRef]
137. Nikuei, P.; Rajaei, M.; Roozbeh, N.; Mohseni, F.; Poordarvishi, F.; Azad, M.; Haidari, S. Diagnostic accuracy of sFlt1/PlGF ratio as a marker for preeclampsia. *BMC Pregnancy Childbirth* **2020**, *20*, 80. [CrossRef]
138. Zeisler, H.; Llurba, E.; Chantraine, F.; Vatish, M.; Staff, A.C.; Sennström, M.; Olovsson, M.; Brennecke, S.P.; Stepan, H.; Allegranza, D.; et al. Predictive Value of the sFlt-1:PlGF Ratio in Women with Suspected Preeclampsia. *N. Engl. J. Med.* **2016**, *374*, 13–22. [CrossRef] [PubMed]
139. Caillon, H.; Tardif, C.; Dumontet, E.; Winer, N.; Masson, D. Evaluation of sFlt-1/PlGF Ratio for Predicting and Improving Clinical Management of Pre-eclampsia: Experience in a Specialized Perinatal Care Center. *Ann. Lab. Med.* **2018**, *38*, 95–101. [CrossRef] [PubMed]
140. Lamarca, B.; Amaral, L.M.; Harmon, A.C.; Cornelius, D.C.; Faulkner, J.L.; Cunningham, M.W., Jr. Placental Ischemia and Resultant Phenotype in Animal Models of Preeclampsia. *Curr. Hypertens. Rep.* **2016**, *18*, 1–9. [CrossRef] [PubMed]
141. Laresgoiti-Servitje, E. A leading role for the immune system in the pathophysiology of preeclampsia. *J. Leukoc. Biol.* **2013**, *94*, 247–257. [CrossRef] [PubMed]
142. Shah, D.M. The role of RAS in the pathogenesis of preeclampsia. *Curr. Hypertens. Rep.* **2006**, *8*, 144–152. [CrossRef] [PubMed]

143. Lumbers, E.R.; Delforce, S.J.; Arthurs, A.; Pringle, K. Causes and Consequences of the Dysregulated Maternal Renin-Angiotensin System in Preeclampsia. *Front. Endocrinol.* **2019**, *10*, 563. [CrossRef] [PubMed]
144. Herse, F.; LaMarca, B. Angiotensin II Type 1 Receptor Autoantibody (AT1-AA)-Mediated Pregnancy Hypertension. *Am. J. Reprod. Immunol.* **2012**, *69*, 413–418. [CrossRef]
145. Saito, S.; Shiozaki, A.; Nakashima, A.; Sakai, M.; Sasaki, Y. The role of the immune system in preeclampsia. *Mol. Asp. Med.* **2007**, *28*, 192–209. [CrossRef]
146. Aggarwal, R.; Jain, A.K.; Mittal, P.; Kohli, M.; Jawanjal, P.; Rath, G. Association of pro- and anti-inflammatory cytokines in preeclampsia. *J. Clin. Lab. Anal.* **2019**, *33*, e22834. [CrossRef]
147. Perez-Sepulveda, A.; Torres, M.J.; Khoury, M.; Illanes, S.E. Innate Immune System and Preeclampsia. *Front. Immunol.* **2014**, *5*, 244. [CrossRef]
148. Liu, L.; Zhao, G.; Fan, H.; Zhao, X.; Li, P.; Wang, Z.; Hu, Y.; Hou, Y. Mesenchymal Stem Cells Ameliorate Th1-Induced Pre-Eclampsia-Like Symptoms in Mice via the Suppression of TNF-α Expression. *PLoS ONE* **2014**, *9*, e88036. [CrossRef] [PubMed]
149. Olaniyi, K.S.; Moodley, J.; Mahabeer, Y.; Mackraj, I. Placental Microbial Colonization and Its Association With Pre-eclampsia. *Front. Cell. Infect. Microbiol.* **2020**, *10*, 413. [CrossRef] [PubMed]
150. Lamarca, B. The role of immune activation in contributing to vascular dysfunction and the pathophysiology of hypertension during preeclampsia. *Minerva Ginecol.* **2010**, *62*, 105–120. [PubMed]
151. Saleh, L.; Verdonk, K.; Visser, W.; van den Meiracker, A.H.; Danser, A.H.J. The emerging role of endothelin-1 in the pathogenesis of pre-eclampsia. *Ther. Adv. Cardiovasc. Dis.* **2016**, *10*, 282–293. [CrossRef] [PubMed]
152. Siddiqui, I.A.; Jaleel, A.; Tamimi, W.; Al Kadri, H.M.F. Role of oxidative stress in the pathogenesis of preeclampsia. *Arch. Gynecol. Obstet.* **2010**, *282*, 469–474. [CrossRef] [PubMed]
153. Aouache, R.; Biquard, L.; Vaiman, D.; Miralles, F. Oxidative Stress in Preeclampsia and Placental Diseases. *Int. J. Mol. Sci.* **2018**, *19*, 1496. [CrossRef]
154. Baker, B.C.; Heazell, A.E.P.; Sibley, C.; Wright, R.; Bischof, H.; Beards, F.; Guevara, T.; Girard, S.; Jones, R.L. Hypoxia and oxidative stress induce sterile placental inflammation in vitro. *Sci. Rep.* **2021**, *11*, 7281. [CrossRef]
155. Silvestro, S.; Calcaterra, V.; Pelizzo, G.; Bramanti, P.; Mazzon, E. Prenatal Hypoxia and Placental Oxidative Stress: Insights from Animal Models to Clinical Evidences. *Antioxidants* **2020**, *9*, 414. [CrossRef]
156. Wu, F.; Tian, F.-J.; Lin, Y. Oxidative Stress in Placenta: Health and Diseases. *BioMed Res. Int.* **2015**, *2015*, 293271. [CrossRef]
157. Guerby, P.; Tasta, O.; Swiader, A.; Pont, F.; Bujold, E.; Parant, O.; Vayssiere, C.; Salvayre, R.; Negre-Salvayre, A. Role of oxidative stress in the dysfunction of the placental endothelial nitric oxide synthase in preeclampsia. *Redox Biol.* **2021**, *40*, 101861. [CrossRef]
158. Sandrim, V.C.; Palei, A.C.; Metzger, I.F.; Gomes, V.A.; Cavalli, R.C.; Tanus-Santos, J.E. Nitric Oxide Formation Is Inversely Related to Serum Levels of Antiangiogenic Factors Soluble Fms-Like Tyrosine Kinase-1 and Soluble Endogline in Preeclampsia. *Hypertension* **2008**, *52*, 402–407. [CrossRef] [PubMed]
159. Tashie, W.; Fondjo, L.A.; Owiredu, W.K.B.A.; Ephraim, R.K.D.; Asare, L.; Adu-Gyamfi, E.A.; Seidu, L. Altered Bioavailability of Nitric Oxide and L-Arginine Is a Key Determinant of Endothelial Dysfunction in Preeclampsia. *BioMed Res. Int.* **2020**, *2020*, 3251956. [CrossRef] [PubMed]
160. Jurado, S.; Saraiva, K.; Marceliano, C.; Souza, V.; Vieira, I. Maternal and Fetal Complications Due to Decreased Nitric Oxide Synthesis during Gestation. Available online: https://www.intechopen.com/chapters/66384 (accessed on 10 December 2021).
161. Sanchez-Aranguren, L.C.; Prada, C.E.; Riãno-Medina, C.E.; Lopez, M. Endothelial dysfunction and preeclampsia: Role of oxidative stress. *Front. Physiol.* **2014**, *5*, 372. [CrossRef]
162. Chen, J.; Sheehan, P.M.; Brennecke, S.; Keogh, R.J. Vessel remodelling, pregnancy hormones and extravillous trophoblast function. *Mol. Cell. Endocrinol.* **2012**, *349*, 138–144. [CrossRef]
163. Handschuh, K.; Guibourdenche, J.; Tsatsaris, V.; Guesnon, M.; Laurendeau, I.; Evain-Brion, D.; Fournier, T. Human Chorionic Gonadotropin Produced by the Invasive Trophoblast But Not the Villous Trophoblast Promotes Cell Invasion and Is Down-Regulated by Peroxisome Proliferator-Activated Receptor-γ. *Endocrinology* **2007**, *148*, 5011–5019. [CrossRef]
164. Atamer, Y.; Erden, A.C.; Demir, B.; Koçyigit, Y.; Atamer, A. The relationship between plasma levels of leptin and androgen in healthy and preeclamptic pregnant women. *Acta Obstet. Gynecol. Scand.* **2004**, *83*, 425–430. [CrossRef] [PubMed]
165. Salustiano, E.; De Pinho, J.C.; Provost, K.; Ruano, R.; Zugaib, M. Maternal Serum Hormonal Factors in the Pathogenesis of Preeclampsia. *Obstet. Gynecol. Surv.* **2013**, *68*, 141–150. [CrossRef]
166. Goswami, D.; Tannetta, D.; Magee, L.; Fuchisawa, A.; Redman, C.; Sargent, I.; von Dadelszen, P. Excess syncytiotrophoblast microparticle shedding is a feature of early-onset pre-eclampsia, but not normotensive intrauterine growth restriction. *Placenta* **2006**, *27*, 56–61. [CrossRef]
167. Raguema, N.; Moustadraf, S.; Bertagnolli, M. Immune and Apoptosis Mechanisms Regulating Placental Development and Vascularization in Preeclampsia. *Front. Physiol.* **2020**, *11*, 98. [CrossRef]
168. Gerasimova, E.M.; Fedotov, S.A.; Kachkin, D.V.; Vashukova, E.S.; Glotov, A.S.; Chernoff, Y.O.; Rubel, A.A. Protein Misfolding during Pregnancy: New Approaches to Preeclampsia Diagnostics. *Int. J. Mol. Sci.* **2019**, *20*, 6183. [CrossRef]
169. López-Otín, C.; Blasco, M.A.; Partridge, L.; Serrano, M.; Kroemer, G. The Hallmarks of Aging. *Cell* **2013**, *153*, 1194–1217. [CrossRef] [PubMed]
170. Manna, S.; McCarthy, C.; McCarthy, F. Placental Ageing in Adverse Pregnancy Outcomes: Telomere Shortening, Cell Senescence, and Mitochondrial Dysfunction. *Oxidative Med. Cell. Longev.* **2019**, *2019*, 3095383. [CrossRef] [PubMed]

171. Mayne, B.T.; Leemaqz, S.; Smith, A.K.; Breen, J.; Roberts, C.; Bianco-Miotto, T. Accelerated placental aging in early onset preeclampsia pregnancies identified by DNA methylation. *Epigenomics* **2017**, *9*, 279–289. [CrossRef]
172. Herzog, E.M.; Eggink, A.J.; Reijnierse, A.; Kerkhof, M.A.; de Krijger, R.R.; Roks, A.J.; Reiss, I.K.; Nigg, A.L.; Eilers, P.H.; Steegers, E.A.; et al. Impact of early- and late-onset preeclampsia on features of placental and newborn vascular health. *Placenta* **2016**, *49*, 72–79. [CrossRef]
173. Lisonkova, S.; Sabr, Y.; Mayer, C.; Young, C.; Skoll, A.; Joseph, K. Maternal Morbidity Associated With Early-Onset and Late-Onset Preeclampsia. *Obstet. Gynecol.* **2014**, *124*, 771–781. [CrossRef]
174. Lisonkova, S.; Joseph, K. Incidence of preeclampsia: Risk factors and outcomes associated with early- versus late-onset disease. *Am. J. Obstet. Gynecol.* **2013**, *209*, 544.e1–544.e12. [CrossRef] [PubMed]
175. Staff, A.C. The two-stage placental model of preeclampsia: An update. *J. Reprod. Immunol.* **2019**, *134–135*, 1–10. [CrossRef] [PubMed]
176. Verlohren, S.; Melchiorre, K.; Khalil, A.; Thilaganathan, B. Uterine artery Doppler, birth weight and timing of onset of pre-eclampsia: Providing insights into the dual etiology of late-onset pre-eclampsia. *Ultrasound Obstet. Gynecol.* **2014**, *44*, 293–298. [CrossRef]
177. Thilaganathan, B. Pre-eclampsia and the cardiovascular-placental axis. *Ultrasound Obstet. Gynecol.* **2018**, *51*, 714–717. [CrossRef]
178. Berg, C.V.D.; Chaves, I.; Herzog, E.M.; Willemsen, S.P.; Van Der Horst, G.T.J.; Steegers-Theunissen, R.P.M. Early- and late-onset preeclampsia and the DNA methylation of circadian clock and clock-controlled genes in placental and newborn tissues. *Chronobiol. Int.* **2017**, *34*, 921–932. [CrossRef]
179. Suhag, A.; Berghella, V. Intrauterine Growth Restriction (IUGR): Etiology and Diagnosis. *Curr. Obstet. Gynecol. Rep.* **2013**, *2*, 102–111. [CrossRef]
180. Unterscheider, J.; Daly, S.; Geary, M.P.; Kennelly, M.M.; McAuliffe, F.M.; O'Donoghue, K.; Hunter, A.; Morrison, J.J.; Burke, G.; Dicker, P.; et al. Optimizing the definition of intrauterine growth restriction: The multicenter prospective PORTO Study. *Am. J. Obstet. Gynecol.* **2013**, *208*, 290.e1–290.e6. [CrossRef] [PubMed]
181. Osuchukwu, O.O.; Reed, D.J. Small for Gestational Age. Available online: https://www.ncbi.nlm.nih.gov/books/NBK563247/ (accessed on 10 December 2021).
182. Beune, I.M.; Bloomfield, F.H.; Ganzevoort, W.; Embleton, N.; Rozance, P.J.; van Wassenaer-Leemhuis, A.G.; Wynia, K.; Gordijn, S.J. Consensus Based Definition of Growth Restriction in the Newborn. *J. Pediatr.* **2018**, *196*, 71–76. [CrossRef] [PubMed]
183. Beune, I.M.; Damhuis, S.E.; Ganzevoort, W.; Hutchinson, J.C.; Khong, T.Y.; Mooney, E.E.; Sebire, N.J.; Gordijn, S.J. Consensus Definition of Fetal Growth Restriction in Intrauterine Fetal Death: A Delphi Procedure. *Arch. Pathol. Lab. Med.* **2020**, *145*, 428–436. [CrossRef]
184. Martín-Estal, I.; de la Garza, R.G.; de Cortázar, I.C. Intrauterine Growth Retardation (IUGR) as a Novel Condition of Insulin-Like Growth Factor-1 (IGF-1) Deficiency. In *Reviews of Physiology, Biochemistry and Pharmacology*; Springer: Cham, Switzerland, 2016; Volume 170. [CrossRef]
185. Saleem, T.; Sajjad, N.; Fatima, S.; Habib, N.; Ali, S.R.; Qadir, M. Intrauterine growth retardation-small events, big consequences. *Ital. J. Pediatr.* **2011**, *37*, 41–44. [CrossRef] [PubMed]
186. Gallego, E.M.; Pujol, A.T.; Bartra, A.J.C.; Roig, M.D.G. Fetal Growth Restriction. Available online: https://www.intechopen.com/chapters/70711 (accessed on 10 December 2021).
187. Salam, R.A.; Das, J.K.; Ali, A.; Lassi, Z.S.; Bhutta, Z.A. Maternal undernutrition and intrauterine growth restriction. *Expert Rev. Obstet. Gynecol.* **2013**, *8*, 559–567. [CrossRef]
188. Radulescu, L.; Munteanu, O.; Popa, F.; Cirstoiu, M. The implications and consequences of maternal obesity on fetal intrauterine growth restriction. *J. Med. Life* **2013**, *6*, 292–298. [PubMed]
189. Parikh, R.M.; Joshi, S.R.; Menon, P.S.; Shah, N.S. Intensive glycemic control in diabetic pregnancy with intrauterine growth restriction is detrimental to fetus. *Med. Hypotheses* **2007**, *69*, 203–205. [CrossRef]
190. Lean, S.C.; Heazell, A.E.P.; Dilworth, M.; Mills, T.; Jones, R.L. Placental Dysfunction Underlies Increased Risk of Fetal Growth Restriction and Stillbirth in Advanced Maternal Age Women. *Sci. Rep.* **2017**, *7*, 9677. [CrossRef]
191. Reeves, S.; Bernstein, I. Effects of maternal tobacco-smoke exposure on fetal growth and neonatal size. *Expert Rev. Obstet. Gynecol.* **2008**, *3*, 719–730. [CrossRef]
192. Garrison, L.; Leeman, L.; Savich, R.D.; Gutierrez, H.; Rayburn, W.F.; Bakhireva, L.N. Fetal Growth Outcomes in a Cohort of Polydrug- and Opioid-Dependent Patients. *J. Reprod. Med.* **2016**, *61*, 311–319. [PubMed]
193. Romo, A.; Carceller, R.; Tobajas, J. Intrauterine Growth Retardation (IUGR): Epidemiology and Etiology—PubMed. *Pediatr. Endocrinol. Rev.* **2009**, *6*, 332–336. [PubMed]
194. Vrachnis, N.; Loukas, N.; Vrachnis, D.; Antonakopoulos, N.; Zygouris, D.; Kolialexi, A.; Pergaliotis, V.; Iavazzo, C.; Mastorakos, G.; Iliodromiti, Z. A Systematic Review of Bisphenol A from Dietary and Non-Dietary Sources during Pregnancy and Its Possible Connection with Fetal Growth Restriction: Investigating Its Potential Effects and the Window of Fetal Vulnerability. *Nutrients* **2021**, *13*, 2426. [CrossRef] [PubMed]
195. Armengaud, J.; Yzydorczyk, C.; Siddeek, B.; Peyter, A.; Simeoni, U. Intrauterine growth restriction: Clinical consequences on health and disease at adulthood. *Reprod. Toxicol.* **2020**, *99*, 168–176. [CrossRef] [PubMed]
196. Choi, H.; Wang, L.; Lin, X.; Spengler, J.D.; Perera, F.P. Fetal Window of Vulnerability to Airborne Polycyclic Aromatic Hydrocarbons on Proportional Intrauterine Growth Restriction. *PLoS ONE* **2012**, *7*, e35464. [CrossRef]

197. Fleiss, B.; Wong, F.; Brownfoot, F.; Shearer, I.K.; Baud, O.; Walker, D.W.; Gressens, P.; Tolcos, M. Knowledge Gaps and Emerging Research Areas in Intrauterine Growth Restriction-Associated Brain Injury. *Front. Endocrinol.* **2019**, *10*, 188. [CrossRef]
198. Cherian, A.; Vijayaselvi, R. Risk assessment of intrauterine growth restriction. *Curr. Med. Issues* **2017**, *15*, 262. [CrossRef]
199. Silver, K.L.; Conroy, A.L.; Leke, R.G.F.; Leke, R.J.I.; Gwanmesia, P.; Molyneux, M.E.; Wallace, D.T.; Rogerson, S.J.; Kain, K.C. Circulating Soluble Endoglin Levels in Pregnant Women in Cameroon and Malawi—Associations with Placental Malaria and Fetal Growth Restriction. *PLoS ONE* **2011**, *6*, e24985. [CrossRef]
200. Gaccioli, F.; Lager, S. Placental Nutrient Transport and Intrauterine Growth Restriction. *Front. Physiol.* **2016**, *7*, 40. [CrossRef]
201. Malhotra, A.; Allison, B.J.; Castillo-Melendez, M.; Jenkin, G.; Polglase, G.R.; Miller, S.L. Neonatal Morbidities of Fetal Growth Restriction: Pathophysiology and Impact. *Front. Endocrinol.* **2019**, *10*, 55. [CrossRef]
202. Sharma, D.; Shastri, S.; Sharma, P. Intrauterine Growth Restriction: Antenatal and Postnatal Aspects. *Clin. Med. Insights Pediatr.* **2016**, *10*, 67–83. [CrossRef] [PubMed]
203. Cinar, B.; Sert, A.; Gokmen, Z.; Aypar, E.; Aslan, E.; Odabas, D. Left ventricular dimensions, systolic functions, and mass in term neonates with symmetric and asymmetric intrauterine growth restriction. *Cardiol. Young* **2013**, *25*, 301–307. [CrossRef] [PubMed]
204. Czernik, C.; Rhode, S.; Metze, B.; Bührer, C.; Schmitz, L. Comparison of left ventricular cardiac dimensions between small and appropriate for gestational age preterm infants below 30 weeks of gestation. *J. Périnat. Med.* **2013**, *41*, 219–226. [CrossRef] [PubMed]
205. Aburawi, E.H.; Malcus, P.; Thuring, A.; Fellman, V.; Pesonen, E. Coronary Flow in Neonates with Impaired Intrauterine Growth. *J. Am. Soc. Echocardiogr.* **2012**, *25*, 313–318. [CrossRef] [PubMed]
206. Rodriguez-Guerineau, L.; Perez-Cruz, M.; Roig, M.D.G.; Cambra, F.J.; Carretero, J.; Prada, F.; Gómez, O.; Crispi, F.; Bartrons, J. Cardiovascular adaptation to extrauterine life after intrauterine growth restriction. *Cardiol. Young* **2017**, *28*, 284–291. [CrossRef]
207. Cosmi, E.; Fanelli, T.; Visentin, S.; Trevisanuto, D.; Zanardo, V. Consequences in Infants That Were Intrauterine Growth Restricted. *J. Pregnancy* **2011**, *2011*, 364381. [CrossRef]
208. Pike, K.; Pillow, J.J.; Lucas, J.S. Long term respiratory consequences of intrauterine growth restriction. *Semin. Fetal Neonatal Med.* **2012**, *17*, 92–98. [CrossRef]
209. Briana, D.D.; Malamitsi-Puchner, A. Perinatal biomarkers implying 'Developmental Origins of Health and Disease' consequences in intrauterine growth restriction. *Acta Paediatr.* **2019**, *109*, 1317–1322. [CrossRef]
210. Shetty, S.; Idell, S. Fibrinolysis: Plasminogen Activator and Plasmin. Available online: https://www.sciencedirect.com/topics/neuroscience/plasminogen-activator (accessed on 10 December 2021).
211. Burton, G.; Jauniaux, E. Pathophysiology of placental-derived fetal growth restriction. *Am. J. Obstet. Gynecol.* **2018**, *218*, S745–S761. [CrossRef]
212. Krishna, U.; Bhalerao, S. Placental Insufficiency and Fetal Growth Restriction. *J. Obstet. Gynecol. India* **2011**, *61*, 505–511. [CrossRef]
213. Zhu, M.Y.; Milligan, N.; Keating, S.; Windrim, R.; Keunen, J.; Thakur, V.; Ohman, A.; Portnoy, S.; Sled, J.G.; Kelly, E.; et al. The hemodynamics of late-onset intrauterine growth restriction by MRI. *Am. J. Obstet. Gynecol.* **2015**, *214*, 367.e1–367.e17. [CrossRef] [PubMed]
214. Romero, R.; Whitten, A.; Korzeniewski, S.; Than, N.G.; Chaemsaithong, P.; Miranda, J.; Dong, Z.; Hassan, S.S.; Chaiworapongsa, T. Maternal Floor Infarction/Massive Perivillous Fibrin Deposition: A Manifestation of Maternal Antifetal Rejection? *Am. J. Reprod. Immunol.* **2013**, *70*, 285–298. [CrossRef] [PubMed]
215. Kim, C.J.; Romero, R.; Chaemsaithong, P.; Kim, J.-S. Chronic inflammation of the placenta: Definition, classification, pathogenesis, and clinical significance. *Am. J. Obstet. Gynecol.* **2015**, *213*, S53–S69. [CrossRef] [PubMed]
216. Martín-Estal, I.; Castilla-Cortázar, I.; Castorena-Torres, F. The Placenta as a Target for Alcohol During Pregnancy: The Close Relation with IGFs Signaling Pathway. *Rev. Physiol. Biochem. Pharmacol.* **2021**, 119–153. [CrossRef]
217. Burd, L.; Roberts, D.; Olson, M.; Odendaal, H. Ethanol and the placenta: A review. *J. Matern. Neonatal Med.* **2007**, *20*, 361–375. [CrossRef]
218. Steane, S.E.; Young, S.L.; Clifton, V.L.; Gallo, L.A.; Akison, L.K.; Moritz, K.M. Prenatal alcohol consumption and placental outcomes: A systematic review and meta-analysis of clinical studies. *Am. J. Obstet. Gynecol.* **2021**, *225*, 607.e1–607.e22. [CrossRef]
219. Sovio, U.; White, I.; Dacey, A.; Pasupathy, D.; Smith, G. Screening for fetal growth restriction with universal third trimester ultrasonography in nulliparous women in the Pregnancy Outcome Prediction (POP) study: A prospective cohort study. *Lancet* **2015**, *386*, 2089–2097. [CrossRef]
220. Mesdaghi-Nia, E.; Behrashi, M.; Saeidi, A.; Abedzadeh-Kalahroudi, M. Association between PAPP-A and placental thickness. *Int. J. Reprod. Biomed.* **2016**, *14*, 421–426. [CrossRef]
221. Ekin, A.; Gezer, C.; Taner, C.E.; Özeren, M. The association between low PAPP-A levels at first trimester and poor pregnancy outcomes. *Périnat. J.* **2014**, *22*, 142–146. [CrossRef]
222. Caliskan, R.; Atis, A.; Aydin, Y.; Acar, D.; Kiyak, H.; Topbas, F. PAPP-A concentrations change in patients with gestational diabetes. *J. Obstet. Gynaecol.* **2019**, *40*, 190–194. [CrossRef]
223. Antwi, E.; Amoakoh-Coleman, M.; Vieira, D.; Madhavaram, S.; Koram, K.A.; Grobbee, D.E.; Agyepong, I.A.; Klipstein-Grobusch, K. Systematic review of prediction models for gestational hypertension and preeclampsia. *PLoS ONE* **2020**, *15*, e0230955. [CrossRef] [PubMed]
224. Fruscalzo, A.; Cividino, A.; Rossetti, E.; Maurigh, A.; Londero, A.P.; Driul, L. First trimester PAPP-A serum levels and long-term metabolic outcome of mothers and their offspring. *Sci. Rep.* **2020**, *10*, 5131. [CrossRef]

225. Westgren, L.; Anneren, R.; Axelsson, O.; Evald, U.; Leb-lanc, K.; Ringden, O.; Winsor, S.; Hornberger, L.; Johnson, J.; Sinai Hospital Toronto, M. Low First-Trimester PAPP-A Identifies Pregnancies Requiring IUGR Screening. *Am. J. Obstet. Gynecol.* **2003**, *189*, S215. [CrossRef]
226. Antsaklis, P.; Fasoulakis, Z.; Theodora, M.; Diakosavvas, M.; Kontomanolis, E.N. Association of Low Maternal Pregnancy-associated Plasma Protein A with Adverse Perinatal Outcome. *Cureus* **2019**, *11*, e4912. [CrossRef] [PubMed]
227. Pascual-Mancho, J.; Pintado-Recarte, P.; Romero-Román, C.; Morales-Camino, J.; Hernández-Martin, C.; Bujan, J.; Ortega, M.; De León-Luis, J. Influence of Cerebral Vasodilation on Blood Reelin Levels in Growth Restricted Fetuses. *Diagnostics* **2021**, *11*, 1036. [CrossRef] [PubMed]
228. Cetin, I.; Mazzocco, M.I.; Giardini, V.; Cardellicchio, M.; Calabrese, S.; Algeri, P.; Martinelli, A.; Todyrenchuk, L.; Vergani, P. PLGF in a clinical setting of pregnancies at risk of Preeclampsia and/or Intrauterine Growth Restriction. *J. Matern. Neonatal Med.* **2016**, *30*, 144–149. [CrossRef] [PubMed]
229. Birdir, C.; Droste, L.; Fox, L.; Frank, M.; Fryze, J.; Enekwe, A.; Köninger, A.; Kimmig, R.; Schmidt, B.; Gellhaus, A. Predictive value of sFlt-1, PlGF, sFlt-1/PlGF ratio and PAPP-A for late-onset preeclampsia and IUGR between 32 and 37 weeks of pregnancy. *Pregnancy Hypertens.* **2018**, *12*, 124–128. [CrossRef]
230. Schoofs, K.; Grittner, U.; Engels, T.; Pape, J.; Denk, B.; Henrich, W.; Verlohren, S. The importance of repeated measurements of the sFlt-1/PlGF ratio for the prediction of preeclampsia and intrauterine growth restriction. *J. Périnat. Med.* **2014**, *42*, 61–68. [CrossRef]
231. Raia-Barjat, T.; Prieux, C.; Gris, J.-C.; Chapelle, C.; Laporte, S.; Chauleur, C. Angiogenic factors for prediction of preeclampsia and intrauterine growth restriction onset in high-risk women: AngioPred study. *J. Matern. Neonatal Med.* **2017**, *32*, 248–257. [CrossRef]
232. Sirikunalai, P.; Wanapirak, C.; Sirichotiyakul, S.; Tongprasert, F.; Srisupundit, K.; Luewan, S.; Traisrisilp, K.; Tongsong, T. Associations between maternal serum free beta human chorionic gonadotropin (β-hCG) levels and adverse pregnancy outcomes. *J. Obstet. Gynaecol.* **2015**, *36*, 178–182. [CrossRef]
233. Massimiani, M.; Salvi, S.; Tiralongo, G.M.; Moresi, S.; Stuhlmann, H.; Valensise, H.; Lanzone, A.; Campagnolo, L. Circulating EGFL7 distinguishes between IUGR and PE: An observational case–control study. *Sci. Rep.* **2021**, *11*, 17919. [CrossRef] [PubMed]
234. Audette, M.C.; Kingdom, J.C. Screening for fetal growth restriction and placental insufficiency. *Semin. Fetal Neonatal Med.* **2018**, *23*, 119–125. [CrossRef] [PubMed]
235. Molvarec, A.; Gullai, N.; Stenczer, B.; Fügedi, G.; Nagy, B.; Jr, J.R. Comparison of placental growth factor and fetal flow Doppler ultrasonography to identify fetal adverse outcomes in women with hypertensive disorders of pregnancy: An observational study. *BMC Pregnancy Childbirth* **2013**, *13*, 161. [CrossRef]
236. Bamfo, J.E.A.K.; Odibo, A.O. Diagnosis and Management of Fetal Growth Restriction. *J. Pregnancy* **2011**, *2011*, 640715. [CrossRef] [PubMed]
237. Kantake, M.; Ikeda, N.; Nakaoka, H.; Ohkawa, N.; Tanaka, T.; Miyabayashi, K.; Shoji, H.; Shimizu, T. IGF1 gene is epigenetically activated in preterm infants with intrauterine growth restriction. *Clin. Epigenet.* **2020**, *12*, 1–9. [CrossRef]
238. Merialdi, M.; Carroli, G.; Villar, J.; Abalos, E.; Guülmezoglu, A.M.; Kulier, R.; De Onis, M. Nutritional Interventions during Pregnancy for the Prevention or Treatment of Impaired Fetal Growth: An Overview of Randomized Controlled Trials. *J. Nutr.* **2003**, *133*, 1626S–1631S. [CrossRef]
239. Clarke, P.E.; Gross, H. Women's behaviour, beliefs and information sources about physical exercise in pregnancy. *Midwifery* **2004**, *20*, 133–141. [CrossRef]
240. Gatford, K.L.; Kaur, G.; Falcão-Tebas, F.; Wadley, G.; Wlodek, M.; Laker, R.C.; Ebeling, P.R.; McConell, G. Exercise as an intervention to improve metabolic outcomes after intrauterine growth restriction. *Am. J. Physiol. Metab.* **2014**, *306*, E999–E1012. [CrossRef]
241. Bhattacharjee, J.; Mohammad, S.; Goudreau, A.D.; Adamo, K.B. Physical activity differentially regulates VEGF, PlGF, and their receptors in the human placenta. *Physiol. Rep.* **2021**, *9*, e14710. [CrossRef]
242. Nascimento, S.L.; Surita, F.; Cecatti, J.G. Physical exercise during pregnancy. *Curr. Opin. Obstet. Gynecol.* **2012**, *24*, 387–394. [CrossRef]
243. Tomić, V.; Sporis, G.; Tomić, J.; Milanović, Z.; Zigmundovac-Klaić, D.; Pantelić, S. The effect of maternal exercise during pregnancy on abnormal fetal growth. *Croat. Med. J.* **2013**, *54*, 362–368. [CrossRef] [PubMed]
244. Melamed, N.; Baschat, A.; Yinon, Y.; Athanasiadis, A.; Mecacci, F.; Figueras, F.; Berghella, V.; Nazareth, A.; Tahlak, M.; McIntyre, H.D.; et al. FIGO (International Federation of Gynecology and Obstetrics) initiative on fetal growth: Best practice advice for screening, diagnosis, and management of fetal growth restriction. *Int. J. Gynecol. Obstet.* **2021**, *152*, 3–57. [CrossRef]
245. Groom, K.M.; David, A. The role of aspirin, heparin, and other interventions in the prevention and treatment of fetal growth restriction. *Am. J. Obstet. Gynecol.* **2018**, *218*, S829–S840. [CrossRef] [PubMed]
246. Pereira, L.; Petitt, M.; Fong, A.; Tsuge, M.; Tabata, T.; Fang-Hoover, J.; Maidji, E.; Zydek, M.; Zhou, Y.; Inoue, N.; et al. Intrauterine Growth Restriction Caused by Underlying Congenital Cytomegalovirus Infection. *J. Infect. Dis.* **2014**, *209*, 1573–1584. [CrossRef] [PubMed]
247. Plotkin, S.A.; Wang, D.; Oualim, A.; Diamond, D.J.; Kotton, C.N.; Mossman, S.; Carfi, A.; Anderson, D.; Dormitzer, P.R. The Status of Vaccine Development Against the Human Cytomegalovirus. *J. Infect. Dis.* **2020**, *221*, S113–S122. [CrossRef] [PubMed]
248. Piazza, G. Varicose Veins. *Circulation* **2014**, *130*, 582–587. [CrossRef]

249. Lurie, F.; Passman, M.; Meisner, M.; Dalsing, M.; Masuda, E.; Welch, H.; Bush, R.L.; Blebea, J.; Carpentier, P.H.; De Maeseneer, M.; et al. The 2020 update of the CEAP classification system and reporting standards. *J. Vasc. Surg. Venous Lymphat. Disord.* **2020**, *8*, 342–352. [CrossRef]
250. Youn, Y.J.; Lee, J. Chronic venous insufficiency and varicose veins of the lower extremities. *Korean J. Intern. Med.* **2019**, *34*, 269–283. [CrossRef]
251. Davies, A.H. The Seriousness of Chronic Venous Disease: A Review of Real-World Evidence—PubMed. *Adv. Ther.* **2019**, *36* (Suppl. S1), 5–12. [CrossRef]
252. Lohr, J.M.; Bush, R.L. Venous disease in women: Epidemiology, manifestations, and treatment. *J. Vasc. Surg.* **2013**, *57*, 37S–45S. [CrossRef]
253. Ortega, M.; Fraile-Martínez, O.; García-Montero, C.; Álvarez-Mon, M.; Chaowen, C.; Ruiz-Grande, F.; Pekarek, L.; Monserrat, J.; Asúnsolo, A.; García-Honduvilla, N.; et al. Understanding Chronic Venous Disease: A Critical Overview of Its Pathophysiology and Medical Management. *J. Clin. Med.* **2021**, *10*, 3239. [CrossRef] [PubMed]
254. Cornu-Thenard, A.; Boivin, P. Chronic Venous Disease during Pregnancy-Servier-Phlebolymphology Servier-Phlebolymphology. *Phlebolymphology* **2014**, *21*, 138–145.
255. Li, X.; Jiang, X.-Y.; Ge, J.; Wang, J.; Chen, G.-J.; Xu, L.; Xie, D.-Y.; Yuan, T.-Y.; Zhang, D.-S.; Zhang, H.; et al. Aberrantly Expressed lncRNAs in Primary Varicose Great Saphenous Veins. *PLoS ONE* **2014**, *9*, e86156. [CrossRef] [PubMed]
256. Ortega, M.A.; Romero, B.; Asúnsolo, Á.; Sola, M.; Álavrez-Rocha, M.J.; Sainz, F.; Álavrez-Mon, M.; Buján, J.; García-Honduvilla, N. Patients with Incompetent Valves in Chronic Venous Insufficiency Show Increased Systematic Lipid Peroxidation and Cellular Oxidative Stress Markers. *Oxidative Med. Cell. Longev.* **2019**, *2019*, 5164576. [CrossRef]
257. Smith, R.K.; Golledge, J. A systematic review of circulating markers in primary chronic venous insufficiency. *Phlebol. J. Venous Dis.* **2013**, *29*, 570–579. [CrossRef]
258. Guss, L.G.; Javvaji, S.; Case, J.; Bs, B.B.; Schaefer, K.N.; Bs, R.G.; Waalen, J.; Greenway, H.T.; Housman, L.B. Differences in Inflammatory Cytokine Levels between Patients with Varying Severity of Chronic Venous Insufficiency. *J. Vasc. Med. Surg.* **2018**, *6*, 1–6. [CrossRef]
259. Raffetto, J.D.; Qiao, X.; Beauregard, K.G.; Khalil, R.A. Estrogen receptor-mediated enhancement of venous relaxation in female rat: Implications in sex-related differences in varicose veins. *J. Vasc. Surg.* **2010**, *51*, 972–981. [CrossRef]
260. Honduvilla, N.G.; Asúnsolo, Á.; Ortega, M.A.; Sainz, F.; Leal, J.; Lopez-Hervas, P.; Pascual, G.; Buján, J. Increase and Redistribution of Sex Hormone Receptors in Premenopausal Women Are Associated with Varicose Vein Remodelling. *Oxidative Med. Cell. Longev.* **2018**, *2018*, 3974026. [CrossRef]
261. Lenković, M.; Cabrijan, L.; Gruber, F.; Batinac, T.; Manestar-Blazić, T.; Stanić Zgombić, Z.; Stasić, A. Effect of Progesterone and Pregnancy on the Development of Varicose Veins. *Acta Derm. Croat.* **2009**, *17*, 263–267.
262. Ciardullo, A.V.; Panico, S.; Bellati, C.; Rubba, P.; Rinaldi, S.; Iannuzzi, A.; Cioffi, V.; Iannuzzo, G.; Berrino, F. High endogenous estradiol is associated with increased venous distensibility and clinical evidence of varicose veins in menopausal women. *J. Vasc. Surg.* **2000**, *32*, 544–549. [CrossRef]
263. Ortega, M.A.; Romero, B.; Asúnsolo, Á.; Martínez-Vivero, C.; Sainz, F.; Bravo, C.; De León-Luis, J.; Álvarez-Mon, M.; Buján, J.; García-Honduvilla, N. Pregnancy-associated venous insufficiency course with placental and systemic oxidative stress. *J. Cell. Mol. Med.* **2020**, *24*, 4157–4170. [CrossRef] [PubMed]
264. Honduvilla, N.G.; Ortega, M.A.; Asúnsolo, Á.; Álvarez-Rocha, M.J.; Romero, B.; De León-Luis, J.; Álvarez-Mon, M.; Buján, J. Placentas from women with pregnancy-associated venous insufficiency show villi damage with evidence of hypoxic cellular stress. *Hum. Pathol.* **2018**, *77*, 45–53. [CrossRef] [PubMed]
265. Ortega, M.; Sánchez-Trujillo, L.; Bravo, C.; Fraile-Martinez, O.; García-Montero, C.; Saez, M.; Alvarez-Mon, M.; Sainz, F.; Alvarez-Mon, M.; Bujan, J.; et al. Newborns of Mothers with Venous Disease during Pregnancy Show Increased Levels of Lipid Peroxidation and Markers of Oxidative Stress and Hypoxia in the Umbilical Cord. *Antioxidants* **2021**, *10*, 980. [CrossRef] [PubMed]
266. Ortega, M.A.; Saez, M.A.; Fraile-Martínez, O.; Asúnsolo, Á.; Pekarek, L.; Bravo, C.; Coca, S.; Sainz, F.; Mon, M.; Buján, J.; et al. Increased Angiogenesis and Lymphangiogenesis in the Placental Villi of Women with Chronic Venous Disease during Pregnancy. *Int. J. Mol. Sci.* **2020**, *21*, 2487. [CrossRef] [PubMed]
267. Ortega, M.A.; Saez, M.A.; Asúnsolo, Á.; Romero, B.; Bravo, C.; Coca, S.; Sainz, F.; Álvarez-Mon, M.; Buján, J.; García-Honduvilla, N. Upregulation of VEGF and PEDF in Placentas of Women with Lower Extremity Venous Insufficiency during Pregnancy and Its Implication in Villous Calcification. *BioMed Res. Int.* **2019**, *2019*, 5320902. [CrossRef]
268. Ortega, M.A.; Fraile-Martínez, O.; Saez, M.A.; Álvarez-Mon, M.A.; Gómez-Lahoz, A.M.; Bravo, C.; Luis, J.A.D.L.; Sainz, F.; Coca, S.; Asúnsolo, Á.; et al. Abnormal proinflammatory and stressor environmental with increased the regulatory cellular IGF-1/PAPP-A/STC and Wnt-1/β-Catenin canonical pathway in placenta of women with Chronic venous Disease during Pregnancy. *Int. J. Med. Sci.* **2021**, *18*, 2814–2827. [CrossRef]
269. Ortega, M.A.; Saez, M.A.; Sainz, F.; Fraile-Martínez, O.; García-Gallego, S.; Pekarek, L.; Bravo, C.; Coca, S.; Mon, M.; Buján, J.; et al. Lipidomic profiling of chorionic villi in the placentas of women with chronic venous disease. *Int. J. Med. Sci.* **2020**, *17*, 2790–2798. [CrossRef]
270. Ortega, M.A.; Asúnsolo, Á.; Álvarez-Rocha, M.J.; Romero, B.; De León-Luis, J.; Álvarez-Mon, M.; Buján, J.; García-Honduvilla, N. Remodelling of collagen fibres in the placentas of women with venous insufficiency during pregnancy. *Histol. Histopathol.* **2018**, *33*, 567–576. [CrossRef]

271. Ortega, M.A.; Asúnsolo, Á.; Fraile-Martínez, O.; Sainz, F.; Saez, M.A.; Bravo, C.; De León-Luis, J.A.; Alvarez-Mon, M.A.; Coca, S.; Álvarez-Mon, M.; et al. An increase in elastogenic components in the placental villi of women with chronic venous disease during pregnancy is associated with decreased EGFL7 expression level. *Mol. Med. Rep.* **2021**, *24*, 556. [CrossRef]
272. Asúnsolo, Á.; Chaowen, C.; Ortega, M.A.; Coca, S.; Borrell, L.N.; De León-Luis, J.; García-Honduvilla, N.; Álvarez-Mon, M.; Buján, J. Association Between Lower Extremity Venous Insufficiency and Intrapartum Fetal Compromise: A Nationwide Cross-Sectional Study. *Front. Med.* **2021**, *8*, 577096. [CrossRef]

cells

MDPI

Article

Primary Human Trophoblasts Mimic the Preeclampsia Phenotype after Acute Hypoxia–Reoxygenation Insult

Barbara Fuenzalida [1], Sampada Kallol [1,2], Jonas Zaugg [1,2], Martin Mueller [3], Hiten D. Mistry [4], Jaime Gutierrez [5], Andrea Leiva [5] and Christiane Albrecht [1,2,*]

1 Institute of Biochemistry and Molecular Medicine, Faculty of Medicine, University of Bern, 3012 Bern, Switzerland; barbara.fuenzalida@ibmm.unibe.ch (B.F.); sampuak@gmail.com (S.K.); jonas.zaugg@ibmm.unibe.ch (J.Z.)
2 Swiss National Center of Competence in Research, NCCR TransCure, University of Bern, 3012 Bern, Switzerland
3 Department of Obstetrics and Gynaecology, Inselspital, University of Bern, 3012 Bern, Switzerland; martin.mueller@insel.ch
4 Department of Women and Children's Health, School of Life Course and Population Health Sciences, King's College London, London WC2R 2LS, UK; hiten.mistry@kcl.ac.uk
5 School of Medical Technology, Faculty of Medicine and Sciences Health, Universidad San Sebastián, Santiago 7510157, Chile; jaime.gutierrez@uss.cl (J.G.); andrea.leiva@uss.cl (A.L.)
* Correspondence: christiane.albrecht@ibmm.unibe.ch; Tel.: +41-31-631-41-08

Abstract: Preeclampsia (PE) is a pregnancy-specific disorder that affects 3 to 5% of pregnancies worldwide and is one of the leading causes of maternal and fetal morbidity and mortality. Nevertheless, how these events occur remains unclear. We hypothesized that the induction of hypoxic conditions in vitro in primary human trophoblast cells would mimic several characteristics of PE found in vivo. We applied and characterized a model of primary cytotrophoblasts isolated from healthy pregnancies that were placed under different oxygen concentrations: ambient O_2 (5% pCO_2, 21%pO_2, 24 h, termed "normoxia"), low O_2 concentration (5% pCO_2, 1.5% pO_2, 24 h, termed "hypoxia"), or "hypoxia/reoxygenation" (H/R: 6 h intervals of normoxia and hypoxia for 24 h). Various established preeclamptic markers were assessed in this cell model and compared to placental tissues obtained from PE pregnancies. Seventeen PE markers were analyzed by qPCR, and the protein secretion of soluble fms-like tyrosine kinase 1 (sFlT-1) and the placenta growth factor (PlGF) was determined by ELISA. Thirteen of seventeen genes associated with angiogenesis, the renin–angiotensin system, oxidative stress, endoplasmic reticulum stress, and the inflammasome complex were susceptible to H/R and hypoxia, mimicking the expression pattern of PE tissue. In cell culture supernatants, the secretion of sFlT-1 was increased in hypoxia, while PlGF release was significantly reduced in H/R and hypoxia. In the supernatants of our cell models, the sFlT-1/PlGF ratio in hypoxia and H/R was higher than 38, which is a strong indicator for PE in clinical practice. These results suggest that our cellular models reflect important pathological processes occurring in PE and are therefore suitable as PE in vitro models.

Keywords: preeclampsia; hypoxia; hypoxia/reoxygenation; sFlT-1/PlGF ratio; primary trophoblast; inflammation; oxidative stress

Citation: Fuenzalida, B.; Kallol, S.; Zaugg, J.; Mueller, M.; Mistry, H.D.; Gutierrez, J.; Leiva, A.; Albrecht, C. Primary Human Trophoblasts Mimic the Preeclampsia Phenotype after Acute Hypoxia–Reoxygenation Insult. *Cells* **2022**, *11*, 1898. https://doi.org/10.3390/cells11121898

Academic Editors: Junie P Warrington, Ana T. Palei, Mark W. Cunningham and Lorena M. Amaral

Received: 28 April 2022
Accepted: 9 June 2022
Published: 11 June 2022

1. Introduction

Preeclampsia (PE) is the most common pregnancy complication and one of the three leading causes of maternal morbidity and mortality worldwide, complicating around 2–8% of pregnancies [1]. PE is characterized by hypertension (systolic blood pressure $\geq$ 140 mm Hg and diastolic blood pressure $\geq$ 90 mm Hg) and proteinuria ($\geq$300 mg/24 h) from 20 weeks of gestation [2]. A critical factor in PE is a poor placental invasion of the uterine vasculature by trophoblast cells and poor differentiation, which leads to impaired placental perfusion, oxidative stress, cellular damage, inflammation, vascular dysfunction, and the release of the antiangiogenic factors soluble fms-like tyrosine kinase 1 (sFlT-1) and soluble endoglin (sENG) components into the maternal circulation [3–8]. It has been observed that the increase in oxidative stress from high ROS production and lipid peroxidation, disturbances in the renin–angiotensin system (RAS), immune maladaptation, and genetic susceptibility may all contribute to the pathogenesis of PE [5,9–13]. Nuclear factor erythroid 2-related factor (NRF2) is the master regulator of cellular oxidative stress responses. It has been associated not only with the pathology of PE but also with the pathogenesis of other disorders such as cancer, Alzheimer's disease, and articular disease [14–17].

In addition, an increase in different inflammation markers in PE has been described, including interleukin (IL)-1β and caspase 1 (CASP1), as well as the activation of the Nod-like receptor protein 3 (NLRP3) inflammasome [18–20].

Based on the findings in different animal and cellular models as well as in preeclamptic patients, some of the differentially expressed proteins were proposed as characteristic markers of PE. These proteins are associated with angiogenic imbalance (FlT-1, PlGF, ENG, and VEGF A), RAS receptors (AGTR1, AGTR4), stress markers (HIF1α, NRF2), endoplasmic reticulum stress (GRP78, GRP94), placenta-specific proteins (PP13, PAPP-A, and RGC-32), and the inflammasome complex (NLRP3, IL-1β, and CASP1).

In the first trimester of pregnancy, low oxygen partial pressures (pO_2) are vital for placental development [21]. This stimulus is required for cytotrophoblast proliferation and differentiation along the extravillous trophoblast (EVT) pathway, as well as for regulation of the invasion of EVT trophoblasts, which is required for uterine arteries' remodeling. During pregnancy, low oxygen partial pressures (<2% pO_2) are observed in the first trimester, which increase to <8% pO_2 as pregnancy progresses during middle and late gestation [22,23]. Dysregulation in this process, such as persistent low O_2 concentrations (hypoxia) or cycles of hypoxia–reoxygenation can lead to pregnancy complications such as PE and intrauterine growth restriction (IUGR) [22–24]. PE, however, is a complex and multifactorial disease associated with an abnormal differentiation of the trophoblast to the invasive EVT occurring in the first trimester. This results in a shallow placental invasion and a failure of the uteroplacental perfusion, triggering the release of antiangiogenic factors to the maternal circulation, damaging the maternal vascular endothelium, and generating a widespread maternal endothelial dysfunction. It is well accepted that several mechanisms contribute to the pathogenesis of PE. However, it is unclear whether the pathways involved in PE are interrelated, have synergistic effects, or act independently.

Since abnormal O_2 concentrations were associated with the dysregulated development and function of the human placenta in PE, we hypothesized that the induction of hypoxic conditions in primary human trophoblasts would mimic several characteristics of PE. Hence, we established a PE model in primary human trophoblast cells obtained from full-term healthy placentae by inducing hypoxia and/or hypoxia/reoxygenation (H/R) and evaluated the cell biological factors dysregulated in PE. Moreover, we assessed the gene expression of various well-established PE markers in these trophoblast cell models, i.e., under ambient O_2 concentration (termed "normoxia"), persistent low O_2 concentration (termed "hypoxia"), and H/R, and compared the expression patterns with placental tissues collected from PE pregnancies.

2. Materials and Methods

2.1. Study Group

Human placentae were obtained from full-term normal pregnancies (38–40 weeks) from the Division of Gynecology and Obstetrics of the Lindenhofgruppe, Bern, Switzerland (n = 9) and from women with PE (32–40 weeks) and full-term normotensive controls (38–40 weeks) pregnancies from Hospital Clínico UC CHRISTUS (n = 29). PE was defined as a systolic blood pressure > 140 mm Hg and a diastolic blood pressure > 90 mm Hg, determined on 2 occasions >4 h apart and arising after 20 weeks of gestation in a previously normotensive woman and de novo proteinuria (protein: creatinine ratio (PCR) > 30; urine protein concentration > 3 g/L in 2 random clean-catch midstream specimens collected >4 h apart) with no evidence of urinary tract infection [2]. The collection of samples was performed according to the principles outlined in the Declaration of Helsinki. All procedures were approved by the local ethics committee (Canton of Bern, Switzerland Basec Nr.2016-00250) and the Faculty of Medicine at Pontificia Universidad Católica de Chile (PUC, ID-180810004). Informed consent and clinical data from patients were obtained. General maternal (i.e., age, height, weight, and blood pressure) and neonatal (i.e., sex, gestational age, weight, height, and the weight of the placenta) variables were obtained from the clinical records.

2.2. Primary Human Trophoblast Cells

Primary cytotrophoblast cells (CTB) were isolated from healthy term placenta villous tissues. In brief, 30 g of placental tissue was carefully minced and digested three times for 30 min at 37 °C in saline Hanks/HEPES solution plus DNase I (Sigma Aldrich, St. Louis, MI, USA) and trypsin (Thermo Fisher Scientific, Waltham, MA, USA), as described previously [25]. After isolation, the cellular pellets were resuspended in DMEM and separated by centrifugation (1500 RCF, 20 °C, 20 min) using a 10–70% Percoll gradient (Sigma Aldrich, St. Louis, MI, USA). CTB cells were obtained from gradient fractions between 35 and 55% Percoll. Cells were plated at a density of 200,000 cell/cm^2 in DMEM high glucose (4.5 g/L glucose) medium supplemented with 10% fetal bovine serum and 1× antibiotic-antimycotic (Gibco, Waltham, MA, USA). After isolation, the cells were cultured for 12 h to allow them to attach before being exposed to different oxygen conditions.

The purity of isolated CTB was evaluated by staining with specific cells markers (an antibody cocktail containing either anti-cytokeratin 7 (mean 93.4% ± 2.01 positivity) direct labeled with Alexa Fluor 488® (Novus Biologicals, Englewood, CO, USA) plus anti-vimentin (mean 2.44% ± 0.87 positivity) labeled with Alexa Fluor 647® (Novus Biologicals, Englewood, Co, USA) or anti-von Willebrand factor (mean 0.28% ± 0.06 positivity) marked with Alexa Fluor 647® (Novus Biologicals, Englewood, Co, USA)). Cells were acquired by flow cytometry (BD FACS LSRII; BD Biosciences, Franklin Lakes, NY, USA). Data acquisition for each staining was based on at least 10,000 events and performed using BD FACSDiva™ software (BD Biosciences, Franklin Lakes, NJ, USA), and the data were analyzed using FlowJo® software version 10 (FlowJo LLC, Ashland, OR, USA).

2.3. Models of Preeclampsia

CTB isolated from healthy term placentae were cultured as described above for 12 h and subsequently kept either at ambient O_2 concentration, defined as a control condition and termed "normoxia" (5% pCO_2, 21% pO_2, 74% N_2, 24 h), at low O_2 concentration, termed as "hypoxia" (5% pCO_2, 1.5% pO_2, 93.5% N_2, 24 h), or at H/R (6 h intervals alternating between normoxia and hypoxia for 24 h) [26].

2.4. Isolation of mRNA and Quantitative RT-PCR

Total RNA was extracted from placental tissues (PE and normotensive control, 100 mg of tissue was homogenized) and CTB after 24 h using TRI reagent (Invitrogen, Oxford, UK). Then, 1 µg of total RNA was reverse transcribed to cDNA using the GoScript™ Reverse Transcriptase System (Promega, Madison, Wi, USA), according to the manufacturer´s instructions. Quantitative reverse transcription-PCR (qRT-PCR) was performed, as previously described [27]. In brief, qRT-PCR was carried out on a CFX qRT-PCR System using the SYBR® Green PCR master mix detection kit (Promega, Madison, Wi, USA). The primer pairs are listed in Table 1. The relative gene expression was calculated using the $2^{-\Delta\Delta Cq}$ method, using Tyrosine 3-monooxygenase/tryptophan 5-monooxygenase (*YWHAZ*) as the reference gene.

Table 1. Characteristics of primers.

Category	Gene	Accession Number	Primer Sequence (5′-3′)
Angiogenic and antiangiogenic markers	*FIT-1*	NM_001160030.1	F-AAGAGCCTGAACTGTATA R-TGATGACGATGATGATGA
	ENG	NM_001114753.2	F-CAAGACCAGGAAGTCCATA R-CGTGTGCGAGTAGATGTA
	PIGF	NM_001293643.1	F-CTCCTAAAGATCCGTTCTG R-CTTTCCGGCTTCATCTTC
	VEGF A	NM_001287044.1	F-TACATCTTCAAGCCATCC R-TTCTTTGGTCTGCATTCA
RAS receptors	*AGTR1*	NM_032049.3	F-TCTCAGCATTGATCGATACC R-TGACTTTGGCTACAAGCATT
	AGTR4	NM_005575.3	F-TATGCCTAAGAAGTCATCAG R-CCAAGTAAGTGCTCATCT
Hypoxia and oxidative stress markers	*HIF1α*	NM_001530.4	F-CGTTGTGAGTGGTATTATTC R-GGCTACTTGTATCTTCTGA
	NRF2	NM_005654.6	F-ATCGTGCTGTTCACGTCAGAC R-TGGCTCCTCACGTACTCCTC
Endoplasmic reticulum (ER) stress markers	*GRP78*	NM_005347.5	F-TGTGCAGCAGGACATCAAGT R-TCCCAAATAAGCCTCAGCGG
	GRP94	NM_003299.3	F-GCCAGTTTGGTGTCGGTTTC R-GGGTAATTGTCGTTCCCCGT
Placenta-specific proteins	*PP13*	NM_013268.3	F- GATATTGCCTTCCGTTTC R-GTAGTCTGTTGTCTCCTC
	PAPP-A	NM_002581.5	F-GCTGTCACATACATCCAT R-GCTGGGTTCATCAATACA
Cell cycle regulator	*RGC-32*	NM_014059.3	F-ATTCTCCAACAGACTCTAC R-CAAGATCAGCAATGAAGG
Syncytial marker	*β-hCG*	NM_033043.2	F-CGGGACATGGGCATCCAA R-GCGCACATCGCGGTAGTT
Inflammasome	*NLRP3*	NM_001079821	F-CACCTGTTGTGCAATCTGAAG R-GCAAGATCCTGACAACATGC
	IL-1β	NM_000576	F-CTGTCCTGCGTGTTGAAAGA R-TTGGGTAATTTTTGGGATCTAC
	CASP1	NM_033295	F-GCCTGTTCCTGTGATGTGGAG R-TGCCCACAGACATTCATACAGTTTC
Reference gene	*YWHAZ*	XM_024447266.1	F-CCGTTACTTGGCTGAGGTTG R-AGTTAAGGGCCAGACCCAGT

2.5. Lipid Peroxidation

As a marker of lipid peroxidation, malondialdehyde (MDA) formation was measured using thiobarbituric acid reactive substances (TBARS) assays in cell culture media, as described in [28]. In brief, cell culture media were collected between 6 and 24 h during normoxia, hypoxia, and H/R conditions. The MDA standard curve was prepared for 0.007–2 nmol/mL. Then, 15% *w*/*v* trichloroacetic acid (TCA, Merck, Darmstadt, Germany), sample or MDA standard (1,1,3,3 tetraethoxypropane), and 0.67% *w*/*v* thiobarbituric acid (TBA, VWR, Dietikon, Switzerland) in 2.5 M HCl were mixed in a 4:5:8 ratio, vortexed, and boiled for 20 min at 95 °C. After cooling the samples to room temperature, 1-butanol was added and gently mixed. (Sigma Aldrich, St. Louis, MI, USA). For phase separation, the samples were centrifuged at 1000 RCF for 1 min. Then, 200 μL of the organic phase was transferred to a black-wall 96-well plate and measured using a Flex Station II fluorescence microplate reader (Thermo Fisher Scientific, Waltham, MA, USA). The MDA combined with TBA to form a fluorescent adduct detected at an excitation/emission wavelength of 530/550 nm. The MDA equivalents were calculated by interpolation to the standard curve.

2.6. Measurement of Protein Secretion by ELISA

The secretion of β-hCG, sFlT-1, and PlGF into CTB culture media after 6–24 h exposure to normoxia, hypoxia, and H/R was measured by using a human hCG (intact) ELISA kit (RAB0092, Sigma Aldrich, St. Louis, MI, USA) and sFlT-1 and PlGF ELISA kits (EA100379 and EA100342, respectively, OriGene, Rockville, MD, USA) following the manufacturer's instructions. The consecutive absorbance measurement was carried out at 450 nm on a Vmax microplate reader (Molecular device, San Jose, CA, USA). The concentrations of hCG, sFlT-1, or PlGF released by CTB were interpolated using the respective standard curves.

2.7. Statistical Analysis

The values for the maternal and neonatal characteristics are presented as the mean ± SD, as described previously [29]. The relative gene expression and protein secretion were presented as the mean ± S.E.M. for the experiments, where n indicates the number of different individual isolated from the CTB cultures or individual placentae. Two or more groups were compared using the student's t-test or ANOVA, respectively. $p < 0.05$ was considered statistically significant. Graphpad Prism 9.2 (GraphPad Software Inc., San Diego, CA, USA) was used for data analysis and for creating the figures.

3. Results

3.1. Maternal and Neonatal Variables in PE and Normotensive Control Pregnancies

The women with normotensive control pregnancy or PE were of comparable age, height, preconceptional weight, delivery, and basal glycemia (Table 2). The systolic and diastolic arterial blood pressure in PE patients was higher than in the normotensive controls. Women with PE presented proteinuria (>300 mg/24 h). Neonates from women with PE had a lower gestational age, birth weight, height, and ponderal index compared with the neonates from normal pregnancies (Table 2). Although the birthweight of the babies born to women with PE were lower, they did not constitute IUGR.

Table 2. Clinical characteristics of pregnant women and newborns.

Pregnancy Characteristics	Normal	Preeclampsia
Numbers	$n = 25$	$n = 13$
Maternal variables		
Age (years)	31.4 ± 4.9 (22–41)	31.7 ± 4.5 (24–42)
Length (cm)	163 ± 6.3 (155–176)	159.6 ± 7 (147–168)
Weight pre-pregnancy (kg)	61.7 ± 9.5 (45–86)	59.8 ± 10.6 (38–85)
Weight at delivery (kg)	73.3 ± 10 (58–102)	72.4 ± 10.8 (46–89)
BMI pre-pregnancy (kg/m^2)	23 ± 2.4 (18.5–27.8)	23.4 ± 3.8 (17.6–33.6)
BMI at delivery (kg/m^2)	27.5 ± 2.3 (23.2–32.9)	28.1 ± 3.8 (21.3–35-2)
Systolic blood pressure at delivery (mm Hg)	112.7 ± 10.9 (90–130)	155 ± 16.9 (138–185) *
Diastolic blood pressure at delivery (mm Hg)	70.9 ± 7 (60–88)	93.1 ± 10.4 (70–113) *
Basal glycemia (mg/dL)	77.6 ± 7.1 (66–92)	79 ± 11.2 (52–87)
Creatinine (mg/dL)	n.d.	115.8 ± 70.9 (26–236)
Proteinuria (mg/dL)	n.d.	2494 ± 1454 (150–19241)
Newborn variables		
Sex (female/male)	15/10	8/5
Gestational age (weeks)	39.1 ± 0.8 (38–40)	36.9 ± 2.1 (32–40) *
Route of delivery (C-section/Labor)	16/9	10/3
Birth weight (g)	3388 ± 367 (2765–4045)	2811 ± 866.2 (1030–3860) *
Height (cm)	50.2 ± 1.6 (48–53)	47.8 ± 4.7 (38–54) *
Ponderal index (g/cm^3 x100)	2.7 ± 0.2 (2.3–3)	2.5 ± 0.3 (1.9–2.8) *

Weight and body mass index (BMI) were determined pregestationally and at the 3rd trimester; blood pressure was determined at delivery. n.d.: not determined. * $p < 0.05$ versus corresponding values in the PE group. Data are presented as the mean ± SD (range).

3.2. In Vitro Models Reflect the Angiogenic and RAS Receptor Expression Patterns Found in PE

We analyzed the mRNA expression of *FlT1*, *ENG*, *VEGF A* (reported to be elevated in PE), and *PlGF* (decreased in PE) in preeclamptic and healthy placentae and in the PE cell models (normoxia, H/R, and hypoxia). As expected, *FlT-1* was increased in PE placentae compared to the controls (2.26 ± 0.49 vs. 1.01 ± 0.01, fold change (FC); $p = 0.0005$), as well as in H/R and hypoxia compared to normoxia (4.07 ± 0.52 and 8.18 ± 1.02 vs. 1.01 ± 0.01 FC, respectively; $p = 0.004$ and $p < 0.0001$) (Figure 1A). The placentae from women with PE showed no differences in *PlGF* expression, and no significant differences were found in H/R (Figure 1B). However, in hypoxia a decrease in *PlGF* compared to normoxia (0.30 ± 0.08 vs. 1.01 ± 0.01 FC; $p = 0.0442$) was observed (Figure 1B). *ENG* was increased in the PE placentae (1.62 ± 0.25 vs. 1.01 ± 0.01 FC; $p = 0.0008$) and in both hypoxic cell models compared to normoxia (3.56 ± 0.91 vs. 1.01 ± 0.01 FC; $p = 0.009$), but only hypoxia alone reached statistical significance (Figure 1C). Similarly, for *VEGF A* we observed an increase in PE placentae (4.83 ± 0.87 vs. 1.01 ± 0.01 FC; $p < 0.0001$) and the hypoxic cell models compared to normoxia, but only hypoxia treatment alone resulted in statistically significant changes (8.04 ± 1.82 vs. 1.01 ± 0.01 FC; $p = 0.0002$) (Figure 1D).

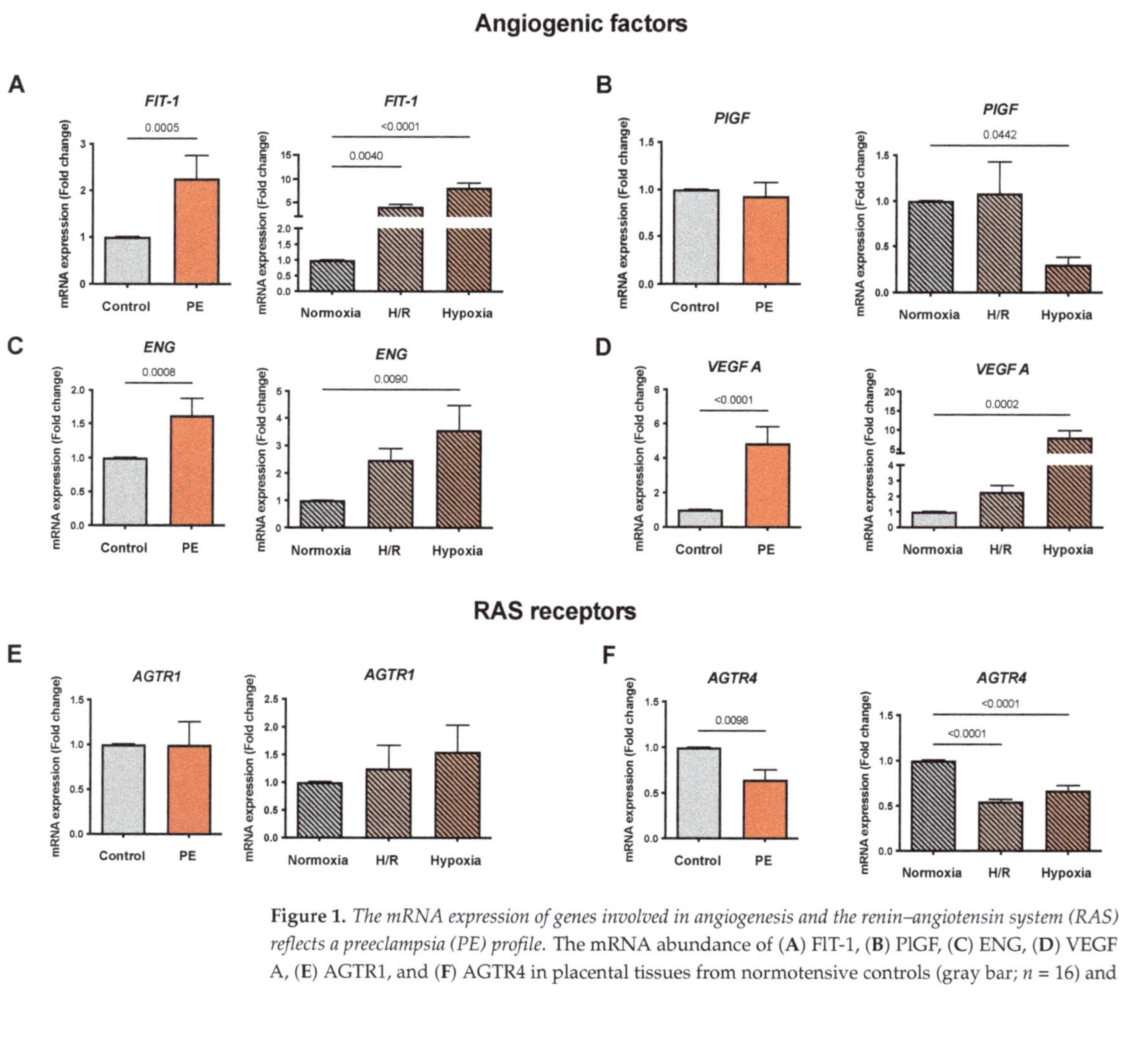

Figure 1. *The mRNA expression of genes involved in angiogenesis and the renin–angiotensin system (RAS) reflects a preeclampsia (PE) profile.* The mRNA abundance of (**A**) FlT-1, (**B**) PlGF, (**C**) ENG, (**D**) VEGF A, (**E**) AGTR1, and (**F**) AGTR4 in placental tissues from normotensive controls (gray bar; $n = 16$) and

preeclamptic (PE: orange bar; n = 13) pregnancies is shown. The cultured primary cytotrophoblasts (CTB) were exposed either to normoxia (24 h: gray bar with diagonal lines; n = 9), hypoxia/reoxygenation (H/R; 6 h intervals each of normoxia and hypoxia for 24 h; light orange bar with diagonal lines; n = 9) and hypoxia (24 h; dark orange bar with diagonal lines; n = 9). The procedures regarding the processing of the placental tissues as well as the isolation of the CTB from the healthy term placentae are described in Materials and Methods. The gene expression was assessed by real-time PCR and normalized to the reference gene YWHAZ. Data are shown as the mean ± S.E.M. and represent the fold change ($2^{-\Delta\Delta Cq}$) of the mRNA expression in comparison to the control placentae and normoxia, respectively. The placental expression patterns between the control and PE were compared using the student's t-test. For comparison of normoxia with hypoxia and H/R, ANOVA was applied. $p < 0.05$ was considered statistically significant. FlT-1: fms-like tyrosine kinase 1; PlGF: placental growth factor; ENG: endoglin; VEGF: vascular endothelial growth factor; ATR: angiotensin receptor; YWHAZ: tyrosine 3-monooxygenase/tryptophan 5- monooxygenase activation protein zeta.

Our study assessed the mRNA expression of the angiotensin receptors, *AGTR1* and *AGTR4,* in the established PE models. The RAS receptor *AGTR1* placental expression was neither altered in our PE cohort nor in the cell models of hypoxia and H/R (Figure 1E). *AGTR4* was reduced in the PE placentae compared to the controls (0.65 ± 0.1 vs. 1.01 ± 0.01 FC; p = 0.0098) as well as in H/R and hypoxia compared to normoxia (0.55 ± 0.03 and 0.67 ± 0.05 vs. 1.01 ± 0.01 FC, respectively; $p < 0.0001$ for both) (Figure 1F).

3.3. *Hypoxia, Oxidative Stress, and Endoplasmic Reticulum Stress Markers Are Selectively Altered in the PE Cell Models*

To determine the markers of hypoxia and oxidative stress in the cell model, we evaluated the *HIF1α* and *NRF2* mRNA levels. Similar to that reported in the literature, *HIF1α* was increased in the PE placentae compared to the controls (2.11 ± 0.23 vs. 1.01 ± 0.01 FC; $p < 0.0001$). In contrast, no significant differences were observed in H/R and hypoxia compared to normoxia (Figure 2A). *NRF2* was reduced in the PE placentae (0.12 ± 0.03 vs. 1.01 ± 0.01 FC; $p < 0.0001$) as well as in the cell models of H/R and hypoxia compared to normoxia (0.47 ± 0.07 and 0.5 ± 0.07 vs. 1.01 ± 0.01 FC, respectively; $p < 0.0001$ for both) (Figure 2B).

We also evaluated selected markers of ER stress, namely *GRP78* and GRP94, which were previously reported to be increased in PE. In our study, *GRP78* placental expression was increased in PE compared to normotensive placentae (2.02 ± 0.42 vs. 1.01 ± 0.01 FC; $p < 0.0001$) and in hypoxia compared to normoxia in cells (1.71 ± 0.27 vs. 1.01 ± 0.01 FC; p = 0.0225) (Figure 2C). The same trend was found for *GRP94* (Figure 2D).

3.4. *mRNA Expression of Placenta-Specific Proteins and Cell Cycle Regulators in PE Placentae and PE Cell Models*

To determine markers of placenta-specific proteins, we evaluated the mRNA expression of *PP13* and *PAPP-A,* which were described to be decreased in PE. The placentae from women with PE showed no differences in PP13 expression compared to the normotensive controls. In contrast, in the cell models, PP13 mRNA abundance was reduced in H/R and hypoxia compared to normoxia (0.5 ± 0.03 and 0.41 ± 0.11 vs. 1.01 ± 0.01 FC, respectively; p = 0.0002 and $p < 0.0001$) (Figure 3A). No significant differences for PAPP-A were found between the PE and normotensive placentae (Figure 3B). *PAPP-A* in H/R was diminished compared to normoxia (0.63 ± 0.1 vs. 1.01 ± 0.01 FC; p = 0.048), but no significant changes were found in hypoxia (Figure 3B).

Following the trend reported in the literature, the cell cycle regulator *RGC32* was decreased in the PE placentae compared to the controls (0.82 ± 0.02 vs. 1.01 ± 0.01 FC; p = 0.0191), but no significant differences were observed in H/R and hypoxia in comparison to normoxia (Figure 3C).

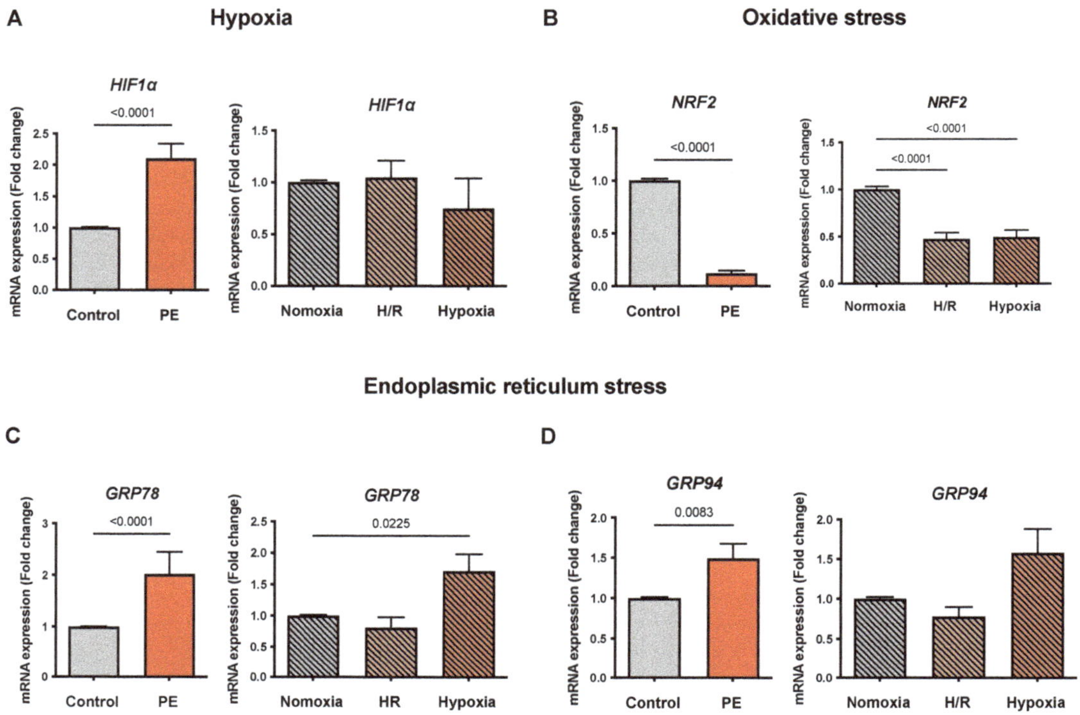

Figure 2. *The mRNA expression of HIF1α, NRF2, GRP78, and GRP94 is selectively altered in the preeclampsia (PE) cell models.* The mRNA abundance of (**A**) hypoxia-inducible factor 1 alpha (HIF1α), (**B**) oxidative stress marker nuclear factor-erythroid 2-related factor 2 (Nrf2), (**C**) endoplasmic reticulum stress marker 78 kDa glucose-regulated protein (GRP78), and (**D**) 94 kDa glucose-regulated protein (GRP94) is shown in the placental tissues of the controls (gray bar) and preeclamptic (PE, orange bar) pregnancies. The cultured CTB were placed under normoxia (24 h, gray bar with diagonal lines), H/R (6 h intervals each of normoxia and hypoxia for 24 h; light orange bar with diagonal lines), and hypoxia (24 h; dark orange bar with diagonal lines). The procedures regarding the processing of the placental tissues as well as the isolation of the CTB from the healthy term placentae are described in Materials and Methods. The gene expression was assessed by real-time PCR and normalized to the reference gene *YWHAZ*. Data are shown as the mean ± S.E.M. (n = 16 Controls; 13 PE; 9 CTB cultured (normoxia, H/R, and hypoxia)) and represent the fold change ($2^{-\Delta\Delta Cq}$) of the mRNA expression as compared to the control or normoxia. The placental expression patterns between the control and PE were compared using the student's t-test. For comparison of normoxia with hypoxia and H/R, ANOVA was applied. $p < 0.05$ was considered statistically significant.

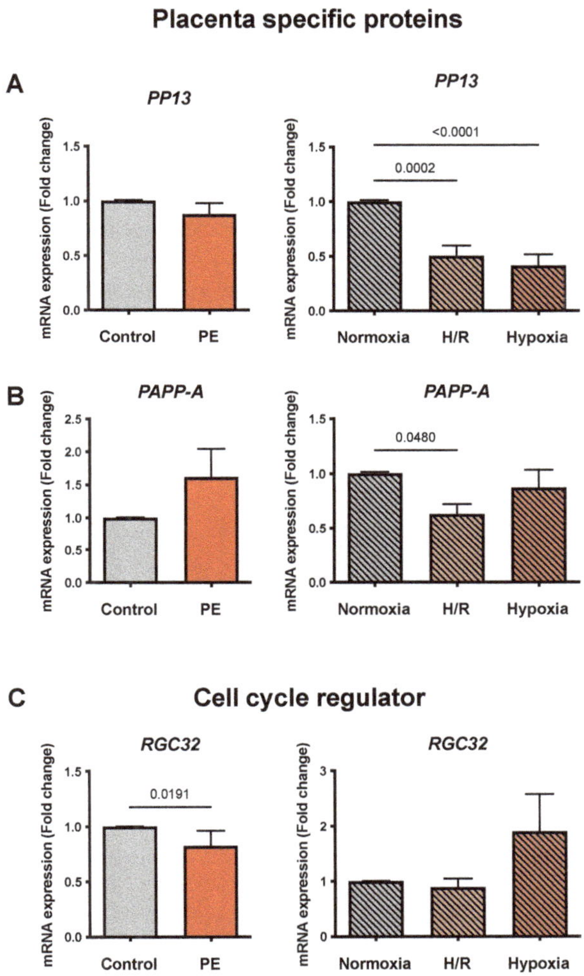

Figure 3. *The mRNA abundance of placenta-specific proteins PP13, PAPP-A, and cell cycle regulation RGC32 is altered in the preeclampsia (PE) cell models.* (**A**) Placental protein 13 (PP13), (**B**) pregnancy-associated plasma protein-A (PAPP-A), and (**C**) the response gene to complement 32 (RGC32), a gene involved in cell cycle regulation. The placental tissues of normotensive controls (gray bar; $n = 16$) and women with PE (orange bar; $n = 13$) were compared. The cultured CTB were placed under normoxia (24 h; gray bar with diagonal lines; $n = 9$), H/R (6 h intervals each of normoxia and hypoxia for 24 h; light orange bar with diagonal lines; $n = 9$), and hypoxia (24 h; dark orange bar with diagonal lines; $n = 9$). The procedures regarding the processing of the placental tissues as well as the isolation of the CTB from the healthy term placentae are described in Materials and Methods. The gene expression was assessed by real-time PCR and normalized to the reference gene YWHAZ. Data are shown as the mean ± S.E.M. and represent the fold change ($2^{-\Delta\Delta Cq}$) of the mRNA expression compared to the control or normoxia. The placental expression patterns between the control and PE were compared using the student's t-test. For comparison of normoxia with hypoxia and H/R, ANOVA was applied. $p < 0.05$ was considered statistically significant.

3.5. Inflammasome Complex Is Altered in PE Placentae and in PE Cell Models

To evaluate the markers of inflammation in PE, we measured the mRNA expression of three molecules of the inflammasome complex. *NLRP3* was increased in PE compared to the control placentae (5.61 ± 1.44 vs. 1.01 ± 0.01 FC; $p < 0.0001$) and in both hypoxic cell models compared to normoxia (3.26 ± 0.66 vs. 1.01 ± 0.01 FC; $p = 0.0174$), but only hypoxic treatment alone reached statistical significance (Figure 4A). *IL-1β* was increased in the PE placentae compared to the controls (2.13 ± 0.53 vs. 1.01 ± 001 FC; $p = 0.0007$), as well as in H/R and hypoxia compared to normoxia (2.44 ± 0.53 and 3.32 ± 0.36 vs. 1.01 ± 0.01 FC, respectively; $p = 0.0229$ and $p = 0.0005$) (Figure 4B). Similarly, *CASP1* was elevated in the PE compared to the control placentae (1.75 ± 0.20 vs. 1.01 ± 0.01 FC; $p = 0.0003$) and in both hypoxic cell models compared to normoxia (4.59 ± 1.02 and 5.07 ± 1.46 vs. 1.01 ± 0.01 FC, respectively; $p = 0.0410$ and $p = 0.0202$) (Figure 4C).

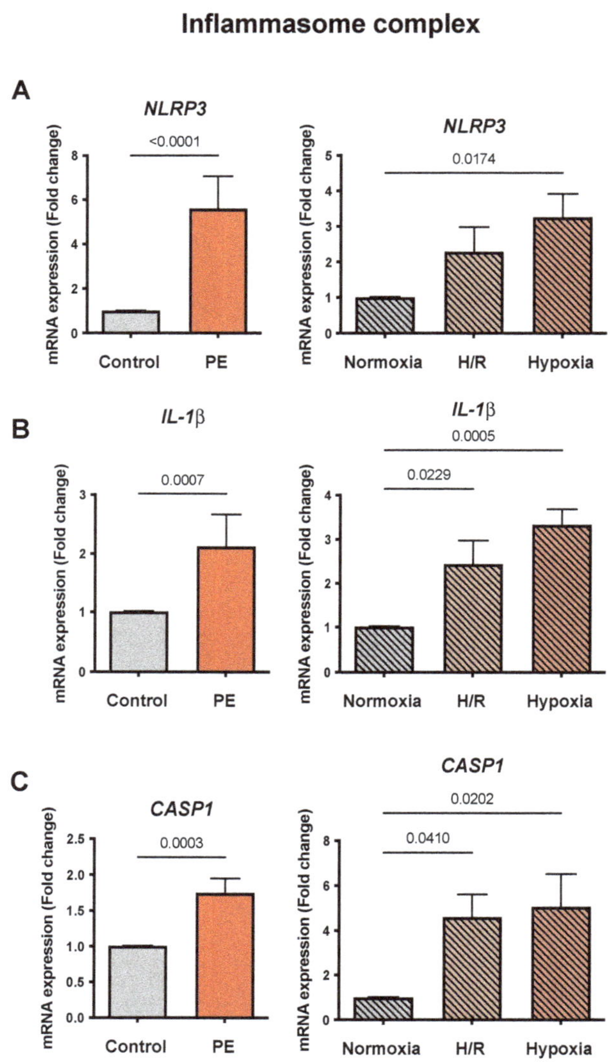

Figure 4. *The mRNA abundances of genes related to the inflammasome complex in the preeclampsia (PE) models reflect the expression patterns of PE placentae.* The gene expressions of (**A**) Nod-like receptor protein 3 (NLRP3), (**B**) interleukin 1β (IL-1β), and (**C**) caspase **1** (CASP1) were assessed by real-time RT-PCR and normalized to the reference gene YWHAZ. The placental tissues of the controls (gray bar; $n = 16$) and preeclamptic (PE, orange bar; $n = 13$) pregnancies were compared. The cultured CTB

were placed under normoxia (24 h: gray bar with diagonal lines; n = 9), H/R (6 h intervals each of normoxia and hypoxia for 24 h; light orange bar with diagonal lines; n = 9), and hypoxia (24 h; dark orange bar with diagonal lines; n = 9). The procedures regarding the processing of placental tissues as well as the isolation of the CTB from the normotensive control placentae are described in Materials and Methods. Data are shown as the mean ± S.E.M. and represent the fold change ($2^{-\Delta\Delta Cq}$) of the mRNA expression compared to the control or normoxia, respectively. The placental expression patterns between the control and PE were compared using the student's t-test. For comparison of normoxia with hypoxia and H/R, ANOVA was applied. $p < 0.05$ was considered statistically significant.

3.6. Human Chorionic Gonadotrophin Is Altered in PE Placentae and the PE Cell Models

We quantified the effect of hypoxia and H/R on the mRNA expression and protein secretion of *hCG* as an indicator for cell differentiation and syncytium formation. We observed an increased *hCG* mRNA expression in the PE placental tissue compared to the controls (2.17 ± 0.34 vs. 1.01 ± 0.01 FC; $p < 0.0001$) (Figure 5A). In contrast, a decrease in *hCG* expression under H/R and hypoxia compared to normoxia was detected (0.61 ± 0.17 and 0.18 ± 0.01 vs. 1.01 ± 0.01 FC, respectively; $p = 0.0258$ and $p < 0.0001$) (Figure 5A). Furthermore, we evaluated the secretion of hCG protein during a 6–24 h time course under the different oxygen conditions. Reduced hCG secretion levels between H/R and normoxia were found after 12 h and 18 h (0.13 ± 0.03 and 0.12 ± 0.04 vs. 0.30 ± 0.07 and 0.28 ± 0.13, respectively; $p = 0.0154$ and $p = 0.0278$) (Figure 5B). hCG secretion was increased after 24 h in hypoxia compared to normoxia (0.42 ± 0.08 vs. 0.23 ± 0.01; $p = 0.0037$)) (Figure 5B).

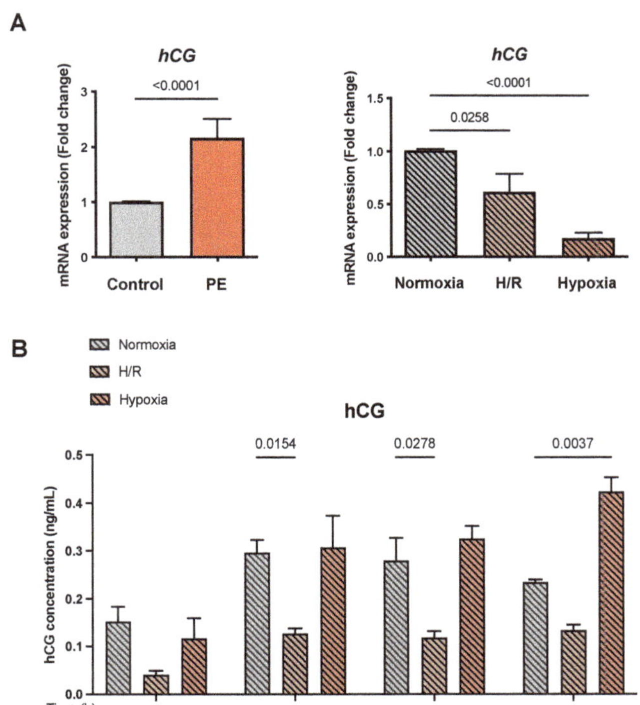

Figure 5. *The mRNA abundance and secretion of hCG.* (**A**) The expression of human chorionic gonadotrophin (hCG) as an indicator for syncytium formation was determined on an mRNA level

by qRT-PCR. (**B**) The secretion of the hCG protein was analyzed during a time course of 6–24 h using ELISA. The placental tissues of the normotensive controls (gray bar; n = 16) and women with PE (orange bar; n = 13) pregnancies were compared. The primary cytotrophoblasts (CTB) were placed under normoxia (24 h: gray bar with diagonal lines; n = 9), H/R (6 h intervals each of normoxia and hypoxia for 24 h; light orange bar with diagonal lines; n = 9), and hypoxia (24 h; dark orange bar with diagonal lines; n = 9). The gene expression was assessed by real-time PCR and normalized to the reference gene YWHAZ. ELISA was performed as described in Material and Methods. Data are shown as the mean ± S.E.M. and represent the fold change ($2^{-\Delta\Delta Cq}$) of the mRNA expression as compared to the control or normoxia. The placental expression patterns between the control and PE were compared using the student's t-test. For comparison of normoxia with hypoxia and H/R, ANOVA was applied. $p < 0.05$ was considered statistically significant.

3.7. Lipid Peroxidation in PE Cell Models

As a functional assessment of oxidative stress, we measured the level of malondialdehyde (MDA) equivalents, a major lipid peroxidation product, in the cell culture supernatants during the 6–24 h exposure to different oxygen conditions. We found significantly increased lipid peroxidation in H/R at 24 h compared with 12–18 h (p = 0.0208 and p= 0.0398) respectively in the same group; no significant changes over time were observed in the normoxia and hypoxia groups (Figure 6).

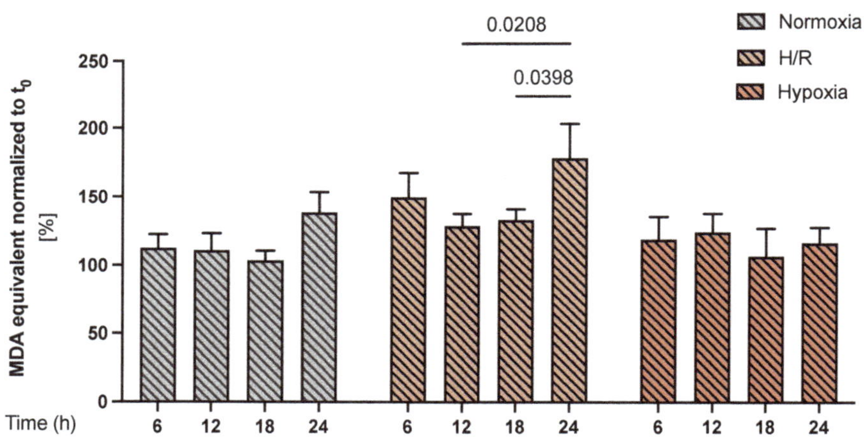

Figure 6. *Lipid peroxidation in the preeclampsia (PE) cell models.* The MDA equivalents were determined in the supernatant of the CTB cultures under normoxia (gray bars with diagonal lines), H/R (light orange bars with diagonal lines), and hypoxia (dark orange bars with diagonal lines) conditions using a thiobarbituric acid reactive substance (TBARS) assay during a time course of 6–24 h. For each condition, the results were normalized to the starting point (i.e., t = 0 h), by setting the time 0 h as 100%. Data are shown as the mean ± S.E.M (n = 9 different CTB isolations cultured under normoxia, H/R, and hypoxia). Statistical analyses were performed by paired 2-way ANOVA. $p < 0.05$ was considered statistically significant.

3.8. Protein Secretion of sFlT-1 and PlGF in PE Cell Models Reflect the Phenotype of PE

Finally, we measured the release of sFlT-1 and PlGF by ELISA in the cell culture supernatants of the three cell models, i.e., under normoxic, H/R, and hypoxic conditions. We observed an increase in sFlT-1 release under hypoxia at 18 and 24 h (8.20 ± 0.38 and 6.30 ± 0.56 FC, (p = 0.0037 and p = 0.0006) respectively) compared to normoxia (Figure 7A). No differences were found for sFlT-1 secretion in H/R (Figure 7A). There was a significant reduction in PlGF secretion in H/R and hypoxia after 6 to 24 h of exposure to hypoxia compared with normoxia (p < 0.0001) (Figure 7B). Furthermore, for the H/R and hypoxia cell models, the calculated sFlT-1 to PlGF ratios exceeded the threshold of 38, a clinically used diagnostic parameter for PE [30,31], already after 12 h of exposure (Figure 7C). In fact, the calculated sFlT-1 to PlGF ratios in the H/R model exceeded the thresholds for early-onset PE (EO-PE; >85) after 18 h and the late-onset PE (LO-PE) thresholds (>110) after 24 h. Even more pronounced was the effect in the cell model under hypoxia alone where both EO-PE and LO-PE thresholds were already exceeded after 12 h of exposure.

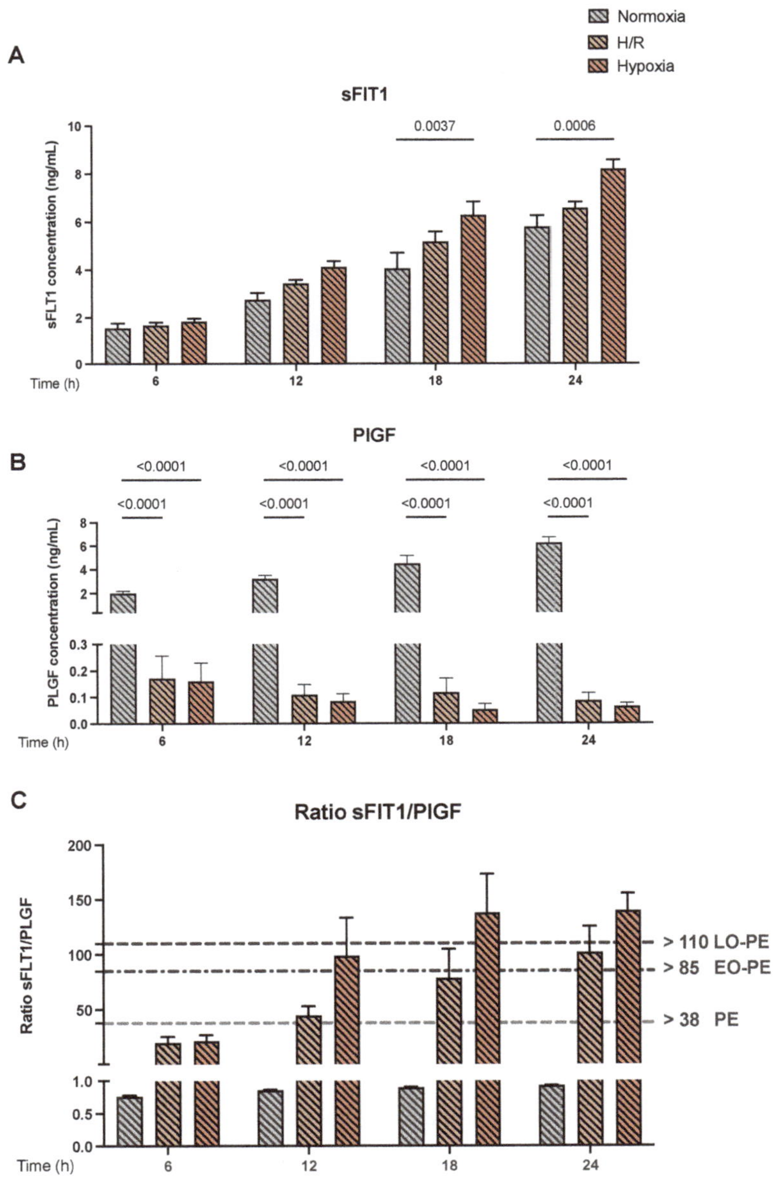

Figure 7. *The protein secretions of sFlT-1 and PlGF into the supernatants of the preeclampsia (PE) cell models.* The protein secretions of (**A**) sFlT-1 and (**B**) PlGF were measured using ELISA kits following the manufacturer´s

instructions. The protein secretion was measured in the media of the CTB cultured under normoxia (gray bar with diagonal lines), H/R (light orange bar with diagonal lines), and hypoxia (dark orange bar with diagonal lines) conditions. (**C**) depicts the calculated sFlT-1/PlGF ratio. The gray dashed line (>38 PE) shows the cutoff value for PE as determined in human serum samples and used for diagnostic purposes in the clinical context. The black dash-dot line (>85) marks the cutoff value for early-onset PE (EO-PE), and the black dotted line (>110) depicts the cutoff value for late-onset PE (LO-PE). The results are presented as mean ± S.E.M. (n = 9 different CTB isolations cultured under normoxia, H/R, and hypoxia). Statistical analyses were performed by paired 2-way ANOVA. $p < 0.05$ was considered statistically significant.

4. Discussion

This novel study presents the thorough characterization of in vitro models for PE using primary cultures of trophoblast cells, exposed to hypoxia or H/R.

It is believed that the alterations characterizing PE placentae are the result of chronic hypoxia [32]. However, comparison with morphological findings in other situations associated with low oxygen tensions such as IUGR and low birth weight suggests that hypoxia alone is insufficient to account for these changes [26]. The fetus and the placenta extract considerable oxygen during middle to late gestation, and the placental tissues will soon become locally hypoxic during periods of vasoconstriction [26,32]. When the maternal blood flow is re-established, there will be a rapid increase in tissue oxygenation. Such fluctuations in oxygen tension could provide the basis for an ischemia–reperfusion type insult [33–35]. Depending on the severity and frequency of these insults, the outcome might range from mild oxidative stress to severe tissue damage [26]. According to this information, we applied hypoxia and H/R in our study to determine if these two models resemble the PE phenotype and to assess the differential effects resulting from the different oxygen concentrations.

The gene expression of the angiogenic factors FlT-1, ENG, and VEGF A were upregulated in the placentae of our PE cohort. These results agreed with other investigations performed under hypoxic conditions induced in isolated CTB [36,37]. The placental PlGF mRNA expression in our cohorts did not differ between the controls and women with PE. While a decrease in circulating PlGF levels in women with PE is manifested, the data for placental tissues are still unclear [38]. Indeed, a decrease due to persistent hypoxia and poor uteroplacental circulation [38,39], an elevation [40], and no changes [41] in PlGF expression in PE placental tissue have been reported. The discrepancies between these findings might be related to the heterogeneity of the women with PE with respect to their genetic background and the temporal onset of the disease. Interestingly, in the context of the measured angiogenic and antiangiogenic factors, the pure hypoxia model turned out to be more sensitive to the expected gene expression changes than H/R. These findings suggest that the trophoblast cells may compensate for the hypoxic damage during the six hours of reoxygenation.

It is well established that the RAS is also affected in PE [9,11,42,43]. Increased renin expression in human PE proposes activation of the uteroplacental RAS, which may lead or contribute to PE [11,42,43]. The results of the present study revealed no significant differences in mRNA abundances for AGTR1 either in the PE placenta samples or in the H/R or hypoxia models. However, it has been described that in PE, AGTR1 increases only in the decidua [44,45]. Considering that decidua was not included in the placental tissues used for this study, this might also explain the lack of differences found in our study. On the other hand, the placental AGTR4 mRNA levels were significantly reduced in the PE samples compared with the controls and in both hypoxia and H/R conditions compared with normoxia. Indeed, in agreement with our results, the placental expression of AGTR4 was found to be reduced in PE at term compared with the normotensive controls [46].

As outlined above, physiological hypoxia or low oxygen tension plays a critical role in early placental development. On the other hand, there is evidence that circulating plasma HIF1α levels are elevated in women with PE [47]. However, in human pregnancy a persistent hypoxic environment due to improper remodeling of the decidual spiral arteries will lead to an imbalanced angiogenic process, especially during the first two trimesters, contributing to the pathogenesis of PE [48,49]. In our experimental in vitro models, the mRNA expression of *HIF1α* remained unchanged under hypoxic conditions. These in vitro findings on *HIF1α* mRNA expression differ from our mRNA results in the PE placentae and other reports showing an increase in mRNA and protein of HIF1α in placental tissue from women with PE [9,49–51]. Though the reason for the lacking response in our experiments is unknown, it cannot be excluded that the effect of hypoxia on HIF1α is time-dependent and, hence, could have been missed in our experimental setup. Hence, the effect on HIF1α protein expression in term villi was previously observed after 7 h of hypoxia, while after 24 h of stimulation it was not detected anymore [52]. Thus, it would be worth analyzing HIF proteins under hypoxia and H/R conditions in tight time course experiments to elucidate the effect of hypoxia and H/R in this context in the future.

To further validate our cell models, we investigated the typical markers of oxidative stress, ER stress, and inflammation. The oxidative stress marker NRF2 was suggested to play a key role in PE [14,53–55] but with conflicting results regarding placental expression in women with PE [56,57]. In our study, the placental gene expression for *NRF2* was significantly reduced in PE and also in our cell models. NRF2 regulates the expression of multiple genes that encode detoxification enzymes and antioxidative proteins and protect the cells from oxidative stress [58]. A previous investigation demonstrated lower placental activation of NRF2 due to oxidative stress in women with PE [57]. This is in line with our findings and suggests that if NRF2 is less activated in trophoblasts, they fail to increase their antioxidant capacity thereby diminishing their cellular protection against oxidative stress.

Recent studies have also shown that the morphology of the ER is markedly altered in PE [59,60]. We detected differential expression levels of the ER stress markers GRP78 and GRP94. In the hypoxia PE cell model compared to normoxia, these findings were also supported by the results of the qPCR measurements in the PE placentae. However, H/R did not affect the expression of the ER stress markers, suggesting that this model reacts quickly and very sensitive to changes in oxygen, thus inducing fast recovery during the episodes of reoxygenation, restoring normal gene expression levels.

Previously, alterations in inflammatory markers such as IL-1β, caspase 1, and NLRP3 inflammasome have been described in PE both in the mother and the placenta [18–20]. Our studies revealed an increase in these three inflammasome markers in our PE cell models as well as in our PE placenta cohort, concurring with a previous placental study [19].

PP13 (also known as galectin-13) was also analyzed, as it has been suggested to be an early biomarker to assess the maternal risk for the subsequent development of pregnancy complications caused by impaired placentation [61]. Decreased placental expression of PP13 and low concentrations in the first trimester maternal sera were associated with an elevated risk of PE [62–64]. We found reduced gene expression of PP13 in hypoxic and H/R conditions compared to normoxia, underlining that the established cell models reflected to a high degree the in vivo situation. Similar to PP13, pregnancy-associated plasma protein (PAPP-A), a large highly glycosylated protein complex synthesized by trophoblasts [62], showed reduced gene expression in H/R. These results agree with the data demonstrating that decreased PAPP-A plasma or serum levels are associated with PE in the first trimester and throughout pregnancy [65–67].

Several studies on RGC32 revealed that this protein participates in cellular processes associated with cell differentiation, angiogenesis, migration, and invasion [68]. In our study, RGC32 was not significantly different in H/R and hypoxia. However, there are conflicting data as to whether PE and the associated hypoxia result in an increased or diminished expression of RGC32. Thus, Wang et al. showed by RT-PCR, western blotting, and immunohistochemistry that placental RGC32 expression was downregulated in PE compared to normotensive controls [68], confirming the reduced RGC32 levels in the PE samples presented in this study.

hCG was evaluated in our study to assess the effect of hypoxia on trophoblast differentiation. Multiple studies have described a relation between high hCG levels and the risk of developing PE, negative effects on fetal development, and increased production of reactive oxygen species [69,70]. Additionally, high hCG levels were associated with increased production of sFlT-1 in women with PE [69]. We detected increased *hCG* mRNA expression in PE tissues, which agrees with other reports [69–71]. In contrast to the diminished *hCG* mRNA levels we found in both the H/R and the hypoxia model, hCG protein secretion was increased under hypoxia after 24 h. It has been previously described in the trophoblast cell lines JEG-3, BeWo, and JAr that hypoxia reduces the mRNA expression of *hCG* [72]. However, the detected changes in hCG secretion in the hypoxia cell model could be related to the de novo synthesis of hCG by trophoblast cells, which can adapt to a hypoxic environment [73], thus mimicking the phenotype of PE.

Endothelial function in PE is affected by the oxidative stress-mediated excess of lipid peroxides [58,74], and increased lipid peroxide products were found in serum samples of preeclamptic women [75,76]. Our data showed continuous elevation and significantly increased MDA equivalents in H/R at 24 h compared to 12 and 18 h but not in the hypoxia model or normoxia control. Considering the levels of the oxidative damage of lipids, the H/R model seems to correspond better to the increased oxidative stress in PE.

Finally, we determined the clinically relevant parameters of sFLT-1 and PIGF in the supernatants of the hypoxic cell models. The protein secretion data showed the same trend as the gene expression results for FlT-1 in H/R and hypoxia and confirmed the significantly reduced PlGF levels in hypoxia in cell supernatants. In this context, Levine et al. showed that five weeks before the development of the clinical symptoms of PE, serum concentrations of sFLT-1 were elevated, and PLGF concentrations were decreased, resulting in an indicative increase in the sFLT-1/PLGF, ratio which can be used to predict the subsequent development of PE [30]. Recent evidence has also emerged that the ratio between sFlT-1 and PlGF in pregnant women's serum could be used to predict early- and late-onset PE. In this context, it has been demonstrated that an sFlT-1/PlGF ratio > 85 for early-onset PE and >110 for late-onset PE represents a marker with a very high specificity of 99.5% and 95.5%, respectively [21]. To date, the sFlT-1/PlGF cutoff value $\leq$ 38 is widely accepted to rule out the assumptions for PE development [77]. Analogous to this clinical assessment, the sFlT-1/PlGF ratios determined in the cell supernatants of the hypoxia and H/R models were above 38 during the entire observation time. These data strongly suggest that the established cell models mimic to a marked extent the pathophysiological characteristics of PE. Therefore, we speculate that the development of novel therapeutic approaches can be envisioned.

In the current study we proposed a cell model for PE using primary CTB isolated from healthy term pregnancies. It is evident that the described in vitro models cannot fully reflect the complex mechanisms occurring during the pathogenesis of PE. Indeed, we focused in our studies on five main pathways known to be dysregulated in PE, but several additional markers could be investigated. Additionally, in PE, several cell types such as extravillous trophoblasts and vascular endothelial cells are supposed to be involved in the pathogenic processes, but we selectively investigated the contribution of the primary trophoblast cells. Moreover, for the disease phenotypes, the differentiation stage of the trophoblasts might play a major role (e.g., the degree of syncytium formation and the differentiation from CTB to EVT). For the sake of clarity and due to technical restrictions, these aspects were currently neglected and represent limitations of our studies.

In conclusion, our in vitro models with primary CTB, isolated from normal term pregnancy placentae and cultured under hypoxia or H/R, exhibited similar characteristics as placentae from women with PE. Thirteen of seventeen genes associated with angiogenesis, the renin–angiotensin system, oxidative stress, ER stress, and the inflammasome complex were susceptible to hypoxia and H/R mimicking the expression pattern of PE tissues. The sFlT-1/PlGF ratio obtained under hypoxia and H/R conditions confirms the results found in the serum from women with PE underlining the suitability of the models. In general, the two different oxygen conditions seem to complement each other and should be chosen depending on the pathways of interest to be investigated in PE. The characterized cell models can serve as a suitable tool for further investigation of PE, enabling the study of the underlying cellular mechanisms of this clinically severe pregnancy disease with still widely unclear pathogenesis.

Author Contributions: Conceptualization, B.F., S.K. and C.A.; methodology, B.F., S.K., J.Z., J.G. and A.L.; data analysis and interpretation of the results, B.F., S.K. and J.Z.; recruitment of patients and provision of patient data, M.M., H.D.M. and A.L.; writing—original draft preparation, B.F.; writing—review and editing, J.Z. and C.A.; supervision and visualization of the project, C.A.; funding acquisition, C.A. All authors have read and agreed to the published version of the manuscript.

Funding: This study was supported by the Swiss National Science Foundation via the National Center of Competence in Research (NCCR) TransCure, University of Bern, Switzerland (grant no. 51NF40-185544; CA), the Swiss 3R Competence Centre (3RCC; grant no OC-2019-019; CA), and the Lindenhof Foundation, Bern, Switzerland (grant no. 17-15-F; CA). HDM received a British Heart Foundation Basic Science Intermediate Fellowship (FS/15/32/31604). AL and JG received support from the Fondo Nacional de Desarrollo Científico y Tecnológico FONDECYT (1190250, 1180935, and 1221362).

Institutional Review Board Statement: The study was performed according to the Declaration of Helsinki, with the approval of the Ethics Committee of the Canton of Bern (Basec Nr. 2016-00250) and from the Faculty of Medicine of the Pontificia Universidad Católica de Chile (PUC, ID-180810004).

Informed Consent Statement: Informed consent was obtained from all subjects involved in the study.

Data Availability Statement: The original contributions presented in the study are included in the article, further inquiries can be directed to the corresponding author.

Acknowledgments: We would like to thank the patients, physicians, and midwives from the Lindenhofgruppe, Bern, Switzerland, for participating in the study. Special thanks in this context to Ruedi Moser, Lindenhofgruppe Bern, for coordinating the placenta sampling process.

Conflicts of Interest: The authors declare no conflict of interest.

References

1. Ghulmiyyah, L.; Sibai, B. Maternal Mortality From Preeclampsia/Eclampsia. *Semin. Perinatol.* **2012**, *36*, 56–59. [CrossRef] [PubMed]
2. Brown, M.A.; Lindheimer, M.D.; de Swiet, M.; Van Assche, A.; Moutquin, J.M. The classification and diagnosis of the hypertensive disorders of pregnancy: Statement from the International Society for the Study of Hypertension in Pregnancy (ISSHP). *Hypertens. Pregnancy* **2001**, *20*, IX–XIV. [CrossRef] [PubMed]
3. Maynard, S.E.; Min, J.-Y.; Merchan, J.; Lim, K.H.; Li, J.; Mondal, S.; Libermann, T.; Morgan, J.P.; Sellke, F.W.; Stillman, I.E.; et al. Excess placental soluble fms-like tyrosine kinase 1 (sFlt1) may contribute to endothelial dysfunction, hypertension, and proteinuria in preeclampsia. *J. Clin. Investig.* **2003**, *111*, 649–658. [CrossRef] [PubMed]
4. Venkatesha, S.; Toporsian, M.; Lam, C.; Hanai, J.I.; Mammoto, T.; Kim, Y.M.; Bdolah, Y.; Lim, K.H.; Yuan, H.T.; Libermann, T.A.; et al. Soluble endoglin contributes to the pathogenesis of preeclampsia. *Nat. Med.* **2006**, *12*, 642–649. [CrossRef]
5. Young, B.C.; Levine, R.J.; Karumanchi, S.A. Pathogenesis of preeclampsia. *Annu. Rev. Pathol. Mech. Dis.* **2010**, *5*, 173–192. [CrossRef]
6. Fantone, S.; Mazzucchelli, R.; Giannubilo, S.R.; Ciavattini, A.; Marzioni, D.; Tossetta, G. AT-rich interactive domain 1A protein expression in normal and pathological pregnancies complicated by preeclampsia. *Histochem. Cell Biol.* **2020**, *154*, 339–346. [CrossRef]
7. Opichka, M.A.; Rappelt, M.W.; Gutterman, D.D.; Grobe, J.L.; McIntosh, J.J. Vascular Dysfunction in Preeclampsia. *Cells* **2021**, *10*, 3055. [CrossRef]
8. Wang, L.J.; Chen, C.P.; Lee, Y.S.; Ng, P.S.; Chang, G.D.; Pao, Y.H.; Lo, H.F.; Peng, C.H.; Cheong, M.L.; Chen, H. Functional antagonism between ΔNp63α and GCM1 regulates human trophoblast stemness and differentiation. *Nat. Commun.* **2022**, *13*, 1626. [CrossRef]
9. Mistry, H.D.; Kurlak, L.O.; Broughton Pipkin, F. The placental renin-angiotensin system and oxidative stress in pre-eclampsia. *Placenta* **2013**, *34*, 182–186. [CrossRef]
10. Poston, L.; Igosheva, N.; Mistry, H.D.; Seed, P.T.; Shennan, A.H.; Rana, S.; Karumanchi, S.A.; Chappell, L.C. Role of oxidative stress and antioxidant supplementation in pregnancy disorders. *Am. J. Clin. Nutr.* **2011**, *94*, 1980S–1985S. [CrossRef]

11. Kurlak, L.O.; Mistry, H.D.; Cindrova-Davies, T.; Burton, G.J.; Pipkin, F.B. Human placental renin-angiotensin system in normotensive and pre-eclamptic pregnancies at high altitude and after acute hypoxia-reoxygenation insult. *J. Physiol.* **2016**, *594*, 1327–1340. [CrossRef] [PubMed]
12. Williams, P.J.; Broughton Pipkin, F. The genetics of pre-eclampsia and other hypertensive disorders of pregnancy. *Best Pract. Res. Clin. Obstet. Gynaecol.* **2011**, *25*, 405–417. [CrossRef] [PubMed]
13. Sargent, I.L.; Borzychowski, A.M.; Redman, C.W. Immunoregulation in normal pregnancy and pre-eclampsia: An overview. *Reprod. Biomed. Online* **2006**, *13*, 680–686. [CrossRef]
14. Kweider, N.; Huppertz, B.; Kadyrov, M.; Rath, W.; Pufe, T.; Wruck, C.J. A possible protective role of Nrf2 in preeclampsia. *Ann. Anat.* **2014**, *196*, 268–277. [CrossRef] [PubMed]
15. Tossetta, G.; Fantone, S.; Montanari, E.; Marzioni, D.; Goteri, G. Role of NRF2 in Ovarian Cancer. *Antioxidants* **2022**, *11*, 663. [CrossRef] [PubMed]
16. Alcaraz, M.J.; Ferrándiz, M.L. Relevance of Nrf2 and heme oxygenase-1 in articular diseases. *Free Radic. Biol. Med.* **2020**, *157*, 83–93. [CrossRef]
17. Kaundal, R.K.; Datusalia, A.K.; Sharma, S.S. Posttranscriptional regulation of Nrf2 through miRNAs and their role in Alzheimer's disease. *Pharmacol. Res.* **2022**, *175*, 106018. [CrossRef]
18. Stødle, G.S.; Silva, G.B.; Tangerås, L.H.; Gierman, L.M.; Nervik, I.; Dahlberg, U.E.; Sun, C.; Aune, M.H.; Thomsen, L.C.V.; Bjørge, L.; et al. Placental inflammation in pre-eclampsia by Nod-like receptor protein (NLRP)3 inflammasome activation in trophoblasts. *Clin. Exp. Immunol.* **2018**, *193*, 84–94. [CrossRef]
19. Weel, I.C.; Romão-Veiga, M.; Matias, M.L.; Fioratti, E.G.; Peraçoli, J.C.; Borges, V.T.; Araujo, J.P.; Peraçoli, M.T. Increased expression of NLRP3 inflammasome in placentas from pregnant women with severe preeclampsia. *J. Reprod. Immunol.* **2017**, *123*, 40–47. [CrossRef]
20. Liu, Z.; Zhao, X.; Shan, H.Y.; Gao, H.; Wang, P. microRNA-520c-3p suppresses NLRP3 inflammasome activation and inflammatory cascade in preeclampsia by downregulating NLRP3. *Inflamm. Res.* **2019**, *68*, 643–654. [CrossRef]
21. Sagrillo-Fagundes, L.; Laurent, L.; Bienvenue-Pariseault, J.; Vaillancourt, C. In Vitro Induction of Hypoxia/Reoxygenation on Placental Cells: A Suitable Model for Understanding Placental Diseases. In Preeclampsia: Methods and Protocols. *Methods Mol. Biol.* **2018**, *1710*, 277–283. [PubMed]
22. Soleymanlou, N.; Jurisica, I.; Nevo, O.; Ietta, F.; Zhang, X.; Zamudio, S.; Post, M.; Caniggia, I. Molecular evidence of placental hypoxia in preeclampsia. *J. Clin. Endocrinol. Metab.* **2005**, *90*, 4299–4308. [CrossRef] [PubMed]
23. Ji, L.; Brkić, J.; Liu, M.; Fu, G.; Peng, C.; Wang, Y.L. Placental trophoblast cell differentiation: Physiological regulation and pathological relevance to preeclampsia. *Mol. Aspects Med.* **2013**, *34*, 981–1023. [CrossRef] [PubMed]
24. Roberts, J.; Hubel, C. The Two Stage Model of Preeclampsia: Variations on the Theme. *Placenta* **2009**, *30*, S32–S37. [CrossRef]
25. Fuenzalida, B.; Cantin, C.; Kallol, S.; Carvajal, L.; Pasten, V.; Contreras-Duarte, S.; Albrecht, C.; Gutierrez, J.; Leiva, A. Cholesterol uptake and efflux are impaired in human trophoblast cells from pregnancies with maternal supraphysiological hypercholesterolemia. *Sci. Rep.* **2020**, *10*, 5264. [CrossRef]
26. Hung, T.H.; Skepper, J.N.; Burton, G.J. In vitro ischemia-reperfusion injury in term human placenta as a model for oxidative stress in pathological pregnancies. *Am. J. Pathol.* **2001**, *159*, 1031–1043. [CrossRef]
27. Huang, X.; Anderle, P.; Hostettler, L.; Baumann, M.U.; Surbek, D.V.; Ontsouka, E.C.; Albrecht, C. Identification of placental nutrient transporters associated with intrauterine growth restriction and pre-eclampsia. *BMC Genom.* **2018**, *19*, 173. [CrossRef]
28. Potter, T.M.; Neun, B.W.; Stern, S.T. Assay to detect lipid peroxidation upon exposure to nanoparticles. *Methods Mol. Biol.* **2011**, *697*, 181–189.
29. Fuenzalida, B.; Sobrevia, B.; Cantin, C.; Carvajal, L.; Salsoso, R.; Gutiérrez, J.; Contreras-Duarte, S.; Sobrevia, L.; Leiva, A. Maternal supraphysiological hypercholesterolemia associates with endothelial dysfunction of the placental microvasculature. *Sci. Rep.* **2018**, *8*, 7690. [CrossRef]
30. Levine, R.J.; Maynard, S.E.; Qian, C.; Lim, K.-H.; England, L.; Yu, K.; Schisterman, B.; Thadhani, R.; Sachs, B.; Epstei, F.; et al. Circulating Angiogenic Factors and the Risk of Preeclampsia. *N. Engl. J. Med.* **2004**, *350*, 673–683. [CrossRef]
31. Verlohren, S.; Herraiz, I.; Lapaire, O.; Schlembach, D.; Zeisler, H.; Calda, P.; Sabria, J.; Markfeld-Erol, F.; Galindo, A.; Schoofs, K.; et al. New gestational phase-specific cutoff values for the use of the soluble fms-like tyrosine kinase-1/placental growth factor ratio as a diagnostic test for preeclampsia. *Hypertension* **2014**, *63*, 346–352. [CrossRef] [PubMed]
32. Burton, G.J.; Jauniaux, E.; Watson, A.L. Influence of oxygen supply on placental structure. In *Fetal Programming: Influences on Development and Disease in Later Life*; O'Brien, P.M.S., Wheeler, T., Barker, D.J.P., Eds.; RCOG Press: London, UK, 1999; pp. 326–341.
33. Burton, G.J.; Hung, T.H. Hypoxia-reoxygenation; a potential source of placental oxidative stress in normal pregnancy and preeclampsia. *Fetal Mat. Med. Rev.* **2003**, *14*, 97–117. [CrossRef]
34. Burton, G.J.; Hung, T.H.; Jauniaux, E. Placental hypoxia, hyperoxia and ischaemia-reperfusion injury in pre-eclampsia. In *Pre-Eclampsia—Etiology and Clinical Practice*; Lyall, F., Belfort, M., Eds.; Cambridge University Press: Cambridge, UK, 2007; pp. 138–151.
35. Neary, P.; Redmond, H.P. Ischaemia-Reperfusion Injury and the Systemic Inflammatory Response Syndrome. In *Ischaemia-Reperfusion Injury*; Grace, P.A., Mathie, R.T., Eds.; Blackwell Science: Oxford, UK, 1999; pp. 123–136.
36. Li, H.; Gu, B.; Zhang, Y.; Lewis, D.F.; Wang, Y. Hypoxia-induced increase in soluble Flt-1 production correlates with enhanced oxidative stress in trophoblast cells from the human placenta. *Placenta* **2005**, *26*, 210–217. [CrossRef] [PubMed]

37. Hung, T.; Burton, G. Hypoxia and reoxygenation: A possible mechanism for placenta oxidative stress in preeclampsia. *Taiwan J. Obstet. Gynecol.* **2006**, *45*, 189–200. [CrossRef]
38. Chau, K.; Hennessy, A.; Makris, A. Placental growth factor and pre-eclampsia. *J. Hum. Hypertens.* **2017**, *31*, 782–786. [CrossRef] [PubMed]
39. Scaife, P.J.; Simpson, A.; Kurlak, L.O.; Briggs, L.V.; Gardner, D.S.; Broughton Pipkin, F.; Jones, C.J.P.; Mistry, H.D. Increased Placental Cell Senescence and Oxidative Stress in Women with Pre-Eclampsia and Normotensive Post-Term Pregnancies. *Int. J. Mol. Sci.* **2021**, *22*, 7295. [CrossRef]
40. Khaliq, A.; Dunk, C.; Jiang, J.; Shams, M.; Li, X.F.; Acevedo, C.; Weich, H.; Whittle, M.; Ahmed, A. Hypoxia down-regulates placenta growth factor, whereas fetal growth restriction up-regulates placenta growth factor expression: Molecular evidence for "placental hyperoxia" in intrauterine growth restriction. *Lab. Investig.* **1999**, *79*, 151–170.
41. Hoeller, A.; Ehrlich, L.; Golic, M.; Herse, F.; Perschel, F.H.; Siwetz, M.; Henrich, W.; Dechend, R.; Huppertz, B.; Verlohren, S. Placental expression of sFlt-1 and PlGF in early preeclampsia vs. early IUGR vs. age-matched healthy pregnancies. *Hypertens. Pregnancy* **2017**, *36*, 151–160. [CrossRef]
42. Shah, D.M. The role of RAS in the pathogenesis of preeclampsia. *Curr. Hypertens. Rep.* **2006**, *8*, 144–152. [CrossRef]
43. Broughton Pipkin, F. The Renin-Angiotensin System in Pre-eclampsia. In *Pre-Eclampsia: Etiology and Clinical Practice*; Lyall, F., Belfort, M., Eds.; Cambridge University Press: Cambridge, UK, 2007; pp. 78–91.
44. Irani, R.A.; Xia, Y. Role of the renin-angiotensin system in pregnancy and preeclampsia. *Placenta* **2008**, *29*, 763–771. [CrossRef]
45. Herse, F.; Dechend, R.; Harsem, N.K.; Wallukat, G.; Janke, J.; Qadri, F.; Hering, L.; Muller, D.N.; Luft, F.C.; Staff, A.C. Dysregulation of the circulating and tissue-based renin-angiotensin system in preeclampsia. *Hypertension* **2007**, *49*, 604–611. [CrossRef] [PubMed]
46. Williams, P.J.; Mistry, H.D.; Innes, B.A.; Bulmer, J.N.; Broughton Pipkin, F. Expression of AT1R, AT2R and AT4R and Their Roles in Extravillous Trophoblast Invasion in the Human. *Placenta* **2010**, *31*, 448–455. [CrossRef] [PubMed]
47. Akhilesh, M.; Mahalingam, V.; Nalliah, S.; Ali, R.M.; Ganesalingam, M.; Haleagrahara, N. Hypoxia-inducible factor-1α as a predictive marker in pre-eclampsia. *Biomed. Rep.* **2013**, *1*, 257–258. [CrossRef]
48. Pennington, K.A.; Schlitt, J.M.; Jackson, D.L.; Schluz, L.C.; Schust, D.J. Preeclampsia: Multiple approaches for multifactorial disease. *Dis. Model Mech.* **2012**, *5*, 9–18. [CrossRef] [PubMed]
49. Kurlak, L.O.; Williams, P.J.; Bulmer, J.N.; Broughton Pipkin, F.; Mistry, H.D. Placental expression of adenosine A(2A) receptor and hypoxia inducible factor-1 alpha in early pregnancy, term and pre-eclamptic pregnancies: Interactions with placental renin-angiotensin system. *Placenta* **2015**, *36*, 611–613. [CrossRef] [PubMed]
50. Harati-Sadegh, M.; Kohan, L.; Teimoori, B.; Mehrabani, M.; Salimi, S. The association of the placental Hypoxia-inducible factor1-α polymorphisms and HIF1-α mRNA expression with preeclampsia. *Placenta* **2018**, *67*, 31–37. [CrossRef]
51. Zhang, Z.; Huang, C.; Wang, P.; Gao, J.; Liu, X.; Li, Y.; Yan, S.; Shi, Y. HIF-1α affects trophoblastic apoptosis involved in the onset of preeclampsia by regulating FOXO3a under hypoxic conditions. *Mol. Med. Rep.* **2020**, *21*, 2484–2492. [CrossRef]
52. Cindrova-Davies, T. Gabor Than Award Lecture 2008: Pre-eclampsia—From Placental Oxidative Stress to Maternal Endothelial Dysfunction. *Placenta* **2009**, *30*, 55–65. [CrossRef]
53. Kweider, N.; Wruck, C.J.; Rath, W. New insights into the pathogenesis of preeclampsia—The role of Nrf2 activators and their potential therapeutic impact. *Geburtshilfe Frauenheilkd* **2013**, *73*, 1236–1240. [CrossRef]
54. Matsubara, K.; Higaki, T.; Matsubara, Y.; Nawa, A. Nitric oxide and reactive oxygen species in the pathogenesis of preeclampsia. *Int. J. Mol. Sci.* **2015**, *16*, 4600–4614. [CrossRef]
55. Du, L.; He, F.; Kuang, L.; Tang, W.; Li, Y.; Chen, D. ENOS/iNOS and endoplasmic reticulum stress-induced apoptosis in the placentas of patients with preeclampsia. *J. Hum. Hypertens.* **2017**, *31*, 49–55. [CrossRef] [PubMed]
56. Li, J.; Zhou, J.; Ye, Y.; Liu, Q.; Wang, X.; Zhang, N.; Wang, X. Increased Heme Oxygenase-1 and Nuclear Factor Erythroid 2-Related Factor-2 in the Placenta Have a Cooperative Action on Preeclampsia. *Gynecol. Obstet. Investig.* **2016**, *81*, 543–551. [CrossRef] [PubMed]
57. Chigusa, Y.; Tatsumi, K.; Kondoh, E.; Fujita, K.; Nishimura, F.; Mogami, H.; Konishi, I. Decreased lectin-like oxidized LDL receptor 1 (LOX-1) and low Nrf2 activation in placenta are involved in preeclampsia. *J. Clin. Endocrinol. Metab.* **2012**, *97*, 1862–1870. [CrossRef]
58. Vaughan, J.E.; Walsh, S.W. Activation of NF-KB in placentas of women with preeclampsia. *Hypertens. Pregnancy* **2012**, *31*, 243–251. [CrossRef] [PubMed]
59. Burton, G.J.; Yung, H.W. Endoplasmic reticulum stress in the pathogenesis of early-onset pre-eclampsia. *Pregnancy Hypertens.* **2011**, *1*, 72–78. [CrossRef] [PubMed]
60. Jones, C.P.; Fox, H. An ultrastructural and ultrahistochemical study of the human placenta in maternal essential hypertension. *Placenta* **1980**, *1*, 61–76. [CrossRef]
61. Anderson, U.D.; Olsson, M.G.; Kristensen, K.H.; Åkerström, B.; Hansson, S.R. Review: Biochemical markers to predict preeclampsia. *Placenta* **2012**, *33*, S42–S47. [CrossRef]
62. Zadeh, M.N.; Naghshvar, F.; Peyvandi, S.; Gheshlaghi, P.; Ehetshami, S. PP13 and PAPP-A in the First and Second Trimesters: Predictive Factors for Preeclampsia? *ISRN Obstet. Gynecol.* **2012**, *2012*, 263871.
63. Romero, R.; Kusanovic, J.P.; Than, N.G.; Erez, O.; Gotsch, F.; Espinoza, J.; Chefetz, I.; Gomez, R.; Nien, J.K.; Kuhnreich, I.; et al. First Trimester Maternal Serum PP13 in the Risk Assessment for Preeclampsia. *Am. J. Obset. Gynecol.* **2008**, *199*, 122.e1–122.e11. [CrossRef]

64. Gadde, R.; Dayanand, C.D.; Sheela, S.R. Placental protein 13: An important biological protein in preeclampsia. *J. Circ. Biomark.* **2018**, *7*, 1–16. [CrossRef]
65. Yaron, Y.; Heifetz, S.; Ochshorn, Y.; Lehavi, O.; Orr-Urtreger, A. Decreased first trimester PAPP-A is a predictor of adverse pregnancy outcome. *Prenat. Diagn.* **2002**, *22*, 778–782. [CrossRef] [PubMed]
66. Luewan, S.; Teja-Intr, M.; Sirichotiyakul, S.; Tongsong, T. Low maternal serum pregnancy-associated plasma protein-A as a risk factor of preeclampsia. *Singap. Med. J.* **2018**, *59*, 55–59. [CrossRef] [PubMed]
67. Grill, S.; Rusterholz, C.; Zanetti-Dällenbach, R.; Tercanli, S.; Holzgreve, W.; Hahn, S.; Lapaire, O. Potential markers of preeclampsia—A review. *Reprod. Biol. Endocrinol.* **2009**, *7*, 70. [CrossRef]
68. Wang, Q.J.; Song, B.F.; Zhang, Y.H.; Ma, Y.Y.; Shao, Q.Q.; Liu, J.; Qu, X. Expression of RGC32 in human normal and preeclamptic placentas and its role in trophoblast cell invasion and migration. *Placenta* **2015**, *36*, 350–356. [CrossRef] [PubMed]
69. Wan, H.; Versnel, M.A.; Cheung, W.Y.; Leenen, P.J.M.; Khan, N.A.; Benner, R.; Kiekens, R.C.M. Chorionic gonadotropin can enhance innate immunity by stimulating macrophage function. *J. Leukoc. Biol.* **2007**, *82*, 926–933. [CrossRef] [PubMed]
70. Lamarca, B. The role of immune activation in contributing to vascular dysfunction and the pathophysiology of hypertension during preeclampsia. *Minerva Ginecol.* **2010**, *62*, 105–120. [PubMed]
71. Barjaktarovic, M.; Korevaar, T.I.M.; Jaddoe, V.W.V.; de Rijke, Y.B.; Peeters, R.P.; Steegers, E.A.P. Human chorionic gonadotropin and risk of pre-eclampsia: Prospective population-based cohort study. *Ultrasound Obstet. Gynecol.* **2019**, *54*, 477–483. [CrossRef]
72. Strohmer, H.; Kiss, H.; Mösl, B.; Egarter, C.; Husslein, P.; Knöfler, M. Hypoxia downregulates continuous and interleukin-1-induced expression of human chorionic gonadotropin in choriocarcinoma cells. *Placenta* **1997**, *18*, 597–604. [CrossRef]
73. Knöfler, M.; Stenzel, M.; Mösl, B.; Strohmer, H.; Husslein, P. Hypoxia-mediated downregulation of human chorionic gonadotrophin subunit mRNA expression in cytotrophoblast JAr cells. *Placenta* **1998**, *19*, 387–402. [CrossRef]
74. Mistry, H.D.; Wilson, V.; Ramsay, M.M.; Symonds, M.E.; Broughton Pipkin, F. Reduced selenium concentrations and glutathione peroxidase activity in preeclamptic pregnancies. *Hypertension* **2008**, *52*, 881–888. [CrossRef]
75. Gupta, S.; Aziz, N.; Sekhon, L.; Agarwal, R.; Mansour, G.; Li, J.; Agarwal, A. Lipid peroxidation and antioxidant status in preeclampsia. A systematic review. *Obstet. Gynecol. Surv.* **2009**, *64*, 750–759. [CrossRef] [PubMed]
76. Serdar, Z.; Gür, E.; Develioğlu, O.; Çolakoğullari, M.; Dirican, M. Placental and decidual lipid peroxidation and antioxidant defenses in preeclampsia—Lipid peroxidation in preeclampsia. *Pathophysiology* **2002**, *9*, 21–25. [CrossRef]
77. Herraiz, I.; Llurba, E.; Verlohren, S.; Galindo, A.; Bartha, J.L.; De La Calle, M.; Delgado, J.L.; De Paco, C.; Escudero, A.I.; Moreno, F.; et al. Update on the Diagnosis and Prognosis of Preeclampsia with the Aid of the sFlt-1/PLGF Ratio in Singleton Pregnancies. *Fetal Diagn. Ther.* **2018**, *43*, 81–89. [CrossRef] [PubMed]

cells MDPI

Article

Functional Analysis of $p21^{Cip1/CDKN1A}$ and Its Family Members in Trophoblastic Cells of the Placenta and Its Roles in Preeclampsia

Nina-Naomi Kreis [1,*], Alexandra Friemel [1], Lukas Jennewein [1], Samira Catharina Hoock [1], Anna Elisabeth Hentrich [1], Thorsten Nowak [2], Frank Louwen [1] and Juping Yuan [1]

1 Obstetrics and Prenatal Medicine, Department of Gynecology and Obstetrics, University Hospital Frankfurt, J. W. Goethe-University Frankfurt, Theodor-Stern-Kai 7, 60590 Frankfurt, Germany; Alexandra.Friemel@kgu.de (A.F.); Lukas.Jennewein@kgu.de (L.J.); SamiraCatharina.Hoock@kgu.de (S.C.H.); AnnaElisabeth.Hentrich@kgu.de (A.E.H.); Louwen@em.uni-frankfurt.de (F.L.); yuan@em.uni-frankfurt.de (J.Y.)
2 Medical Practice for Gynecology and Obstetrics, Mainzer Landstr. 265, 60326 Frankfurt, Germany; info@dr-nowak.net
* Correspondence: Nina-Naomi.Kreis@kgu.de; Tel.: +49-69-6301-83426

Abstract: Preeclampsia (PE), a gestational hypertensive disease originating from the placenta, is characterized by an imbalance of various cellular processes. The cell cycle regulator $p21^{Cip1/CDKN1A}$ (p21) and its family members p27 and p57 regulate signaling pathways fundamental to placental development. The aim of the present study was to enlighten the individual roles of these cell cycle regulators in placental development and their molecular involvement in the pathogenesis of PE. The expression and localization of p21, phospho-p21 (Thr-145), p27, and p57 was immunohistochemically analyzed in placental tissues from patients with early-onset PE, early-onset PE complicated by the HELLP (hemolysis, elevated liver enzymes and low platelet count) syndrome as well as late-onset PE compared to their corresponding control tissues from well-matched women undergoing caesarean sections. The gene level was evaluated using real-time quantitative PCR. We demonstrate that the delivery mode strongly influenced placental gene expression, especially for *CDKN1A* (p21) and *CDKN1B* (p27), which were significantly upregulated in response to labor. Cell cycle regulators were highly expressed in first trimester placentas and impacted by hypoxic conditions. In support of these observations, p21 protein was abundant in trophoblast organoids and hypoxia reduced its gene expression. Microarray analysis of the trophoblastic BeWo cell line depleted of p21 revealed various interesting candidate genes and signaling pathways for the fusion process. The level of p21 was reduced in fusing cytotrophoblasts in early-onset PE placentas and depletion of p21 led to reduced expression of fusion-related genes such as syncytin-2 and human chorionic gonadotropin (β-hCG), which adversely affected the fusion capability of trophoblastic cells. These data highlight that cell cycle regulators are important for the development of the placenta. Interfering with p21 influences multiple pathways related to the pathogenesis of PE.

Keywords: $p21^{Cip1/CDKN1A}$; trophoblasts; preeclampsia; hypoxia; trophoblast organoids; fusion

Citation: Kreis, N.-N.; Friemel, A.; Jennewein, L.; Hoock, S.C.; Hentrich, A.E.; Nowak, T.; Louwen, F.; Yuan, J. Functional Analysis of $p21^{Cip1/CDKN1A}$ and Its Family Members in Trophoblastic Cells of the Placenta and Its Roles in Preeclampsia. *Cells* **2021**, *10*, 2214. https://doi.org/10.3390/cells10092214

Academic Editors: Junie P Warrington, Ana T. Palei, Mark W. Cunningham and Lorena M. Amaral

Received: 19 July 2021
Accepted: 25 August 2021
Published: 27 August 2021

1. Introduction

Pregnancy can be affected by various health problems, of which preeclampsia (PE) is the most common. PE is a multisystemic gestational disease with a global prevalence of up to 8% [1]. It is characterized by concurrent hypertension and proteinuria or any other sign of end organ damage including liver or brain, occurring after 20 weeks of gestation [2]. PE is a consequence of diverse pathophysiological processes linked to maternal endothelial dysfunction and systemic inflammation, which can result in multiorgan failure, if the fetus and placenta are not delivered [1–3]. It can be subdivided into an early-onset (<34th week of gestation) and a late-onset ($\geq$34th week of gestation) form [4]. Since early-onset PE mainly originates from the placenta, it is also referred to as placental PE, where it is tightly associated with defective trophoblast invasion and inadequate spiral artery remodeling, while late-onset PE is rather linked to preexisting maternal risk factors like obesity [5]. This classification has prognostic importance, since early-onset PE entails a greater risk of maternal and fetal complications [6]. PE is often complicated by the HELLP syndrome, which is an acronym for hemolysis, elevated liver enzymes, and low platelet count [7]. Despite intensive research in the last decades, prevention and treatment options are very limited. Delivery is currently the only definitive treatment for PE patients. Consequently, it remains one of the leading causes of maternal and perinatal mortality and morbidity, and is associated with increased rates of cardiovascular and metabolic diseases in later life [1,8].

Defective placentation is considered to be causative for the development of pregnancy-related disorders like PE, fetal growth restriction [9], and the HELLP syndrome [10], suggesting that trophoblasts, which constitute the major placental cell type, are deficient and the origin of gestational diseases. The human placenta is a rather complex organ structured as villous trees consisting of an inner layer of proliferative progenitor cells, called villous cytotrophoblasts (CTBs). On one hand, these CTBs further differentiate and then fuse into the non-proliferative and multinucleated syncytiotrophoblast (STB), responsible for nutrient exchange and hormone production [11]. Of importance, high attention is paid to a small portion of villous CTBs, so called "fusing cytotrophoblasts" (fCTBs), which are characterized by a marked reduction in E-cadherin expression during the differentiation process to the STB [12,13] and ready to fuse to the STB, thus, also referred to as intermediate CTBs [14]. On the other hand, CTBs grow into the proliferative cell columns and differentiate to an invasive and growth-arrested phenotype, termed extravillous trophoblasts (EVTs) [11]. CTBs are therefore considered as progenitors for the STB as well as EVTs [15]. EVTs invade into the maternal decidua and colonize the lumen of spiral arteries, which are remodeled for sufficient blood supply to the embryo [16]. PE is associated with profound cellular dysfunctions including impaired placentation, deregulated proliferation and differentiation manifested by shallow EVT invasion with incomplete spiral artery remodeling, defective fusion of CTBs, increased STB microparticles due to deregulated apoptosis, and a hypoxic environment resulting in elevated oxidative stress, angiogenic imbalance, and inflammation at the maternal–fetal interface [1,17–20].

The rapid development of the human placenta requires strictly regulated and coordinated cell proliferation and differentiation processes, crucial for tissue homeostasis [21]. While a successful placentation depends on precise trophoblast regulation, the roles of cell cycle regulators are understudied in the human placenta. $p21^{Cip1}$, encoded by the gene *CDKN1A* (cyclin-dependent kinase inhibitor 1A), is a pivotal broad cell cycle regulator with heterogeneous roles and the founding member of the Cip/Kip (cyclin-dependent kinase interacting protein/kinase inhibitory protein) family including $p27^{Kip1}$ (*CDKN1B*) and $p57^{Kip2}$ (*CDKN1C*) [22]. Interestingly, these regulators are evolutionally highly conserved and able to substitute each other [23]. Apart from its role in cell cycle regulation including mitosis, p21 is involved in cell differentiation, reprogramming of induced pluripotent stem cells, transcription, DNA repair, migration, apoptosis, autophagy, and the onset of senescence [24]. p21 acts either as a positive or negative regulator of the same pathway, often related to its cytoplasmic localization, expression level, and posttranslational modifications [23,24]. Since p21 participates in a diversity of cellular activities in cancer cells and stem cells, it is reasonable to assume that p21 is also multifunctional in placental trophoblasts coordinating various cellular processes. In fact, we have recently shown that the deficiency of p21 lowers the migration and invasion capability of cancer and trophoblastic cells [25]. In the present work, we investigated the expression and localization of p21 and its family members in the placenta, and their potential involvement in placental development and the pathogenesis of PE.

2. Materials and Methods

2.1. Placental Tissue Collection

This study was approved by the Ethics Committee at the University Hospital (reference number: 375/11), Goethe-University Frankfurt. Written informed approval was obtained from all donors. PE was diagnosed as an occurrence of hypertension after 20 weeks of gestation with a blood pressure ≥140/90 mmHg and proteinuria with ≥300 mg in 24 h. The HELLP syndrome was defined as the presence of hemolysis, elevated liver enzymes, and thrombocytopenia (low platelet count <100,000/μL). The reasons for delivery of early-onset control groups were PE-irrelevant like breech presentation, premature rupture of membranes, premature placental abruption, non-reassuring fetal heart rate, or umbilical cord prolapse. Tissue samples were taken from placentas within 30 min post-delivery, formalin-fixed and paraffin-embedded (FFPE) for immunohistochemistry staining (IHC), or frozen immediately in liquid nitrogen for mRNA and protein extraction, which were stored at −80 °C until usage. Clinical information of all participants is shown in (Tables 1–4). For IHC staining, Prof. Dr. Qi Chen, Department of Obstetrics and Gynecology, University of Auckland, and Fudan University, Shanghai, China, kindly provided us with six first trimester placental FFPE samples (six to nine weeks of gestation, age 20 to 33 years). The sample collection was approved by the Ethics Committee of the Hospital of Obstetrics and Gynecology of Fudan University (reference number 2018-62), China. Written consent was obtained from healthy donors undergoing elective surgical terminations of pregnancy. For trophoblast organoid formation, primary villous cytotrophoblasts or placental mesenchymal stem/stromal cells were isolated from first trimester placentas. Written informed consent was obtained from patients undergoing elective terminations of normal pregnancies (seven to 12 weeks of gestation, age 25 to 41 years) at the Medical Practice for gynecology and obstetrics led by Dr. Thorsten Nowak with ethical approval from the Ethics Committee at the University Hospital, Goethe-University Frankfurt (reference number 19-455).

Table 1. Clinical information of patients for delivery mode analysis. Mean value ± standard deviation is shown. CS, caesarean section; eCS, emergency caesarean section after the onset of labor; VD, vaginal delivery; opVD, operative vaginal delivery; ns, not significant.

	n	Gestational Age (Weeks)	Body Mass Index (BMI)	Age	Birth Weight (g)
CS	5	38.2 ± 0.8	21.9 ± 1.4	28.4 ± 0.9	2960 ± 634
eCS	5	38.0 ± 1.0	21.5 ± 1.9	27.0 ± 2.2	3020 ± 370
VD	5	38.0 ± 1.0	21.7 ± 2.3	27.0 ± 1.9	3002 ± 302
opVD	5	38.0 ± 1.0	22.8 ± 1.0	26.4 ± 2.2	3008 ± 302
p-value		ns	ns	ns	ns

Table 2. Clinical information of patients with early-onset preeclampsia (ePE) and matched controls. Mean value ± standard deviation is shown. BP, blood pressure; n.d., not determined.

	n	Age	Gestational Age (Weeks)	Body Mass Index (BMI)	Birth Weight (g)	Systolic BP (mmHg)	Diastolic BP (mmHg)	Proteinuria (mg/24 h)
control	20	32.6 ± 4.6	29.7 ± 2.6	24.9 ± 3.9	1284 ± 710	118 ± 13	71 ± 11	n.d.
early-onset PE	20	32.4 ± 5.9	29.6 ± 2.6	25.7 ± 4.4	1072 ± 387	167 ± 22	102 ± 12	4153 ± 4569
p-value		0.876	0.330	0.256	0.114	0.00000011	0.0000038	

Table 3. Clinical information of patients with late-onset preeclampsia (PE) and matched controls. Mean value ± standard deviation is shown. BP, blood pressure; n.d., not determined.

	n	Age	Gestational Age (Weeks)	Body Mass Index (BMI)	Birth Weight (g)	Systolic BP (mmHg)	Diastolic BP (mmHg)	Proteinuria (mg/24 h)
control	10	30.9 ± 3.4	37.7 ± 1.3	23.4 ± 4.0	2914 ± 524	119 ± 7	73 ± 11	n.d.
late-onset PE	10	30.9 ± 2.8	37.7 ± 1.3	24.4 ± 2.0	2413 ± 461	153 ± 16	96 ± 13	1794 ± 1901
p-value		0.989	1.0	0.273	0.020	0.00034	0.00020	

Table 4. Clinical information of patients with early-onset preeclampsia (ePE) complicated by the HELLP (hemolysis, elevated liver enzymes, and low platelet count) syndrome and matched controls. Mean value ± standard deviation is shown. BP, blood pressure; n.d., not determined.

	n	Age	Gestational Age (Weeks)	Body Mass Index (BMI)	Birth Weight (g)	Systolic BP (mmHg)	Diastolic BP (mmHg)	Proteinuria (mg/24 h)
control	17	29.6 ± 9.1	30.2 ± 7.9	23.7 ± 6.7	1751 ± 920	109 ± 29	66 ± 20	n.d.
ePE + HELLP	16	31.4 ± 4.5	31.7 ± 2.4	25.3 ± 4.3	1389 ± 443	175 ± 21	107 ± 9	4815 ± 4856
p-value		0.589	0.164	0.973	0.0035	0.0000001	0.00000004	

2.2. Formation of Trophoblast Organoids from Human First Trimester Placental Tissue

The protocol was adapted from Sheridan et al. [26] with some modifications. In brief, to obtain trophoblast-enriched cell suspensions, villi from first trimester placental tissue were washed with PBS (Thermo Fisher, Waltham, MA, USA) containing 1% penicillin/streptomycin (Sigma-Aldrich, Taufkirchen, Germany), further minced into small pieces and sequentially digested with 0.2% trypsin/EDTA in PBS, then with 1.0 mg/mL collagenase V (Sigma-Aldrich, Taufkirchen, Germany) in PBS containing 0.1% BSA (Carl Roth, Karlsruhe, Germany). The digestion steps were performed in a shaker with 120 rpm for 5 min at 37 °C. Both digestion steps were stopped with Ham's F12 medium (Life Technologies, Carlsbad, CA, USA) containing 20% fetal bovine serum (FBS) (Biowest, Riverside, CA, USA), filtered with a 100 μm cell strainer (Corning, New York, NY, USA), pooled and washed with PBS. Erythrocytes were lysed by the addition of lysis buffer (155 mM NH_4Cl, 10 mM $KHCO_3$, 0.1 mM EDTA) [27], incubated for 10 min at 37 °C, centrifuged (1000× *g*, 5 min), and washed twice with PBS. Cells were plated onto 24 well Corning Costar ultra-low attachment plates (Corning; Berlin, Germany) or embedded in Matrigel (Trevigen®, Gaithersburg, MD, USA). The trophoblast organoid medium (TOM) contained Advanced DMEM/F12 supplemented with 1× N2 (Life Technologies), 1× B27 (Life Technologies), 100 μg/mL primocin (Invivogen, San Diego, CA, USA), 1.25 nM *N*-Acetyl-L-cysteine (Sigma-Aldrich, Taufkirchen, Germany), 2 mM L-glutamine (Life Technologies, Carlsbad, CA, USA), 1 mM A83-01 (Tocris, Bristol, UK), 1.5 μM CHIR99021 (Selleck Chemicals Llc., Houston, TX, USA), 50 ng/mL EGF (epidermal growth factor) (PeproTech, Rocky Hill, NJ, USA), 100 ng/mL R-spondin 1 (PeproTech, Rocky Hill, NJ, USA), 100 ng/mL FGF-2 (fibroblast growth factor) (PeproTech, Rocky Hill, NJ, USA), 50 ng/mL HGF (hepatocyte growth factor) (PeproTech, Rocky Hill, NJ, USA), 2 μM Y-27632 (Selleckchem Llc., Houston, TX, USA), and 2.5 μM prostaglandin E2 (R&D Systems, Minneapolis, MN, USA) [28]. Cultures were maintained in 5% CO_2 in a humidified incubator at 37 °C. Medium was replaced every 2–3 days. Small organoid clusters became visible around day 10 (Figure S1A).

2.3. Preparation of Organoids and Placental Tissues for IHC-IF

For paraffin embedding, organoids embedded in Matrigel were gently collected into cold medium, washed with PBS containing 0.1% BSA (Carl Roth, Karlsruhe, Germany), fixed in ROTI® Histofix solution (Carl Roth, Karlsruhe, Germany) for 30 min on ice, washed twice with PBS containing 0.1% BSA and stored at least overnight in 70% ethanol at 4 °C.

For visualization, organoids were stained with a drop of hematoxylin for 5–10 min, washed gently with Aqua dest., mixed with 100 µL Histowax (Leica Biosytems, Nussloch, Germany), and incubated for 30 min on ice for embedding. Five µm sections were prepared. After deparaffinization, the slides were incubated with the target retrieval solution from DAKO EnVision™ FLEX Kit (DAKO, Hamburg, Germany) for 15 min, or 30 min in the case of placental tissue sections in a water bath (100 °C), blocked with peroxidase for 5 min, and incubated with primary antibodies for 60 min at room temperature. Following primary antibodies were used: rabbit monoclonal p21 (#2947; Cell Signaling Technology, Danvers, MA, USA), mouse monoclonal cytokeratin 7 (#M7018; DAKO, Hamburg, Germany), mouse monoclonal E-cadherin (#610181; BD Transduction Laboratories, San Jose, CA, USA), rabbit monoclonal epidermal growth factor receptor (EGFR) (#4267; Cell Signaling, Technology, Danvers, MA, USA), mouse monoclonal human leukocyte antigen G (HLA-G) (#11-291-C100; Exbio Praha, a.s., Vestec, Czech), and mouse monoclonal Ki67 (#M7240; DAKO, Hamburg, Germany). This was followed by the incubation with secondary antibodies for 30 min at room temperature: goat anti-rabbit or goat anti-mouse Alexa Fluor® 594 (#ab150080 or #ab150116; Abcam, Cambridge, UK) and goat-anti mouse DyLight® 488 or goat anti-rabbit Alexa Fluor® 488 (#ab96879 or #ab150077; Abcam, Cambridge, UK). DAPI (4′,6-diamidino-2-phenylindole-dihydrochloride; Roche, Mannheim, Germany) was used to stain the DNA content. Slides were examined with an AxioObserver.Z1 microscope (Zeiss, Göttingen, Germany) equipped with an AxioCam MRm camera (Zeiss, Göttingen, Germany).

2.4. Isolation of Primary Villous Cytotrophoblasts from Human First Trimester Placenta

Isolation of villous cytotrophoblasts from first trimester placenta was performed according to Haider et al. [29] and Vondra et al. [30] using three consecutive digestion steps followed by Percoll density gradient centrifugation (5–70%; Sigma-Aldrich, Taufkirchen, Germany). Placental tissues were washed with PBS and placental villi were minced. The digestion steps were performed in 50 mL tubes at 37 °C in a shaker with 200 rpm for 8 min, 15 min, and 15 min. The digestion buffer (10× HBSS (Hank's Balanced Salt) [31], 7.5% $NaHCO_3$, 1 M HEPES (4-(2-hydroxyethyl)-1-piperazineethanesulfonic acid)) was supplemented with 0.25% trypsin (Gibco, Life Technologies, Carlsbad, CA, USA) and 1.25 mg/mL DNase I (Sigma-Aldrich, Taufkirchen, Germany). The second and third digestion steps were pooled and purified with a Percoll gradient as detailed for term placental cytotrophoblasts isolation (see below).

2.5. Isolation of Placental Mesenchymal Stem/Stromal Cells from First Trimester Placental Tissue

The isolation protocol by Stiegman et al. [32] was modified. Villous like structures were washed twice, minced, and mixed with 15 mL collagenase II (275 U/mL; Worthington, Columbus, OH, USA), diluted in HBSS, and 500 µL Dispase (90 U/mL; Roche, Mannheim, Germany). After 75 min at 37 °C and 200 rpm, the enzyme activity was stopped with medium containing 10% FBS (Biochrome, Berlin, Germany) and filtered through a 100 µm mesh. Erythrocytes were lysed with lysis buffer and resuspended in DMEM (Gibco, Carlsbad, CA, USA) containing 20% FBS (Biochrome, Berlin, Germany), 1% penicillin/streptomycin, 1 µg/mL amphotericin B (Sigma-Aldrich, Taufkirchen, Germany), and 5 ng/mL hFGF (Promega GmbH, Walldorf, Germany). The medium was changed after 24 h to remove non-adherent cells.

2.6. Isolation of Primary Villous Cytotrophoblasts from Human Term Placental Tissue

Villous cytotrophoblast cell isolation and purification was carried out according to Petroff et al. [31]. In brief, approximately 50 g of villous placental tissue free of calcification or hematoma was finely minced within 30 min after delivery, rinsed with 0.9% NaCl, and digested with 0.25% trypsin (Thermo Fisher Scientific, Dreieich, Germany) and 300 U/mL DNase I (Sigma-Aldrich, Taufkirchen, Germany) for 20 min shaking with 200 rpm at 37 °C. After digestion, the supernatant was transferred into tubes containing 1.5 mL FBS (Merck

Millipore, Darmstadt, Germany) and centrifuged (1000× g, 15 min). The digestion, transfer, and centrifugation steps were repeated two more times. The pellet was resuspended in DMEM (Thermo Fisher Scientific, Dreieich, Germany) and filtered with a 100 μm cell strainer (Corning, NY, USA). The cells were centrifuged (1000× g, 10 min), resuspended in Ca/Mg-free Hank's balanced salt solution, and stratified on two Percoll gradients (5–70%; Sigma-Aldrich, Taufkirchen, Germany). The gradients were centrifuged without brake (1200× g, 20 min). The fractions between 35 and 50% of the gradients were used, pooled, and diluted in pre-warmed medium for centrifugation (1000× g, 5 min). The cell pellet was resuspended in erythrocyte lysis buffer (155 mM NH_4Cl, 10 mM $KHCO_3$, and 0.1 mM EDTA) [27], incubated for 10 min at 37 °C, centrifuged (1000× g, 5 min), and washed twice with PBS. The remaining cells were seeded onto 15 cm cell culture plates in DMEM/F12 (Life Technologies) containing 20% FBS (Biowest, Riverside, CA, USA), 100 μg/mL streptomycin, 100 U/mL penicillin, and 1 μg/mL amphotericin B (Sigma-Aldrich, Taufkirchen, Germany) for 45 min to remove adherent stromal cells. Non-adherent trophoblastic cells were collected, seeded onto collagen-coated plates (Greiner Bio-One, Frickenhausen, Germany), and cultured under standard cell culture conditions.

2.7. Immunohistochemistry of Placental Tissue

A standard staining procedure with DAKO EnVision™FLEX Kit (#K8000; DAKO, Hamburg, Germany) was used to stain FFPE placental tissue sections from PE patients and matched controls, as stated [33]. The following antibodies were used: mouse monoclonal antibody against p21 (#2946; 1:25, incubation 1 h at 37 °C), rabbit monoclonal antibody against p27 (#3686; 1:30, incubation 1 h), rabbit polyclonal antibody against p57 (#2557; 1:30, incubation 1 h; Cell Signaling Technology, Danvers, MA, USA), and p-p21 (Thr-145, #AF3290; 1:30, incubation 1 h at 37 °C; Affinity Biosciences, Cincinnati, OH, USA). Slides were counterstained with hematoxylin and analyzed using an AxioObserver.Z1 microscope (Zeiss, Göttingen, Germany). Negative controls included samples stained with control immunoglobulin G (IgG) lacking primary antibody. Evaluation was carried out without knowing the diagnosis. The average of the percentage of positive cells was determined. In the case of syncytiotrophoblast (STB), the positive area per visual field was estimated. Ten fields per sample were counted. The slides were further evaluated by the semi-quantitative H-score method, which takes the staining intensity into account. The H-score is determined by adding the results of multiplying the percentage of positive stained cells with their staining intensity (scored as 0 for no signal, 1 = weak, 2 = moderate and 3 = strong): [1 × (% cells 1) + 2 × (% cells 2) + 3 × (% cells 3)]. The highest possible value is 300 [34,35].

2.8. RNA Extraction and Real-Time Quantitative PCR

Total RNAs were extracted with EXTRACTME Total RNA Kit, with DNase digestion for tissue samples or without DNase digestion for cell lines, according to manual instructions (7Bioscience GmbH, Neuenburg, Rhein, Germany). In the case of first trimester samples, total RNAs were extracted from FFPE tissues using the ReliaPrep™ FFPE Total RNA Miniprep System as instructed (Promega GmbH, Walldorf, Germany). Reverse transcription was performed using the Go Script Reverse Transcription Mix as instructed (Promega GmbH, Walldorf, Germany). A StepOnePlus Real-time PCR System (Applied Biosystems, Darmstadt, Germany) was used to perform real-time quantitative PCR and data were analyzed with StepOne Software v2.3 (Applied Biosystems, Darmstadt, Germany). To analyze primary placental tissue, the mean value of expression levels of *SDHA* (succinate dehydrogenase complex, subunit A), *TBP* (TATA box-binding protein), and *YWHAZ* (tyrosine 3-monooxygenase/tryptophan 5-monooygenase activation protein, zeta polypeptide) or *TBP* alone served as the endogenous control [36,37]. For gene evaluation from cultured cells *GAPDH* (glyceraldehyde 3-phosphate dehydrogenase) was used as the endogenous control. The primers and probes for *GAPDH* (Hs_02786624), *CDKN1A* (Hs00355782_m1), *CDKN1B* (Hs_00153277_m1), *CDKN1C* (Hs00175938_m1),

TP53 (Hs01034249_m1), *HERV-FRD* (Hs01942443_s1), *CGbeta5* (Hs00361224_gH), *GCM1* (Hs_00961601_m1), *GATA3* (Hs00231122_m1), *TFAP2A* (Hs01029413_m1), *TFAP2C* (Hs00231476_m1), *ELF5* (Hs01063023_g1), *SDHA* (Hs00188166_m1), *TBP* (Hs99999910_m1), and *YWHAZ* (Hs00237047_m1) were obtained from Applied Biosystems (Darmstadt, Germany). All results were shown as relative quantification (RQ) [38].

2.9. Cell Culture, Transfection and Treatment

The HTR-8/SVneo cell line (referred to as HTR) was kindly provided by Prof. Dr. Charles Graham [39] and the SGHPL-4 cell line by Prof. Dr. Guy Whitley [40]. BeWo (Sigma-Aldrich, Taufkirchen, Germany) and JEG-3 cells (ATCC, Wesel, Germany) were cultured as instructed. The 1% oxygen atmosphere was supplied in a special hypoxic O_2 incubator (Galaxy 48 R, Eppendorf, Hamburg, Germany).

siRNA targeting p21 (sense: ACACCUCCUCAUGUACAUA and antisense: UAUGUACAUGAGGAGGUGU; designated as sip21) was manufactured by Sigma-Aldrich (Taufkirchen, Germany). A different siRNA against p21 (referred to as sip21 #2), containing a mixed pool of siRNAs, was obtained from Santa Cruz (Heidelberg, Germany; sc-29427). Control siRNA was obtained from Qiagen (Hilden, Germany; #1027281). siRNAs (30 nM). Cells were transiently transfected with siRNA with Oligofectamine™ (Thermo Fisher Scientific, Dreieich, Germany) [41].

Trophoblast fusion was induced by treating cells with 25 μM forskolin (Sigma-Aldrich, Taufkirchen, Germany) for indicated time points. An equal amount of dimethyl sulfoxide (DMSO, Sigma-Aldrich, Taufkirchen, Germany) was used as the vehicle control.

2.10. Western Blot Analysis

Western blot analysis was performed as reported [42]. Cells were harvested with RIPA buffer (50 mM Tris pH 8.0, 150 mM NaCl, 1% NP-40, 0.5% Na-desoxycholate, 0.1% SDS, 1 mM NaF, phosphatase and protease inhibitor cocktail tablets (Roche, Mannheim, Germany)). The following antibodies were used: mouse monoclonal antibody against GAPDH (GTX627408; GeneTex, Eching, Germany) and p53 (DO-1, sc-126; Santa Cruz, Heidelberg, Germany); rabbit monoclonal antibody against p21 (#2947) and p27 (#3686); and rabbit polyclonal antibody against p57 (#2557; Cell Signaling Technology, Danvers, MA, USA). ImageJ 1.48v software (National Institutes of Health, Bethesda, MD, USA) was used for densitometric measurements of western blot analysis.

2.11. Immunofluorescence Staining

Indirect immunofluorescence was performed as reported [25,41]. Following primary antibodies were used: rabbit monoclonal antibody against epidermal growth factor receptor (EGFR) (#4267; Cell Signaling, Technology, Danvers, MA, USA), mouse monoclonal antibody against E-cadherin (#610181; BD Transduction Laboratories, San Jose, CA, USA), rabbit polyclonal antibody against human chorionic gonadotropin (β-hCG) (#SAB4500168; Sigma-Aldrich, Taufkirchen, Germany), mouse monoclonal antibody against cytokeratin 7 (#M7018; DAKO, Hamburg, Germany), rabbit monoclonal antibody against cytokeratin 18 (#ab32118; Abcam, Cambridge, UK), mouse monoclonal antibody against human leukocyte antigen G (HLA-G) (#11-291-C100; Exbio Praha, a.s., Vestec, Czech), and mouse monoclonal antibody against vimentin (#M7020; DAKO, Hamburg, Germany). FITC- and Cy3-conjugated secondary antibodies were obtained from Jackson Immunoresearch (Cambridgeshire, UK). DAPI (Roche, Mannheim, Germany) was used to stain the DNA content. Slides were examined with an AxioObserver.Z1 microscope (Zeiss, Göttingen, Germany) equipped with an AxioCam MRm camera (Zeiss, Göttingen, Germany).

2.12. Luciferase Assay

BeWo or JEG-3 cells were treated with scrambled siRNA (sicon) or siRNA against the 3′-untranslated region (UTR) of p21 (sip21) or mixed siRNAs against the coding region of p21 (sip21 #2). After 24 h, the syncytin-2 (2 μg) promoter plasmid [43] was transfected with

FuGENE® HD transfection reagent (Promega GmbH, Walldorf, Germany) for 48 h. Cells were harvested with cell culture lysis reagent from Promega GmbH (Walldorf, Germany; #E1531) and the assays were performed with the luciferase assay system from Promega GmbH (Walldorf, Germany; #E1501).

2.13. Microarray Analysis

BeWo cells were treated with control siRNA or siRNA targeting the UTR of p21 (sip21) for 48 h. Cells from three independent experiments were harvested and the total RNA was isolated using RNeasy Kits (Qiagen, Hilden, Germany). The expression was assessed using Human HT-12 v4 Beadchip (Illumina, San Diego, CA, USA), a direct hybridization whole-gene expression array. The expression profiling service from the German Cancer Research Center (DKFZ Microarray Core Facility, Heidelberg, Germany) was used. The most significant genes with a *p*-value (Student's *t*-test) less than 0.05 were selected. A full gene list is available as Supplementary Table S1, which shows data from the whole gene-expression array with a *p*-value smaller than 0.05.

2.14. Statistical Analysis

Outliers were detected with Grubbs' test (GraphPath QuickCalcs, San Diego, CA, USA). Data distribution normality was analyzed with the Shapiro–Wilk test and statistical significance was analyzed with the Student's *t* test, or, if not Gaussian distributed, with the non-parametric Wilcoxon-test (paired) or the Mann–Whitney U test (unpaired samples). Difference was defined as statistically significant when $p < 0.05$.

3. Results

3.1. Cell Cycle Regulators Are Affected by the Delivery Mode, and Specifically Expressed in Trophoblast Organoids and Placental Tissues

The rapid expansion of the human placenta is attributed to spatiotemporally regulated cell proliferation, for which the cell cycle regulators are indispensable. A previous study showed that the mode of delivery affected the expression of certain genes including *CDKN1C* (p57), which was elevated over 2-fold in placental tissues from labor deliveries compared to elective caesarean sections [44]. We started to verify and extend this very important issue in terms of the impact of the delivery mode on gene expression of *CDKN1A* (p21), *CDKN1B* (p27), and *CDKN1C* (p57). Placental tissues were collected from healthy women with different delivery modes without significant differences in gestational age, maternal age, and maternal body mass index (BMI) (Table 1). Total RNAs were extracted for gene analysis. We compared the gene expression levels of placental tissues from elective caesarean section (CS), emergency caesarean section after the onset of labor (eCS), vaginal delivery (VD), and operative vaginal delivery (opVD, mainly forceps delivery). Indeed, as shown in Figure 1A, the delivery mode strongly affected the gene expression of cell cycle regulators in the human placenta, especially for *CDKN1A* and *CDKN1B*, where a significant increase of about 2-fold was obtained after VD and opVD delivery. For *CDKN1C*, a significant 2-fold elevation was observed by opVD. Based on these data, we decided to collect placental tissues from caesarean sections without mechanical compression caused by uterine contractions or additional stress factors.

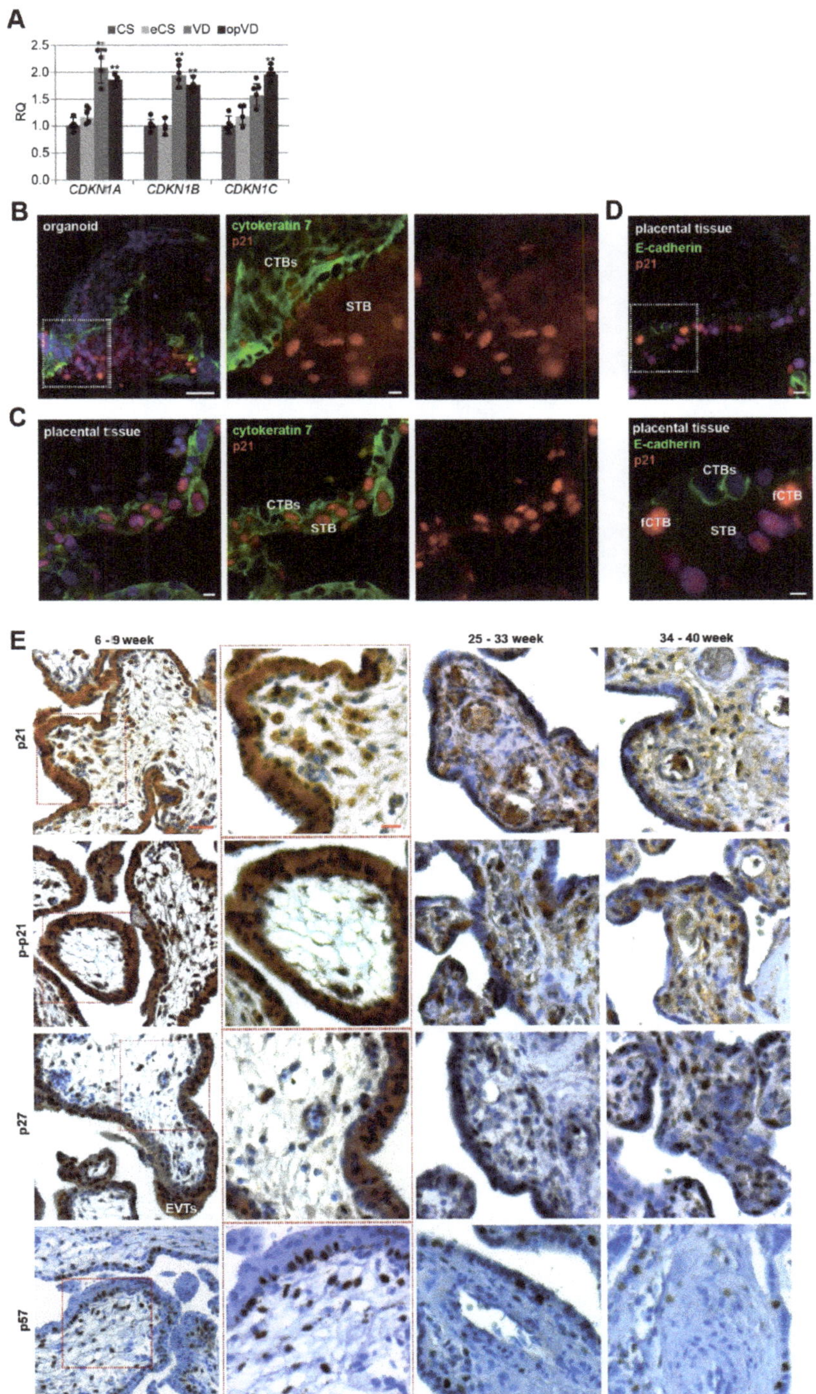

Figure 1. Cell cycle regulators are affected by the delivery mode, and expressed in trophoblast organoids and placental tissues. (**A**) The relative amount of the gene levels of *CDKN1A* (p21), *CDKN1B*

(p27), and *CDKN1C* (p57) was analyzed in placental tissues from different delivery modes ($n = 5$). The results are presented as relative quantification (RQ) with minimum and maximum range. The mean value of the expression levels of *SDHA* (succinate dehydrogenase complex, subunit A), *TBP* (TATA box–binding protein), and *YWHAZ* (tyrosine 3–monooxygenase/tryptophan 5–monooxygenase activation protein, zeta polypeptide) served as the endogenous control. CS, caesarean section; eCS, emergency caesarean section after the onset of labor; VD, vaginal delivery; opVD, operative vaginal delivery. Unpaired Student's t-test was used for statistical analysis, ** $p < 0.01$. (**B**) Representative immunohistochemistry-immunofluorescence (IHC–IF) image of a trophoblast organoid (8 weeks of gestation) stained for cytokeratin 7 (green), p21 (red), and nuclei (DAPI, blue) is shown. Scale: 50 µm. Inset scale: 10 µm. Villous cytotrophoblasts (CTBs) and the syncytiotrophoblast (STB) are indicated. (**C**) Representative IHC-IF image of placental tissue (8 weeks of gestation) stained for cytokeratin 7 (green), p21 (red), and nuclei (DAPI, blue) is presented. Scale: 10 µm. (**D**) Representative IHC–IF image of placental tissue (7 weeks of gestation) stained for E-cadherin (green), p21 (red), and nuclei (DAPI, blue) is shown. Scale: 10 µm. Inset scale: 5 µm. CTBs ongoing to fuse to the STB (fCTBs), CTBs and the STB are indicated. (**E**) Formalin-fixed and paraffin-embedded (FFPE) tissue sections were immunohistochemically stained with p21 (first row), p–p21 (second row), p27 (third row) or p57 (fourth row) antibody (brown), respectively, and counterstained with hematoxylin (blue). Scale: 50 µm. Inset scale: 20 µm. EVTs (extravillous cytotrophoblasts) are indicated.

To gain insight into the possible roles of cell cycle regulators in placental development, we generated human trophoblast 3D organoids (ORGs) according to well-established protocols [26,28,29]. Villous cytotrophoblasts (CTBs) were purified from first trimester placental tissues and embedded in Matrigel or plated on ultra-low attachment plates (Figure S1A). The trophoblast origin was corroborated by immunohistochemistry-immunofluorescence (IHC-IF) staining of the presence of the trophoblast marker cytokeratin 7 and epidermal growth factor receptor (EGFR), the proliferation marker Ki67, and the absence of human leukocyte antigen G (HLA-G) (Figure S1B), and by analyzing the gene expression of *GATA3* (GATA binding protein 3), *TFAP2A* (transcription factor AP-2 alpha), *TFAP2C* (transcription factor AP-2 gamma), and *ELF5* (E74 like ETS transcription factor 5) (Figure S1C), previously described trophoblast identity criteria [28]. The data revealed similar gene expression profiles between first trimester organoids (ORGs) and isolated primary cytotrophoblasts (pCTBs); both displayed higher expression levels of *GATA3, TFAP2A, TFAP2C,* and *ELF5* compared to placental mesenchymal stem/stromal cells (pMSCs) isolated from first trimester placental tissues. Additionally, relative to first trimester pCTBs, trophoblast organoids expressed very high levels of *CGbeta5* (β-hCG) (Figure S1D). IHC-IF staining for cytokeratin 7 revealed that organoids had the previously described inside-out structure [28,29], where the CTBs formed the outer layer and fused toward the center to generate the STB (Figure 1B and Figure S1B). In comparison, the IHC-IF of first trimester placenta showed the familiar structure (Figure 1C). Interestingly, in both trophoblast organoids and placental tissue sections, p21 was expressed in CTBs and the STB (Figure 1B,C), in the nucleus and the cytoplasm.

For further analyses, we used sections of placental tissues with gestational ages between six and nine weeks, between 25 and 33 weeks as well as between 34 and 40 weeks from normal donors. Using IHC, placental sections were stained for p21, p–p21 (Thr-145), p27 or p57, and counterstained with hematoxylin. Whereas p21's role in cell cycle arrest is attributed to its nuclear localization, its phosphorylation at Thr-145 by distinct kinases is described as a marker for cytoplasmic translocation or enhanced protein stability contributing to cell cycle progression [23]. Placental tissue was highly positive for p21, p–p21, and p27 (Figure 1E). The positive staining was found in the cytoplasm as well as the nucleus of trophoblastic cells of the placenta, especially in the proliferative villous CTBs, in particular, CTBs ongoing to fuse to the STB, so called fCTBs marked by partial loss of E-cadherin staining, suggesting the breakdown of apical and lateral plasma membranes (fCTBs, Figure 1D), the terminally differentiated, non-proliferative, and multinucleated

STB, the migrating EVTs in proliferative cell columns, and villous stromal cells throughout gestation. The positive staining of p57 was predominantly present in the nucleus of fCTBs, stromal cells, and EVTs (Figure 1E, last panel).

3.2. Cell Cycle Regulators Are Highly Expressed during the First Trimester of Gestation

We analyzed the positively stained CTBs, fCTBs, and the STB area for p21, p–p21, p27, and p57 in tissue sections from healthy donors with gestational ages between six and nine weeks (n = 6), between 25 and 33 weeks (n = 20; Table 2, also served as early-onset control group), and between 34 and 40 weeks (n = 10; Table 3, also served as late-onset control group) (representatives are shown in Figure 1E). In the first trimester sections, there were not enough EVTs for a reliable quantification. For all staining, first trimester sections showed the highest percentage of positive CTBs and positive STB area with a significant decline in the early- and late-onset control group (Figure 2A,B; named 25–33 or 34–40 week). There was a significant difference in p–p21 positive staining of CTBs and the STB area between early- and late-onset control samples (Figure 2A,B, middle panel). Interestingly, fCTBs displayed high percentages of positive staining of p21, p–p21, p27, and p57 throughout gestation (Figure 2C), suggesting that this cell population is highly active in proliferation and differentiation. Next, the gene expression of *CDKN1A* (p21), *CDKN1B* (p27), and *CDKN1C* (p57) was evaluated in placental tissues from the first trimester samples (named 6–9 week) compared to early- and late-onset controls (Figure 2D). The relative amounts of *CDKN1A*, *CDKN1B*, and *CDKN1C* in early-onset controls were significantly reduced by 92%, 79%, and 90%, respectively, compared to the first trimester group. Particularly, *CDKN1A* was significantly reduced in early- as well as late-onset controls compared to the first trimester samples (Figure 2D, left). In addition, a moderate increase was observed in the gene expression of all cell cycle regulators in late-onset controls compared to early-onset controls (Figure 2D).

3.3. p21 Expression Is Reduced in fCTBs of Early-Onset PE Placental Samples

To address if cell cycle regulators were altered in preeclamptic placentas, the expression levels of p21, p–p21, p27, or p57 were compared between twenty early-onset PE placental tissues and twenty samples from well-matched control donors (Table 2) by the semi-quantitative H-score method, which combines the percentage of stained cells/area and their staining intensity. There was no apparent difference in cell cycle regulators in the H-score of CTBs or the STB, and the percentage of positive CTBs or in the positive stained area per visual field in the STB in early-onset PE samples compared to their respective control counterparts (Figure 3A,B). Interestingly, a significant reduction of p21 in the H-score of fCTBs as well as in the percentage of positive fCTBs were observed in early-onset preeclamptic placental samples, in comparison to the matched control tissues (Figure 3C, left graph). While p–p21 and p27 were almost comparable (Figure 3C, 2nd and 3rd graph), the percentage of positive p57 fCTBs declined in early-onset PE samples compared to the control tissues (Figure 3C, right graph, bottom). Moreover, the percentage of p57 positive stained EVTs was significantly reduced, whereas p21, p–p21, and p27 were hardly altered (Figure 3D). To underscore these results, we next evaluated the expression with western blot analyses from whole tissue samples. While the expression of p21, p27, and p57 was decreased, only p57 showed a significant reduction (Figure 3E). A limitation of our study is that the p–p21 antibody did not work for tissue western blot analysis. Further gene analysis showed reduced levels of *CDKN1A* (p21) and *CDKN1B* (p27) (Figure 3F).

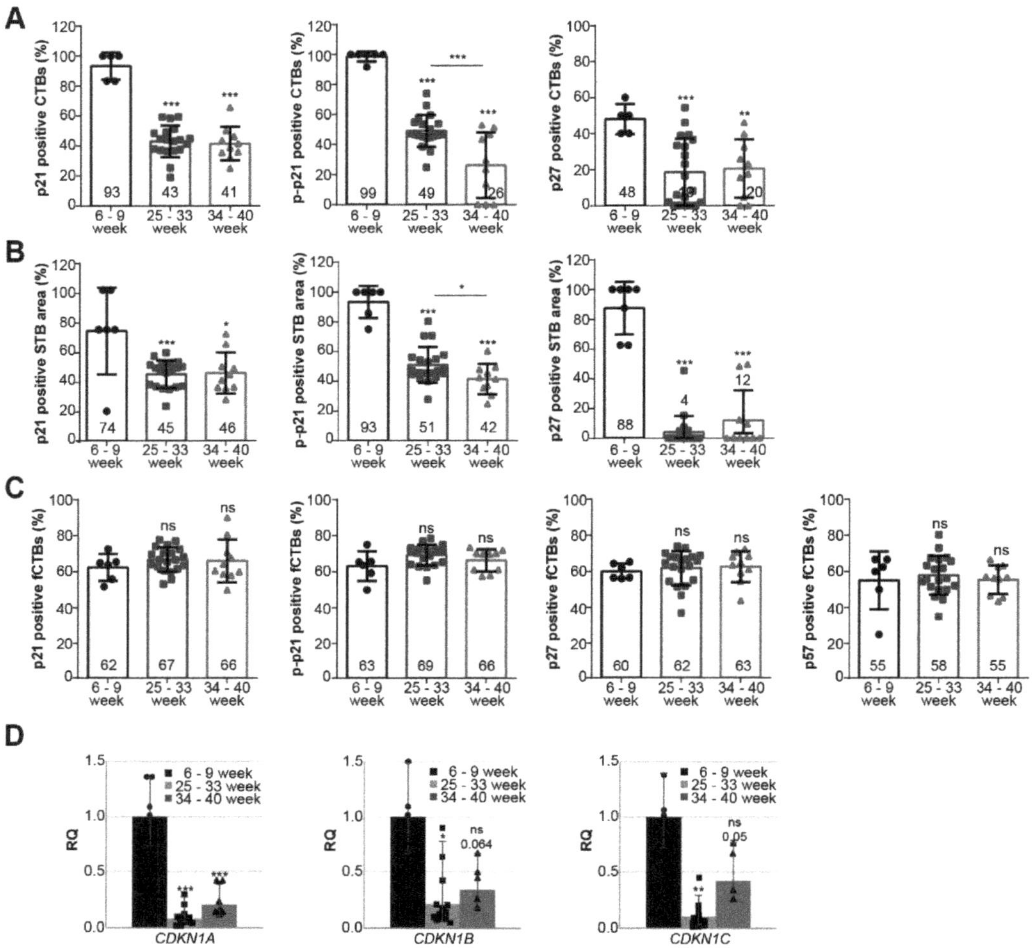

Figure 2. Cell cycle regulators are highly expressed during the first trimester of gestation. (**A–C**) Evaluation of positive cells in first trimester placental sections ($n = 6$; named 6–9 week), early-onset control ($n = 20$; named 25–33 week), and late-onset control samples ($n = 10$; named 34–40 week). The results are presented as bar and scatter plots showing the mean value ± SD. (**A**) Quantification of p21 (left panel), p–p21 (middle panel), and p27 (right panel) positive CTBs in %. (**B**) Quantification of p21 (left panel), p-p21 (middle panel), and p27 (right panel) positive stained STB area in %. (**C**) Quantification of p21 (left panel), p–p21 (second panel), p27 (third panel), and p57 (right panel) positive fCTBs in %. (**D**) The relative amount of the gene levels was analyzed from placental tissues: left panel *CDKN1A* (p21), middle panel *CDKN1B* (p27), and right panel *CDKN1C* (p57). The results are presented as relative quantification (RQ) with minimum and maximum range. *TBP* was used as the endogenous control. Unpaired Student's *t*-test or Mann–Whitney U test referring to first trimester samples was used for statistical analysis, * $p < 0.05$, ** $p < 0.01$, *** $p < 0.001$. CTBs, cytotrophoblasts; fCTBs, cytotrophoblasts ongoing to fuse; STB, syncytiotrophoblast.

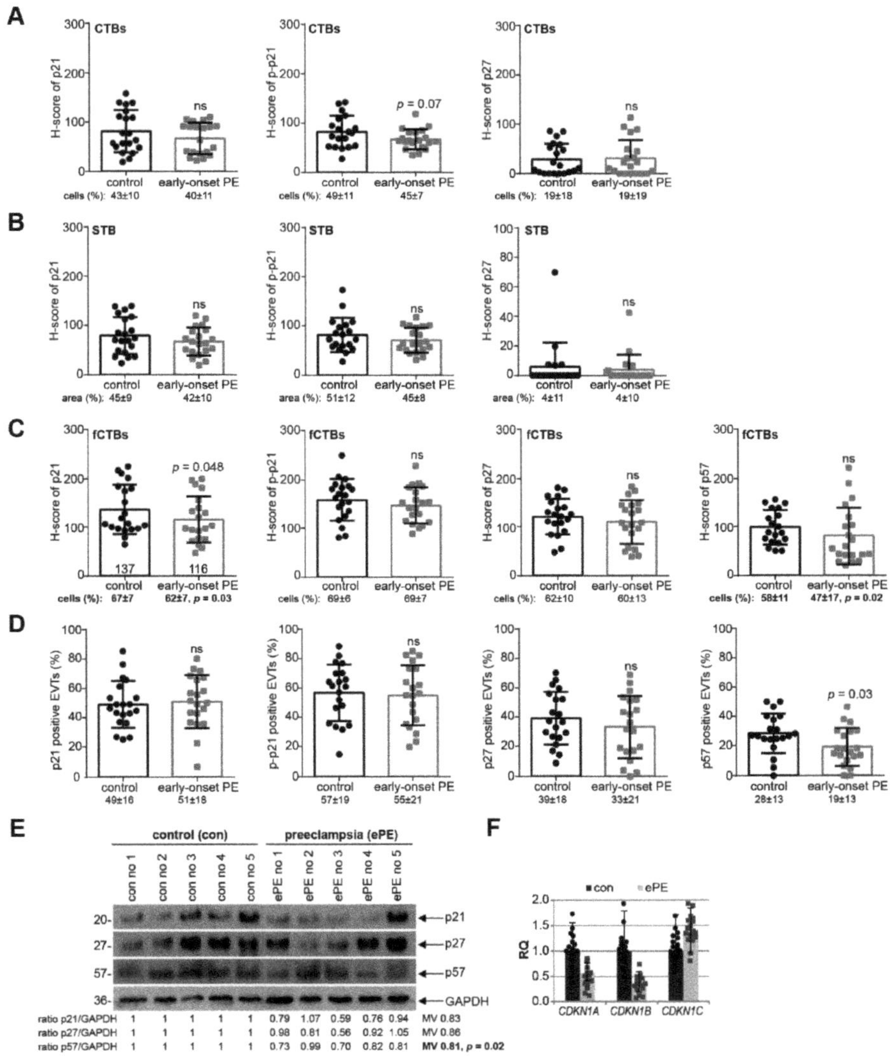

Figure 3. p21 expression is reduced in fCTBs of early-onset PE samples. (**A–C**) Quantification of cell cycle regulators in placental sections of control donors (control, n = 20) and patients with early-onset PE (ePE, n = 20) using the H-score method. The results are presented as bar and scatter plots showing the mean value with SD. The percentage of positive stained cells/area is shown under each graph. (**A**) H-score of p21 (left panel), p–p21 (middle panel), and p27 (right panel) for CTBs. (**B**) H-score of p21 (left panel), p–p21 (middle panel), and p27 (right panel) for the STB area. (**C**) H-score of p21 (left panel), p–p21 (second panel), p27 (third panel), and p57 (right panel) for fCTBs. (**D**) Quantification of p21 positive (left panel), p–p21 positive (second panel), p27 positive (third panel), and p57 positive EVTs (right panel) in %. (**E**) Western blot analysis with extracts from placental tissues is shown. Glyceraldehyde-3-phosphate dehydrogenase (GAPDH) served as the loading control. (**F**) The relative amount of the gene levels of *CDKN1A* (p21), *CDKN1B* (p27), and *CDKN1C* (p57) was analyzed from placental tissues. The results are presented as relative quantification (RQ) with minimum and maximum range. *TBP* was used as the endogenous control. Paired Student's t-test or Wilcoxon-test was used for statistical analysis. CTBs, cytotrophoblasts; fCTBs, cytotrophoblasts ongoing to fuse; STB, syncytiotrophoblast; EVT, extravillous cytotrophoblasts; no, number; MV, mean value.

Ten late-onset PE samples and ten control tissues from well-matched donors (Table 3) were also systematically analyzed. Obvious change in p21, p–p21, p27, and p57 expression was not observed at protein as well as at gene level (Figure S2). Interestingly, the H-score of p–p21 was significantly elevated in the STB (Figure S2B, middle graph). In sum, these data support the notion that early- and late-onset PE derive from distinct pathogenesis and should be considered separately from each other.

3.4. Reduced p21 Protein Expression in Early-Onset PE with HELLP Syndrome

Since early-onset PE is often complicated by the HELLP syndrome (ePEH) [1], placental samples were also collected from patients with ePEH and their well-matched controls (Table 4). The percentage of positive cells/area and the staining intensity of CTBs, the STB, and fCTBs were evaluated for p21, p–p21, p27, and p57 (Figure 4A–C). There was no obvious difference of cell cycle regulators in the H-score of CTBs or the STB, and the percentage of positive CTBs or in the positive stained area of the STB in ePEH compared to the well-matched controls (Figure 4A,B). Interestingly, there was a significant increase in the H-score of p–p21 in fCTBs (Figure 4C). The percentage of positive stained EVTs was comparable (Figure 4D). Western blot analyses with cellular lysates from whole tissue samples showed that the expression of p21, p27, and p57 was decreased, and only p21 showed a significant reduction (Figure 4E). Further gene analysis showed reduced levels of *CDKN1A* (p21) and *CDKN1B* (p27) (Figure 4F), as observed in early-onset PE (Figure 3F).

3.5. p21 Expression Is Decreased in Trophoblastic Cell Lines and in Isolated Primary Cytotrophoblasts under Hypoxic Conditions

PE is associated with chronic hypoxia of the placenta through defective trophoblast invasion and inadequate remodeling of the maternal spiral arteries [5,17,45,46]. To mimic the situation in PE, immortalized first trimester trophoblast cell lines SGHPL-4 [40] and HTR [39] were grown under normal (21.4% O_2) or hypoxic conditions (1% O_2) for 48 h. The gene levels of *CDKN1A* (p21), *CDKN1B* (p27), *CDKN1C* (p57), and *TP53* (p53), the master regulator of p21, were measured. *CDKN1A* was significantly reduced under hypoxic conditions in SGHPL-4 cells, whereas *CDKN1B*, *CDKN1C*, and *TP53* were decreased but not significantly (Figure 5A). Moreover, the p21 protein was also reduced, while the p53 protein expression was not affected by hypoxia (Figure 5B). Comparable results were also observed with HTR cells, which showed a significantly lowered gene expression of *CDKN1A* and *CDKN1B* (Figure 5C) as well as decreased p21 protein expression under hypoxic conditions (Figure 5D). To investigate cell cycle regulators in a more physiological setting, trophoblast organoids (ORGs) derived from first trimester placenta were generated and cultured under normal (21.4% O_2) or under hypoxic conditions (1% O_2) for 48 h. The expression of cell cycle regulator genes was also affected, showing a significant decline in *CDKN1A*, *CDKN1B*, *CDKN1C*, and *TP53* under hypoxia (Figure 5E). The choriocarcinoma cell line BeWo, a widely used cell culture model mimicking CTB fusion and differentiation [47], was also grown under normoxia and hypoxia for gene analysis. *CDKN1A*, *CDKN1B*, and *TP53* were significantly reduced under low oxygen supply (Figure 5F).

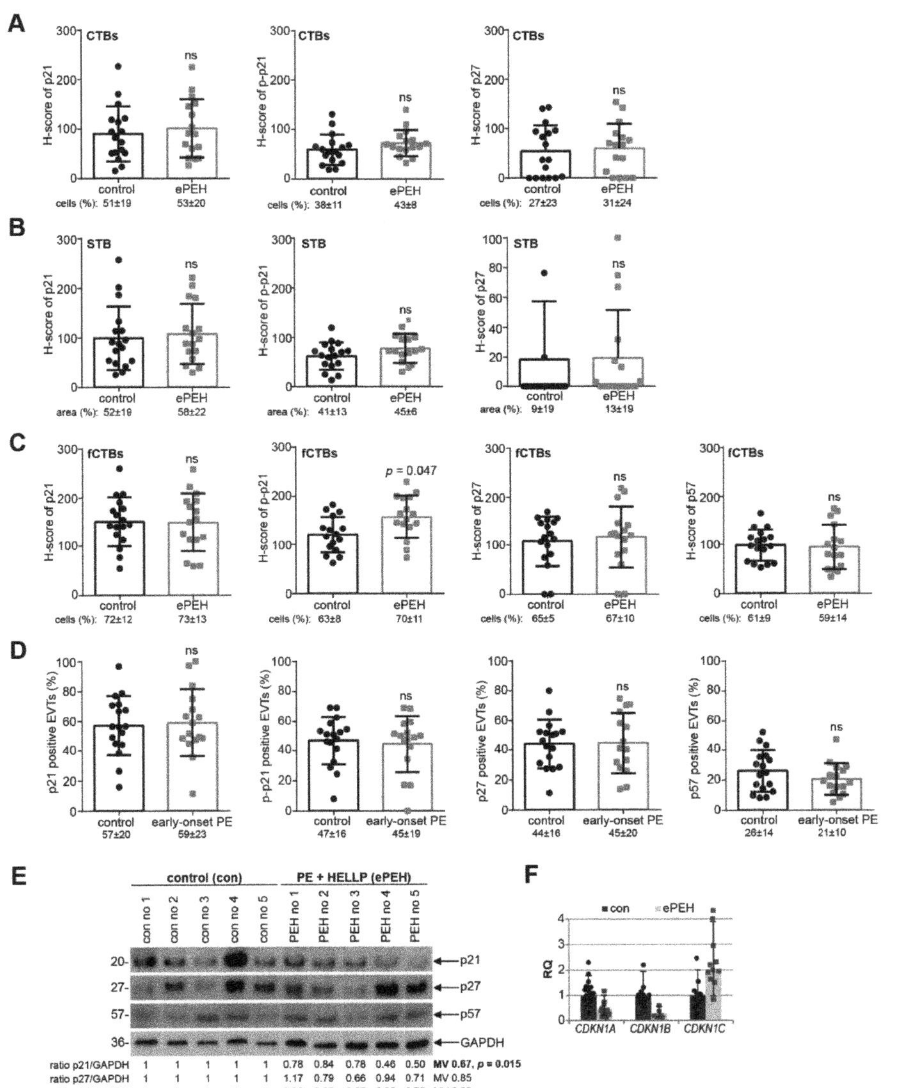

Figure 4. p21 expression is reduced in early-onset PE complicated by the HELLP syndrome. (**A–C**) Quantification of cell cycle regulators in placental sections of control donors (control, n = 17) and placental tissues from early-onset PE complicated by the HELLP syndrome (ePEH, n = 16) using the H-score method. The results are presented as bar and scatter plots showing the mean value with SD. The percentage of positive stained cells/area is shown under each graph. (**A**) H-score of p21 (left panel), p–p21 (middle panel), and p27 (right panel) for CTBs. (**B**) H-score of p21 (left panel), p–p21 (middle panel), and p27 (right panel) for the STB area. (**C**) H-score of p21 (left panel), p–p21 (second panel), p27 (third panel), and p57 (right panel) for fCTBs. (**D**) Quantification of p21 positive (left panel), p–p21 positive (second panel), p27 positive (third panel), and p57 positive EVTs (right panel) in %. (**E**) Western blot analysis with extracts from placental tissues is shown. GAPDH served as the loading control. (**F**) The relative amount of the gene levels of *CDKN1A* (p21), *CDKN1B* (p27) and *CDKN1C* (p57) was analyzed with placental tissues. The results are presented as relative quantification (RQ) with minimum and maximum range. *TBP* was used as the endogenous control. Paired Student's t-test or Wilcoxon-test was used for statistical analysis. CTBs, cytotrophoblasts; fCTBs, cytotrophoblasts ongoing to fuse; STB, syncytiotrophoblast; EVT, extravillous cytotrophoblasts; no, number; MV, mean value.

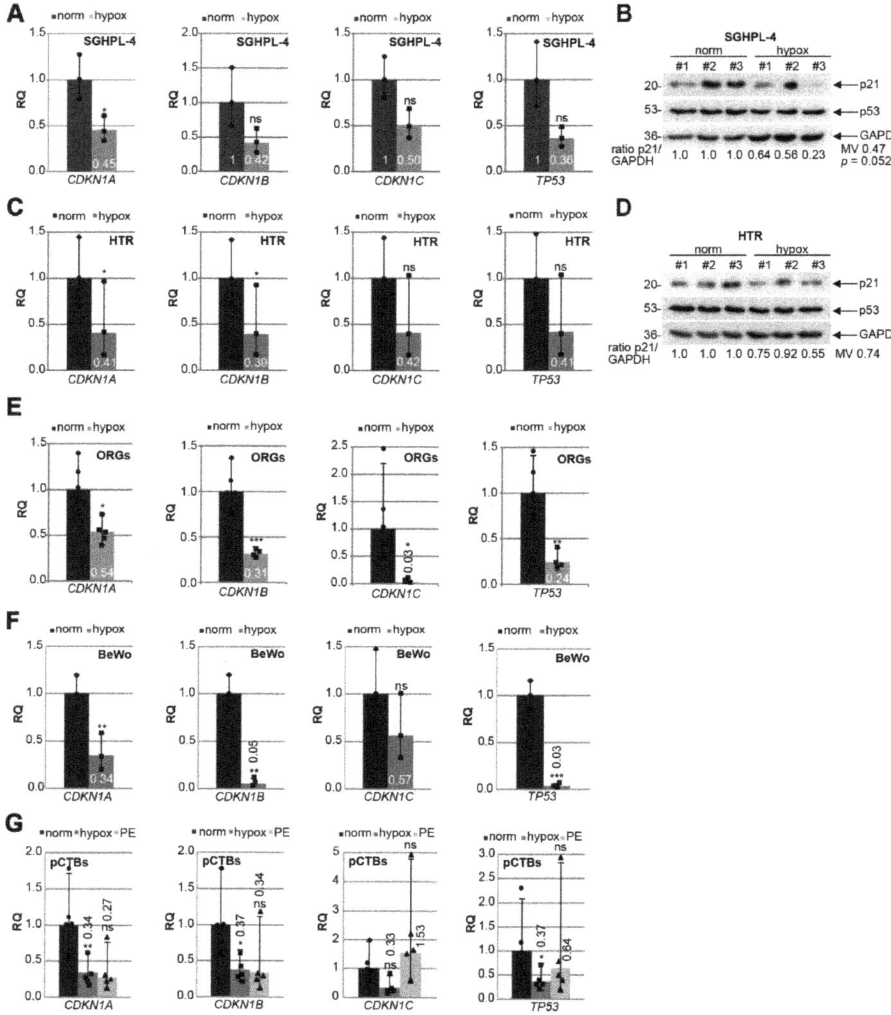

Figure 5. p21 expression is decreased in trophoblastic cell lines and in isolated primary trophoblasts under hypoxic conditions. Cells were grown under normoxia (norm, 21.4% O_2) or hypoxia (hypox, 1% O_2) for 48 h prior to RNA extraction or western blot analysis. (**A**) SGHPL-4 cells. Gene analysis of *CDKN1A* (p21), *CDKN1B* (p27), *CDKN1C* (p57) and *TP53* (p53) is shown. The results ($n = 3$) are presented as relative quantification (RQ) with minimum and maximum range. *GAPDH* was used as the endogenous control. (**B**) Western blot analysis with cellular lysates from SGHPL-4 cells ($n = 3$). GAPDH served as the loading control. (**C**) HTR cells. Gene analysis of *CDKN1A* (p21), *CDKN1B* (p27), *CDKN1C* (p57), and *TP53* (p53) is shown. The results ($n = 3$) are presented as relative quantification (RQ) with minimum and maximum range. *GAPDH* was used as the endogenous control. (**D**) Western blot analysis with cellular lysates from HTR cells ($n = 3$). GAPDH served as the loading control. (**E**) Organoids (ORGs) cultured under normoxia (norm; $n = 5$) or hypoxia (hypox; $n = 4$). Relative gene levels of *CDKN1A* (p21), *CDKN1B* (p27), *CDKN1C* (p57), and *TP53* (p53) are shown. The results are presented as relative quantification (RQ) with minimum and maximum range. *GAPDH* was used as endogenous control. (**F**) BeWo cells. Relative gene levels of *CDKN1A* (p21), *CDKN1B* (p27), *CDKN1C* (p57), and *TP53* (p53) are shown. The results ($n = 3$) are presented as the relative quantification (RQ) with minimum and maximum range. *GAPDH* was used as the endogenous control. (**G**) Primary trophoblasts (pCTB) were isolated from healthy and preeclamptic (PE) donors. pCTBs from healthy donors were cultured under normoxia (norm; dot) or hypoxia (hypox, square), PE (triangle) pCTBs were grown under normoxia. The gene level of *CDKN1A* (p21), *CDKN1B* (p27), *CDKN1C* (p57), and *TP53* (p53) was evaluated ($n = 5$). *GAPDH* was used as the housekeeping gene control. Student's *t*-test, * $p < 0.05$, ** $p < 0.01$, *** $p < 0.001$. ns, not significant; MV, mean value.

To further underline the observations, human primary cytotrophoblasts (pCTBs) were isolated from five term placentas of healthy donors, which were grown under normoxia or hypoxia and compared to pCTBs isolated from late-onset PE patients grown under normoxic conditions (Table 5). The characterization of isolated trophoblastic cells was performed with positive staining of EGFR, E-cadherin, β-hCG, cytokeratin 7, cytokeratin 18, and negative markers HLA-G and vimentin (Figure S3). Compared to normoxia, the gene levels of *CDKN1A* (p21), *CDKN1B* (p27), and *TP53* (p53) were significantly reduced in primary cytotrophoblasts under hypoxic conditions (Figure 5G). Interestingly, a decrease in *CDKN1A* and *CDKN1B* expression was also observed in the pCTBs from late-onset PE placentas, albeit not significant due to the small sample size (Figure 5G).

Table 5. Clinical information of preeclamptic patients and matched donors, whose placentas were used for primary trophoblast isolation (term). Mean value ± standard deviation is shown. BP, blood pressure; n.d., not determined.

	n	Age	Gestational Age (Weeks)	Body Mass Index (BMI)	Birth Weight (g)	Systolic BP (mmHg)	Diastolic BP (mmHg)	Proteinuria (mg/24 h)
control	5	32.6 ± 5.3	40 ± 1.2	22.6 ± 3.7	3407 ± 448	114 ± 9	74 ± 11	n.d.
PE	5	34.1 ± 5	36.8 ± 4.0	23.7 ± 2.7	2462 ± 974	150 ± 11	97 ± 5	2763 ± 3459
p-value		0.666	0.123	0.604	0.084	0.00042	0.0031	

3.6. Knockdown of p21 Impairs the Fusion Ability of Trophoblastic BeWo Cells

Since the expression of p21 was decreased in fCTBs, a cell population ready to fuse into the STB of placental tissues from early-onset PE patients (Figure 3C, left graph), we focused on the role of p21 in cell differentiation and fusion. To look at possible mechanisms by which reduced p21 could cause defects in cell differentiation, total RNAs were extracted from BeWo cells depleted of p21 for RNA microarray analysis (Figure 6A, Table S1). Differently expressed genes were compared between BeWo cells treated with the control siRNA (sicon) and siRNA specifically targeting p21. The heatmap depicts genes with a *p*-value smaller than 0.05, and a fold change greater than 1 (red color code) and below 1 (blue color code), respectively (Figure 6A, left panel).

Interestingly, as reported for HTR cells depleted of p21 [25], the extracellular signal-regulated kinase 3 (ERK3), encoded by the gene *MAPK6*, is among the top three of altered genes and strongly reduced upon p21-depletion (hit number 1, *CDKN1A*). ERK3 is a distantly related member of the mitogen-activated protein kinase (MAPK) superfamily [48], which is known to be involved in cell differentiation [49]. Further analysis revealed that the fusion-related genes *HERV-FRD* (syncytin-2), *ERVWE1* (syncytin-1), *CGbeta5* (β-hCG), *KLF6* (Krüppel-like factor 6), and *GCM1* (glial cells missing transcription factor 1) were reduced (Figure 6A, right upper panel) upon p21 depletion (Figure 6A, right lower panel). This was corroborated for *HERV-FRD*, *CGbeta5*, and *GCM1* by quantitative PCR analysis with BeWo (Figure 6B) and JEG-3 cells (Figure 6C). To further address this issue, BeWo or JEG-3 cells depleted of p21 with two different siRNAs (sip21 and sip21 #2; Figures 6F and S3G) were stimulated to fuse with forskolin up to 48 h. The amount of β-hCG, which is upregulated upon cell fusion and induced by forskolin [47], was reduced upon p21 depletion, visualized by immunofluorescence staining compared to the control cells (Figures 6D and S3E). Moreover, the luciferase assay showed that reduced p21 significantly decreased the expression of syncytin-2 in a promoter-dependent manner in BeWo and JEG-3 cells, respectively (Figures 6E and S3F).

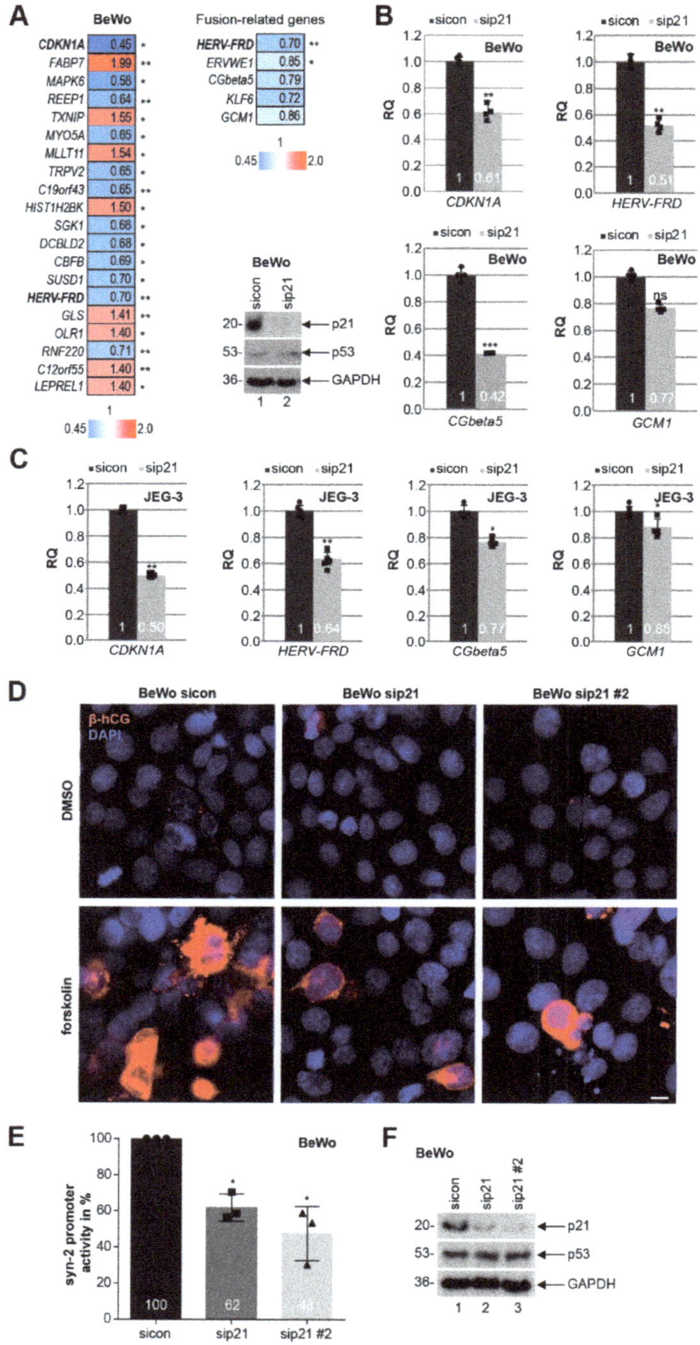

Figure 6. Knockdown of p21 impairs the fusion ability of BeWo and JEG-3 cells. (**A**) Left panel: genome-wide profiling of BeWo cells treated with sicon or sip21 for 48 h. Total RNAs were extracted from three independent experiments. Gene expression was analyzed using HumanHT-12 v4 beadchip

array. Genes with a p-value < 0.05 are included. Heatmap of the most differently expressed genes are shown (greater than 1, red color code; below 1, blue color code). Right upper panel: heatmap of fusion-related genes is depicted. Right lower panel: western blot analysis was performed with cellular lysates from BeWo cells treated with scrambled siRNA (sicon) or siRNA targeting the untranslated region (UTR) of p21 (sip21). GAPDH was taken as the loading control. (**B**) Gene analysis of BeWo cells depleted of p21. The mRNA levels of p21 (*CDKN1A*), syncytin-2 (*HERV-FRD*), β-hCG (*CGbeta5*), and *GCM1* are shown. *GAPDH* was used as the housekeeping gene control. The results are presented as RQ with minimum and maximum range (n = 4). (**C**) Gene analysis of JEG-3 cells depleted of p21. The mRNA levels of p21 (*CDKN1A*), syncytin-2 (*HERV-FRD*), β-hCG (*CGbeta5*), and *GCM1* are shown. *GAPDH* was used as the housekeeping gene control. The results are presented as RQ with the minimum and maximum range (n = 5). (**D**) BeWo cells, treated with sicon, sip21, or mixed siRNAs against the coding region of p21 (sip21 #2) for 24 h, were incubated with forskolin or DMSO for another 48 h. Treated BeWo cells were stained for the fusion marker β-hCG (red) and DNA (DAPI, blue). Examples are shown. Scale: 10 μm. (**E**) BeWo cells were treated with sicon, sip21, or sip21 #2. After 24 h, the syncytin-2 promoter plasmid was transfected for 48 h. Luciferase assays of BeWo cells for syncytin-2 promoter activity is shown as mean value with SD (n = 3). Dot, square, and triangle show the individual data points of sicon, sip21 and sip21 #2, respectively. (**F**) Western blot analysis as transfection control. GAPDH was used as the loading control. Student's t-test, * $p < 0.05$, ** $p < 0.01$, *** $p < 0.001$. ns, not significant.

4. Discussion

The cell cycle regulator p21 is a key player in various cellular events including cell differentiation, migration, stem cell maintenance, and gene transcription [24], processes important for placental development and altered in its pathogenesis. However, p21's expression and roles in the placenta and its diseases are still contradictory, despite numerous studies and intensive work. To clarify these issues, in the present work, we collected early-onset PE, early-onset PE complicated by the HELLP syndrome, and late-onset PE placental samples, and their well-matched control tissues, and systematically examined the expression of p21 (*CDKN1A*) and its family members p27 *(CDKN1B)* and p57 *(CDKN1C)* as well as their roles in normal pregnancy and hypertensive disorders. We show here that these cell cycle regulators are highly expressed in first trimester placentas, while their expression generally decreases in the second and third trimester placenta at the protein as well as gene level. Notably, fCTBs, a special portion of CTBs ongoing to fuse into the STB, express high levels of p21 and its family members, which remain high throughout gestation. Importantly, the level of p21 is only reduced in fCTBs of early-onset PE placental tissues and its overall protein expression is decreased in early-onset PE complicated by the HELLP syndrome. These data strongly support the notion that early- and late-onset PE derive from distinct pathogenesis. Moreover, cell cycle regulators are decreased upon hypoxic conditions and depletion of p21 leads to reduced expression of fusion-related genes and an impaired fusion capacity of trophoblastic cells. In addition, the expression of these cell cycle regulators is dependent on the delivery mode, underscoring the importance of optimization and standardization of patient selection and placental collection for proper interpretation of existing data on cell cycle regulators.

We show that overall p21 was reduced in early-onset PE and early-onset PE complicated by the HELLP syndrome, whereas it was unaltered in late-onset PE, suggesting its potential involvement in the development of early-onset PE. Several studies provide insight into p21′s roles in PE and are, however, accompanied by inconsistency and divergence: one study reported that the gene and protein expression of p21 was increased in preeclamptic placentas [50], whereby another study showed that p21 expression was decreased via IHC staining [51]. In other investigations, its protein or gene expression was unchanged [52,53]. The contradictory expression of p21 in PE might be explained by the highly variable study design concerning gestational age without the differentiation in early- and late-onset PE or concomitant diseases like the HELLP syndrome, sample number, and mode of delivery. In further support of our results, a recent study showed a decrease in p21 in early-onset PE, whereas it was unchanged in late-onset PE at the gene and protein level [54]. The reduced p21 is proposed to lead to elevated Cdk2/cyclin E levels and reduced phospho-retinoblastoma protein (RB) critical for cell cycle progression, which hampered the differentiation and fusion process of CTBs to the STB [54].

PE, especially early-onset PE, is associated with constant hypoxia and oxidative stress of the placenta [17]. Interestingly, we show that hypoxia significantly decreased mRNA levels of *CDKN1A* (p21) and *CDKN1B* (p27) in diverse trophoblastic cell lines, organoids derived from first trimester placenta, and isolated primary cytotrophoblasts. Others reported that the p21 protein was downregulated in CTBs upon 2% oxygen supply for 72 h, which was 3.8-fold higher in normoxia cell extracts isolated from explants of anchoring villi of first trimester placenta [55]. Moreover, we observed that hypoxia also decreased the mRNA level of *TP53* (p53), the upstream regulator of p21. However, the p53 protein expression from whole cellular lysates was not affected. p53 downregulation was reported in the STB of primary isolated term trophoblasts upon 1% hypoxia for 24 h, associated with the induction of apoptosis [56]. Evaluation of nuclear p53 and its post-modifications is required to clarify whether p53 is responsible for reduced p21. Interestingly, transcriptional repressors of p21 like BCL6 (B cell lymphoma 6) [57] or TFAP2C [58] are known to be enhanced in PE [52,59]. In addition, the miRNA family miR-130 including miR-130a/b and miR-301a/b repressed *CDKN1A* in human pulmonary artery smooth muscle cells upon hypoxia [60]. Interestingly, miR-130a and miR-301a have been reported to be overexpressed in PE placenta [61]. These regulators may be responsible for the reduced expression of p21 under hypoxia, bridging its reduction to the pathogenesis of PE.

PE is associated with profound cellular dysfunctions including reduced differentiation and fusion ability [62], and p21 is of crucial importance for differentiation [24]. Indeed, p21 protein levels increased during spontaneous differentiation and fusion of term CTBs [63] as well as *CDKN1A* during differentiation of mouse trophoblast stem cells and fusion of BeWo cells upon forskolin treatment [64], whereas p53 levels were reduced during BeWo cell differentiation [65]. We report here that p21 levels were significantly decreased in fCTBs, a special portion of CTBs ongoing to fuse to the STB, in early-onset PE samples, suggestive of its involvement in CTB differentiation and fusion. Moreover, in early-onset PE complicated by the HELLP syndrome, the H-score of overall p21 was reduced, whereas phosphorylated-p21 was significantly elevated in fCTBs, possibly a compensatory mechanism in a crisis situation induced by the HELLP syndrome, by which the stress protein p21 is trying to be stabilized and possibly translocated to the cytoplasm, where it could exert its anti-apoptotic function but hardly affect the transcription of fusion-related genes. In fact, our data revealed that the fusion-related genes *HERV-FRD* (syncytin-2), *CGbeta5* (β-hCG), and *GCM1* were reduced upon p21 depletion. Importantly, decreased p21 significantly reduced the expression of syncytin-2 in a promoter-dependent manner in BeWo and JEG-3 cells. Indeed, p21 has been reported to interact as a co-activator with GCM1 to bind to the promoter region of syncytin-2, regulating its transcription [66]. Our findings are further supported by previous studies. Reduced p21 was reported to be involved in impaired fusion mediated by KLF6 silencing in BeWo cells [63]. Recently, it has been revealed that p21, but not its family member p27, coordinates trophoblast fusion contributing to G0 arrest and terminal differentiation [66]. Of significance, the reduced expression of the fusion-related genes/proteins syncytin-1 and syncytin-2 correlates with the severity of PE [67]. In sum, our data clearly suggest that reduced p21 in fCTBs compromises the expression of fusion-related genes, contributing to impaired differentiation and fusion of trophoblasts, an important hallmark of early-onset PE.

The studies concerning p27 and p57 in PE are rather limited: in one study, p27 and p57 were significantly increased in preeclamptic placentas [68]. However, in that study, the averaged gestational age of PE patients was lower than in the control group [68]. p57 has been reported to be important in trophoblast fusion [69,70], and in migration and invasion [71]. In early-onset PE samples, we observed a significant reduction in the percentage of p57 positive fCTBs, suggestive of its potential contribution to compromised

CTB differentiation and fusion, possibly in collaboration with reduced p21, in early-onset PE placentas. Additionally, we detected a reduced p57 level in EVTs, indicative of a possible involvement in impaired cell motility, a hallmark of PE. In fact, the amount of p57 was also decreased in western blot analyses of early-onset PE placental samples, whereas we did not observe significant differences in p27 expression. Interestingly, mutant mice, defective only for the maternal p57 allele, displayed clinical manifestations of PE including proteinuria and elevated blood pressure [72]. However, a further study failed to reproduce these PE manifestations using the same mouse model, although mice exhibited placental abnormalities [73]. p57-mutant murine placentas demonstrated significant alterations of transcripts coding for a variety of molecules involved in blood pressure regulation, inflammation, and apoptosis [74]. Moreover, loss of p57 seems to be associated with placentomegaly due to unrestricted endoreduplication [75]. The role of p57 in placental development needs further investigation.

It has been shown that the co-expression of p53 and p21 can lead to cell cycle arrest via suppression of G2 phase and mitotic (M) genes by the DREAM (*D*imerization partner, *R*B-like proteins, *E*2Fs *A*nd *M*ulti-vulval B) multiprotein complex [76,77]. In cancer cells, loss of p21 reduced the DREAM binding to the cell cycle-dependent element (CDE) and cell cycle genes homology region (CHR) of the promoters, resulting in the expression of G2/M genes [77]. However, microarray analysis using BeWo cells depleted of p21 did not show an increase in G2/M genes, which could be attributed to their long cell doubling time [78]. Remarkably, it revealed other interesting candidate genes such as the elevated gene *TXNIP*, which encodes the protein thioredoxin interacting protein, a major regulator of the cellular redox signaling, protecting cells from oxidative stress [79]. Interestingly, TXNIP was expressed in the STB, CTBs, and endothelial cells of the human placenta, where its expression was increased in the second and third trimester [80]. TXNIP loss significantly increased p21 protein expression levels in ARPE-19 cells, inhibiting cell growth [81], whereas TXNIP stabilized p27 protein indirectly in fibroblasts [82]. Further investigations are required to study the relationship of p21 with these candidate genes and their functions in PE.

5. Conclusions

Taken together, we show that p21 and its family members p27 and p57 are highly expressed in the first trimester of pregnancy, pointing to their importance in early placental development. The expression of p21 in CTBs and the STB is further highlighted with 3D trophoblast organoids, which have been reported to mimic the placental villi structurally, phenotypically, metabolically, and endocrinologically [28,29]. Moreover, we report that p21 is reduced in early-onset PE fCTBs ongoing to fuse to the STB. This finding is further underscored by our observation that hypoxia reduces p21 in trophoblastic cells and organoids. Depletion of p21 decreases the expression of fusion-related genes such as syncytin-2 and impairs the fusion capability of trophoblasts, characteristic of PE. Moreover, the percentages of positive p57 fCTBs and EVTs are significantly reduced in early-onset PE placentas, indicating its potential involvement in CTB fusion and EVT migration. These data strongly suggest that p21 deficiency, in collaboration with reduced p57, is likely to contribute to the pathogenesis of early-onset PE.

Supplementary Materials: The following are available online at https://www.mdpi.com/article/10.3390/cells10092214/s1, Figure S1: Characterization of long-term trophoblast organoid culture, Figure S2: Expression of cell cycle regulators in late-onset PE, Figure S3: Characterization of primary cytotrophoblasts. Figure S4: Raw data of all western blots. Table S1: Data from whole gene-expression array with a p-value smaller than 0.05.

Author Contributions: Conceptualization, N.-N.K.; Methodology, A.F. and N.-N.K.; Software, N.-N.K.; Validation, N.-N.K.; Formal analysis, A.F. and N.-N.K.; Investigation, A.F. and N.-N.K.; Resources, L.J., S.C.H., A.E.H., T.N., and F.L.; Data curation, N.-N.K.; Writing—original draft preparation, N.-N.K.; Writing—review and editing, J.Y. and N.-N.K.; Visualization, N.-N.K.; Supervision,

N.-N.K. and J.Y.; Project administration, N.-N.K.; Funding acquisition, N.-N.K. All authors have read and agreed to the published version of the manuscript.

Funding: This research was funded by the Deutsche Forschungsgemeinschaft (DFG, German Research Foundation, project number 390921723).

Institutional Review Board Statement: The collection of placental tissues was approved by the Ethics Committee at the University Hospital (reference number: 375/11), Goethe-University Frankfurt. Clinical information of all patients is listed in Tables 1–5. First trimester samples for IHC staining were kindly provided by Prof. Dr. Qi Chen, Department of Obstetrics and Gynecology, University of Auckland, and Fudan University, Shanghai, China, approved by the Ethics Committee of the Hospital of Obstetrics and Gynecology of Fudan University (reference number 2018-62), China. First trimester placentas for trophoblast organoid formation and primary trophoblast isolation were provided by Dr. Thorsten Nowak, Medical Practice for Gynecology and Obstetrics in Frankfurt, approved by the Ethics Committee of the University Hospital, Goethe-University Frankfurt (reference number: 19-455). Written informed approval was obtained from all patients.

Informed Consent Statement: Informed consent was obtained from all subjects involved in the study.

Data Availability Statement: The datasets generated and/or analyzed during the current study are available from the corresponding author on reasonable request.

Acknowledgments: We are grateful to our patients and the clinical team for making this study possible. We thank Qi Chen, University of Auckland, New Zealand, and Fudan University, Shanghai, China, for supporting this study with six FFPE first trimester placental tissues. We also thank Charles Graham, Queen's University at Kingston, for kindly providing us with the cell line HTR-8/SVneo and Guy Whitley, St George's hospital medical school, University of London, for the SGHPL-4 cell line. We thank Hungwen Chen, Institute of Biological Chemistry, Academia Sinica in Taiwan, for kindly providing us with the syncytin-2 promoter plasmid.

Conflicts of Interest: The authors declare no conflict of interest The funders had no role in the design of the study; in the collection, analyses, or interpretation of data; in the writing of the manuscript, or in the decision to publish the results.

References

1. Steegers, E.A.; von Dadelszen, P.; Duvekot, J.J.; Pijnenborg, R. Pre-eclampsia. *Lancet* **2010**, *376*, 631–644. [CrossRef]
2. Phipps, E.A.; Thadhani, R.; Benzing, T.; Karumanchi, S.A. Pre-eclampsia: Pathogenesis, novel diagnostics and therapies. *Nat Rev. Nephrol.* **2019**, *15*, 275–289. [CrossRef]
3. Roberts, J.M.; Taylor, R.N.; Musci, T.J.; Rodgers, G.M.; Hubel, C.A.; McLaughlin, M.K. Preeclampsia: An endothelial cell disorder. *Am. J. Obstet. Gynecol.* **1989**, *161*, 1200–1204. [CrossRef]
4. Myatt, L.; Roberts, J.M. Preeclampsia: Syndrome or Disease? *Curr. Hypertens. Rep.* **2015**, *17*, 83. [CrossRef] [PubMed]
5. Robillard, P.Y.; Dekker, G.; Chaouat, G.; Scioscia, M.; Iacobelli, S.; Hulsey, T.C. Historical evolution of ideas on eclampsia/preeclampsia: A proposed optimistic view of preeclampsia. *J. Reprod. Immunol.* **2017**, *123*, 72–77. [CrossRef]
6. Phipps, E.; Prasanna, D.; Brima, W.; Jim, B. Preeclampsia: Updates in Pathogenesis, Definitions, and Guidelines. *Clin. J. Am. Soc. Nephrol.* **2016**, *11*, 1102–1113. [CrossRef]
7. Weinstein, L. Syndrome of hemolysis, elevated liver enzymes, and low platelet count: A severe consequence of hypertension in pregnancy. *Am. J. Obstet. Gynecol.* **1982**, *142*, 159–167. [CrossRef]
8. Rana, S.; Lemoine, E.; Granger, J.P.; Karumanchi, S.A. Preeclampsia: Pathophysiology, Challenges, and Perspectives. *Circ. Res.* **2019**, *124*, 1094–1112. [CrossRef] [PubMed]
9. Turco, M.Y.; Moffett, A. Development of the human placenta. *Development* **2019**, *146*, dev163428. [CrossRef] [PubMed]
10. Haram, K.; Mortensen, J.H.; Nagy, B. Genetic aspects of preeclampsia and the HELLP syndrome. *J. Pregnancy* **2014**, *2014*, 910751. [CrossRef] [PubMed]
11. Knofler, M.; Pollheimer, J. Human placental trophoblast invasion and differentiation: A particular focus on Wnt signaling. *Front Genet.* **2013**, *4*, 190. [CrossRef] [PubMed]
12. Coutifaris, C.; Kao, L.C.; Sehdev, H.M.; Chin, U.; Babalola, G.O.; Blaschuk, O.W.; Strauss, J.F., 3rd. E-cadherin expression during the differentiation of human trophoblasts. *Development* **1991**, *113*, 767–777. [CrossRef] [PubMed]
13. Gauster, M.; Huppertz, B. Fusion of cytotrophoblast with syncytiotrophoblast in the human placenta: Factors involved in syncytialization. *J. Reprod. Med. Endocrinol.* **2008**, *5*, 76–82.
14. Aplin, J.D.; Jones, C.J.P. Cell dynamics in human villous trophoblast. *Hum. Reprod. Update* **2021**, *27*, 904–922. [CrossRef]
15. Knofler, M.; Haider, S.; Saleh, L.; Pollheimer, J.; Gamage, T.; James, J. Human placenta and trophoblast development: Key molecular mechanisms and model systems. *Cell Mol. Life Sci.* **2019**, *76*, 3479–3496. [CrossRef]

16. O'Tierney-Ginn, P.F.; Lash, G.E. Beyond pregnancy: Modulation of trophoblast invasion and its consequences for fetal growth and long-term children's health. *J. Reprod. Immunol.* **2014**, *104–105*, 37–42. [CrossRef]
17. Sanchez-Aranguren, L.C.; Prada, C.E.; Riano-Medina, C.E.; Lopez, M. Endothelial dysfunction and preeclampsia: Role of oxidative stress. *Front. Physiol.* **2014**, *5*, 372. [CrossRef]
18. Heazell, A.E.; Lacey, H.A.; Jones, C.J.; Huppertz, B.; Baker, P.N.; Crocker, I.P. Effects of oxygen on cell turnover and expression of regulators of apoptosis in human placental trophoblast. *Placenta* **2008**, *29*, 175–186. [CrossRef]
19. Huppertz, B.; Kadyrov, M.; Kingdom, J.C. Apoptosis and its role in the trophoblast. *Am. J. Obstet. Gynecol.* **2006**, *195*, 29–39. [CrossRef]
20. Fisher, S.J. The placental problem: Linking abnormal cytotrophoblast differentiation to the maternal symptoms of preeclampsia. *Reprod. Biol. Endocrinol.* **2004**, *2*, 1–4. [CrossRef]
21. Ruijtenberg, S.; van den Heuvel, S. Coordinating cell proliferation and differentiation: Antagonism between cell cycle regulators and cell type-specific gene expression. *Cell Cycle* **2016**, *15*, 196–212. [CrossRef]
22. Warfel, N.A.; El-Deiry, W.S. p21WAF1 and tumourigenesis: 20 years after. *Curr. Opin. Oncol.* **2013**, *25*, 52–58. [CrossRef]
23. Kreis, N.N.; Louwen, F.; Yuan, J. Less understood issues: p21(Cip1) in mitosis and its therapeutic potential. *Oncogene* **2015**, *34*, 1758–1767. [CrossRef] [PubMed]
24. Kreis, N.N.; Louwen, F.; Yuan, J. The Multifaceted p21 (Cip1/Waf1/CDKN1A) in Cell Differentiation, Migration and Cancer Therapy. *Cancers* **2019**, *11*, 1220. [CrossRef] [PubMed]
25. Kreis, N.N.; Friemel, A.; Ritter, A.; Roth, S.; Rolle, U.; Louwen, F.; Yuan, J. Function of p21 (Cip1/Waf1/CDKN1A) in Migration and Invasion of Cancer and Trophoblastic Cells. *Cancers* **2019**, *11*, 989. [CrossRef] [PubMed]
26. Sheridan, M.A.; Fernando, R.C.; Gardner, L.; Hollinshead, M.S.; Burton, G.J.; Moffett, A.; Turco, M.Y. Establishment and differentiation of long-term trophoblast organoid cultures from the human placenta. *Nat. Protoc.* **2020**, *15*, 3441–3463. [CrossRef]
27. Ritter, A.; Friemel, A.; Fornoff, F.; Adjan, M.; Solbach, C.; Yuan, J.; Louwen, F. Characterization of adipose-derived stem cells from subcutaneous and visceral adipose tissues and their function in breast cancer cells. *Oncotarget* **2015**, *6*, 34475–34493. [CrossRef]
28. Turco, M.Y.; Gardner, L.; Kay, R.G.; Hamilton, R.S.; Prater, M.; Hollinshead, M.S.; McWhinnie, A.; Esposito, L.; Fernando, R.; Skelton, H.; et al. Trophoblast organoids as a model for maternal-fetal interactions during human placentation. *Nature* **2018**, *564*, 263–267. [CrossRef]
29. Haider, S.; Meinhardt, G.; Saleh, L.; Kunihs, V.; Gamperl, M.; Kaindl, U.; Ellinger, A.; Burkard, T.R.; Fiala, C.; Pollheimer, J.; et al. Self-Renewing Trophoblast Organoids Recapitulate the Developmental Program of the Early Human Placenta. *Stem Cell Rep.* **2018**, *11*, 537–551. [CrossRef]
30. Vondra, S.; Kunihs, V.; Eberhart, T.; Eigner, K.; Bauer, R.; Haslinger, P.; Haider, S.; Windsperger, K.; Klambauer, G.; Schutz, B.; et al. Metabolism of cholesterol and progesterone is differentially regulated in primary trophoblastic subtypes and might be disturbed in recurrent miscarriages. *J. Lipid Res.* **2019**, *60*, 1922–1934. [CrossRef]
31. Petroff, M.G.; Phillips, T.A.; Ka, H.; Pace, J.L.; Hunt, J.S. Isolation and culture of term human trophoblast cells. *Methods Mol. Med.* **2006**, *121*, 203–217.
32. Steigman, S.A.; Fauza, D.O. Isolation of Mesenchymal Stem Cells from Amniotic Fluid and Placenta. *Curr. Protoc. Stem Cell Biol.* **2007**, *1*, 1E.2.1–1E.2.12. [CrossRef] [PubMed]
33. Rodel, F.; Steinhauser, K.; Kreis, N.N.; Friemel, A.; Martin, D.; Wieland, U.; Rave-Frank, M.; Balermpas, P.; Fokas, E.; Louwen, F.; et al. Prognostic impact of RITA expression in patients with anal squamous cell carcinoma treated with chemoradiotherapy. *Radiother. Oncol. J. Eur. Soc. Ther. Radiol. Oncol.* **2018**, *126*, 214–221. [CrossRef]
34. Detre, S.; Saclani Jotti, G.; Dowsett, M. A "quickscore" method for immunohistochemical semiquantitation: Validation for oestrogen receptor in breast carcinomas. *J. Clin. Pathol.* **1995**, *48*, 876–878. [CrossRef]
35. McClelland, R.A.; Finlay, P.; Walker, K.J.; Nicholson, D.; Robertson, J.F.; Blamey, R.W.; Nicholson, R.I. Automated quantitation of immunocytochemically localized estrogen receptors in human breast cancer. *Cancer Res.* **1990**, *50*, 3545–3550. [PubMed]
36. Meller, M.; Vadachkoria, S.; Luthy, D.A.; Williams, M.A. Evaluation of housekeeping genes in placental comparative expression studies. *Placenta* **2005**, *26*, 601–607. [CrossRef] [PubMed]
37. Louwen, F.; Muschol-Steinmetz, C.; Reinhard, J.; Reitter, A.; Yuan, J. A lesson for cancer research: Placental microarray gene analysis in preeclampsia. *Oncotarget* **2012**, *3*, 759–773. [CrossRef]
38. Muschol-Steinmetz, C.; Friemel, A.; Kreis, N.N.; Reinhard, J.; Yuan, J.; Louwen, F. Function of survivin in trophoblastic cells of the placenta. *PLoS ONE* **2013**, *8*, e73337. [CrossRef]
39. Graham, C.H.; Hawley, T.S.; Hawley, R.G.; MacDougall, J.R.; Kerbel, R.S.; Khoo, N.; Lala, P.K. Establishment and characterization of first trimester human trophoblast cells with extended lifespan. *Exp. Cell Res.* **1993**, *206*, 204–211. [CrossRef]
40. Dash, P.R.; Cartwright, J.E.; Whitley, G.S. Nitric oxide inhibits polyamine-induced apoptosis in the human extravillous trophoblast cell line SGHPL-4. *Hum. Reprod.* **2003**, *18*, 959–968. [CrossRef]
41. Kreis, N.N.; Friemel, A.; Zimmer, B.; Roth, S.; Rieger, M.A.; Rolle, U.; Louwen, F.; Yuan, J. Mitotic p21Cip1/CDKN1A is regulated by cyclin-dependent kinase 1 phosphorylation. *Oncotarget* **2016**, *7*, 50215–50228. [CrossRef]
42. Kreis, N.N.; Sommer, K.; Sanhaji, M.; Kramer, A.; Matthess, Y.; Kaufmann, M.; Strebhardt, K.; Yuan, J. Long-term downregulation of Polo-like kinase 1 increases the cyclin-dependent kinase inhibitor p21(WAF1/CIP1). *Cell Cycle* **2009**, *8*, 460–472. [CrossRef]
43. Liang, C.Y.; Wang, L.J.; Chen, C.P.; Chen, L.F.; Chen, Y.H.; Chen, H. GCM1 regulation of the expression of syncytin 2 and its cognate receptor MFSD2A in human placenta. *Biol. Reprod.* **2010**, *83*, 387–395. [CrossRef] [PubMed]

44. Janssen, A.B.; Tunster, S.J.; Savory, N.; Holmes, A.; Beasley, J.; Parveen, S.A.; Penketh, R.J.; John, R.M. Placental expression of imprinted genes varies with sampling site and mode of delivery. *Placenta* **2015**, *36*, 790–795. [CrossRef] [PubMed]
45. Roberts, J.M.; Hubel, C.A. The two stage model of preeclampsia: Variations on the theme. *Placenta* **2009**, *30*, S32–S37. [CrossRef] [PubMed]
46. Verlohren, S.; Geusens, N.; Morton, J.; Verhaegen, I.; Hering, L.; Herse, F.; Dudenhausen, J.W.; Muller, D.N.; Luft, F.C.; Cartwright, J.E.; et al. Inhibition of trophoblast-induced spiral artery remodeling reduces placental perfusion in rat pregnancy. *Hypertension* **2010**, *56*, 304–310. [CrossRef]
47. Orendi, K.; Gauster, M.; Moser, G.; Meiri, H.; Huppertz, B. The choriocarcinoma cell line BeWo: Syncytial fusion and expression of syncytium-specific proteins. *Reproduction* **2010**, *140*, 759–766. [CrossRef]
48. Coulombe, P.; Meloche, S. Atypical mitogen-activated protein kinases: Structure, regulation and functions. *Biochim. Biophys. Acta* **2007**, *1773*, 1376–1387. [CrossRef] [PubMed]
49. Cargnello, M.; Roux, P.P. Activation and function of the MAPKs and their substrates, the MAPK-activated protein kinases. *Microbiol. Mol. Biol. Rev.* **2011**, *75*, 50–83. [CrossRef]
50. Sharp, A.N.; Heazell, A.E.; Baczyk, D.; Dunk, C.E.; Lacey, H.A.; Jones, C.J.; Perkins, J.E.; Kingdom, J.C.; Baker, P.N.; Crocker, I.P. Preeclampsia is associated with alterations in the p53-pathway in villous trophoblast. *PLoS ONE* **2014**, *9*, e87621. [CrossRef]
51. Cobellis, L.; Mastrogiacomo, A.; Federico, E.; Schettino, M.T.; De Falco, M.; Manente, L.; Coppola, G.; Torella, M.; Colacurci, N.; De Luca, A. Distribution of Notch protein members in normal and preeclampsia-complicated placentas. *Cell Tissue Res.* **2007**, *330*, 527–534. [CrossRef]
52. Louwen, F.; Muschol-Steinmetz, C.; Friemel, A.; Kampf, A.K.; Tottel, E.; Reinhard, J.; Yuan, J. Targeted gene analysis: Increased B-cell lymphoma 6 in preeclamptic placentas. *Hum. Pathol.* **2014**, *45*, 1234–1242. [CrossRef] [PubMed]
53. Londero, A.P.; Orsaria, M.; Marzinotto, S.; Grassi, T.; Fruscalzo, A.; Calcagno, A.; Bertozzi, S.; Nardini, N.; Stella, E.; Lelle, R.J.; et al. Placental aging and oxidation damage in a tissue micro-array model: An immunohistochemistry study. *Histochem. Cell Biol.* **2016**, *146*, 191–204. [CrossRef] [PubMed]
54. Duan, L.; Schimmelmann, M.; Wu, Y.; Reisch, B.; Faas, M.; Kimmig, R.; Winterhager, E.; Koninger, A.; Gellhaus, A. CCN3 Signaling Is Differently Regulated in Placental Diseases Preeclampsia and Abnormally Invasive Placenta. *Front Endocrinol.* **2020**, *11*, 1–18. [CrossRef]
55. Genbacev, O.; Zhou, Y.; Ludlow, J.W.; Fisher, S.J. Regulation of human placental development by oxygen tension. *Science* **1997**, *277*, 1669–1672. [CrossRef] [PubMed]
56. Chen, B.; Longtine, M.S.; Sadovsky, Y.; Nelson, D.M. Hypoxia downregulates p53 but induces apoptosis and enhances expression of BAD in cultures of human syncytiotrophoblasts. *Am. J. Physiol. Cell Physiol.* **2010**, *299*, C968–C976. [CrossRef]
57. Phan, R.T.; Saito, M.; Basso, K.; Niu, H.; Dalla-Favera, R. BCL6 interacts with the transcription factor Miz-1 to suppress the cyclin-dependent kinase inhibitor p21 and cell cycle arrest in germinal center B cells. *Nat. Immunol.* **2005**, *6*, 1054–1060. [CrossRef]
58. Sharma, N.; Kubaczka, C.; Kaiser, S.; Nettersheim, D.; Mughal, S.S.; Riesenberg, S.; Holzel, M.; Winterhager, E.; Schorle, H. Tpbpa-Cre-mediated deletion of TFAP2C leads to deregulation of Cdkn1a, Akt1 and the ERK pathway, causing placental growth arrest. *Development* **2016**, *143*, 787–798. [CrossRef] [PubMed]
59. Kotani, T.; Iwase, A.; Ino, K.; Sumigama, S.; Yamamoto, E.; Hayakawa, H.; Nagasaka, T.; Itakura, A.; Nomura, S.; Kikkawa, F. Activator protein-2 impairs the invasion of a human extravillous trophoblast cell line. *Endocrinology* **2009**, *150*, 4376–4385. [CrossRef]
60. Brock, M.; Haider, T.J.; Vogel, J.; Gassmann, M.; Speich, R.; Trenkmann, M.; Ulrich, S.; Kohler, M.; Huber, L.C. The hypoxia-induced microRNA-130a controls pulmonary smooth muscle cell proliferation by directly targeting CDKN1A. *Int. J Biochem. Cell Biol.* **2015**, *61*, 129–137. [CrossRef]
61. Yang, S.; Li, H.; Ge, Q.; Guo, L.; Chen, F. Deregulated microRNA species in the plasma and placenta of patients with preeclampsia. *Mol. Med. Rep.* **2015**, *12*, 527–534. [CrossRef]
62. Kudo, Y.; Boyd, C.A.; Sargent, I.L.; Redman, C.W. Hypoxia alters expression and function of syncytin and its receptor during trophoblast cell fusion of human placental BeWo cells: Implications for impaired trophoblast syncytialisation in pre-eclampsia. *Biochim. Biophys. Acta* **2003**, *1638*, 63–71. [CrossRef]
63. Racca, A.C.; Ridano, M.E.; Camolotto, S.; Genti-Raimondi, S.; Panzetta-Dutari, G.M. A novel regulator of human villous trophoblast fusion: The Kruppel-like factor 6. *Mol. Hum. Reprod.* **2015**, *21*, 347–358. [CrossRef]
64. Ullah, R.; Naz, A.; Akram, H.S.; Ullah, Z.; Tariq, M.; Mithani, A.; Faisal, A. Transcriptomic analysis reveals differential gene expression, alternative splicing, and novel exons during mouse trophoblast stem cell differentiation. *Stem. Cell Res. Ther.* **2020**, *11*, 1–17. [CrossRef] [PubMed]
65. Gauster, M.; Maninger, S.; Siwetz, M.; Deutsch, A.; El-Heliebi, A.; Kolb-Lenz, D.; Hiden, U.; Desoye, G.; Herse, F.; Prokesch, A. Downregulation of p53 drives autophagy during human trophoblast differentiation. *Cell Mol. Life Sci.* **2018**, *75*, 1839–1855. [CrossRef]
66. Lu, X.; Wang, R.; Zhu, C.; Wang, H.; Lin, H.Y.; Gu, Y.; Cross, J.C.; Wang, H. Fine-Tuned and Cell-Cycle-Restricted Expression of Fusogenic Protein Syncytin-2 Maintains Functional Placental Syncytia. *Cell Rep.* **2017**, *21*, 1150–1159. [CrossRef]
67. Vargas, A.; Toufaily, C.; LeBellego, F.; Rassart, E.; Lafond, J.; Barbeau, B. Reduced expression of both syncytin 1 and syncytin 2 correlates with severity of preeclampsia. *Reprod. Sci.* **2011**, *18*, 1085–1091. [CrossRef] [PubMed]

68. Unek, G.; Ozmen, A.; Mendilcioglu, I.; Simsek, M.; Korgun, E.T. The expression of cell cycle related proteins PCNA, Ki67, p27 and p57 in normal and preeclamptic human placentas. *Tissue Cell* **2014**, *46*, 198–205. [CrossRef]
69. Song, H.L.; Liu, T.H.; Wang, Y.H.; Li, F.F.; Ruan, L.L.; Adu-Gyamfi, E.A.; Hu, S.C.; Chen, X.M.; Ding, Y.B.; Fu, L.J. Appropriate expression of P57kip2 drives trophoblast fusion via cell cycle arrest. *Reproduction* **2021**, *161*, 633–644. [CrossRef]
70. Takahashi, K.; Yoneyama, Y.; Koizumi, N.; Utoguchi, N.; Kanayama, N.; Higashi, N. Expression of p57(KIP2) reduces growth and invasion, and induces syncytialization in a human placental choriocarcinoma cell line, BeWo. *Placenta* **2021**, *104*, 168–178. [CrossRef] [PubMed]
71. He, G.Q.; Liu, G.Y.; Xu, W.M.; Liao, H.J.; Liu, X.H.; He, G.L. p57KIP2mediated inhibition of human trophoblast apoptosis and promotion of invasion in vitro. *Int. J. Mol. Med.* **2019**, *44*, 281–290. [CrossRef]
72. Kanayama, N.; Takahashi, K.; Matsuura, T.; Sugimura, M.; Kobayashi, T.; Moniwa, N.; Tomita, M.; Nakayama, K. Deficiency in p57Kip2 expression induces preeclampsia-like symptoms in mice. *Mol. Hum. Reprod.* **2002**, *8*, 1129–1135. [CrossRef] [PubMed]
73. Falcao, S.; Solomon, C.; Monat, C.; Berube, J.; Gutkowska, J.; Lavoie, J.L. Impact of diet and stress on the development of preeclampsia-like symptoms in p57kip2 mice. *Am. J. Physiol. Heart Circ. Physiol.* **2009**, *296*, H119–H126. [CrossRef] [PubMed]
74. Knox, K.S.; Baker, J.C. Genome-wide expression profiling of placentas in the p57Kip2 model of pre-eclampsia. *Mol. Hum. Reprod.* **2007**, *13*, 251–263. [CrossRef]
75. Ullah, Z.; Kohn, M.J.; Yagi, R.; Vassilev, L.T.; DePamphilis, M.L. Differentiation of trophoblast stem cells into giant cells is triggered by p57/Kip2 inhibition of CDK1 activity. *Genes Dev.* **2008**, *22*, 3024–3036. [CrossRef]
76. Engeland, K. Cell cycle arrest through indirect transcriptional repression by p53: I have a DREAM. *Cell Death Differ.* **2018**, *25*, 114–132. [CrossRef]
77. Fischer, M.; Quaas, M.; Steiner, L.; Engeland, K. The p53-p21-DREAM-CDE/CHR pathway regulates G2/M cell cycle genes. *Nucleic Acids Res.* **2016**, *44*, 164–174. [CrossRef]
78. Rothbauer, M.; Patel, N.; Gondola, H.; Siwetz, M.; Huppertz, B.; Ertl, P. A comparative study of five physiological key parameters between four different human trophoblast-derived cell lines. *Sci. Rep.* **2017**, *7*, 1–11. [CrossRef]
79. Yoshihara, E.; Masaki, S.; Matsuo, Y.; Chen, Z.; Tian, H.; Yodoi, J. Thioredoxin/Txnip: Redoxisome, as a redox switch for the pathogenesis of diseases. *Front Immunol.* **2014**, *4*, 514. [CrossRef] [PubMed]
80. Mogami, H.; Yura, S.; Kondoh, E.; Masutani, H.; Yodoi, J.; Konishi, I. Differential expression of thioredoxin binding protein-2/Txnip in human placenta: Possible involvement of hypoxia in its suppression during early pregnancy. *J. Obstet. Gynaecol. Res.* **2017**, *43*, 50–56. [CrossRef]
81. Ji Cho, M.; Yoon, S.J.; Kim, W.; Park, J.; Lee, J.; Park, J.G.; Cho, Y.L.; Hun Kim, J.; Jang, H.; Park, Y.J.; et al. Oxidative stress-mediated TXNIP loss causes RPE dysfunction. *Exp. Mol. Med.* **2019**, *51*, 1–13. [CrossRef] [PubMed]
82. Jeon, J.H.; Lee, K.N.; Hwang, C.Y.; Kwon, K.S.; You, K.H.; Choi, I. Tumor suppressor VDUP1 increases p27(kip1) stability by inhibiting JAB1. *Cancer Res.* **2005**, *65*, 4485–4489. [CrossRef] [PubMed]

cells

MDPI

Article

Melatonin, a Potential Therapeutic Agent for Preeclampsia, Reduces the Extrusion of Toxic Extracellular Vesicles from Preeclamptic Placentae

Yunhui Tang [1,2,*], Katie Groom [3,4], Larry Chamley [2,5] and Qi Chen [2,5,*]

1 Department of Family Planning, The Hospital of Obstetrics & Gynaecology, Fudan University, Shanghai 200081, China
2 Department of Obstetrics and Gynaecology, The University of Auckland, 1142 Auckland, New Zealand; l.chamley@auckland.ac.nz
3 National Women's Health, Auckland City Hospital, 1142 Auckland, New Zealand; k.groom@auckland.ac.nz
4 Liggins Institute, The University of Auckland, 1142 Auckland, New Zealand
5 Hub for Extracellular Vesicle Investigations, The University of Auckland, 1142 Auckland, New Zealand
* Correspondence: tangyunhui1786@fckyy.org.cn (Y.T.); q.chen@auckland.ac.nz (Q.C.)

Abstract: Preeclampsia, characterised by maternal endothelial cell activation, is triggered by toxic factors, such as placental extracellular vesicles (EVs) from a dysfunctional placenta. The increased oxidative stress seen in the preeclamptic placenta links to endoplasmic reticulum (ER) stress. The ER regulates protein folding and trafficking. When the ER is stressed, proteins are misfolded, and misfolded proteins are toxic. Misfolded proteins can be exported from cells, via EVs which target to other cells where the misfolded proteins may also be toxic. Melatonin is a hormone and antioxidant produced by the pineal gland and placenta. Levels of melatonin are reduced in preeclampsia. In this study we investigated whether melatonin treatment can change the nature of placental EVs that are released from a preeclamptic placenta. EVs were collected from preeclamptic ($n = 6$) and normotensive ($n = 6$) placental explants cultured in the presence or absence of melatonin for 18 h. Misfolded proteins were measured using a fluorescent compound, Thioflavin-T (ThT). Endothelial cells were exposed to placental EVs overnight. Endothelial cell activation was measured by the quantification of cell-surface ICAM-1 using a cell-based ELISA. EVs from preeclamptic placentae carried significantly ($p < 0.001$) more misfolded proteins than normotensive controls. Incubating preeclamptic placental explants in the presence of melatonin (1 μM and 10 μM) significantly ($p < 0.001$) reduced the misfolded proteins carried by EVs. Culturing endothelial cells in the presence of preeclamptic EVs significantly increased the expression of ICAM-1. This increased ICAM-1 expression was significantly reduced when the endothelial cells were exposed to preeclamptic EVs cultured in the presence of melatonin. This study demonstrates that melatonin reduces the amount of misfolded proteins carried by EVs from preeclamptic placentae and reduces the ability of these EVs to activate endothelial cells. Our study provides further preclinical support for the use of melatonin as a treatment for preeclampsia.

Keywords: melatonin; extracellular vesicle; exosome; preeclampsia; misfold proteins; endothelial cell activation

Citation: Tang, Y.; Groom, K.; Chamley, L.; Chen, Q. Melatonin, a Potential Therapeutic Agent for Preeclampsia, Reduces the Extrusion of Toxic Extracellular Vesicles from Preeclamptic Placentae. *Cells* **2021**, *10*, 1904. https://doi.org/10.3390/cells10081904

Academic Editors: Junie P Warrington, Ana T. Palei, Mark W. Cunningham, Lorena M. Amaral and Alexander Kalyuzhny

Received: 7 June 2021
Accepted: 23 July 2021
Published: 27 July 2021

1. Introduction

Preeclampsia, a human specific pregnancy disorder, is clinically characterised by high blood pressure after 20 weeks of gestation accompanied by one or more of a spectrum of signs of organ dysfunction [1,2]. It affects 2–8% of all pregnancies worldwide [1,2]. The clinical signs of preeclampsia are preceded by an exaggerated inflammatory response and generalised maternal endothelial cell dysfunction, which are fundamental components of the pathogenesis of preeclampsia [3]. Dysfunctional endothelial cells consequently prevent the normal adaptation of the maternal vasculature seen in pregnancy, with resulting hypertension and other signs and symptoms of preeclampsia. Although the underlying mechanisms of preeclampsia are still not fully understood, it is recognised that one or more toxic factors released from the placenta trigger the maternal endothelial dysfunction. Extracellular vesicles (EVs) are among the placental factors that are increasingly recognised as being able to contribute to the endothelial cell dysfunction of preeclampsia [4–6].

EVs are lipid-enclosed packages of cellular contents that are extruded from all cells. Most eukaryote cells extrude both micro- and nano-EVs. While there are varying definitions, generally micro-EVs range in size from 200 to 1000 nm, while nano-EVs range from 10 to 200 nm. A subpopulation of nano-EVs are exosomes. The maternal-facing aspect of the human placenta is covered entirely by a single multinucleated cell, the syncytiotrophoblast, which extrudes vast numbers of both micro- and nano-EVs directly into the maternal circulation. These placental EVs are carried around the maternal body and can interact with multiple maternal cells and organs, including endothelial cells [7,8]. Like EVs from other cell types, placental EVs contain biologically active proteins, nucleic acids, lipids and RNAs, that they can transfer to recipient cells and organs, affecting their function [9–12]. We and others have previously reported that placental EVs derived from preeclamptic placentae activated maternal endothelial cells in vitro and shown that nano-EVs alter the maternal vascular tone in pregnant mice [10,13–15].

Oxidative stress is significantly increased in the preeclamptic placenta [16], and there is a strong connection between oxidative stress and endoplasmic reticulum (ER) stress [17]. The ER regulates protein folding and trafficking. When the ER is stressed, proteins are misfolded, and misfolded proteins can then be exported from cells or tissues, via EVs which target to other cells where the misfolded proteins may be toxic. We have recently shown that preeclamptic placentae contained significantly more aggregated transthyretin than normotensive placentae and that this transthyretin is specifically packaged into nano-EVs [18].

Melatonin is a lipid-soluble hormone with antioxidant activities originally identified as being of primarily pineal origin. However, melatonin is also produced in large quantities by the placenta [19,20]. In addition to its endocrine functions, melatonin also has direct free radical scavenging and indirect antioxidant activities [21–24]. The peak melatonin levels in pregnancy are 80–100 pg/mL, twice as high as those in men or non-pregnant women. Much of the melatonin is of placental origin, suggesting a particular importance of melatonin in pregnancy [19,25–28]. However, women with preeclampsia have significantly decreased serum levels of melatonin [25,26], as well as decreased levels of the enzymes responsible for melatonin synthesis and decreased levels of melatonin receptors [29]. Given the action of melatonin in modulating free radicals/oxidative stress and a potential role in the treatment of preeclampsia [30–32], we undertook this study to investigate whether melatonin supplementation has a preventative effect on the production of toxic placental EVs from preeclamptic placentae.

2. Methods

This study was approved by the Northern X Health and Disabilities Ethics Committee, New Zealand (NTX/12/06/057/AM06), and conforms to the principles outlined in the Declaration of Helsinki. All patient-derived tissues were obtained following informed written consent.

3. Collection of Preeclamptic Placentae

Six placentae were collected from women with preeclampsia (two of them in an early onset and four of them in a late onset form), and ten term placentae were collected from normotensive pregnancies from National Women's Health, Auckland City Hospital, New Zealand. The clinical parameters of the study cohort are summarised in Table 1. We were not able to collect gestation-matched placentae from normotensive pregnancies due to ethical issues. There was no difference in the maternal age (p = 0.388) between the two groups. All women with preeclampsia received medication, either labetalol or nifedipine. The mean BMI in preeclampsia was 27.08 ± 5.2 kg/m^2.

Table 1. Clinical parameters of the study cohort.

	Preeclampsia (n = 6)	Normotensive (n = 10)
Maternal age (years, mean/SD)	30.2 ± 3.1	32 ± 5.1
Onset week (mean/SD)	33 + 6 ± 4	N/A
Delivery week	35 + 2 ± 4	39 ± 1
Birthweight (g, mean/SD)	2365 ± 338	3350 ± 180
Systolic blood pressure (mmHg, mean/SD)	156 ± 7	N/A
Diastolic blood pressure (mmHg, mean/SD)	98 ± 11	N/A

Preeclampsia was defined as a maternal systolic blood pressure ≥140 mmHg and/or diastolic blood pressure ≥90 mmHg on two occasions separated by 6 h, and proteinuria >300 mg in a 24 h period, a protein-creatinine ratio >30 mg/mmol or impaired liver function after 20 weeks of gestation, in accordance with the guideline of the Society of Obstetric Medicine of Australia and New Zealand (SOMANZ), which are consistent with the international guideline from the International Society for the Study of Hypertension in Pregnancy (ISSHP) [33].

4. Collection of Placental EVs from Preeclamptic Placentae

Approximately 400 mg wet weight placental explants were dissected from either preeclamptic or normotensive placentae, as previously described [13,34,35]. Four quadruplicate explants from each placenta were cultured in Netwell™ culture inserts and suspended in 12 well culture plates, for 24 h at 37 °C, in 3 mL Advanced DMEM/F12 containing 2.5% fetal bovine serum in an ambient oxygen atmosphere in the presence or absence of melatonin (1 μM and 10 μM, dissolved in PBS) (Sigma-Aldrich, Auckland, New Zealand). The concentration of melatonin followed that of our previous study [36]. As our previous study [36] showed no dose response on the effect of melatonin, in this study normotensive placental explants were treated with a higher dose of melatonin (10 μM). The conditioned media were then collected and centrifuged at 2000× g for 5 min for the removal of culture debris. The supernatant was centrifuged at 20,000× g for 1 h for the collection of micro-EVs. The supernatant was further centrifuged at 100,000× g for 1 h for collection of nano-EVs (Avanti J30 I Ultracentrifuge, JA 30.50 fixed angle rotor, Beckman Coulter, New Zealand).

The relevant cultured explants with melatonin treatment were then collected and immediately embedded in an optimal cutting temperature (OCT) compound, and then sectioned at 5 μm using a cryostat for the measurement of misfolded proteins.

5. Quantification of Micro-EVs and Nano-EVs

The amount of micro- and nano-EVs collected from preeclamptic and healthy placentae were quantified using a Nanosight™ NS300 nanoparticle tracking device (Nanosight, UK). Placental micro-EVs and nano-EVs were re-suspended in 1 mL PBS after ultra-centrifugation and analysed. All automatic settings were applied, with the viscosity setting at 0.95 cP and the temperature at 25 °C. A single measurement consists of three 30-s videos, and each sample was measured five times at camera level 10. The detection threshold was set at 10, and data acquisition and processing were performed using the NTA3.2 software (Nanosight). Only recordings with over 1000 valid tracks/vesicles were included in the analysis. The number of placental EVs was expressed as the number of EVs per mL.

6. Measurement of Endothelial Cell Activation by Cell-Based ELISA

A human microvascular endothelial cell line, composed of HMEC-1 cells (ATCC, CRL-3243), was grown until confluent in 96 well culture plates in MCDB 131 media. The endothelial cells (HMEC-1) were exposed to placental micro- or nano-EVs from preeclamptic placental explants or from normotensive placental explants which had been treated with or without melatonin for 24 h. After the removal of the remaining placental EVs by washing with PBS, the cell surface expression of ICAM-1 by HMEC-1 monolayers was determined by a cell-based ELISA, as described previously [13]. Each measurement of ICAM-1 was conducted in quadruplicate. The data are presented as the median and 5th and 95th percentiles of the fold changes relative to untreated controls.

7. Measurement of Misfolded Proteins

Misfolded proteins in preeclamptic and healthy placentae and in the preeclamptic placental explants that had been treated with melatonin were measured using a fluorescent compound, Thioflavin-T (ThT, Sigma-Aldrich, Sydney, Australia), as described previously [37,38]. Briefly, frozen placental sections (5 μm) were fixed with 4% paraformaldehyde (PFA) for 5 min at room temperature and washed with PBS. The sections were then stained with ThT (500 μM dissolved in PBS) for 3 min at room temperature, followed by counter-staining with DAPI for 1 min. The sections were then mounted with Citifluor and examined using a fluorescent microscope (Nikon, ECLISPE Ni-E, Tokyo, Japan). Sections from which ThT was omitted were used as negative controls.

Misfolded proteins in micro- or nano-EVs collected from preeclamptic or healthy placental explant cultures that had been treated with melatonin were stained with ThT (5 μM) for 10 min and read in a fluorescent plate reader at 485 nm (Synergie 2, BioTek, Auckland, New Zealand), following previous reports [37,38].

8. Semiquantitative Analysis of Immunofluorescent Staining

All images were converted into a 32-bit RGB colour format. Subsequently, a representative mean grey value integral was quantified from each image, which served as an index for fluorescence intensity and allowed for relative comparisons of protein expression between different images and samples. Each image was opened using ImageJ software, and five lines were randomly drawn on the images. The fluorescent density of each line on the images was calculated by Matlab software (R2019 A version). The resultant five integral values were averaged to yield a mean grey value integral [39].

9. Statistical Analysis

The concentrations of placental micro- and nano-EVs were expressed as a median and 95% confidence interval (CI). The statistical analysis of the concentration of placental micro- or nano-EVs, the levels of expression of cell surface ICAM-1 by endothelial cells or the levels of misfolded proteins in placental EVs were assessed with an ANOVA test, or a Mann–Whitney U-test, when appropriate. The GraphPad Prism software package (version 8.0) was used, and a $p < 0.05$ was considered as statistically significant.

10. Results

10.1. The Amount of Placental EVs Extruded from Preeclamptic Placentae Was Not Changed by Treatment with Melatonin

We confirmed previous reports [40] that the amount of micro- and nano-EVs extruded from preeclamptic placentae was significantly increased compared to the number of EVs released from normotensive placental explants (Figure 1A, $p = 0.016$, or Figure 1B, $p < 0.01$, respectively). The treatment with melatonin (either 1 μM or10 μM) did not change the amount of micro-EVs (Figure 1A, $p = 0.954$, ANOVA) or nano-EVs (Figure 1B, $p = 0.776$, ANOVA) extruded from preeclamptic placental explant cultures.

10.2. Treatment with Melatonin Prevented the Production of Toxic Placental EVs from Preeclamptic Placentae

We confirmed that placental EVs from preeclamptic placentae activated endothelial cells (Figure 2A,B, $p < 0.001$), which was previously reported [41]. Both the micro-EVs and nano-EVs released from preeclamptic placental explants that had been incubated with melatonin did not induce an increase in endothelial cell activation (as measured by cell surface ICAM-1 expression, Figure 2A, $p = 0.0014$, ANOVA or Figure 2B, $p = 0.0002$,

ANOVA). There was no dose response in the reduction of ICAM-1 levels ($p > 0.05$) across the concentrations of melatonin tested.

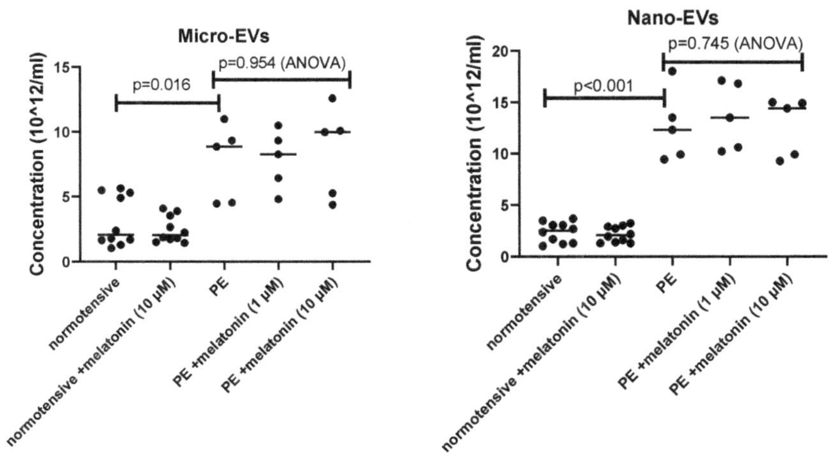

(A) (B)

Figure 1. The numbers of micro- (**A**) or nano-EVs (**B**) extruded from preeclamptic placentae (PE) was quantified by NTA, and the statistical analyses were assessed with an ANOVA test or a Mann–Whitney U-test, when appropriate.

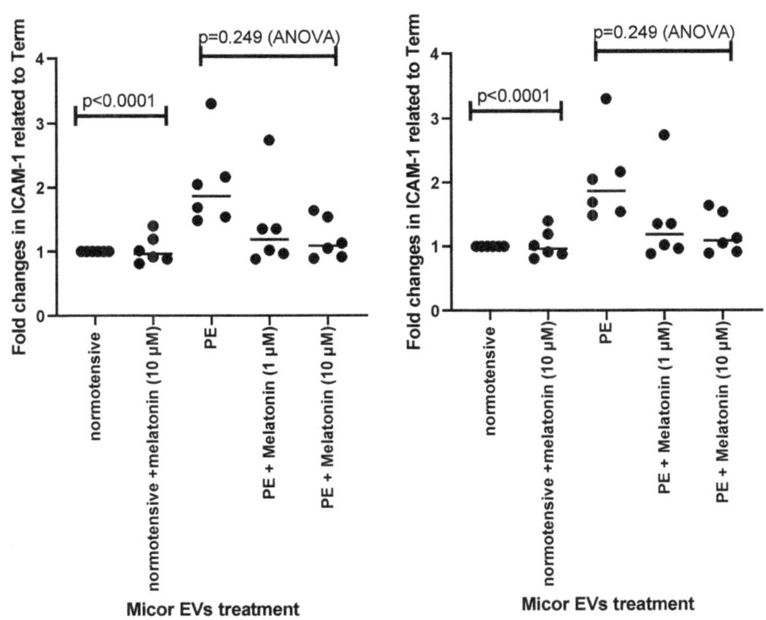

(A) (B)

Figure 2. The activation of endothelial cells after being exposed to micro- (**A**) or nano-EVs (**B**) extruded from preeclamptic (PE) or normotensive placentae was measured by cell-surface ICAM-1 levels. The statistical analysis was assessed by an ANOVA test or a Mann–Whitney U-test, when appropriate.

10.3. Treatment with Melatonin Reduced the Levels of Misfolded Proteins in Preeclamptic Placentae and in Placental EVs from Preeclamptic Placentae

There were significantly increased levels of misfolded proteins in micro- (Figure 3A, $p = 0.0001$) and nano-EVs (Figure 3B, $p = 0.0001$) collected from preeclamptic placentae compared to placental EVs collected from normotensive term placental explants, measured by the fluorescent intensity of ThT (measured at 485 nm). However, this increased fluorescent intensity of ThT was significantly reduced when placental EVs were collected from melatonin-treated (1 or 10 μM) preeclamptic placental explant cultures (Figure 3A,B, $p = 0.0001$, ANOVA). There was no dose response in the reduction of the fluorescent intensity of ThT.

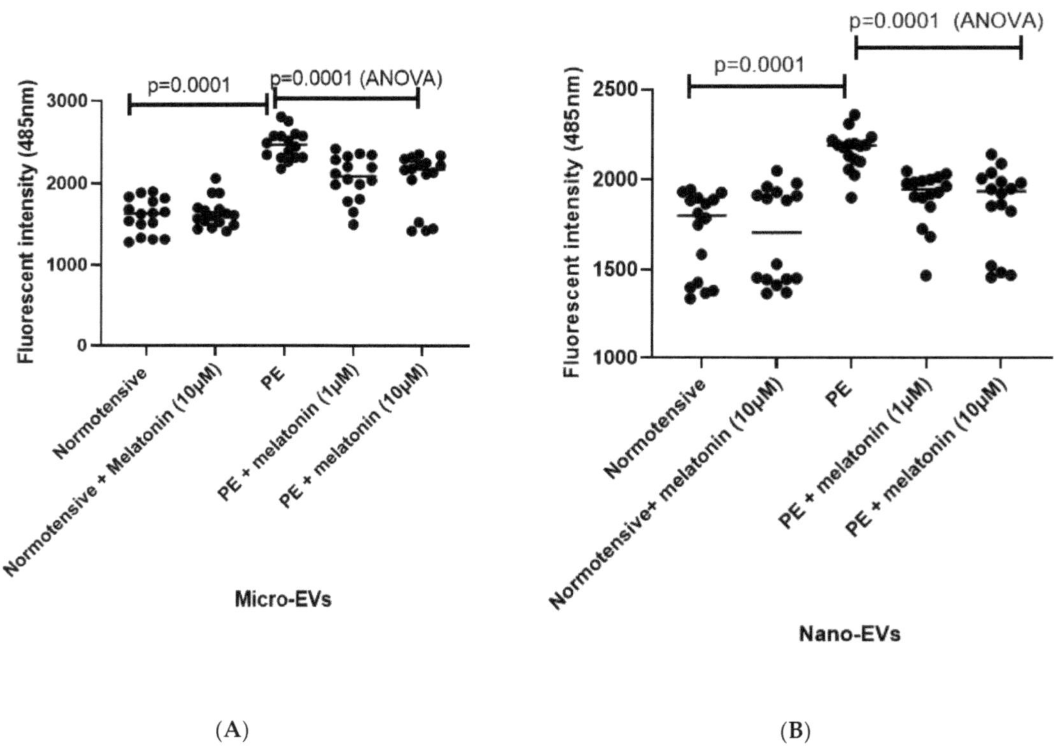

Figure 3. The misfolded proteins in placental micro-EVs (**A**) and nano-EVs (**B**) collected from either preeclamptic (PE) or normotensive placentae were measured by ThT fluorescent intensity. The statistical analysis was assessed by an ANOVA test or a Mann–Whitney U-test, when appropriate.

In addition, there were significantly increased levels of misfolded proteins in preeclamptic placentae (Figure 4A), compared to normotensive placentae (Figure 4B), measured by the increased fluorescent intensity levels of ThT in a semi-quantitative assay (Figure 4C, $p = 0.0079$). However, the treatment of the preeclamptic placentae with melatonin (1 μM Figure 5B or 10 μM Figure 5C) significantly reduced the fluorescent intensity of ThT in preeclamptic placentae, compared to the untreated ones (Figure 5A), measured by a semi-quantitative assay (Figure 5E, $p < 0.0001$). There was also no dose response in the reduction of the fluorescent intensity of ThT (Figure 5E, $p > 0.05$).

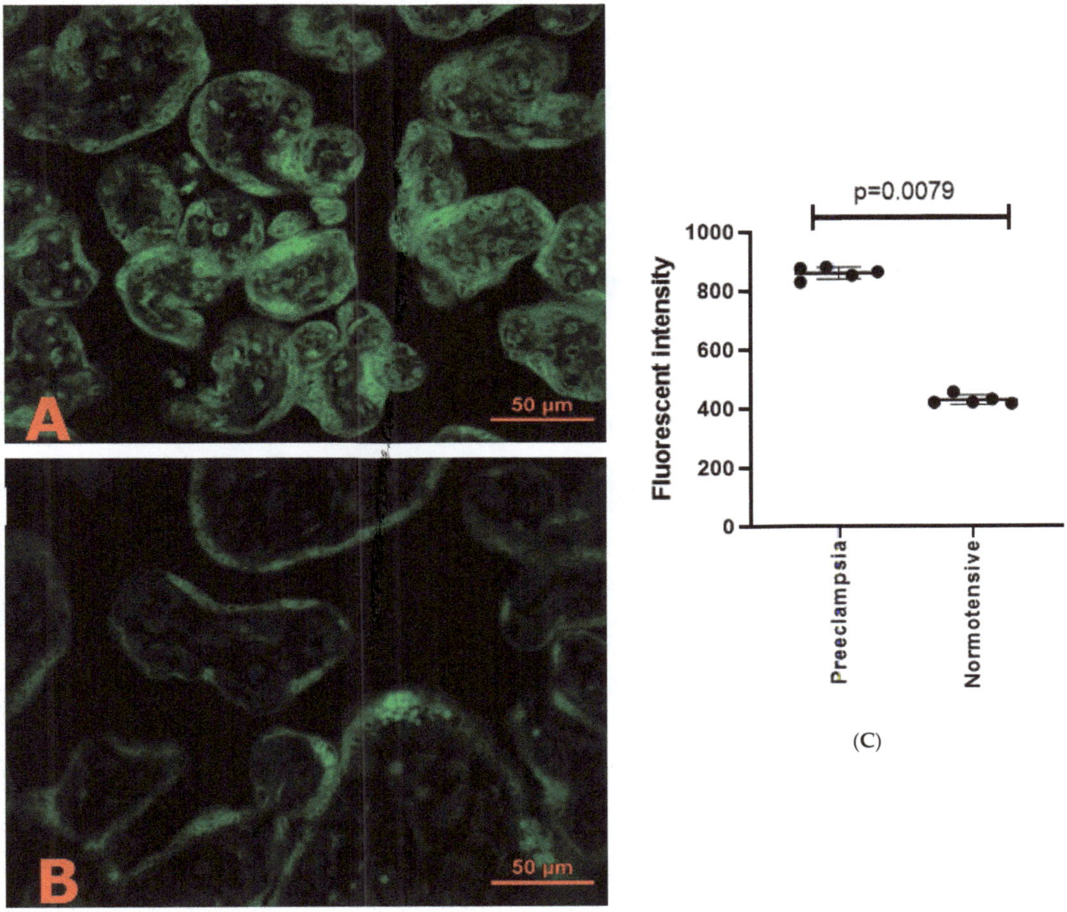

Figure 4. The levels of misfolded proteins in preeclamptic placentae, especially in the syncytiotrophoblast (**A**) and in normotensive term placentae (**B**), were measured by ThT fluorescent intensity. The statistical difference was confirmed by a semi-quantitative assay (**C**).

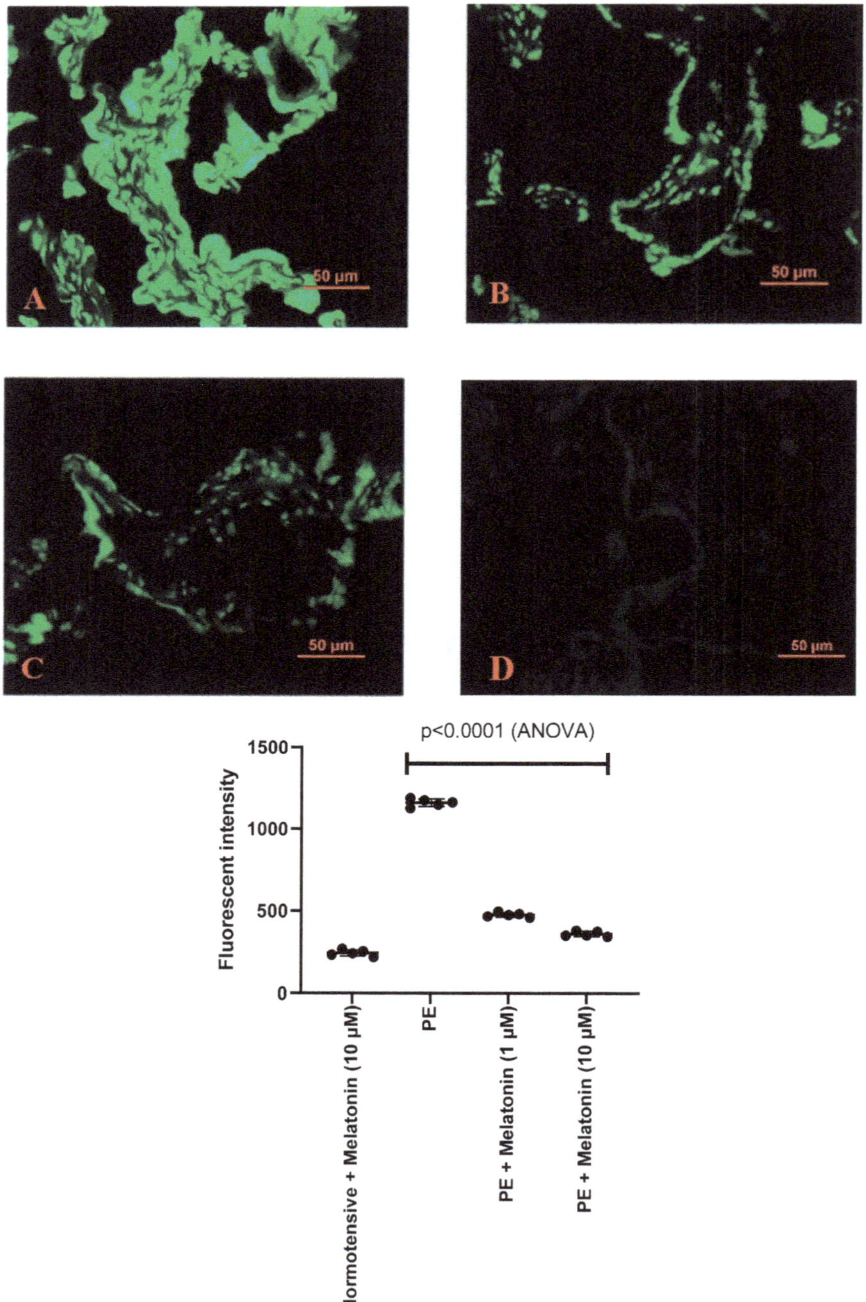

(E)

Figure 5. Misfolded proteins in preeclamptic placental explants (PE) that had been treated with 1 µM (**B**) or 10 µM (**C**) melatonin or without melatonin (**A**) were measured by the fluorescent intensity of ThT. In comparison, misfold proteins in normotensive placental explants that had been treated with 10 µM (**D**) were also measured. The statistical difference was confirmed by a semi-quantitative assessment and assessed by an ANOVA test (**E**).

11. Discussion

In this in vitro study, we found that the treatment with melatonin did not reduce the amount of EVs produced by preeclamptic placentae but did alter the nature (function) of the EVs such that they were not toxic and did not active endothelial cells. Reducing the increased levels of ER stress in preeclamptic placentae and in EVs that were exported from preeclamptic placentae could be one of the underlying mechanisms of this protective effect of endothelial cell activation.

Although the exact causes of preeclampsia are still unclear, there is increasing evidence suggesting that EVs extruded from the placenta are associated with the pathogenesis of preeclampsia [13]. These placental EVs from preeclamptic placentae may contribute to triggering the disease by inducing maternal systemic endothelial cell dysfunction [42], as they carry several anti-angiogenic factors, including soluble fms-like tyrosine kinase-1 (sFlt1) and soluble Endoglin (sEng), cytokines and oxidants, as well as multiple "danger" signals [43]. Oxidative stress results from an imbalance in oxidant and antioxidant molecules [44] and is reported to play an important role in the development of preeclampsia [45].

Melatonin, typically thought of as a lipid hormone that regulates circadian rhythms, has been shown to have direct free-radical scavenging properties and to induce the expression of antioxidants [46]. During pregnancy, the placenta is a major source of melatonin, producing higher levels than the pineal gland [20], and placental melatonin production does not appear to be regulated in a circadian fashion [47]. Studies have suggested that alterations of circadian rhythms increase the risk of developing preeclampsia [48]. In women with preeclampsia there are significantly reduced levels of circulating melatonin [25,26], accompanied by a reduction in the levels of the synthetic enzymes and melatonin receptors in preeclamptic placentae [29]. Therefore, it has been suggested that exogenous melatonin may have potential benefits in preeclampsia [30–32,48,49]. Here, we confirmed that EVs from preeclamptic placentae induce endothelial cell activation, which is a hallmark of preeclampsia, and found that EVs from preeclamptic placental explants that had been treated with melatonin did not induce endothelial cell activation. This finding fits with our previous report that melatonin reversed the production of toxic placental EVs extruded from normal placental explants that had been treated with either preeclamptic serum or antiphospholipid antibodies, which are a major maternal risk factor for preeclampsia [50]. In our previous report, we demonstrated that treating placental explants with melatonin reduced the markers of oxidative stress that were induced by the preeclamptic sera and antiphospholipid antibodies [36]. Others have recently shown that melatonin induces the expression of antioxidants in the placenta [51]. Taken together, our data further suggest that exogenous melatonin could directly change the production of toxic EVs by the preeclamptic placenta.

The molecular signalling pathways that change the production of EVs in preeclampsia are largely unknown. A number of factors found in preeclampsia, such as the inflammatory cytokines IL-6 and TNFα [52], disrupt placental function, resulting in the production of toxic EVs [53]. Exactly how those cytokines induce the production of toxic EVs is still unclear, but more is known about how antiphospholipid antibodies induce the production of toxic EVs [54,55]. These autoantibodies are internalised into the syncytiotrophoblast where they cause mitochondrial dysfunction, oxidative damage, the production of both necroptosis and apoptosis cell death pathway proteins and ER stress, characterised by an excess of misfolded proteins [38,56]. ER stress is a feature of preeclamptic placentae that is characterised by increased protein misfolding and aggregation (reviewed in [57]). These misfolded and aggregated proteins are toxic and can be deported from the placenta into the maternal circulation via EVs. For example, we have previously shown increased levels of aggregated transthyretin in EVs extruded from preeclamptic placentae [18]. Removing misfolded proteins via EVs may be a survival mechanism for the syncytiotrophoblast, but these misfolded proteins are potentially damaging to the maternal cells that take up the EVs containing an excess of misfolded proteins. Here, we confirmed an increased level of misfolded proteins in preeclamptic placentae. In addition, we also found an increased level of misfolded proteins in both micro- and nano-EVs extruded from preeclamptic placentae. These EVs are targeted to several maternal organs, such as the liver and kidneys, that are often effected in preeclampsia [9,10]. The uptake of micro-EVs and/or nano-EVs that contain toxic misfolded proteins may be harmful to these organs. The key finding of our current study is that treating explants from preeclamptic placentae with melatonin reduces the levels of misfolded proteins both in the placenta and in the EVs extruded from the placentae. The reduction of the load of misfolded proteins delivered to maternal organs by EVs from melatonin-treated preeclamptic placentae may reduce the harm to those organs. The clinical utility of melatonin for the treatment of preeclampsia has recently been suggested by a phase I clinical trial that demonstrated the safety of administering melatonin to women with preeclampsia and also demonstrated a prolongation of the time from diagnosis to delivery [58].

We acknowledge that, due to the small clinical samples used in this study, we were not able to analyse whether the protective effect on producing toxic placental EVs by melatonin is associated with the severity or time of onset of preeclampsia. Future study is required to confirm our findings.

In conclusion: this study demonstrates that exogenous melatonin prevents the production of endothelial-activating EVs from preeclamptic placentae. Reducing the amount of misfolded proteins carried by EVs from preeclamptic placentae exporting to the maternal circulation may be one of the underlying mechanisms to prevent endothelial cell activation. Our study provides further preclinical support for the use of melatonin for the treatment of preeclampsia.

Author Contributions: All authors were involved in the drafting, editing and approval of the manuscript for publication. In addition to this, each author contributed to the following work: Y.T. and Q.C.: experiments performance and analysis; L.C. and Q.C.: data analysis; K.G., L.C. and Q.C.: designed the study and wrote the manuscript draft. All authors have read and agreed to the published version of the manuscript.

Funding: This research was supported by grants from the Performance Based Research Foundation (PBRF), the School of Medicine, The University of Auckland, and the Health Research Council of New Zealand (grant number 18/408). Y.T. was funded by training grants from The Hospital of Obstetrics and Gynaecology, Fudan University, China. The APC was funded by The Hospital of Obstetrics and Gynaecology, Fudan University, China.

Institutional Review Board Statement: This study was approved by the Northern X Health and Disabilities Ethics Commit-tee, New Zealand (NTX/12/06/057/AM06).

Informed Consent Statement: All patient-derived tissues were obtained following informed written consent.

Data Availability Statement: All the data will be available upon to request.

Acknowledgments: The authors thank the patients and staff of Auckland City Hospital for the donation and collection of placentae.

Conflicts of Interest: None of the authors has a conflict of interest to declare.

References

1. Sibai, B.; Dekker, G.; Kupferminc, M. Pre-eclampsia. *Lancet* **2005**, *365*, 785–799. [CrossRef]
2. Gestational Hypertension and Preeclampsia: ACOG Practice Bulletin, Number 222. *Obstet. Gynecol.* **2020**, *135*, e237–e260. [CrossRef] [PubMed]
3. Roberts, J.M. Endothelial dysfunction in preeclampsia. *Semin. Reprod. Endocrinol.* **1998**, *16*, 5–15. [CrossRef] [PubMed]
4. Schmorl, G. *Pathologisch-Anatomische Untersuchungen Uber Puerperal-Eklampsie*; Verlag von FC Vogel: Leipzig, Germany, 1893.
5. Attwood, H.; Park, W. Embolism to the lungs by trophoblast. *BJOG Int. J. Obstet. Gynaecol.* **1961**, *68*, 611–617. [CrossRef]
6. Jaameri, K.E.; Koivuniemi, A.P.; Carpen, E.O. Occurrence of trophoblasts in the blood of toxaemic patients. *Gynaecologia* **1965**, *160*, 315–320. [CrossRef]
7. Tong, M.; Stanley, J.L.; Chen, Q.; Wise, M.R.; Stone, P.R.; James, J.L.; Chamley, L.W. In Vivo Targets of Human Placental Extracellular Vesicles. *Reprod. Sci.* **2016**, *23*, 153a.

8. Cronqvist, T.; Tannetta, D.; Mörgelin, M.; Belting, M.; Sargent, I.; Familari, M.; Hansson, S.R. Syncytiotrophoblast derived extracellular vesicles transfer functional placental miRNAs to primary human endothelial cells. *Sci. Rep.* **2017**, *7*, 4558. [CrossRef]
9. Tong, M.; Chen, Q.; James, J.L.; Wise, M.R.; Stone, P.R.; Chamley, L.W. In vivo targets of human placental micro-vesicles vary with exposure time and pregnancy. *Reproduction* **2017**, *153*, 835–845. [CrossRef]
10. Tong, M.; Stanley, J.L.; Chen, Q.; James, J.L.; Stone, P.R.; Chamley, L.W. Placental Nano-vesicles Target to Specific Organs and Modulate Vascular Tone In Vivo. *Hum. Reprod.* **2017**, *32*, 2188–2198. [CrossRef] [PubMed]
11. Wei, J.; Blenkiron, C.; Tsai, P.; James, J.L.; Chen, Q.; Stone, P.R.; Chamley, L.W. Placental trophoblast debris mediated feto-maternal signalling via small RNA delivery: Implications for preeclampsia. *Sci. Rep.* **2017**, *7*, 14681. [CrossRef] [PubMed]
12. Wei, J.; Lau, S.Y.; Blenkiron, C.; Chen, Q.; James, J.L.; Kleffmann, T.; Wise, M.; Stone, P.R.; Chamley, L.W. Trophoblastic debris modifies endothelial cell transcriptome in vitro: A mechanism by which fetal cells might control maternal responses to pregnancy. *Sci. Rep.* **2016**, *6*, 30632. [CrossRef]
13. Chen, Q.; Stone, P.R.; McCowan, L.M.; Chamley, L.W. Phagocytosis of necrotic but not apoptotic trophoblasts induces endothelial cell activation. *Hypertension* **2006**, *47*, 116–121. [CrossRef]
14. Chen, Q.; Stone, P.; Ching, L.M.; Chamley, L. A role for interleukin-6 in spreading endothelial cell activation after phagocytosis of necrotic trophoblastic material: Implications for the pathogenesis of pre-eclampsia. *J. Pathol.* **2009**, *217*, 122–130. [CrossRef] [PubMed]
15. Chen, Q.; Guo, F.; Liu, S.; Xiao, J.; Wang, C.; Snowise, S.; Stone, P.R.; Chamley, L.W. Calcium channel blockers prevent endothelial cell activation in response to necrotic trophoblast debris: Possible relevance to pre-eclampsia. *Cardiovasc. Res.* **2012**, *96*, 484–493. [CrossRef] [PubMed]
16. Burton, G.J.; Jauniaux, E. Oxidative stress. *Best Pract. Res. Clin. Obstet. Gynaecol.* **2011**, *25*, 287–299. [CrossRef]
17. Kritsiligkou, P.; Rand, J.D.; Weids, A.J.; Wang, X.; Kershaw, C.J.; Grant, C.M. Endoplasmic reticulum (ER) stress-induced reactive oxygen species (ROS) are detrimental for the fitness of a thioredoxin reductase mutant. *J. Biol. Chem.* **2018**, *293*, 11984–11995. [CrossRef] [PubMed]
18. Tong, M.; Cheng, S.B.; Chen, Q.; DeSousa, J.; Stone, P.R.; James, J.L.; Chamley, L.W.; Sharma, S. Aggregated transthyretin is specifically packaged into placental nano-vesicles in preeclampsia. *Sci. Rep.* **2017**, *7*, 6694. [CrossRef] [PubMed]
19. Kivela, A. Serum melatonin during human pregnancy. *Acta Endocrinol.* **1991**, *124*, 233–237. [CrossRef]
20. Lanoix, D.; Beghdadi, H.; Lafond, J.; Vaillancourt, C. Human placental trophoblasts synthesize melatonin and express its receptors. *J. Pineal Res.* **2008**, *45*, 50–60. [CrossRef]
21. Miller, S.L.; Yawno, T.; Alers, N.O.; Castillo-Melendez, M.; Supramaniam, V.G.; VanZyl, N.; Sabaretnam, T.; Loose, J.M.; Drummond, G.R.; Walker, D.W. Antenatal antioxidant treatment with melatonin to decrease newborn neurodevelopmental deficits and brain injury caused by fetal growth restriction. *J. Pineal Res.* **2014**, *56*, 283–294. [CrossRef]
22. Miller, S.L.; Wallace, E.M.; Walker, D.W. Antioxidant therapies: A potential role in perinatal medicine. *Neuroendocrinology* **2012**, *96*, 13–23. [CrossRef]
23. Hardeland, R. Antioxidative protection by melatonin: Multiplicity of mechanisms from radical detoxification to radical avoidance. *Endocrine* **2005**, *27*, 119–130. [CrossRef]
24. Storm, W.F.; Eddy, D.R.; Welch, C.B.; Hickey, P.A.; Fischer, J.; Cardenas, R. Cognitive performance following premature awakening from zolpidem or melatonin induced daytime sleep. *Aviat. Space Environ. Med.* **2007**, *78*, 10–20. [PubMed]
25. Zeng, K.; Gao, Y.; Wan, J.; Tong, M.; Lee, A.C.; Zhao, M.; Chen, Q. The reduction in circulating levels of melatonin may be associated with the development of preeclampsia. *J. Hum. Hypertens.* **2016**, *30*, 666–671. [CrossRef]
26. Nakamura, Y.; Tamura, H.; Kashida, S.; Takayama, H.; Yamagata, Y.; Karube, A.; Sugino, N.; Kato, H. Changes of serum melatonin level and its relationship to feto-placental unit during pregnancy. *J. Pineal Res.* **2001**, *30*, 29–33. [CrossRef]
27. Rosell, R.; Danenberg, K.D.; Alberola, V.; Bepler, G.; Sanchez, J.J.; Camps, C.; Provencio, M.; Isla, D.; Taron, M.; Diz, P.; et al. Ribonucleotide reductase messenger RNA expression and survival in gemcitabine/cisplatin-treated advanced non-small cell lung cancer patients. *Clin. Cancer Res.* **2004**, *10*, 1318–1325. [CrossRef]
28. Sae-Teaw, M.; Johns, J.; Johns, N.P.; Subongkot, S. Serum melatonin levels and antioxidant capacities after consumption of pineapple, orange, or banana by healthy male volunteers. *J. Pineal Res.* **2013**, *55*, 58–64. [CrossRef]
29. Lanoix, D.; Guerin, P.; Vaillancourt, C. Placental melatonin production and melatonin receptor expression are altered in preeclampsia: New insights into the role of this hormone in pregnancy. *J. Pineal Res.* **2012**, *53*, 417–425. [CrossRef] [PubMed]
30. Reiter, R.J.; Tan, D.X.; Korkmaz, A.; Rosales-Corral, S.A. Melatonin and stable circadian rhythms optimize maternal, placental and fetal physiology. *Hum. Reprod. Update* **2014**, *20*, 293–307. [CrossRef] [PubMed]
31. Chuffa, L.G.A.; Lupi, L.A.; Cucielo, M.S.; Silveira, H.S.; Reiter, R.J.; Seiva, F.R.F. Melatonin Promotes Uterine and Placental Health: Potential Molecular Mechanisms. *Int. J. Mol. Sci.* **2020**, *21*, 300. [CrossRef] [PubMed]
32. Marseglia, L.; D'Angelo, G.; Manti, S.; Reiter, R.J.; Gitto, E. Potential Utility of Melatonin in Preeclampsia, Intrauterine Fetal Growth Retardation, and Perinatal Asphyxia. *Reprod. Sci.* **2016**, *23*, 970–977. [CrossRef] [PubMed]
33. Brown, M.A.; Magee, L.A.; Kenny, L.C.; Karumanchi, S.A.; McCarthy, F.P.; Saito, S.; Hall, D.R.; Warren, C.E.; Adoyi, G.; Ishaku, S. The hypertensive disorders of pregnancy: ISSHP classification, diagnosis & management recommendations for international practice. *Pregnancy Hypertens.* **2018**, *13*, 291–310. [PubMed]
34. Abumaree, M.; Stone, P.; Chamley, L. An in vitro model of human placental trophoblast deportation/shedding. *Mol. Hum. Reprod.* **2006**, *12*, 687–692. [CrossRef] [PubMed]

35. Tong, M.; Kleffmann, T.; Pradhan, S.; Johansson, C.L.; DeSousa, J.; Stone, P.R.; James, J.L.; Chen, Q.; Chamley, L.W. Proteomic characterization of macro-, micro- and nano-extracellular vesicles derived from the same first trimester placenta: Relevance for feto-maternal communication. *Hum. Reprod.* **2016**, *31*, 687–699. [CrossRef]
36. Zhao, M.; Li, Y.; Xu, L.; Hickey, A.; Groom, K.; Stone, P.R.; Chamley, L.W.; Chen, Q. Melatonin prevents preeclamptic sera and antiphospholipid antibodies inducing the production of reactive nitrogen species and extrusion of toxic trophoblastic debris from first trimester placentae. *Placenta* **2017**, *58*, 17–24. [CrossRef] [PubMed]
37. Beriault, D.R.; Werstuck, G.H. Detection and quantification of endoplasmic reticulum stress in living cells using the fluorescent compound, Thioflavin T. *Biochim. Biophys. Acta* **2013**, *1833*, 2293–2301. [CrossRef]
38. Tang, Y.; Chen, Y.; Nursalim, Y.; Groom, K.; Hickey, A.; Chamley, L.; Chen, Q. Endoplasmic reticulum stress occurs in association with the extrusion of toxic extracellular vesicles from human placentae treated with antiphospholipid antibodies. *Clin. Sci.* **2020**, *134*, 459–472. [CrossRef] [PubMed]
39. Eggert, N.; Lund, J. The trapezoidal rule for analytic functions of rapid decrease. *J. Comput. Appl. Math.* **1989**, *27*, 389–406. [CrossRef]
40. Tong, M.; Chen, Q.; James, J.L.; Stone, P.R.; Chamley, L.W. Micro- and Nano-vesicles from First Trimester Human Placentae Carry Flt-1 and Levels Are Increased in Severe Preeclampsia. *Front. Endocrinol.* **2017**, *8*, 174. [CrossRef]
41. Shen, F.; Wei, J.; Snowise, S.; DeSousa, J.; Stone, P.; Viall, C.; Chen, Q.; Chamley, L. Trophoblast debris extruded from preeclamptic placentae activates endothelial cells: A mechanism by which the placenta communicates with the maternal endothelium. *Placenta* **2014**, *35*, 839–847. [CrossRef] [PubMed]
42. Colombo, M.; Raposo, G.; Thery, C. Biogenesis, secretion, and intercellular interactions of exosomes and other extracellular vesicles. *Annu. Rev. Cell Dev. Biol.* **2014**, *30*, 255–289. [CrossRef]
43. Sadovsky, Y.; Ouyang, Y.; Powell, J.S.; Li, H.; Mouillet, J.F.; Morelli, A.E.; Sorkin, A.; Margolis, L. Placental small extracellular vesicles: Current questions and investigative opportunities. *Placenta* **2020**, *102*, 34–38. [CrossRef]
44. Espinosa-Diez, C.; Miguel, V.; Mennerich, D.; Kietzmann, T.; Sanchez-Perez, P.; Cadenas, S.; Lamas, S. Antioxidant responses and cellular adjustments to oxidative stress. *Redox Biol.* **2015**, *6*, 183–197. [CrossRef]
45. Goulopoulou, S.; Davidge, S.T. Molecular mechanisms of maternal vascular dysfunction in preeclampsia. *Trends Mol. Med.* **2015**, *21*, 88–97. [CrossRef] [PubMed]
46. Socaciu, A.I.; Ionut, R.; Socaciu, M.A.; Ungur, A.P.; Barsan, M.; Chiorean, A.; Socaciu, C.; Rajnoveanu, A.G. Melatonin, an ubiquitous metabolic regulator: Functions, mechanisms and effects on circadian disruption and degenerative diseases. *Rev. Endocr. Metab. Disord.* **2020**, *21*, 465–478. [CrossRef] [PubMed]
47. Nakazawa, K.; Kanakura, Y.; Kometani, K.; Iwasaki, S.; Yosimura, Y. Study on melatonin in human and rat placental tissue. *Placenta* **1999**, *20* (Suppl. S1), 467–474. [CrossRef]
48. Valenzuela, F.J.; Vera, J.; Venegas, C.; Pino, F.; Lagunas, C. Circadian System and Melatonin Hormone: Risk Factors for Complications during Pregnancy. *Obstet. Gynecol. Int.* **2015**, *2015*, 825802. [CrossRef]
49. Hobson, S.R.; Wallace, E.M.; Kingdom, J.C.; Hodges, R.J. A Randomized Double-Blinded Placebo-Controlled Intervention Trial of Melatonin for the Prevention of Preeclampsia in Moderate- and High-Risk Women: The MELPOP Trial. *Methods Mol. Biol.* **2018**, *1710*, 347–352. [PubMed]
50. Duckitt, K.; Harrington, D. Risk factors for pre-eclampsia at antenatal booking: Systematic review of controlled studies. *BMJ* **2005**, *330*, 565. [CrossRef] [PubMed]
51. Hannan, N.J.; Binder, N.K.; Beard, S.; Nguyen, T.-V.; Kaitu'u-Lino, T.J.; Tong, S. Melatonin enhances antioxidant molecules in the placenta, reduces secretion of soluble fms-like tyrosine kinase 1 (sFLT) from primary trophoblast but does not rescue endothelial dysfunction: An evaluation of its potential to treat preeclampsia. *PLoS ONE* **2018**, *13*, e0187082. [CrossRef]
52. Lau, S.Y.; Guild, S.J.; Barrett, C.J.; Chen, Q.; McCowan, L.; Jordan, V.; Chamley, L.W. Tumor Necrosis Factor-Alpha, Interleukin-6, and Interleukin-10 Levels are Altered in Preeclampsia: A Systematic Review and Meta-Analysis. *Am. J. Reprod. Immunol.* **2013**, *70*, 412–427. [CrossRef]
53. Chen, L.; Liu, B.; Zhao, H.; Stone, P.; Chen, Q.; Chamley, L. IL-6, TNF [alpha] and TGF [beta] Promote Nonapoptotic Trophoblast Deportation and Subsequently Causes Endothelial Cell Activation. *Placenta* **2010**, *31*, 75–80. [CrossRef] [PubMed]
54. Chen, Q.; Viall, C.; Kang, Y.; Liu, B.; Stone, P.; Chamley, L. Anti-phospholipid antibodies increase non-apoptotic trophoblast shedding: A contribution to the pathogenesis of pre-eclampsia in affected women? *Placenta* **2009**, *30*, 767–773. [CrossRef] [PubMed]
55. Tong, M.; Tsai, B.W.; Chamley, L.W. Antiphospholipid antibodies and extracellular vesicles in pregnancy. *Am. J. Reprod. Immunol.* **2021**, *85*, e13312. [CrossRef]
56. Viall, C.A.; Chen, Q.; Liu, B.; Hickey, A.; Snowise, S.; Salmon, J.E.; Stone, P.R.; Chamley, L.W. Antiphospholipid antibodies internalised by human syncytiotrophoblast cause aberrant cell death and the release of necrotic trophoblast debris. *J. Autoimmun.* **2013**, *47*, 45–57. [CrossRef] [PubMed]
57. Gerasimova, E.M.; Fedotov, S.A.; Kachkin, D.V.; Vashukova, E.S.; Glotov, A.S.; Chernoff, Y.O.; Rubel, A.A. Protein Misfolding during Pregnancy: New Approaches to Preeclampsia Diagnostics. *Int. J. Mol. Sci.* **2019**, *20*, 6183. [CrossRef] [PubMed]
58. Hobson, S.R.; Gurusinghe, S.; Lim, R.; Alers, N.O.; Miller, S.L.; Kingdom, J.C.; Wallace, E.M. Melatonin improves endothelial function in vitro and prolongs pregnancy in women with early-onset preeclampsia. *J. Pineal Res.* **2018**, *65*, e12508. [CrossRef]

cells MDPI

Article

Cerebral Biomarkers and Blood-Brain Barrier Integrity in Preeclampsia

Therese Friis [1,*], Anna-Karin Wikström [1], Jesenia Acurio [2,3], José León [2,3,4], Henrik Zetterberg [5,6,7,8,9], Kaj Blennow [5,6], Maria Nelander [1], Helena Åkerud [10], Helena Kaihola [10], Catherine Cluver [11], Felipe Troncoso [2], Pablo Torres-Vergara [3,12], Carlos Escudero [2,3,†] and Lina Bergman [1,11,13,†]

1 Department of Women's and Children's Health, Uppsala University, 75185 Uppsala, Sweden; anna-karin.wikstrom@kbh.uu.se (A.-K.W.); maria.nelander@kbh.uu.se (M.N.); lina.bergman@kbh.uu.se (L.B.)
2 Vascular Physiology Laboratory, Department of Basic Sciences, Faculty of Sciences, University of Bío-Bío, Chillán 3810178, Chile; jeseniacurio2@gmail.com (J.A.); joseman1115@gmail.com (J.L.); fetronc@gmail.com (F.T.); cescudero@ubiobio.cl (C.E.)
3 Group of Research and Innovation in Vascular Health (GRIVAS Health), Chillán 3810178, Chile; pabltorr@udec.cl
4 Escuela de Enfermería, Facultad de Salud, Universidad Santo Tomás, Los Ángeles 4441171, Chile
5 Department of Psychiatry and Neurochemistry, Institute of Neuroscience and Physiology, The Sahlgrenska Academy at the University of Gothenburg, 43180 Mölndal, Sweden; henrik.zetterberg@clinchem.gu.se (H.Z.); kaj.blennow@neuro.gu.se (K.B.)
6 Clinical Neurochemistry Laboratory, Sahlgrenska University Hospital, 43180 Mölndal, Sweden
7 Department of Neurodegenerative Disease, UCL Queen Square Institute of Neurology, Queen Square, London WC1E 6BT, UK
8 UK Dementia Research Institute at UCL, London WC1E 6BT, UK
9 Hong Kong Center for Neurodegenerative Diseases, Clear Water Bay, Hong Kong, China
10 Department of Immunology, Genetics and Pathology, Uppsala University, 75185 Uppsala, Sweden; helena.akerud@gynhalsan.se (H.Å.); helena.kaihola@outlook.com (H.K.)
11 Department of Obstetrics and Gynecology, Stellenbosch University, Cape Town 7500, South Africa; cathycluver@hotmail.com
12 Department of Pharmacy, Faculty of Pharmacy, University of Concepción, Concepción 4070386, Chile
13 Department of Obstetrics and Gynecology, Gothenburg University, 41650 Gothenburg, Sweden
* Correspondence: therese.friis@kbh.uu.se; Tel.: +46-18-611-6613
† These authors contributed equally to this work.

Citation: Friis, T.; Wikström, A.-K.; Acurio, J.; León, J.; Zetterberg, H.; Blennow, K.; Nelander, M.; Åkerud, H.; Kaihola, H.; Cluver, C.; et al. Cerebral Biomarkers and Blood-Brain Barrier Integrity in Preeclampsia. *Cells* **2022**, *11*, 789. https://doi.org/10.3390/cells11050789

Academic Editors: Junie P. Warrington, Ana T. Palei, Mark W. Cunningham and Lorena M. Amaral

Received: 27 December 2021
Accepted: 20 February 2022
Published: 24 February 2022

Abstract: Cerebral complications in preeclampsia contribute substantially to maternal mortality and morbidity. There is a lack of reliable and accessible predictors for preeclampsia-related cerebral complications. In this study, plasma from women with preeclampsia (n = 28), women with normal pregnancies (n = 28) and non-pregnant women (n = 16) was analyzed for concentrations of the cerebral biomarkers neurofilament light (NfL), tau, neuron-specific enolase (NSE) and S100B. Then, an in vitro blood–brain barrier (BBB) model, based on the human cerebral microvascular endothelial cell line (hCMEC/D3), was employed to assess the effect of plasma from the three study groups. Transendothelial electrical resistance (TEER) was used as an estimation of BBB integrity. NfL and tau are proteins expressed in axons, NSE in neurons and S100B in glial cells and are used as biomarkers for neurological injury in other diseases such as dementia, traumatic brain injury and hypoxic brain injury. Plasma concentrations of NfL, tau, NSE and S100B were all higher in women with preeclampsia compared with women with normal pregnancies (8.85 vs. 5.25 ng/L, $p < 0.001$; 2.90 vs. 2.40 ng/L, $p < 0.05$; 3.50 vs. 2.37 µg/L, $p < 0.001$ and 0.08 vs. 0.05 µg/L, $p < 0.01$, respectively). Plasma concentrations of NfL were also higher in women with preeclampsia compared with non-pregnant women ($p < 0.001$). Higher plasma concentrations of the cerebral biomarker NfL were associated with decreased TEER ($p = 0.002$) in an in vitro model of the BBB, a finding which indicates that NfL could be a promising biomarker for BBB alterations in preeclampsia.

Keywords: blood-brain barrier; preeclampsia; pregnancy; in vitro studies; cerebral biomarkers; NfL; tau; NSE; S100B

1. Introduction

Preeclampsia affects 3–5% of all pregnancies and is one of the most common causes of maternal and perinatal morbidity and mortality [1]. Preeclampsia is defined as de novo hypertension after 20 weeks of gestation accompanied by signs of maternal organ dysfunction, such as renal insufficiency, liver dysfunction, neurological features, hematological complications or fetal growth restriction as a sign of uteroplacental dysfunction [2,3]. Annually, more than 70,000 maternal deaths are associated with hypertensive disorders of pregnancy, where cerebral complications due to preeclampsia, such as eclampsia, cerebral edema and cerebral hemorrhage are leading causes of maternal death [3–5]. In the majority of cases, eclampsia is preceded by hypertension and in some cases by neurological symptoms; however, eclampsia can occur before the onset of hypertension and often in the absence of premonitory symptoms [6]. There is a lack of objective biomarkers for cerebral complications to preeclampsia and the underlying pathophysiology remains partly unknown.

The cerebral biomarkers neurofilament light (NfL), tau, neuron-specific enolase (NSE) and S100B are all present within cells of the central nervous system, the last of these in the astrocytic endfeet contributing to the neurovascular unit of the blood–brain barrier (BBB) [7]. They have all been extensively studied in the context of hypoxic or traumatic brain injury, different types of dementia and epilepsy. NfL and tau are both axonal proteins used as biomarkers for neurodegenerative disease [8,9]. In patients with traumatic brain injury, increased concentrations of NfL have been detected in both cerebrospinal fluid (CSF) and peripheral blood, compared with controls [10]. NSE is a glycolytic enzyme mainly found in neurons [11], which has proved useful in prognostication of patients with cardiac arrest and hypoxic ischemic encephalopathy [12]. Similarly, tau has been identified to predict six-month outcomes regarding cerebral symptoms after cardiac arrest [13]. S100B and NSE have both been used as predictors for poor neurological outcome after traumatic brain injury [14]. S100B might enter the circulation after an isolated BBB injury with loss of BBB integrity, even without injury to the brain parenchyma [15].

Previous studies exploring cerebral biomarkers for preeclampsia have reported increased plasma concentrations before the onset of disease [16–18], during disease [19,20] and one year postpartum [21] in women with preeclampsia compared with women with normal pregnancies. Studies have also reported higher S100B plasma concentrations in women with severe preeclampsia and eclampsia compared with women with preeclampsia without severe features [20,22]. In addition, neurological symptoms, such as visual disturbances, have been correlated with increased S100B plasma concentrations [19,23].

The BBB is a highly restrictive and specialized neurovascular network that isolates and protects the brain parenchyma from potential harmful molecules present in the systemic circulation. Disruption of the BBB in preeclampsia and eclampsia has been studied in both laboratory and clinical settings, mainly with animal models [24], or ex vivo animal models exposed to plasma from women with preeclampsia. Cerebral edema, commonly seen in eclampsia and sometimes in severe preeclampsia, may partly be due to disruption of the BBB, resulting in increased BBB permeability and passage of fluid into the brain parenchyma [25,26]. Up until recently there has been a paucity of studies assessing the human BBB in preeclampsia, due to difficulties in studying the BBB in a clinical setting. However, our research group has recently presented promising results on the human cerebral microvascular endothelial cell line (hCMEC/D3) [27–29] as a new in vitro model of the BBB in preeclampsia research [30,31].

It is still not known if cerebral biomarkers are useful in reflecting BBB alterations in preeclampsia and to our knowledge, cerebral biomarkers have never before been correlated to any measures of BBB integrity in preeclampsia. Therefore, the aim of this study was to correlate plasma concentrations of cerebral biomarkers in women with preeclampsia, women with normal pregnancies and non-pregnant women with BBB integrity, measured as changes in transendothelial electrical resistance (TEER) in the hCMEC/D3 in vitro model.

2. Materials and Methods

2.1. Study Population

The study population consisted of pregnant women with preeclampsia (n = 28) diagnosed according to the International Society for Studies on Hypertension in Pregnancy (ISSHP) 2018 guidelines [3]. Diagnostic criteria were de novo hypertension (systolic blood pressure (SBP) > 140 and/or diastolic blood pressure (DBP) > 90 mmHg) in combination with significant proteinuria (protein level > 300 mg/24 h or urine dipstick > 1+) after 20 gestational weeks. Although clinical preeclampsia can be diagnosed in the absence of proteinuria, with other signs of maternal organ dysfunction present, it has been recommended that proteinuria is used for patients enrolled in scientific research to ensure more specificity around the diagnosis [2]. These were also the common criteria used for preeclampsia diagnosis in the obstetric clinic at the time when cases were enrolled in the study.

Severe preeclampsia was defined according to the guidelines from the International Society for the study of Hypertension in Pregnancy (ISSHP) [32]. Criteria were an SBP $\geq$ 160 and/or DBP $\geq$ 110, development of HELLP syndrome, eclampsia or other severe organ manifestations.

As controls, women with normal pregnancies (n = 28), matched for maternal age and gestational length at inclusion, and non-pregnant women (=16) were recruited. The definition of a normal pregnancy required that the woman remained normotensive throughout her pregnancy. The pregnancy also had to result in term delivery (gestational week $\geq$ 37) of an infant with normal birth weight ($\pm$2 standard deviations of the mean birth weight for gestational age and sex) [33]. Women with prior hypertensive disorder in pregnancy were not included in any of the control groups. Further, none of the study groups included women with chronic hypertension, diabetes mellitus or chronic kidney disease.

The women were all recruited from the obstetric ward or the outpatient clinic at Uppsala University Hospital, Sweden between 2013 and 2016 [34].

Uppsala Ethical Review Board approved the study and informed consent was obtained from all participants.

2.2. Sample Collection

Plasma samples were collected in Vacutainer tubes with lithium heparin (Becton, Dickinson, Franklin Lakes, NJ, USA) within four hours of study inclusion. The samples were centrifuged for 10 min at 1500 *g* and the plasma was immediately frozen at −70 °C for later analysis. Thawed plasma was used for analysis of the cerebral biomarkers. NfL and tau concentrations were measured in Mölndal, Sweden, whereas NSE and S100B concentrations were measured in Uppsala, Sweden. Further, frozen plasma was shipped to Chillán, Chile, thawed and added to the BBB model for TEER measurements.

2.3. Biomarker Assay

Plasma NfL concentration was measured with an in-house single molecule array (Simoa) method, whilst plasma tau concentration was measured with the Human Total Tau 2.0 kit and the Simoa platform (Quanterix, Billerica, MA, USA), both previously described in detail [8,35]. Laboratory technicians, who were blinded to clinical data, performed measurements in one round of experiments, using one batch of reagents. Two quality-control samples were run in duplicates in the beginning and the end of each run, showing coefficients of variance (CVs) for intermediate precision of 6.0% at 8.5 pg/mL and 5.1% at 121 pg/mL for NfL, whereas CVs were 7.3% at 32.2 pg/mL and 7.0% at 7.5 pg/mL for tau.

Plasma NSE and S100B concentrations were measured by an enzyme-linked immunosorbent assay (ELISA). A commercially available kit (Sangtec 100 Elisa, Diasorin, MN, USA) was used and the samples were run according to the manufacturer's recommendation. The intra- and inter-assay coefficients of variation were 2.8% and 4.3% and 4.6% and 3.1%, respectively, for NSE and S100B.

2.4. hCMEC/D3 In Vitro Model

The hCMEC/D3 cell line (Merck Millipore, Darmstadt, Germany) [27] was used for the in vitro experiments. Monolayers of cells were seeded on semipermeable plates coated with rat-tail type I collagen (Discovery Labware, Bedford, MA, USA) at a density of 20,000 cells/well. Medium EndoGro MV Supplement Kit (Merck Millipore) was used as a culturing medium, and cells were incubated at 37 °C, 5% CO_2. Once cells reached 100% of confluence, and a TEER value larger than 20 Ωcm^2, they were used for experiments. In addition, six hours prior to experiments the culture medium was replaced by a medium without growth supplements. Cultured hCMEC/D3 cells were treated (1%, v/v, 12 h) with thawed plasma either from women with preeclampsia, women with normal pregnancies or non-pregnant women. An epithelial Volt/Ohm meter (EVOM2, World Precision Instruments, Sarasota, FL, USA) with two electrodes in each compartment was used to measure the TEER. Measurements were performed both before adding plasma (baseline), and after incubation with plasma. Delta-TEER values were calculated by subtracting basal TEER values from TEER values after exposure to plasma.

Confirmatory experiments of TEER and cell permeability to high-molecular weight fluorescent dye (Fluorescein-5-isothiocyanate FITC-dextran 70 kDa) were performed as previously reported [30] using randomly selected plasmas from women with preeclampsia (n = 12), and women with normal pregnancies (n = 13).

The hCMEC/D3 cell line was used in passages 5 to 10. Individual plasmas were used in duplicate experimental replicates. None of the plasmas were excluded at the final analysis. More detailed experimental conditions are described in a previous publication [30].

2.5. Statistical Analyses

Background characteristics were presented as medians with interquartile range (IQR) and numbers with percentage (%) as appropriate. Groups were compared by one-way ANOVA, Chi-Square or Kruskal–Wallis test as appropriate, and the significance level was set at 0.05.

Plasma concentrations of the cerebral biomarkers (NfL, tau, NSE and S100B), TEER and permeability were presented as medians with interquartile range (IQR). Differences between groups were compared by non-parametric analysis by Kruskal–Wallis test and pairwise comparisons by Mann-Whitney U-test. In case of statistical significance, a Bonferroni post-hoc test was used.

Associations between concentrations of the cerebral biomarkers NfL, tau, NSE and S100B, and TEER values were analyzed with a cumulative probability model [36] allowing for different, possibly non-linear, associations in the three groups of women. The model was further adjusted for baseline TEER, and for the confounders maternal age, parity and BMI. The confounders were identified with a DAG (directed acyclic graph). For this fully flexible model a p-value for all non-linear terms of 0.835 was calculated. The model was subsequently adjusted to a model where all associations were linear. Data and statistical analyses were performed with IBM SPSS Statistics for Windows, Version 25.0 (IBM Corp, Armonk, NY, USA), GraphPad Prism 6.00 (GraphPad Software, San Diego, CA, USA) and R version 3.6.1 with the add-on package rms [37,38].

Subgroup analyses of the women with preeclampsia were performed with regards to clinical symptoms in association with cerebral biomarkers and change in TEER, i.e., the difference between measured TEER before and after adding plasma to the model. Analyses were performed with the Mann-Whitney U-test.

3. Results

3.1. Background Characteristics

Clinical characteristics of the study population are described in Tables 1 and 2. There was no difference in maternal age or gestational length at inclusion between the pregnant study groups. Women with preeclampsia had a higher early-pregnancy body mass index (BMI) and were more often nulliparous compared with women with normal pregnancies and non-pregnant women. In the group of women with preeclampsia most women had antihypertensive treatment (79%), and at inclusion approximately one third of them had severe headaches and/or visual disturbances. At delivery, 57% of the women had developed severe preeclampsia, with a recorded SBP ≥160 and/or DBP ≥110 [32]. None of the women developed eclampsia, neurological complications or other severe organ manifestations, and none required magnesium sulphate prophylaxis against imminent eclampsia.

Table 1. Clinical characteristics of the study population.

Clinical Characteristics	Preeclampsia (n = 28)	Normal Pregnancy (n = 28)	Non-Pregnant (n = 16)	p-Values
Maternal age (years)	28 (25–32)	33 (29–35)	27 (24–36)	n.s.
Nulliparous	23 (82%)	10 (36%)	9 (56%)	<0.001
BMI	26 (23–29)	24 (22–26)	22 (20–25)	<0.001
At inclusion				
Gestational week	35 (29–37)	35 (27–38)		n.s.
Blood pressure (mmHg)				
Systolic	150 (140–160)	110 (110–120)	110 (110–118)	<0.001
Diastolic	98 (86–100)	70 (60–75)	70 (65–70)	<0.001
MAP	113 (107–120)	83 (77–90)	83 (80–86)	<0.001
Neurological symptoms (yes)				
Headache	18 (64%)	4 (14%)	3 (19%)	<0.001
Severe headache (VAS ≥ 5)	10 (36%)	0	0	<0.001
Visual disturbances	10 (36%)	0	0	<0.001
Headache & visual disturbances	10 (36%)	0	0	<0.001
Any neurological symptom	20 (71%)	4 (14%)	3 (19%)	<0.001
TEER (Ωcm^2)				
Baseline value	34.9(29.7–38.8)	33.7 (29.7–42.6)	29.2 (25.4–38.8)	n.s.
After plasma exposure	22.9 (18.1–27.5)	27.1 (18.8–35.5)	23.8 (21.6–29.3)	n.s.
Δ-TEER	11.9 (8.5–14.8)	7.6 (3.7–11.9)	5.8 (2.0–8.0)	<0.001

Data are presented as medians (IQR) or numbers (%). Abbreviations: BMI, body mass index; MAP, mean arterial pressure; VAS, visual analogue scale; TEER, transendothelial electrical resistance; Δ-TEER, the difference between TEER values before and after exposure to plasma; n.s., non-significant.

Table 2. Characteristics of the women with preeclampsia.

Clinical Characteristics	Preeclampsia (n = 28)
Gestational week at preeclampsia diagnosis	35 (22–41)
Severe preeclampsia at inclusion, n	10 (36%)
Blood pressure medication at inclusion, n	22 (79%)
Magnesium treatment, n	0 (0%)
Gestational week at delivery	35 (25–41)
Severe preeclampsia at delivery, n	16 (57%)

Numbers are presented as median (range) or numbers (%). Severe preeclampsia is defined according to the guidelines from the International Society for the study of Hypertension in Pregnancy (ISSHP).

3.2. Plasma Concentrations of NfL, Tau, NSE and S100B

Compared with women with normal pregnancies, plasma concentrations of the cerebral biomarkers were higher in plasma from women with preeclampsia; NfL (5.25 ng/L, IQR 3.93–7.63 ng/L vs. 8.85 ng/L, IQR 6.78–12.65 ng/L, $p < 0.001$); tau (2.40 ng/L, IQR 1.80–2.58 ng/L vs. 2.90 ng/L, IQR 2.40–4.35 ng/L, $p < 0.05$); NSE (2.37 μg/L, IQR 1.93–2.85 μg/L vs. 3.50 μg/L, IQR 2.84–4.55 μg/L, $p < 0.001$) and S100B (0.05 μg/L, IQR 0.03–0.08 μg/L vs. 0.08 μg/L, IQR 0.05–0.10 μg/L, $p < 0.01$) (Figure 1a–d). In addition, plasma concentrations of NfL were also higher in women with preeclampsia compared with non-pregnant women (8.85 ng/L, IQR 6.78–12.65 ng/L vs. 5.65 ng/L, IQR 4.83–6.40 ng/L, $p < 0.001$) (Figure 1a).

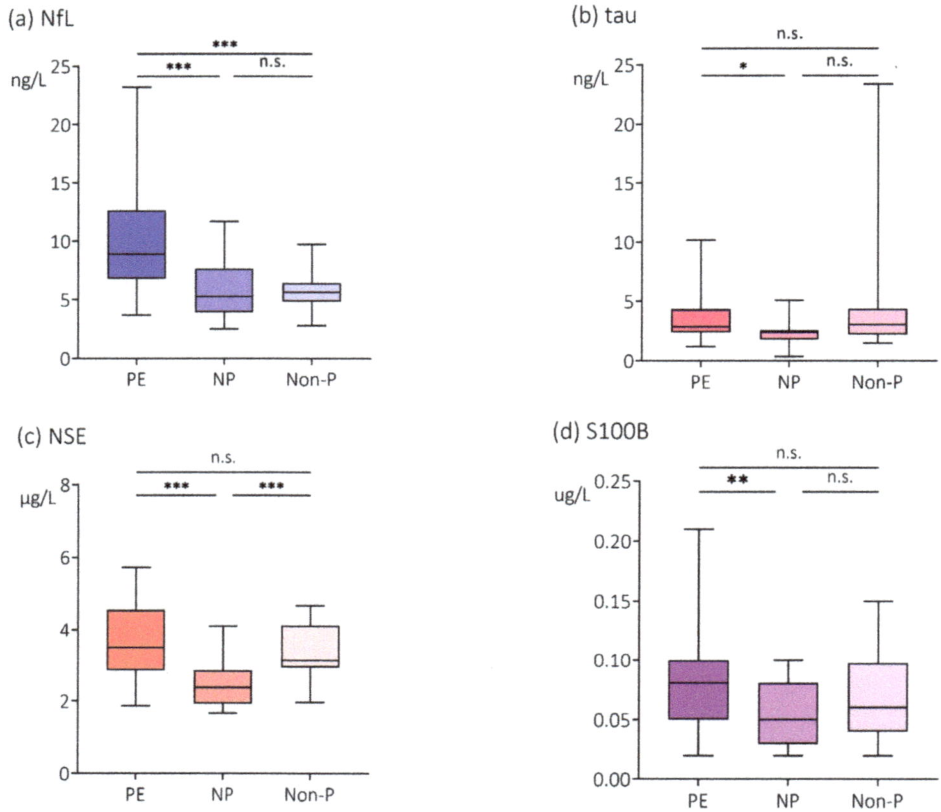

Figure 1. Plasma concentrations of (**a**) NfL, (**b**) tau, (**c**) NSE and (**d**) S100B. Plasma concentrations of the cerebral biomarkers (**a**) neurofilament light (NfL), (**b**) tau, (**c**) neuron specific enolase (NSE) and (**d**) S100B in women with preeclampsia (PE), women with normal pregnancies (NP) and non-pregnant women (Non-P). Values are represented by medians with interquartile range (IQR). Pairwise comparisons by Mann-Whitney U-test, Bonferroni correction. * $p < 0.05$; ** $p < 0.01$; *** $p < 0.001$; n.s. = non-significant.

3.3. *Effects of Plasma on TEER in an In Vitro BBB Model*

The effects of plasma on changes in TEER, measured in the hCMEC/D3 in vitro model, in this population have been published previously [30]. Baseline TEER (i.e., TEER prior to exposure to plasma) was not statistically different among cell monolayers. After exposure to plasma, a significantly larger reduction in TEER was detected in cell monolayers exposed to plasma from women with preeclampsia compared with plasma from women with normal pregnancies and from non-pregnant women [30].

For this manuscript a re-confirmation analysis of previously published data was performed, where plasma was randomly included from the group of women with preeclampsia (PE, n = 12), and the women with normal pregnancies (NP, n = 13). Results are shown in Figure 2.

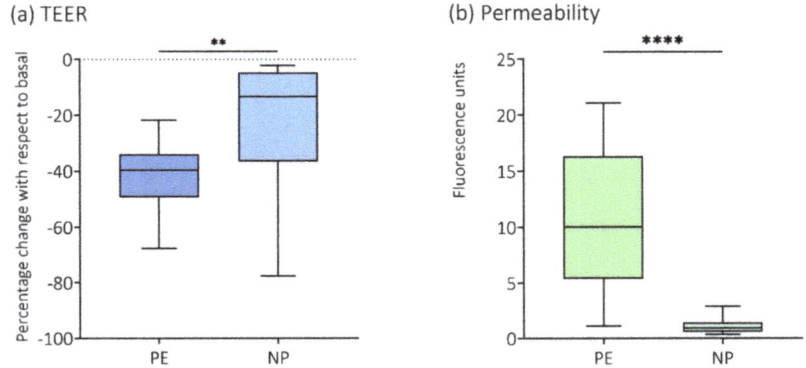

Figure 2. Re−confirmation analysis of TEER and permeability to 70 kDa Dextran. Plasma from a small, randomly chosen sample of the women with preeclampsia (PE, n = 12) and the women with normal pregnancies (NP, n = 13) was analyzed for re−confirmation of previously published data on (**a**) transendothelial electrical resistance (TEER) and (**b**) permeability to 70 kDa fluorescein isothiocyanate (FITC)−dextran [30]. ** $p < 0.01$; **** $p < 0.0001$.

3.4. Association of NfL, Tau, NSE and S100B with TEER

The associations between cerebral biomarkers and TEER are presented in Figure 3. NfL was the only biomarker that showed a negative linear association with TEER, i.e., higher plasma concentrations of NfL correlated with a larger reduction in TEER ($p < 0.01$) after exposure to plasma. Data did not support associations between tau, NSE or S100B and TEER.

3.5. Subgroup Analyses of Neurological Symptoms in the Women with Preeclampsia

In the group of women with preeclampsia, 10 out of 28 women reported severe headaches (VAS $\geq$ 5), and 10 women reported visual disturbances.

Higher plasma concentrations of NfL were found in the women with preeclampsia that expressed severe headache than in the women with mild or no headache (11.65 ng/L, IQR 9.58–15.63 ng/L vs. 7.40 ng/L, IQR 5.93–9.93 ng/L, $p = 0.024$). The plasma concentrations of the other biomarkers (tau, NSE and S100B) did not relate to severity of headache. No associations were found between reported visual disturbances and any of the biomarkers. When comparing the women who had a combination of severe headache and visual disturbances with women who did not have this combination of symptoms, there were no differences in plasma concentrations of the respective biomarkers.

Women with preeclampsia that expressed severe headache (VAS $\geq$ 5) exhibited a greater reduction of TEER than the women with mild or no headache (10.17, IQR 7.32–10.45 vs. 7.74, IQR 5.65–9.42, $p = 0.040$). No associations were found between reported visual disturbances and change in TEER. However, a greater reduction of TEER was seen in the women who had a combination of severe headache and visual disturbances, compared with the women without this combination of symptoms (10.32, IQR 7.82–11.72 vs. 7.81, IQR 5.77–9.56, $p < 0.021$).

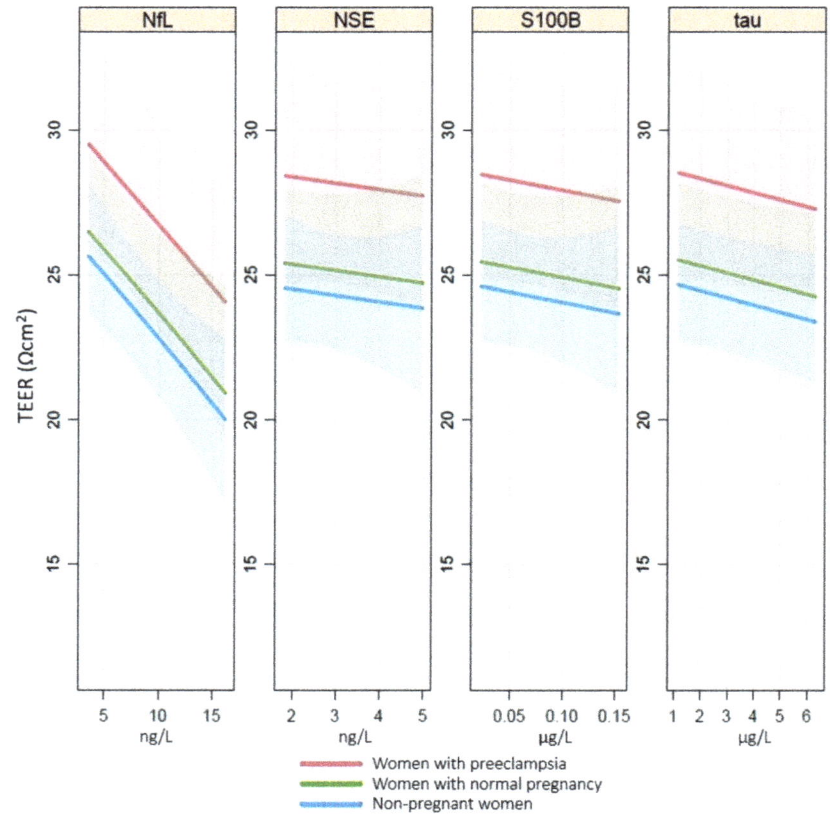

Figure 3. Association of TEER with NfL, tau, NSE and S100B. Associations between transendothelial electrical resistance (TEER (Ωcm^2)) and cerebral biomarkers neurofilament light (NfL (ng/L)), tau (ng/L), neuron-specific enolase (NSE (µg/L)) and S100B (µg/L) in plasma analyzed by a cumulative probability model stratified by group and adjusted for baseline TEER, body mass index, parity and maternal age. NfL $p < 0.01$, tau p = n.s., NSE p = n.s., S100B p = n.s.

4. Discussion

4.1. Main Findings

In this study we demonstrated that circulating concentrations of NfL was associated with decreased TEER. In addition, we corroborated previous findings of higher plasma concentrations of the cerebral biomarkers NfL, tau, NSE and S100B in women with preeclampsia compared with women with normal pregnancies. Concentrations of NfL were also higher in women with preeclampsia compared with non-pregnant women. Further, women with preeclampsia that expressed severe headache had higher concentrations of plasma NfL and a greater reduction of TEER, compared with women with lesser symptoms. There was also a greater reduction of TEER in women with the combination of severe headache and visual disturbances.

4.2. Research Implications

Several previous studies have reported increased circulating concentrations of cerebral biomarkers in preeclampsia [16–18,20,22], findings that are also supported in this study. A novel finding, however, is that plasma concentrations of NfL were also significantly higher in plasma in women with preeclampsia compared with non-pregnant women. We did not find a difference in plasma concentrations of any of the biomarkers comparing women with normal pregnancies with non-pregnant women. Hence, pregnancy alone does not seem to have a major impact on circulating levels of these cerebral biomarkers.

NfL, tau, NSE and S100B have repeatedly been reported useful as circulating biomarkers of cerebral injury in other neurological disorders such as neurodegenerative disease, traumatic brain injury and epilepsy [8–10,14,15,39]. Previous studies of these biomarkers in relation to BBB injury have, however, most often evaluated biomarker concentrations in male or mixed populations with other neurological conditions than preeclampsia. The results from these studies seem to be somewhat conflicting regarding whether they support a BBB injury or not [40,41]. Our findings in a female population are suggestive of BBB alterations in terms of loss of ionic tightness, and possibly neuroaxonal injury, reflected by higher plasma concentrations of NfL in women with preeclampsia. We propose this by an association between higher circulating NfL concentrations in plasma and decreased TEER in an in vitro model of the BBB. However, this finding could have different underlying causes [15]. It could mean that intracerebral concentrations of NfL are normal, but are secreted over the BBB to a greater extent due to increased permeability, and are subsequently detected in higher concentrations in plasma. It could also reflect that NfL is released in higher concentrations in the central nervous system (CNS) following neuroaxonal injury, which might in turn cause the BBB to become more permeable or render increased NfL concentrations in plasma through alternative transport mehcanisms [15]. Alternatively, it may reflect a secondary breakdown of the BBB caused by circulating harmful molecules associated with preeclampsia, e.g., anti-angiogenic or inflammatory factors, including small extracellular vesicles [31], that allow NfL to cross from the brain to the circulation. It could also be caused by a combination of these proposed theories.

However, our findings could not support that higher plasma concentrations of NSE, S100B and tau had a significant association to decreased TEER. Thus, alternative explanations, such as other transport pathways across the BBB, dose-response-dependent passage, increased extracerebral existence or hemolysis in serum/plasma samples, have to be considered as causes of the higher concentrations of these biomarkers detected in women with preeclampsia. Furthermore, an in vitro model also has its own limitations and may not perfectly reflect in vivo conditions.

Recently published results from our research group reported increased concentrations of NfL in both plasma and CSF in women with preeclampsia compared with normotensive pregnant women. In addition, there was a strong correlation between concentrations of NfL in plasma and CSF [42]. Previous studies of women with preeclampsia have not been able to establish whether a BBB impairment exists in these women or not [43]; however, human studies are few on this subject.

Models for studying the brain and BBB function in preeclampsia that have found evidence of BBB injury have traditionally included in vitro models using animal endothelial cells, along with studies of brain hemodynamic regulation in humans and different preeclampsia-mimicking models in animals in vivo [24–26,29,44–46]. Recently our research group introduced the use of the hCMEC/D3 cell line as a means of studying the effect of preeclampsia on the BBB. We demonstrated a reduction of TEER, and increased permeability to 70 kDa FITC-dextran over the BBB, when a monolayer of cells was exposed to plasma from women with preeclampsia and compared with plasma from women with normal pregnancies and non-pregnant controls [30]. The expression of tight junction proteins (zonula occludens-1 (ZO-1) and occludin) and phosphorylation of two tyrosine residues of VEGFR2 (pY951 and pY1175) were also explored, with no changes in the mRNA expression of the tight junction proteins. However, changes in their localization morphology or function were not investigated [30].

For this study we re-confirmed findings of a reduction in TEER and increased permeability in cells exposed to plasma from women with preeclampsia (Figure 2). In addition, we examined expression of tight junction protein claudin-5. This was performed as a sole experiment with only four randomly chosen plasma samples. Brain endothelial cells exposed to plasma from women with preeclampsia demonstrated a reduced protein abundance of claudin-5 in the cell membrane, whereas it was enhanced in the cytoplasmic fraction (Supplementary Material Figure S1).Characteristically, brain endothelial cells express higher levels of tight junction proteins than peripheral endothelial cells [47]. The key components of intercellular tight junctions are the transmembrane proteins occludin, claudin and junctional adhesion molecules (JAMs), which form complex strands that govern the permeability characteristics of the paracellular route [48,49]. Among them, claudins are directly responsible for the selective permeability of tight junctions [50,51]. There are 27 different claudin isoforms found in mammals [52]. Present in the cell membrane, claudin-5 increases TEER, primarily by decreasing the permeability of cations [52,53]. In order to maintain this restricted diffusion pathway, tight junctions are linked to the cytoplasmic zonula occludens proteins that provide a structural bridge to the actin cytoskeleton. Consequently, the degree of tightness is determined by the interactions of tight junction family members on endothelial cells.

Pre-clinical studies show conflicting results regarding the involvement of these tight junction proteins on the increased BBB permeability seen in preeclampsia models [54–56]. One study found that despite exhibiting increased water content and increased BBB permeability in the anterior cerebrum in a preeclampsia-like syndrome in rats, the protein expression of claudin-1, ZO-1 and occludin was not altered in that part of the brain. However, an upregulation of claudin-1 was detected in the posterior cerebrum of these animals, where brain water content was not altered [56].

In this manuscript, we report a reduction in the availability of claudin-5 in the cell membrane of brain endothelial cells exposed to plasma from women with preeclampsia. This may constitute an underlying mechanism for the reduction in TEER under these experimental conditions.

To our knowledge, no previous studies have explored the association between circulating concentrations of cerebral biomarkers and BBB alterations in women with preeclampsia. However, several studies have established a correlation between serum or plasma concentrations of S100B, which as of now is the most explored of the cerebral biomarkers in preeclampsia, and severity of disease and/or neurological symptoms [19,20,22,23].

4.3. Strengths and Limitations

Strengths in our study include the use of a human in vitro model of the BBB, as opposed to strictly animal-based models or the use of human plasma in animal models [24], to correlate BBB alterations in preeclampsia with peripheral biomarkers. This might reflect the human BBB to a higher extent compared with previous animal studies. Another strength is the well-characterized cohort of women and simultaneous measurements of cerebral biomarkers and BBB-assessment in the same woman.

A limitation of our study is that only one third of the women with preeclampsia in our cohort reported neurological symptoms, in the form of headaches or visual disturbances, and none developed severe cerebral complications [34]. The women in this study also underwent brain MRI scans, as part of previously published studies [34,57,58]. None of the women exhibited cerebral edema on their scans, which might reflect the fact that none of them suffered from severe neurological complications to preeclampsia, such as overt cerbral edema. Nevertheless, there were findings of decreased concentrations of magnesium (Mg) on magnetic spectroscopy and decreased perfusion in the caudate nucleus in the women with preeclampsia. Since the study population did not fully reflect the degree of CNS pathophysiology that can occur in preeclampsia, our results are only generalizable to preeclampsia without severe neurological engagement.

Unfortunately, there were no CSF samples available for this population. Measurements of NfL in CSF, and also the CSF/serum albumin ratio, in the same women, could possibly have helped address what the plasma NfL finding represents: neuroaxonal injury and/or leakage across the BBB.

For TEER measurements in this study we used a static in vitro model based on a monoculture. Future studies should include adaptations of the in vitro model to enhance the tightness of the BBB in order to better reflect in vivo conditions [27]. For example, co-culturing of hCMEC/D3 monolayers with astrocytes has been shown to increase TEER from 30 to 60 Ωcm^2 [59], whereas exposure of hCMEC/D3 monolayers to a pulsatile flow in a capillary cartridge system resulted in TEER values of 1000–1200 Ωcm^2 [60].

In this last regard, physical and technical parameters, such as setup, culture medium viscosity, electrodes and other instruments used, have been demonstrated to influence the measurement of TEER in BBB models. This may result in great variation of TEER results. Drawbacks of methods for TEER measurements have been further discussed by Vigh et al. [61]. To avoid these potential experimental pitfalls, all in vitro experiments were run in a blinded manner for the groups of plasma. In addition, analyses were performed head-to-head using plasmas from the three different groups, as previously reported [30].

4.4. Future Perspectives

Manifestations of acute cerebral complications of preeclampsia, such as eclampsia and cerebral hemorrhage or edema, are difficult to predict. A circulating biomarker, alone or in combination with other clinical data, that could identify women with preeclampsia at risk of such events, may in the future contribute to the prevention of severe cerebral complications. This would allow for a more customized management of women at high risk, e.g., by suggesting more aggressive treatment with antihypertensive drugs, administration of neuroprotective treatment such as $MgSO_4$ [46], and/or aid in decisions about timing of delivery. Here we suggest that NfL could be a promising biomarker for BBB alterations and/or axonal injury in preeclampsia. However, it needs to be further explored in a population presenting with cerebral complications of the disease, and additional methods to assess BBB permeability and function should also be considered.

5. Conclusions

We demonstrated that plasma concentrations of NfL had a negative association with TEER in an in vitro model of the BBB. This finding supports the role of NfL as a promising biomarker to reflect cerebral involvement in women with preeclampsia. Further, in women with preeclampsia we found that those with severe headache, a possible sign of cerebral involvement, had both higher concentrations of plasma NfL and a greater reduction of TEER, compared with those without severe headache.

Prospective studies should be conducted to confirm the role of NfL as a biomarker for BBB alterations and/or neuroaxonal injury in women with preeclampsia, preferably in a population with manifest neurological complications.

Supplementary Materials: The following supporting information can be downloaded at: https://www.mdpi.com/article/10.3390/cells11050789/s1, Figure S1: Reduced protein abundance of claudin-5 in cell membrane of brain endothelial cells exposed to plasma from women with preeclampsia [62].

Author Contributions: C.E. and L.B. conceptualized the study. L.B., T.F., C.E. and A.-K.W. conducted the statistical analyses. T.F., L.B., A.-K.W. and C.E. wrote the manuscript. M.N. collected the data. J.A. and J.L. performed the BBB experiments. P.T.-V. was a consultant for the in vitro BBB experiments. H.Z., K.B., H.Å. and H.K. performed the biomarker analyses. F.T. performed confirmatory experiments regarding TEER and permeability, and claudin-5. C.C. helped supervise the project. All authors have read and agreed to the published version of the manuscript.

Funding: This research was funded by the Swedish foundation for international cooperation in research and higher education (STINT) for collaboration between Chile and Sweden (Saving maternal brains; MG2019-8462); The Swedish Research Council (Saving maternal brains; 2020-01640); The Swedish Brain Foundation (Saving maternal brains; FO2019-0128). L.B. is supported by the Swedish Society of Medicine (SLS-934559). C.E. is financially supported by Fondecyt (1200250). T.F. received funding from the Birth Foundation.

Institutional Review Board Statement: Ethical approval was obtained from Uppsala Ethical Review Board (protocol number 2012/087, approval of amendment: protocol number 2012/087/1).

Informed Consent Statement: Informed consent was obtained from all subjects involved in the study.

Data Availability Statement: The data presented in this study will be anonymized and made available on request from the corresponding author, after approval from the Uppsala Ethical Review Board. The data are not publicly available due to secrecy.

Acknowledgments: The authors would like to thank the researchers from GRIVAS health for their valuable input. We would also like to thank Erik Lampa, Uppsala Clinical Research Center, for valuable statistical support.

Conflicts of Interest: H.Z. has served at scientific advisory boards and/or as a consultant for Abbvie, Alector, Eisai, Denali, Roche, Wave, Samumed, Siemens Healthineers, Pinteon Therapeutics, Nervgen, AZTherapies, CogRx and Red Abbey Labs, has given lectures in symposia sponsored by Cellectricon, Fujirebio, Alzecure and Biogen, and is a co-founder of Brain Biomarker Solutions in Gothenburg AB (BBS), which is a part of the GU Ventures Incubator Program. The funders had no role in the design of the study; in the collection, analyses, or interpretation of data; in the writing of the manuscript, or in the decision to publish the results. The other authors declare no conflict of interest.

References

1. Chappell, L.C.; Cluver, C.A.; Kingdom, J.; Tong, S. Pre-eclampsia. *Lancet* **2021**, *398*, 341–354. [CrossRef]
2. Tranquilli, A.L.; Dekker, G.; Magee, L.; Roberts, J.; Sibai, B.M.; Steyn, W.; Zeeman, G.G.; Brown, M.A. The classification, diagnosis and management of the hypertensive disorders of pregnancy: A revised statement from the ISSHP. *Pregnancy Hypertens.* **2014**, *4*, 97–104. [CrossRef] [PubMed]
3. Brown, M.A.; Magee, L.A.; Kenny, L.C.; Karumanchi, S.A.; McCarthy, F.P.; Saito, S.; Hall, D.R.; Warren, C.E.; Adoyi, G.; Ishaku, S.; et al. Hypertensive Disorders of Pregnancy: ISSHP Classification, Diagnosis, and Management Recommendations for International Practice. *Hypertension* **2018**, *72*, 24–43. [CrossRef] [PubMed]
4. MacKay, A.P.; Berg, C.J.; Atrash, H.K. Pregnancy-related mortality from preeclampsia and eclampsia. *Obstet. Gynecol.* **2001**, *97*, 533–538. [CrossRef] [PubMed]
5. Say, L.; Chou, D.; Gemmill, A.; Tuncalp, O.; Moller, A.B.; Daniels, J.; Gulmezoglu, A.M.; Temmerman, M.; Alkema, L. Global causes of maternal death: A WHO systematic analysis. *Lancet Glob. Health* **2014**, *2*, e323–e333. [CrossRef]
6. Hastie, R.; Brownfoot, F.C.; Cluver, C.A.; Walker, S.P.; Hesselman, S.; Tong, S.; Bergman, L. Predictive Value of the Signs and Symptoms Preceding Eclampsia: A Systematic Review. *Obstet. Gynecol.* **2019**. [CrossRef]
7. Donato, R.; Cannon, B.R.; Sorci, G.; Riuzzi, F.; Hsu, K.; Weber, D.J.; Geczy, C.L. Functions of S100 proteins. *Curr. Mol. Med.* **2013**, *13*, 24–57. [CrossRef]
8. Rohrer, J.D.; Woollacott, I.O.; Dick, K.M.; Brotherhood, E.; Gordon, E.; Fellows, A.; Toombs, J.; Druyeh, R.; Cardoso, M.J.; Ourselin, S.; et al. Serum neurofilament light chain protein is a measure of disease intensity in frontotemporal dementia. *Neurology* **2016**, *87*, 1329–1336. [CrossRef]
9. Avila, J.; Lucas, J.J.; Perez, M.; Hernandez, F. Role of tau protein in both physiological and pathological conditions. *Physiol. Rev.* **2004**, *84*, 361–384. [CrossRef]
10. Bogoslovsky, T.; Gill, J.; Jeromin, A.; Davis, C.; Diaz-Arrastia, R. Fluid Biomarkers of Traumatic Brain Injury and Intended Context of Use. *Diagnostics* **2016**, *6*, 37. [CrossRef]
11. Marchi, N.; Rasmussen, P.; Kapural, M.; Fazio, V.; Kight, K.; Mayberg, M.R.; Kanner, A.; Ayumar, B.; Albensi, B.; Cavaglia, M.; et al. Peripheral markers of brain damage and blood-brain barrier dysfunction. *Restor. Neurol. Neurosci.* **2003**, *21*, 109–121. [PubMed]
12. Chou, S.H.; Robertson, C.S.; the Participants in the International Multi-disciplinary Consensus Conference on the Multimodality Monitoring. Monitoring biomarkers of cellular injury and death in acute brain injury. *Neurocrit. Care* **2014**, *21* (Suppl. S2), S187–S214. [CrossRef] [PubMed]
13. Randall, J.; Mortberg, E.; Provuncher, G.K.; Fournier, D.R.; Duffy, D.C.; Rubertsson, S.; Blennow, K.; Zetterberg, H.; Wilson, D.H. Tau proteins in serum predict neurological outcome after hypoxic brain injury from cardiac arrest: Results of a pilot study. *Resuscitation* **2013**, *84*, 351–356. [CrossRef] [PubMed]
14. Stefanovic, B.; Duric, O.; Stankovic, S.; Mijatovic, S.; Doklestic, K.; Stefanovic, B.; Jovanovic, B.; Marjanovic, N.; Kalezic, N. Elevated Serum Protein S100B and Neuron Specific Enolase Values as Predictors of Early Neurological Outcome After Traumatic Brain Injury. *J. Med. Biochem.* **2017**, *36*, 314–321. [CrossRef]
15. Marchi, N.; Cavaglia, M.; Fazio, V.; Bhudia, S.; Hallene, K.; Janigro, D. Peripheral markers of blood-brain barrier damage. *Clin. Chim. Acta* **2004**, *342*, 1–12. [CrossRef]
16. Evers, K.S.; Atkinson, A.; Barro, C.; Fisch, U.; Pfister, M.; Huhn, E.A.; Lapaire, O.; Kuhle, J.; Wellmann, S. Neurofilament as Neuronal Injury Blood Marker in Preeclampsia. *Hypertension* **2018**, *71*, 1178–1184. [CrossRef]

17. Bergman, L.; Zetterberg, H.; Kaihola, H.; Hagberg, H.; Blennow, K.; Akerud, H. Blood-based cerebral biomarkers in preeclampsia: Plasma concentrations of NfL, tau, S100B and NSE during pregnancy in women who later develop preeclampsia—A nested case control study. *PLoS ONE* **2018**, *13*, e0196025. [CrossRef]
18. Bergman, L.; Akerud, H. Plasma Levels of the Cerebral Biomarker, Neuron-Specific Enolase, are Elevated During Pregnancy in Women Developing Preeclampsia. *Reprod. Sci.* **2016**, *23*, 395–400. [CrossRef]
19. Bergman, L.; Akhter, T.; Wikstrom, A.K.; Wikstrom, J.; Naessen, T.; Akerud, H. Plasma levels of S100B in preeclampsia and association with possible central nervous system effects. *Am. J. Hypertens.* **2014**, *27*, 1105–1111. [CrossRef]
20. Vettorazzi, J.; Torres, F.V.; de Avila, T.T.; Martins-Costa, S.H.; Souza, D.O.; Portela, L.V.; Ramos, J.G. Serum S100B in pregnancy complicated by preeclampsia: A case-control study. *Pregnancy Hypertens.* **2012**, *2*, 101–105. [CrossRef]
21. Bergman, L.; Akerud, H.; Wikstrom, A.K.; Larsson, M.; Naessen, T.; Akhter, T. Cerebral Biomarkers in Women With Preeclampsia Are Still Elevated 1 Year Postpartum. *Am. J. Hypertens.* **2016**, *29*, 1374–1379. [CrossRef] [PubMed]
22. Schmidt, A.P.; Tort, A.B.; Amaral, O.B.; Schmidt, A.P.; Walz, R.; Vettorazzi-Stuckzynski, J.; Martins-Costa, S.H.; Ramos, J.G.; Souza, D.O.; Portela, L.V. Serum S100B in pregnancy-related hypertensive disorders: A case-control study. *Clin. Chem.* **2004**, *50*, 435–438. [CrossRef] [PubMed]
23. Artunc-Ulkumen, B.; Guvenc, Y.; Goker, A.; Gozukara, C. Maternal Serum S100-B, PAPP-A and IL-6 levels in severe preeclampsia. *Arch. Gynecol Obs.* **2015**, *292*, 97–102. [CrossRef] [PubMed]
24. Cipolla, M.J. Cerebrovascular function in pregnancy and eclampsia. *Hypertension* **2007**, *50*, 14–24. [CrossRef]
25. Hammer, E.S.; Cipolla, M.J. Cerebrovascular Dysfunction in Preeclamptic Pregnancies. *Curr. Hypertens. Rep.* **2015**, *17*, 64. [CrossRef]
26. Amburgey, O.A.; Chapman, A.C.; May, V.; Bernstein, I.M.; Cipolla, M.J. Plasma from preeclamptic women increases blood-brain barrier permeability: Role of vascular endothelial growth factor signaling. *Hypertension* **2010**, *56*, 1003–1008. [CrossRef]
27. Weksler, B.; Romero, I.A.; Couraud, P.O. The hCMEC/D3 cell line as a model of the human blood brain barrier. *Fluids Barriers CNS* **2013**, *10*, 16. [CrossRef]
28. Rahman, N.A.; Rasil, A.; Meyding-Lamade, U.; Craemer, E.M.; Diah, S.; Tuah, A.A.; Muharram, S.H. Immortalized endothelial cell lines for in vitro blood-brain barrier models: A systematic review. *Brain Res.* **2016**, *1642*, 532–545. [CrossRef]
29. Deli, M.A.; Abraham, C.S.; Kataoka, Y.; Niwa, M. Permeability studies on in vitro blood-brain barrier models: Physiology, pathology, and pharmacology. *Cell. Mol. Neurobiol.* **2005**, *25*, 59–127. [CrossRef]
30. Bergman, L.; Acurio, J.; Leon, J.; Gatu, E.; Friis, T.; Nelander, M.; Wikstrom, J.; Larsson, A.; Lara, E.; Aguayo, C.; et al. Preeclampsia and increased permeability over the blood brain barrier—A role of vascular endothelial growth receptor 2. *Am. J. Hypertens.* **2020**. [CrossRef]
31. León, J.; Acurio, J.; Bergman, L.; López, J.; Wikström, A.-K.; Torres-Vergara, P.; Troncoso, F.; Castro, F.O.; Vatish, M.; Escudero, C. Disruption of the Blood-Brain Barrier by Extracellular Vesicles From Preeclampsia Plasma and Hypoxic Placentae: Attenuation by Magnesium Sulfate. *Hypertension* **2021**. [CrossRef]
32. Tranquilli, A.L.; Brown, M.A.; Zeeman, G.G.; Dekker, G.; Sibai, B.M. The definition of severe and early-onset preeclampsia. Statements from the International Society for the Study of Hypertension in Pregnancy (ISSHP). *Pregnancy Hypertens.* **2013**, *3*, 44–47. [CrossRef] [PubMed]
33. Marsal, K.; Persson, P.H.; Larsen, T.; Lilja, H.; Selbing, A.; Sultan, B. Intrauterine growth curves based on ultrasonically estimated foetal weights. *Acta Paediatr.* **1996**, *85*, 843–848. [CrossRef] [PubMed]
34. Nelander, M.; Weis, J.; Bergman, L.; Larsson, A.; Wikstrom, A.K.; Wikstrom, J. Cerebral Magnesium Levels in Preeclampsia; A Phosphorus Magnetic Resonance Spectroscopy Study. *Am. J. Hypertens.* **2017**, *30*, 667–672. [CrossRef]
35. Mattsson, N.; Zetterberg, H.; Janelidze, S.; Insel, P.S.; Andreasson, U.; Stomrud, E.; Palmqvist, S.; Baker, D.; Tan Hehir, C.A.; Jeromin, A.; et al. Plasma tau in Alzheimer disease. *Neurology* **2016**, *87*, 1827–1835. [CrossRef] [PubMed]
36. Liu, Q.; Shepherd, B.E.; Li, C.; Harrell, F.E., Jr. Modeling continuous response variables using ordinal regression. *Stat. Med.* **2017**, *36*, 4316–4335. [CrossRef] [PubMed]
37. R: A Language and Environment for Statistical Computing. Available online: https://www.R-project.org/ (accessed on 26 December 2021).
38. Harrell, F.E., Jr. rms: Regression Modeling Strategies. R Package Version 6.2-0. Available online: https://CRAN.R-project.org/package=rms (accessed on 26 December 2021).
39. Simani, L.; Sadeghi, M.; Ryan, F.; Dehghani, M.; Niknazar, S. Elevated Blood-Based Brain Biomarker Levels in Patients with Epileptic Seizures: A Systematic Review and Meta-analysis. *ACS Chem. Neurosci.* **2020**, *11*, 4048–4059. [CrossRef]
40. Kalm, M.; Bostrom, M.; Sandelius, A.; Eriksson, Y.; Ek, C.J.; Blennow, K.; Bjork-Eriksson, T.; Zetterberg, H. Serum concentrations of the axonal injury marker neurofilament light protein are not influenced by blood-brain barrier permeability. *Brain Res.* **2017**, *1668*, 12–19. [CrossRef]
41. Uher, T.; McComb, M.; Galkin, S.; Srpova, B.; Oechtering, J.; Barro, C.; Tyblova, M.; Bergsland, N.; Krasensky, J.; Dwyer, M.; et al. Neurofilament levels are associated with blood-brain barrier integrity, lymphocyte extravasation, and risk factors following the first demyelinating event in multiple sclerosis. *Mult. Scler.* **2021**, *27*, 220–231. [CrossRef]
42. Andersson, M.; Oras, J.; Thorn, S.E.; Karlsson, O.; Kalebo, P.; Zetterberg, H.; Blennow, K.; Bergman, L. Signs of neuroaxonal injury in preeclampsia-A case control study. *PLoS ONE* **2021**, *16*, e0246786. [CrossRef]

43. Burwick, R.M.; Togioka, B.M.; Speranza, R.J.; Gaffney, J.E.; Roberts, V.H.J.; Frias, A.E.; Rincon, M. Assessment of blood-brain barrier integrity and neuroinflammation in preeclampsia. *Am. J. Obs. Gynecol.* **2019**, *221*, 269.e1–269.e8. [CrossRef] [PubMed]
44. Helms, H.C.; Abbott, N.J.; Burek, M.; Cecchelli, R.; Couraud, P.O.; Deli, M.A.; Forster, C.; Galla, H.J.; Romero, I.A.; Shusta, E.V.; et al. In vitro models of the blood-brain barrier: An overview of commonly used brain endothelial cell culture models and guidelines for their use. *J. Cereb. Blood Flow Metab.* **2016**, *36*, 862–890. [CrossRef] [PubMed]
45. van Veen, T.R.; Panerai, R.B.; Haeri, S.; Singh, J.; Adusumalli, J.A.; Zeeman, G.G.; Belfort, M.A. Cerebral autoregulation in different hypertensive disorders of pregnancy. *Am. J. Obs. Gynecol.* **2015**, *212*, 513.e1–513.e7. [CrossRef] [PubMed]
46. Johnson, A.C.; Tremble, S.M.; Chan, S.L.; Moseley, J.; LaMarca, B.; Nagle, K.J.; Cipolla, M.J. Magnesium sulfate treatment reverses seizure susceptibility and decreases neuroinflammation in a rat model of severe preeclampsia. *PLoS ONE* **2014**, *9*, e113670. [CrossRef]
47. Uwamori, H.; Ono, Y.; Yamashita, T.; Arai, K.; Sudo, R. Comparison of organ-specific endothelial cells in terms of microvascular formation and endothelial barrier functions. *Microvasc. Res.* **2019**, *122*, 60–70. [CrossRef] [PubMed]
48. Haseloff, R.F.; Dithmer, S.; Winkler, L.; Wolburg, H.; Blasig, I.E. Transmembrane proteins of the tight junctions at the blood-brain barrier: Structural and functional aspects. *Semin. Cell Dev. Biol.* **2015**, *38*, 16–25. [CrossRef]
49. Keaney, J.; Campbell, M. The dynamic blood-brain barrier. *FEBS J.* **2015**, *282*, 4067–4079. [CrossRef]
50. Koval, M. Claudins—Key pieces in the tight junction puzzle. *Cell Commun. Adhes.* **2006**, *13*, 127–138. [CrossRef]
51. Krause, G.; Winkler, L.; Mueller, S.L.; Haseloff, R.F.; Piontek, J.; Blasig, I.E. Structure and function of claudins. *Biochim. Biophys. Acta* **2008**, *1778*, 631–645. [CrossRef]
52. Angelow, S.; Ahlstrom, R.; Yu, A.S. Biology of claudins. *Am. J. Physiol. Ren. Physiol.* **2008**, *295*, F867–F876. [CrossRef]
53. Wen, H.; Watry, D.D.; Marcondes, M.C.; Fox, H.S. Selective decrease in paracellular conductance of tight junctions: Role of the first extracellular domain of claudin-5. *Mol. Cell Biol.* **2004**, *24*, 8408–8417. [CrossRef] [PubMed]
54. Marra, A.; Vargas, M.; Striano, P.; Del Guercio, L.; Buonanno, P.; Servillo, G. Posterior reversible encephalopathy syndrome: The endothelial hypotheses. *Med. Hypotheses.* **2014**, *82*, 619–622. [CrossRef] [PubMed]
55. Clayton, A.M.; Shao, Q.; Paauw, N.D.; Giambrone, A.B.; Granger, J.P.; Warrington, J.P. Postpartum increases in cerebral edema and inflammation in response to placental ischemia during pregnancy. *Brain Behav. Immun.* **2018**, *70*, 376–389. [CrossRef]
56. Warrington, J.P.; Fan, F.; Murphy, S.R.; Roman, R.J.; Drummond, H.A.; Granger, J.P.; Ryan, M.J. Placental ischemia in pregnant rats impairs cerebral blood flow autoregulation and increases blood-brain barrier permeability. *Physiol. Rep.* **2014**, *2*. [CrossRef] [PubMed]
57. Nelander, M.; Hannsberger, D.; Sundström-Poromaa, I.; Bergman, L.; Weis, J.; Åkerud, H.; Wikström, J.; Wikström, A.-K. Assessment of cerebral perfusion and edema in preeclampsia with intravoxel incoherent motion MRI. *AOGS* **2018**. [CrossRef]
58. Nelander, M.; Wikström, A.-K.; Weis, J.; Bergman, L.; Larsson, A.; Sundström-Poromaa, I.; Wikström, J. Cerebral Osmolytes and Plasma Osmolality in Pregnancy and Preeclampsia: A Proton Magnetic Resonance Spectroscopy Study. *Am. J. Hypertens.* **2018**, *31*, 847–853. [CrossRef]
59. Hatherell, K.; Couraud, P.O.; Romero, I.A.; Weksler, B.; Pilkington, G.J. Development of a three-dimensional, all-human in vitro model of the blood-brain barrier using mono-, co-, and tri-cultivation Transwell models. *J. Neurosci. Methods* **2011**, *199*, 223–229. [CrossRef]
60. Cucullo, L.; Couraud, P.O.; Weksler, B.; Romero, I.A.; Hossain, M.; Rapp, E.; Janigro, D. Immortalized human brain endothelial cells and flow-based vascular modeling: A marriage of convenience for rational neurovascular studies. *J. Cereb. Blood Flow Metab.* **2008**, *28*, 312–328. [CrossRef]
61. Vigh, J.P.; Kincses, A.; Ozgür, B.; Walter, F.R.; Santa-Maria, A.R.; Valkai, S.; Vastag, M.; Neuhaus, W.; Brodin, B.; Dér, A. Transendothelial Electrical Resistance Measurement across the Blood–Brain Barrier: A Critical Review of Methods. *Micromachines* **2021**, *12*, 685. [CrossRef]
62. Escudero, C.; Bertoglia, P.; Hernadez, M.; Celis, C.; Gonzalez, M.; Aguayo, C.; Acurio, J. Impaired A2A adenosine receptor/nitric oxide/VEGF signaling pathway in fetal endothelium during late- and early-onset preeclampsia. *Purinergic Signal.* **2013**, *9*, 215–226. [CrossRef]

cells

Article

Evidence of Neuroinflammation and Blood–Brain Barrier Disruption in Women with Preeclampsia and Eclampsia

Lina Bergman [1,2,3,*], Roxanne Hastie [2,4,5], Henrik Zetterberg [6,7,8,9,10], Kaj Blennow [6,7], Sonja Schell [1], Eduard Langenegger [1], Ashley Moodley [1], Susan Walker [4,5], Stephen Tong [4,5] and Catherine Cluver [1,4,5]

1 Department of Obstetrics and Gynecology, Stellenbosch University, Cape Town 7500, South Africa; sonjaschell@sun.ac.za (S.S.); langen@sun.ac.za (E.L.); ashleymoodley69@gmail.com (A.M.); cathycluver@hotmail.com (C.C.)
2 Department of Women's and Children's Health, Uppsala University, 75309 Uppsala, Sweden; hastie.r@unimelb.edu.au
3 Department of Obstetrics and Gynecology, Institute of Clinical Sciences, Sahlgrenska Academy, University of Gothenburg, 41685 Gothenburg, Sweden
4 Translational Obstetrics Group, Department of Obstetrics and Gynecology, University of Melbourne, Heidelberg, VIC 3084, Australia; spwalker@unimelb.edu.au (S.W.); stong@unimelb.edu.au (S.T.)
5 Mercy Perinatal, Mercy Hospital for Women, Heidelberg, VIC 3084, Australia
6 Institute of Neuroscience and Physiology, Department of Psychiatry and Neurochemistry, Sahlgrenska Academy, University of Gothenburg, 41345 Mölndal, Sweden; henrik.zetterberg@clinchem.gu.se (H.Z.); kaj.blennow@neuro.gu.se (K.B.)
7 Clinical Neurochemistry Laboratory, Sahlgrenska University Hospital, 41385 Mölndal, Sweden
8 UCL Institute of Neurology, Department of Neurodegenerative Disease, Queen Square, London WC1N3BG, UK
9 UK Dementia Research Institute, London WC1E6BT, UK
10 Hong Kong Center for Neurodegenerative Diseases, Clear Water Bay, Kowloon, Hong Kong 999077, China
* Correspondence: lina.bergman.2@gu.se; Tel.: +46-7079-207-80

Citation: Bergman, L.; Hastie, R.; Zetterberg, H.; Blennow, K.; Schell, S.; Langenegger, E.; Moodley, A.; Walker, S.; Tong, S.; Cluver, C. Evidence of Neuroinflammation and Blood–Brain Barrier Disruption in Women with Preeclampsia and Eclampsia. *Cells* **2021**, *10*, 3045. https://doi.org/10.3390/cells10113045

Academic Editors: Junie P. Warrington, Ana T. Palei, Mark W. Cunningham, Lorena M. Amaral and Alexander E. Kalyuzhny

Received: 6 September 2021
Accepted: 4 November 2021
Published: 5 November 2021

Publisher's Note: MDPI stays neutral with regard to jurisdictional claims in published maps and institutional affiliations.

Abstract: Cerebral complications in preeclampsia are leading causes of maternal mortality. Animal models suggest that an injured blood–brain barrier and neuroinflammation may be important but there is paucity of data from human studies. Therefore, we aimed to evaluate this in women with preeclampsia and eclampsia. We included women recruited to the South African Preeclampsia Obstetric Adverse Events (PROVE) biobank. Blood and cerebrospinal fluid (CSF) were collected around delivery. CSF was analyzed for neuroinflammatory markers interleukin 1β, interleukin 6, interleukin-8 and tumor necrosis factor alpha (TNF-alpha). The CSF to plasma albumin ratio was measured to assess blood–brain barrier function. Women with eclampsia ($n = 4$) showed increased CSF concentrations of all pro-inflammatory cytokines and TNF-alpha compared to women with normotensive pregnancies ($n = 7$) and also for interleukin-6 and TNF-alpha compared to women with preeclampsia ($n = 4$). Women with preeclampsia also showed increases in pro-inflammatory cytokines IL-6 and IL-8 but not TNF-alpha in the CSF compared to women with normotensive pregnancies. In particular, women with eclampsia but also women with preeclampsia showed an increase in the CSF to plasma albumin ratio compared to normotensive women. In conclusion, women with preeclampsia and eclampsia show evidence of neuroinflammation and an injured blood–brain barrier. These findings are seen in particular among women with eclampsia.

Keywords: eclampsia; neuroinflammation; blood–brain barrier; preeclampsia; cerebral edema

1. Introduction

Preeclampsia is a multisystem disorder that complicates 4–6% of all pregnancies [1,2]. The cerebral complications of preeclampsia, which include eclampsia, cerebral edema and stroke, are a leading cause of severe maternal morbidity and mortality [3,4]. Additionally, long-term neurological consequences of preeclampsia and its complications include an increased risk for white matter lesions, stroke, seizure disorders and vascular dementia later in life [5–7].

Magnesium sulfate is the drug of choice to prevent and treat eclampsia and has shown effects in reducing maternal mortality and morbidity [4,8]. The mechanism of action is not completely understood but animal studies suggest that magnesium sulfate may decrease neuroinflammation and potentially protect the functionality of the blood–brain barrier [9–12].

Animal models have demonstrated an impaired blood–brain barrier and increased neuroinflammation in preeclampsia. In these studies, animals with induced preeclampsia demonstrated an increased number of activated microglial cells and an increased cerebrospinal fluid (CSF) concentration of cytokines. In addition, the CSF/plasma albumin ratio was increased in preeclampsia as a measurement of blood–brain barrier impairment [10,12–14]. So far, there has been no evidence from pregnant women with preeclampsia to support this finding, with only one study having found increased levels of interleukins in CSF in preeclampsia, although other studies have not [15–18]. To our knowledge, no studies have assessed neuroinflammation and the blood–brain barrier in women with eclampsia.

We, therefore, evaluated whether there is evidence of blood–brain barrier disruption and neuroinflammation in women with eclampsia compared to those with preeclampsia and normotensive women.

2. Materials and Methods

2.1. Study Cohort

We included women who were recruited into the South African Preeclampsia Obstetric Adverse Events (PROVE) biobank and database that had a cesarean section delivery. The PROVE biobank is an ongoing collaborative project that facilitates research in the field of preeclampsia, with a focus on phenotyping severe disease. Women diagnosed with preeclampsia and normotensive controls are enrolled in the biobank at admission to Tygerberg Hospital (Stellenbosch University, Cape Town, South Africa). Biological samples and clinical data are collected at inclusion or delivery, during the hospital stay and postpartum [19]. Tygerberg Hospital is the largest referral hospital in the Western Cape Provence of South Africa. In 2018 there were 32,422 deliveries in the referral area, of which 8067 were considered high risk and delivered at Tygerberg. PROVE biobank includes the majority of women with eclampsia presenting at the hospital, where around 50 women with eclampsia are recruited yearly [19].

For this study, we included women with singleton pregnancies. Exclusion criteria were known neurological and cardiac diseases. For normotensive women, additional exclusion criteria were chronic hypertension and diabetes mellitus. Preeclampsia and eclampsia were defined according to the American College of Obstetricians and Gynecologists (ACOG) 2020 Practice Bulletin [20]. A woman was considered normotensive if she had no documented systolic blood pressure greater than or equal to 140 mmHg or a diastolic blood pressure greater than or equal to 90 mmHg during her pregnancy until discharge postpartum. All women with eclampsia were recruited after the first generalized seizure and all women with preeclampsia were recruited after diagnosis. All women were recruited before delivery.

Baseline data were obtained by interview and extraction from medical records. All data were entered and stored using REDCap (Research Electronic Data Capture) tools hosted at Stellenbosch University [21]. Electronic data were double-checked for accuracy and cross-referenced with original data collection forms collected by PROVE research midwives.

2.2. Sample Collection

Blood was drawn on the day before delivery (normotensive control = 1), at delivery (normotensive controls = 2, preeclampsia = 1, eclampsia = 3), one day postpartum (normotensive controls = 3, eclampsia = 1), two days postpartum (normotensive control = 1, preeclampsia = 1), three days postpartum (preeclampsia = 1) and four days postpartum (preeclampsia = 1). CSF was collected at the time of spinal anesthesia at delivery. CSF was analyzed for concentrations of neuroinflammatory markers interleukin 1 (IL-1β), interleukin 6 (IL-6), interleukin-8 (IL-8) and tumor necrosis factor alpha (TNF-alpha). The CSF/plasma albumin ratio was measured to assess blood–brain barrier integrity.

Blood samples were collected in a 9 mL ethylenediaminetetraacetic acid (EDTA) tube at time of inclusion in the study. CSF samples were collected at cesarean section delivery at the time of spinal anesthesia. All samples were centrifuged for 10 min at 2800× g and plasma and CSF were aliquoted in 2 mL cryotubes and subsequently frozen at −80 °C until the analysis.

2.3. Biomarker Assays

All CSF measurements and plasma/serum concentrations were assessed in one round of experiments using a single batch of reagents for each assay by laboratory technicians who were blinded to clinical data and groups.

CSF and plasma albumin concentrations were measured by immunonephelometry on a Beckman Immage Immunohistochemistry system (Beckman Instruments, Beckman Coulter, Brea, CA, USA). The albumin ratio was calculated as CSF albumin (mg/L)/serum albumin (g/L) and was used as a measure of the blood–brain barrier integrity [22]. CSF concentrations of IL-1β, IL-6, IL-8 and TNF-α were measured using the Meso Scale Discovery 4-plex Proinflammatory Panel II according to instructions from the kit manufacturer (Meso Scale Discovery, Rockville, MD). The individual assays in the multiplexed inflammation marker panel (the Meso Scale Discovery 4-plex Proinflammatory Panel II) have been validated for research use in human CSF, as previously described (https://www.mesoscale.com/~/media/files/scientific%20poster/measuring-low-concentrations-cytokines-csf-assays-aai-2017-msd.pdf?la=en, accessed on 5 November 2021). The measurement ranges were 0.284–1166 pg/mL for IL-1β, 0.324–1326 pg/mL for IL-6, 0.28–1150 pg/mL for IL-8 and 0.165–676 pg/mL for TNF-α. All measurements were performed in one round of experiments using one batch of reagents. The intra-assay coefficient of variation for internal control samples within the linear range was <10%.

2.4. Statistics

Demographic and clinical characteristics are presented as means with standard deviations (SD) and percentages. Albumin quotient and neuroinflammatory markers are presented as medians with ranges and compared between groups as fold-changes with 95% confidence intervals.

Due to the small sample sizes, p-values for differences in concentrations of inflammatory markers in CSF and albumin ratios were calculated using an exact permutation test for the mean difference on log-transformed variables. Corresponding confidence intervals were calculated by test inversion.

In all hypothesis tests, a two-sided p-value of less than 0.05 was considered statistically significant. Data and statistical analyses were performed using SPSS version 26.0 (SPSS; PASW statistics, IBM corp) and SAS version 9.4 (SAS Institute INC, Cary, NC, USA).

2.5. Ethical Permission and Registration Details

Ethical approval was obtained (protocol number N18/03/034, Federal Wide assurance number 00001372, Institutional Review Board number IRB0005239) and all included participants signed informed consent before being enrolled in the Preeclampsia Obstetric Adverse Events (PROVE) biobank. The biobank is registered at ISRCTN with registration number ISRCTN10623443 and the protocol has been published previously [19].

3. Results

We included women enrolled in the PROVE biobank between April 2018 and March 2020. During this time, 233 women with biological samples were included in the PROVE

Biobank and 15 of these had CSF collected. Seven women were normotensive, four had preeclampsia and four had eclampsia (Figure 1).

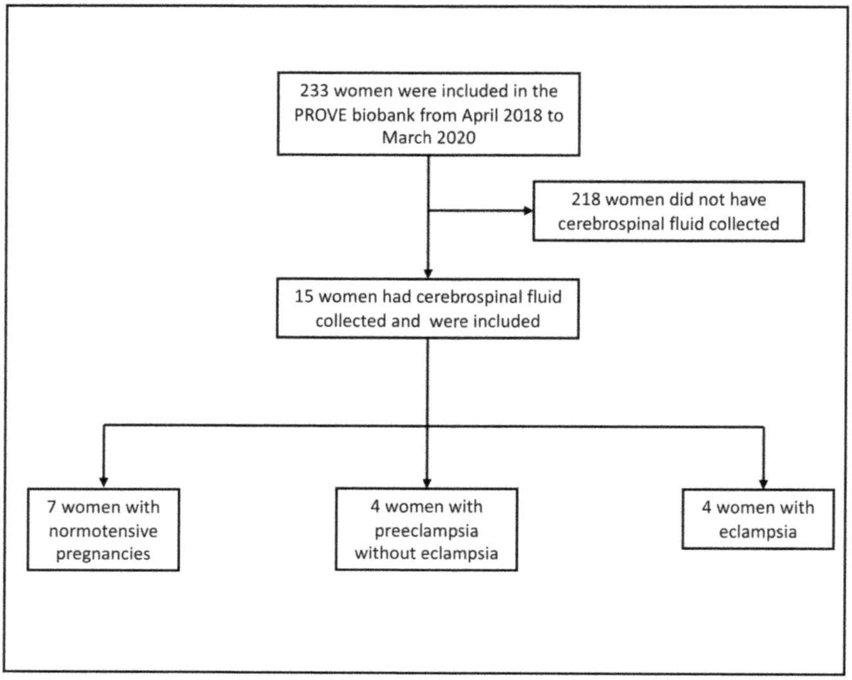

Figure 1. Background characteristics.

Maternal characteristics and pregnancy outcomes are presented in Table 1.

Table 1. Background characteristics of the population.

	Normotensiv Pregnancy	Preeclampsia	Eclampsia
n	7	4	4
At baseline			
Maternal age (years)	30.9 (5.0)	27.3 (6.4)	18.3 (4.2)
Nulliparous (%)	1 (14)	2 (50)	3 (75)
HIV positive(%)	1 (14)	1 (25)	0 (0)
Smoking (%)	1 (14)	1 (25)	0 (0)
Diabetes (%)	0 (0)	0 (0)	0 (0)
Chronic hypertension (%)	0 (0)	2 (50)	0 (0)
BMI (kg/m^2)	31.6 (5.7)	19.8 (2.3) *	24.6 (4.8)
After inclusion			
GA at delivery (weeks)	38.4 (0.8)	33.3 (3.8)	34.3 (3.6)
Magnesium sulfate (%)	0 (0)	3 (75)	4 (100)
Hours before CSF sample	N/A	3.5 (1.2–25.5)	23 (12.5–30)

Table 1. *Cont.*

	Normotensiv Pregnancy	Preeclampsia	Eclampsia
n	7	4	4
Mode of delivery (%)			
Vaginal delivery	0 (0)	0 (0)	0 (0)
Elective CS	7 (100)	0 (0)	0 (0)
Emergency CS	0 (0)	4 (100)	4 (100)
Liveborn (%)	7 (100)	4 (100)	4 (100)
Birthweight (g)	3279.3 (242.7)	1713.8 (693.9)	2235.0 (1087.3)
Maternal complications			
Recurrent eclampsia	0 (0)	0 (0)	3 (75)
Severe hypertension	0 (0)	1 (25)	1 (25)

Abbreviations: HIV = human immune deficiency virus. * Missing data on one participant. Severe hypertension was defined as a systolic blood pressure of >160 mm Hg and/or a diastolic blood pressure of >110 mm Hg despite intravenous treatment. No women in this study reported alcohol or methamphetamine use during pregnancy.

Compared to normotensive women, those who had eclampsia were younger and more often nulliparous. Women with preeclampsia or eclampsia delivered at an earlier gestation and had lower birthweight infants. All women with preeclampsia and eclampsia gave birth by emergency cesarean section. Women in the normotensive group all had elective cesarean sections. One woman with eclampsia and one with preeclampsia had severe hypertension. There were no other neurological complications such as stroke and no women were admitted to the general intensive care unit, had a low Glasgow coma scale or needed intubation. None experienced pulmonary edema or hemolysis, elevated liver enzyme and low platelet (HELLP) syndrome. All women with eclampsia and preeclampsia were treated with magnesium sulfate until after delivery. All women in the study received cephalosporin antibiotics at the time of cesarean section.

3.1. *Neuroinflammatory Markers*

When compared to normotensive pregnancies, women with eclampsia showed evidence of increased neuroinflammation. Those with eclampsia showed an 18.0-fold increase in CSF concentrations of IL-6 (95% CI 3.17–121.15, p = 0.006), a 7.8-fold increase in IL-8 (95% CI 3.58–18.75, p = 0.006) and a 3.7-fold increase in TNF-alpha (95% CI 1.53–7.70, p = 0.006). There was also evidence of increased neuroinflammation among those who had eclampsia compared to women with preeclampsia: a 4.6-fold increase in IL-8 (95% CI 1.66–40.92, p = 0.029), a 3.2-fold increase in TNF-alpha (95% CI 1.10–22.53, p = 0.029) and a non-significant 8.5-fold increase in IL-6 (95% CI 0.86–521.37, p = 0.057).

When women with preeclampsia were compared to normotensive women there was a 2.1-fold increase in IL-6 (95% CI 1.18–3.85, p = 0.024) and a 1.7-fold increase in IL-8 (95% CI 1.03–2.79, p = 0.036) (Figure 2 and Tables 2 and 3).

Table 2. Cerebrospinal fluid concentrations of neuroinflammatory markers and CSF/plasma albumin ratios.

	Normotensive	Preeclampsia	Eclampsia
n	7	4	4
IL-6 (pg/mL)	0.54 (0.30–0.76)	1.13 (0.59–1.90)	4.23 (1.63–305.00)
IL-8 (pg/mL	31.36 (16.66–38.75)	53.88 (28.10–74.45)	137.81 (123.43–1150.00)
TNF-alpha (pg/mL)	0.17 (0.17–0.30)	0.20 (0.17–0.28)	0.40 (0.31–3.72)
CSF/plasma albumin ratio	3.15 (2.41–4.05)	5.21 (2.83–14.79)	10.42 (8.23–23.17)

CSF, cerebrospinal fluid; IL-6, interleukin 6; IL-8, interleukin 8; n, number' TNF-alpha, tumor necrosis factor-alpha. Numbers are presented as medians with ranges.

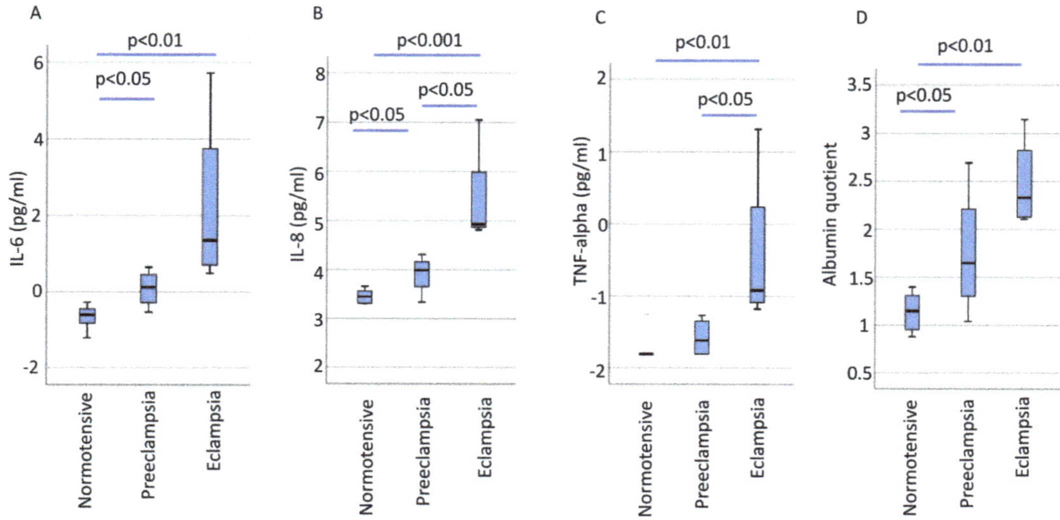

Figure 2. Log-transformed concentrations of neuroinflammatory markers and albumin quotient in CSF samples. Boxplot showing log-transformed CSF concentrations (IL-6, IL-8 and TNF-alpha) and log-transformed CSF/plasma albumin ratios as medians with interquartile ranges for IL-6 (**A**), IL-8 (**B**), TNF alpha (**C**) and albumin (**D**). Normotensive pregnancy (n = 7), preeclampsia (n = 4) and eclampsia (n = 4). IL-6, interleukin 6; IL-8, interleukin 8; TNF-alpha, tumor necrosis factor alpha.

Table 3. Fold changes of cerebrospinal fluid concentrations of cerebral biomarkers, neuroinflammatory markers and CSF/plasma albumin ratios.

Biomarker	Preeclampsia vs. Normotensive	Eclampsia vs. Normotensive	Eclampsia vs. Preeclampsia
IL-6 (pg/mL)	2.11 (1.18–3.85) p = 0.024	18.03 (3.17–121.15) p = 0.006	8.53 (0.86–521.37) p = 0.057
IL-8 (pg/mL)	1.69 (1.03–2.79) p = 0.036	7.76 (3.58–18.75) p = 0.006	4.59 (1.66–40.92) p = 0.029
TNF-alpha (pg/mL)	1.16 (0.75–1.58) p = 0.56	3.65 (1.53–7.70) p = 0.006	3.16 (1.10–22.53) p = 0.029
CSF/plasma albumin ratio	1.86 (1.01–3.65) p = 0.048	3.83 (2.33–6.48) p = 0.006	2.05 (0.56–8.19) p = 0.14

Data are presented as fold changes with 95% confidence intervals; p-values were calculated using exact permutation test for the mean difference on log-transformed variables. Corresponding confidence intervals were calculated by test inversion. Abbreviations: CSF, cerebrospinal fluid; IL-6, interleukin 6; IL-8, interleukin 8; TNF-alpha, tumor necrosis factor alpha.

IL-1β CSF concentrations were also measured, although concentrations were generally below the limit of detection. One woman with eclampsia showed concentrations of the inflammatory markers 10 times higher than the others in the eclampsia group. Excluding the results of this participant did not significantly change the findings.

3.2. Albumin Quotient

There was also evidence of significant disruption of blood–brain barrier integrity among those who had eclampsia and even preeclampsia. Women with eclampsia showed a 3.8-fold increase in the CSF/plasma albumin ratio (95% CI 2.33–6.48, p = 0.006) when compared to normotensive women. Women with preeclampsia showed a 1.9-fold increase in the CSF/plasma albumin ratio (95% CI 1.01–3.65, p = 0.048) compared to normotensive women (Figure 2 and Table 3). When restricting the analyses by matching eclampsia and normotensive controls for day of plasma sampling (n = 4 in each group), the plasma/albumin ratio in women with eclampsia remained increased compared to controls (median, range, 10.4, 8.2–23.2 vs. 3.3, 2.8–4.1, $p = 0.029$).

4. Discussion

4.1. Principal Findings

Inflammatory markers (IL-6, IL-8 and TNF-alpha) were increased in the CSF of women with eclampsia, compared to those with preeclampsia and women with normotensive pregnancies. Furthermore, there was evidence of injury to the blood–brain barrier with an increased CSF/plasma albumin ratio among women with eclampsia compared to normotensive women. There was also evidence of injury among those with preeclampsia, though it was to a lesser extent. This study provides evidence of neuroinflammation and blood–brain barrier injury in women with eclampsia and preeclampsia.

4.2. Results in Context

To the best of our knowledge, there are no previous studies reporting on CSF findings among women with eclampsia. There have been a few studies that have measured protein content in CSF from women with preeclampsia. One study that used SOMAscan proteomics with CSF samples from 13 women with preeclampsia and 14 women with normotensive pregnancies demonstrated increased levels of IL-8 and IL-9 in CSF from women with preeclampsia, although only related levels were reported and no information was given about absolute concentrations [17]. Another study analyzed CSF samples using liquid chromatography tandem mass spectrometry proteomics in 43 women with severe preeclampsia and 55 normotensive women [18]. Of the 457 proteins analyzed, 25 were different between those with preeclampsia and normotensive controls. There were eight proteins that were more abundant in preeclampsia and 18 that were less abundant. The most significantly abundant protein was protein alpha-1-microglobulin/bikunin precursor (AMBP), a precursor of alpha-1-microglobulin (A1M) that is a heme scavenger and potentially protective against oxidative stress caused by free hemoglobin (Hb). The other proteins that were more or less abundant in preeclampsia included acute phase proteins and proteins related to angiogenesis. Their analyses did not include interleukins or TNF-alpha [18]. The CSF/plasma albumin ratio was not assessed in these studies [17,18].

The CSF/plasma albumin ratio is an expression of the integrity of the blood–brain barrier, where the plasma-derived protein albumin should be found in very low concentrations in the CSF in normal conditions, resulting in a low CSF/plasma albumin ratio. In one study that had previously investigated the CSF/plasma albumin ratio, 15 women with preeclampsia and 15 women with normotensive pregnancies were included. There were no women with eclampsia in this study [15]. One woman in the preeclampsia group had HELLP syndrome but none had eclampsia or other severe end organ complications. In the normotensive group, the CSF/plasma albumin ratio was 2.73, similar to our findings of 3.1. Women with preeclampsia had a median CSF/plasma albumin ratio of 3.02 [15]. In our study, the median CSF/plasma albumin ratio in women with preeclampsia was 5.2. The only study we could find that investigated both neuroinflammation and blood–brain barrier integrity in preeclampsia assessed the concentrations of C5a, C5b-9, TNF-alpha and IL-6 in CSF and the CSF/plasma albumin ratio in preeclampsia cases with or without severe features ($n = 16$), compared to hypertensive disease ($n = 16$) without proteinuria and normotensive controls ($n = 16$) [16]. In this study, severe disease was mostly due to severe hypertension, with only one case of HELLP syndrome and one case of pulmonary edema. No difference was found between groups for any of the inflammatory markers or the CSF/plasma albumin ratio [16]. In this study, the overall CSF concentrations of IL-6 were higher than in our study but not different between preeclampsia and normotensive women. The CSF concentrations of TNF-alpha and the CSF/plasma albumin ratio in normotensive women were similar to our study. Interestingly, women with preeclampsia had a lower CSF/plasma albumin ratio of 3.5 (IQR 2.9–5.1) compared to 5.21 (range 2.83–14.79) in our study. These differences may be due to different assays with variable intra- and inter-coefficients or differences in the disease phenotype. The previous study recruited women from a high-income setting where preeclampsia may be identified earlier, potentially resulting in a shorter time of exposure to harmful molecules that could injure the blood–brain barrier compared to our study where women often presented late.

Perhaps owing to the difficulty of studying blood–brain barrier integrity and pathophysiology in women with preeclampsia, several animal models have been established. The reduced uterine perfusion pressure (RUPP) model in rats or mice is one of the most commonly used. Studies have reported increased permeability of the blood–brain barrier to the larger molecule dextran [14] and others only to the smaller molecule sodium fluorescein [10]. RUPP rats also had increased cerebral water content two months postpartum and this was thought to be a good proxy of persistent cerebral edema [23]. Neuroinflammatory activity measured using immunohistochemistry in this RUPP model also showed increased neuroinflammatory activity, which was interestingly reversed by magnesium sulfate treatment [10].

Our group previously reported on in vitro studies of the human brain endothelium, exposed to plasma from women with preeclampsia [24]. We demonstrated increased permeability and a lower trans endothelial electrical resistance over the in vitro blood–brain barrier after exposure to plasma from women with preeclampsia compared to normotensive women, implicating that plasma from women with preeclampsia induces injury to the blood–brain barrier [24], supporting the findings from animal studies.

This study supports data from preclinical studies and suggests that neuroinflammation and blood–brain barrier injury may be important in the pathophysiology of eclampsia and preeclampsia.

4.3. Clinical Implications

In this study, women with eclampsia had a median albumin quotient of 10.4 with a range of 8.23–23.17. These are all above the reference value of <6.5 for corresponding age and gender [25–27]. This suggests that eclampsia represents a pathological state severely affecting blood–brain barrier integrity.

The novel findings of blood–brain barrier injury and increased neuroinflammation in preeclampsia and eclampsia in the acute phase may be important for short-term complications and the risk for long-term neurological consequences of preeclampsia [5–7,28].

The presence of an injured blood–brain barrier and neuroinflammation in other medical conditions such as traumatic brain injury have been suggested as underlying factors of long-term neurological outcome, such as cognitive dysfunction and chronic traumatic encephalopathy [29,30]. There are similar hypotheses in the field of neurodegenerative disease with an initial insult to the blood–brain barrier and subsequent neuroinflammation as an underlying trigger to the development of later neurodegenerative disease [31].

Treatments that restore or protect blood–brain barrier integrity and decrease neuroinflammation could be important in preventing and treating eclampsia and may possibly protect women from long-term cerebral adverse outcomes associated with preeclampsia, such as cognitive decline, epilepsy and dementia [5,6,32].

4.4. Research Implications

Larger studies should be done to validate our findings and further evaluate the effects of magnesium sulfate and other neuroprotective drugs on the degree of neuroinflammation and blood–brain barrier integrity in a clinical cohort of women with preeclampsia and eclampsia. If confirmed, future research should focus on neuroprotective treatment for women with preeclampsia and neurological complications to decrease neuroinflammation and restore blood–brain barrier integrity with short- and long-term effects.

4.5. Strengths and Limitations

We have collected unique CSF samples in a cohort of women with eclampsia. It is a difficult task to obtain CSF samples in eclampsia, as these women are often delivered in an emergency situation after hours and they often require general anesthesia due to other complications including HELLP syndrome, which precludes CSF sampling. Collecting these samples was only possible due to the high incidence of eclampsia at our hospital and the infrastructure of the PROVE biobank. Even so, we were only able to collect CSF samples from four women with eclampsia with corresponding controls with a total of 233 women in the biobank, which highlights the difficulty of obtaining CSF samples at delivery in eclampsia. A limitation is that not all plasma samples were drawn on the day of delivery, inducing potential bias in the interpretation of the CSF/plasma albumin ratio. However, when restricting the analyses by matching eclampsia and normotensive controls for day of plasma sampling, the plasma/albumin ratio in women with eclampsia remained increased compared to controls. Another limitation is that our sample size was small. Despite this, we were still able to demonstrate significant differences in neuroinflammation and blood–brain barrier integrity between the groups, which implies that the differences are very significant. If a larger sample size could be obtained in future studies, it would be of value to perform adjusted models correcting for potential confounders such as BMI, maternal age and parity.

5. Conclusions

Women with eclampsia show evidence of neuroinflammation and an injured blood–brain barrier. This provides important knowledge in understanding the underlying pathophysiological mechanisms in eclampsia and may help direct therapeutic interventions to prevent and treat neurological complications of preeclampsia (such as anti-inflammatory agents after an eclamptic seizure).

Author Contributions: Conceptualization, L.B., S.W., S.T. and C.C.; Data curation, L.B. and S.S.; Formal analysis, L.B., R.H., H.Z., K.B., S.W., S.T. and C.C.; Funding acquisition, L.B., H.Z., K.B., S.W. and S.T.; Methodology, L.B. and C.C.; Project administration, L.B., S.S. and C.C.; Resources, L.B., H.Z., E.L. and C.C.; Supervision, E.L., A.M. and C.C.; Validation, H.Z. and C.C.; Visualization, L.B.; Writing—original draft, L.B. and C.C.; Writing—review & editing, L.B., R.H., H.Z., K.B., S.S., E.L., A.M., S.W., S.T. and C.C. All authors have read and agreed to the published version of the manuscript.

Funding: The study was supported by the Swedish Medical Society, Märta Lundqvist Foundation, Swedish Foundation for International Cooperation in Research and Higher Education (STINT), Jane and Dan Olssons Foundation, Mercy Perinatal, the Swedish Research Council (VR), Center for Clinical Research Dalarna and the Preeclampsia Foundation. L.B. is supported by the Swedish Society for Medical Research (SSMF). C.C. receives salary support from the Mercy Health Foundation. R.H. and S.T. receive salary support from the National Health and Medical Research Council of Australia (NHMRC). H.Z. is a Wallenberg Scholar supported by grants from the Swedish Research Council (#2018-02532), the European Research Council (#681712), Swedish State Support for Clinical Research (#ALFGBG-720931), the Alzheimer Drug Discovery Foundation (ADDF), USA (#201809-2016862), the AD Strategic Fund and the Alzheimer's Association (#ADSF-21-831376-C, #ADSF-21-831381-C and #ADSF-21-831377-C), the Olav Thon Foundation, the Erling-Persson Family Foundation, Stiftelsen för Gamla Tjänarinnor, Hjärnfonden, Sweden (#FO2019-0228), the European Union's Horizon 2020 research and innovation programme under the Marie Skłodowska-Curie grant agreement No 860197 (MIRIADE) and the UK Dementia Research Institute at UCL.

Institutional Review Board Statement: The study was conducted according to the guidelines of the Declaration of Helsinki, and approved by the Institutional Review Board IRB0005239 of Stellenbosch University, Federal-Wide assurance number 00001372 (protocol code N18/03/034 approval date 01-06-2018). All included participants signed informed consent before being enrolled in the Preeclampsia Obstetric Adverse Events (PROVE) biobank. The biobank is registered at ISRCTN with registration number ISRCTN10623443 and the protocol has been published previously [19].

Informed Consent Statement: Informed consent was obtained from all subjects involved in the study.

Data Availability Statement: Anonymized data not published within this article will be made available on request from any qualified investigator after approval, as described in the PROVE protocol [19].

Acknowledgments: We thank all study participants and the staff at Tygerberg Hospital and Stellenbosch University for their support. In particular, we thank the anesthetic team for assisting with the collection of the CSF samples. We would also like to thank Henrik Imberg, Statistika Konsultgruppen, Gothenburg, for valuable statistical support.

Conflicts of Interest: H.Z. has served on scientific advisory boards or as a consultant for Alector, Eisai, Denali, Roche Diagnostics, Wave, Samumed, Siemens Healthineers, Pinteon Therapeutics, Nervgen, AZTherapies, CogRx and Red Abbey Labs; has given lectures in symposia sponsored by Cellectricon, Fujirebio, Alzecure and Biogen; and is a co-founder of Brain Biomarker Solutions in Gothenburg AB (BBS), which is a part of the GU Ventures Incubator Program. K.B. has served as a consultant or at advisory boards for Abcam, Axon, Biogen, Lilly, MagQu, Novartis and Roche Diagnostics, and is a co-founder of Brain Biomarker Solutions in Gothenburg AB (BBS), which is a part of the GU Ventures Incubator Program. L.B. is a part of a steering group in a study investigating first trimester prediction for preeclampsia where Roche, Termo Fischer and Perkin Elmer provide free reagents for PlGF. The remaining authors report no conflict of interest. The funders had no role in the design of the study; in the collection, analyses, or interpretation of data; in the writing of the manuscript, or in the decision to publish the results.

References

1. Chappell, L.C.; Cluver, C.A.; Kingdom, J.; Tong, S. Pre-eclampsia. *Lancet* **2021**, *398*, 341–354. [CrossRef]
2. Abalos, E.; Cuesta, C.; Grosso, A.L.; Chou, D.; Say, L. Global and regional estimates of preeclampsia and eclampsia: A systematic review. *Eur. J. Obstet. Gynecol. Reprod. Biol.* **2013**, *170*, 1–7. [CrossRef]
3. Duley, L. The Global Impact of Pre-eclampsia and Eclampsia. *Semin. Perinatol.* **2009**, *33*, 130–137. [CrossRef] [PubMed]
4. Fishel Bartal, M.; Sibal, B.M. Eclampsia in the 21st century. *Am. J. Obstet. Gynecol.* **2020**. [CrossRef]
5. Basit, S.; Wohlfahrt, J.; A Boyd, H. Pre-eclampsia and risk of dementia later in life: Nationwide cohort study. *BMJ* **2018**, *363*, k4109. [CrossRef]
6. Nerenberg, K.A.; Park, A.L.; Vigod, S.N.; Saposnik, G.; Berger, H.; Hladunewich, M.A.; Gandhi, S.; Silversides, C.K.; Ray, J.G. Long-term Risk of a Seizure Disorder After Eclampsia. *Obstet. Gynecol.* **2017**, *130*, 1327–1333. [CrossRef] [PubMed]
7. McDonald, S.D.; Malinowski, A.; Zhou, Q.; Yusuf, S.; Devereaux, P. Cardiovascular sequelae of preeclampsia/eclampsia: A systematic review and meta-analyses. *Am. Heart J.* **2008**, *156*, 918–930. [CrossRef]
8. Altman, D.; Carroli, G.; Duley, L.; Farrell, B.; Moodley, J.; Neilson, J.; Smith, D.; Magpie Trial Collaboration Group. Do women with pre-eclampsia, and their babies, benefit from magnesium sulphate? The Magpie Trial: A randomised placebo-controlled trial. *Lancet* **2002**, *359*, 1877–1890. [CrossRef]
9. Euser, A.; Bullinger, L.; Cipolla, M.J. Magnesium sulphate treatment decreases blood-brain barrier permeability during acute hypertension in pregnant rats. *Exp. Physiol.* **2007**, *93*, 254–261. [CrossRef]
10. Johnson, A.C.; Tremble, S.M.; Chan, S.-L.; Moseley, J.; Lamarca, B.; Nagle, K.J.; Cipolla, M.J. Magnesium Sulfate Treatment Reverses Seizure Susceptibility and Decreases Neuroinflammation in a Rat Model of Severe Preeclampsia. *PLoS ONE* **2014**, *9*, e113670. [CrossRef] [PubMed]
11. Li, X.; Han, X.; Yang, J.; Bao, J.; Di, X.; Zhang, G.; Liu, H. Magnesium Sulfate Provides Neuroprotection in Eclampsia-Like Seizure Model by Ameliorating Neuroinflammation and Brain Edema. *Mol. Neurobiol.* **2016**, *54*, 7938–7948. [CrossRef] [PubMed]
12. Zhang, L.W.; Warrington, J.P. Magnesium Sulfate Prevents Placental Ischemia-Induced Increases in Brain Water Content and Cerebrospinal Fluid Cytokines in Pregnant Rats. *Front. Neurosci.* **2016**, *10*, 561. [CrossRef]
13. Ambugey, O.A.; Chapman, A.C.; May, V.; Bernstein, I.M.; Cipolla, M.J. Plasma from preeclamptic women increases blood-brain barrier permeability: Role of vascular endothelial growth factor signaling. *Hypertension* **2010**, *56*, 1003–1008.
14. Warrington, J.P.; Fan, F.; Murphy, S.R.; Roman, R.J.; Drummond, H.A.; Granger, J.P.; Ryan, M.J. Placental ischemia in pregnant rats impairs cerebral blood flow autoregulation and increases blood-brain barrier permeability. *Physiol. Rep.* **2014**, *2*, e12134. [CrossRef] [PubMed]
15. Andersson, M.; Oras, J.; Thörn, S.E.; Karlsson, O.; Kälebo, P.; Zetterberg, H.; Blennow, K.; Bergman, L. Signs of neuroaxonal injury in preeclampsia—A case control study. *PLoS ONE* **2021**, *16*, e0246786. [CrossRef]
16. Burwick, R.M.; Togioka, B.M.; Speranza, R.J.; Gaffney, J.E.; Roberts, V.H.; Frias, A.E.; Rincón, M. Assessment of blood-brain barrier integrity and neuroinflammation in preeclampsia. *Am. J. Obstet. Gynecol.* **2019**, *221*, 1–8. [CrossRef]
17. Ciampa, E.; Li, Y.; Dillon, S.; Lecarpentier, E.; Sorabella, L.; Libermann, T.A.; Karumanchi, S.A.; Hess, P.E. Cerebrospinal Fluid Protein Changes in Preeclampsia. *Hypertension* **2018**, *72*, 219–226. [CrossRef] [PubMed]
18. Van Den Berg, C.B.; Duvekot, J.J.; Guzel, C.; Hansson, S.R.; de Leeuw, T.G.; Steegers, E.A.P.; Versendaal, J.; Luider, T.M.; Stoop, M.T. Elevated levels of protein AMBP in cerebrospinal fluid of women with preeclampsia compared to normotensive pregnant women. *Proteom. Clin. Appl.* **2016**, *11*, 1600082. [CrossRef] [PubMed]
19. Bergman, L.; Bergman, K.; Langenegger, E.; Moodley, A.; Griffith-Richards, S.; Wikström, J.; Hall, D.; Joubert, L.; Herbst, P.; Schell, S.; et al. PROVE—Pre-Eclampsia Obstetric Adverse Events: Establishment of a Biobank and Database for Pre-Eclampsia. *Cells* **2021**, *10*, 959. [CrossRef]

20. Gestational Hypertension and Preeclampsia: ACOG Practice Bulletin, Number 222. *Obstet. Gynecol.* **2020**, *135*, 237–260. [CrossRef]
21. Harris, P.A.; Taylor, R.; Minor, B.L.; Elliott, V.; Fernandez, M.; O'Neal, L.; McLeod, L.; Delacqua, G.; Delacqua, F.; Kirby, J.; et al. The REDCap consortium: Building an international community of software platform partners. *J. Biomed. Inform.* **2019**, *95*, 103208. [CrossRef]
22. Tibbling, G.; Link, H.; Ohman, S. Principles of albumin and IgG analyses in neurological disorders. I. Establishment of reference values. *Scand. J. Clin. Lab. Invest.* **1977**, *37*, 385–390.
23. Clayton, A.M.; Shao, Q.; Paauw, N.D.; Giambrone, A.B.; Granger, J.P.; Warrington, J.P. Postpartum increases in cerebral edema and inflammation in response to placental ischemia during pregnancy. *Brain Behav. Immun.* **2018**, *70*, 376–389. [CrossRef]
24. Bergman, L.; Acurio, J.; Leon, J.; Gatu, E.; Friis, T.; Nelander, M.; Wikström, J.; Larsson, A.; Lara, E.; Aguayo, C.; et al. Preeclampsia and Increased Permeability Over the Blood–Brain Barrier: A Role of Vascular Endothelial Growth Receptor 2. *Am. J. Hypertens.* **2020**, *34*, 73–81. [CrossRef] [PubMed]
25. Hegen, H.; Auer, M.; Zeileis, A.; Deisenhammer, F. Upper reference limits for cerebrospinal fluid total protein and albumin quotient based on a large cohort of control patients: Implications for increased clinical specificity. *Clin. Chem. Lab. Med.* **2016**, *54*, 285–292. [CrossRef] [PubMed]
26. Parrado-Fernandez, C.; Blennow, K.; Hansson, M.; Leoni, V.; Cedazo-Minguez, A.; Bjorkhem, I. Evidence for sex difference in the CSF/plasma albumin ratio in ~20 000 patients and 335 healthy volunteers. *J. Cell Mol. Med.* **2018**, *22*, 5151–5154. [CrossRef] [PubMed]
27. Blennow, K.; Fredman, P.; Wallin, A.; Gottfries, C.G.; Karlsson, I.; Långström, G.; Skoog, I.; Svennerholm, L.; Wikkelsö, C. Protein analysis in cerebrospinal fluid. II. Reference values derived from healthy individuals 18-88 years of age. *Eur. Neurol.* **1993**, *33*, 129–133. [CrossRef]
28. Aukes, A.M.; Wessel, I.; Dubois, A.M.; Aarnoudse, J.G.; Zeeman, G.G. Self-reported cognitive functioning in formerly eclamptic women. *Am. J. Obstet. Gynecol.* **2007**, *197*, 1–6. [CrossRef]
29. Rengel, K.F.; Hayhurst, C.J.; Pandharipande, P.; Hughes, C.G. Long-term Cognitive and Functional Impairments After Critical Illness. *Anesth. Analg.* **2019**, *128*, 772–780. [CrossRef]
30. Ling, H.; Hardy, J.; Zetterberg, H. Neurological consequences of traumatic brain injuries in sports. *Mol. Cell. Neurosci.* **2015**, *66*, 114–122. [CrossRef] [PubMed]
31. Stolp, H.B.; Dziegielewska, K.M. Review: Role of developmental inflammation and blood-brain barrier dysfunction in neurodevelopmental and neurodegenerative diseases. *Neuropathol. Appl. Neurobiol.* **2009**, *35*, 132–146. [CrossRef] [PubMed]
32. Elharram, M.; Dayan, N.; Kaur, A.; Landry, T.; Pilote, L. Long-Term Cognitive Impairment After Preeclampsia: A Systematic Review and Meta-analysis. *Obstet. Gynecol.* **2018**, *132*, 355–364. [CrossRef] [PubMed]

cells

Article

Acid Sensing Ion Channel 2a Is Reduced in the Reduced Uterine Perfusion Pressure Mouse Model and Increases Seizure Susceptibility in Pregnant Mice

Maria Jones-Muhammad [1], Qingmei Shao [2], Loretta Cain-Shields [3], James P. Shaffery [4] and Junie P. Warrington [2,5,*]

[1] Program in Neuroscience, University of Mississippi Medical Center, Jackson, MS 39216, USA; mmjones@umc.edu
[2] Department of Neurology, University of Mississippi Medical Center, Jackson, MS 39216, USA; qshao2@umc.edu
[3] Department of Data Sciences, University of Mississippi Medical Center, Jackson, MS 39216, USA; lcain@umc.edu
[4] Department of Psychiatry, University of Mississippi Medical Center, Jackson, MS 39216, USA; jshaffery@umc.edu
[5] Department of Neurobiology and Anatomical Sciences, University of Mississippi Medical Center, Jackson, MS 39216, USA
* Correspondence: jpwarrington@umc.edu; Tel.: +1-601-815-8969

Abstract: Eclampsia is diagnosed in pregnant women who develop novel seizures. Our laboratory showed that the reduced uterine perfusion pressure (RUPP) rat model of preeclampsia displays reduced latency to drug-induced seizures. While acid sensing ion channels (ASIC1a and 3) are important for reducing seizure longevity and severity, the role of ASIC2a in mediating seizure sensitivity in pregnancy has not been investigated. We hypothesized that 1) RUPP reduces hippocampal ASIC2a, and 2) pregnant mice with reduced ASIC2a (ASIC2a+/−) have increased seizure sensitivity. On gestational day 18.5, hippocampi from sham and RUPP C57BL/6 mice were harvested, and ASIC2a was assessed using Western blot. Pregnant wild-type and ASIC2a+/− mice received 40 mg/kg of pentylenetetrazol (i.p.) and were video recorded for 30 min. Behaviors were scored using a modified Racine scale (0–7: 0 = no seizure; 7 = respiratory arrest/death). Seizure severity was classified as mild (score = 1–3) or severe (score = 4–7). RUPP mice had reduced hippocampal and placental ASIC2a protein. ASIC2a+/− mice had reduced latency to seizures, increased seizure duration, increased severe seizure duration, and higher maximum seizure scores. Reduced hippocampal ASIC2a in RUPP mice and increased seizure activity in pregnant ASIC2a+/− mice support the hypothesis that reduced ASIC2a increases seizure sensitivity associated with the RUPP.

Keywords: eclampsia; placental ischemia; seizure; ASIC2a; pregnancy; RUPP; pentylenetetrazol

Citation: Jones-Muhammad, M.; Shao, Q.; Cain-Shields, L.; Shaffery, J.P.; Warrington, J.P. Acid Sensing Ion Channel 2a Is Reduced in the Reduced Uterine Perfusion Pressure Mouse Model and Increases Seizure Susceptibility in Pregnant Mice. *Cells* **2021**, *10*, 1135. https://doi.org/10.3390/cells10051135

Academic Editor: Anthony Ashton

Received: 23 March 2021
Accepted: 5 May 2021
Published: 8 May 2021

1. Introduction

Preeclampsia, a hypertensive disorder of pregnancy that affects approximately 5% of pregnancies in the US, can advance to eclampsia, where the mother displays new-onset seizures [1]. Of the 5% of women diagnosed with preeclampsia, about 1.4% are subsequently diagnosed with eclampsia [1], which is responsible for roughly 10% of maternal deaths worldwide. The number of (pre)eclampsia cases is growing [2–4]. Women who survive preeclampsia/eclampsia complications are at a higher risk for cognitive decline later in life [5], and their children are at a higher risk of developing high blood pressure and other cardiovascular diseases, in addition to neurodevelopmental disorders [6,7]. In some cases, eclampsia can be avoided if the fetus and placenta are delivered early [8]. However, there are cases of women experiencing eclampsia during the postpartum period [9] and early delivery increases the number of premature births, which is also associated with adverse outcomes. Due to the detrimental impacts of eclampsia on the mother and fetus, it is imperative that the underlying mechanisms are identified.

Why some women, with otherwise healthy pregnancies, go on to have novel seizures is still unknown. In the clinic, women diagnosed with preeclampsia with severe symptoms are given magnesium sulfate prophylactically to prevent the progression to eclampsia. While magnesium sulfate is effective in preventing seizures, it cannot be used for chronic treatment, and the mechanism of magnesium sulfate's anti-seizure activity is not fully understood. Moreover, treatment with magnesium sulfate does not work in all patients [10]. In order to develop better treatments for eclampsia, a greater understanding of the mechanisms underlying the transition from preeclampsia to eclampsia is needed.

In preclinical experimental studies, seizures are induced in animal models of preeclampsia to produce eclampsia models. Pentylenetetrazol (PTZ), a GABA receptor antagonist previously demonstrated to produce temporal lobe epilepsy [11], is commonly used in preclinical eclampsia models [12–14]. Previous research from our group used PTZ to induce seizures in RUPP rats and showed reduced latency to seizure activity. This implies increased seizure susceptibility in RUPP rats compared to normal pregnant controls [13]. The mechanisms by which the RUPP procedure increases seizure susceptibility have not been fully elucidated.

One potential mechanism could be reduced acid sensing ion channel (ASIC) expression and activity. ASICs, proton-gated ion channels found in neural and peripheral tissue, are activated by reductions in the extracellular pH, a physiological response to ischemia and epilepsy. ASIC1a and 3 have important roles in mediating seizure activity [15]. The role of ASIC2a in seizure activity is not fully known, given that it is activated at a lower threshold of acidity (pH 4.0–4.9) [15]. Interestingly, Wu et al. reported that following induced seizures in male rats, ASIC2a expression was upregulated [16]. Moreover, overexpression of ASIC2a in the hippocampus of male rats increased seizure susceptibility [17]. Whether ASIC2a expression is altered in the RUPP model and whether the knockdown of ASIC2a in pregnant mice induces increased seizure severity and duration are not known.

Thus, the objectives of the current study were to: (1) determine whether RUPP surgery in mice induces decreased ASIC2a protein expression in the hippocampus and placenta; (2) determine whether ASIC2a heterozygous knockout mice have increased seizure severity and longevity during late pregnancy.

2. Materials and Methods

2.1. Animals

In this study, 10- to 12-week-old timed-pregnant C57BL/6 female mice ($n = 11$) from Charles River Laboratories were used for establishing the RUPP model in mice and for analysis of placental and hippocampal ASIC2a expression. Heterozygous B6.129-Asic2tm1Wsh/J mice were obtained from Jackson Laboratory (JAX Mice stock number: 013126) [18] and bred in the Lab Animal Facilities at the University of Mississippi Medical Center (UMMC) to generate genotypes for the experiments. Mice were fed a standard rodent diet (Teklad 22/5 Rodent Diet 8640) and water ad libitum until breeding, when they were switched to a breeders' diet (Teklad global 19% protein extruded rodent diet 2019) from mating until euthanasia at gestational day (GD) 18.5. Mice were housed at 2–5 per cage at a constant temperature of 74 °F and a 12 h light and 12 h dark cycle. All procedures were approved by the UMMC Institutional Animal Care and Use Committee (Protocol: 1434, 1434A, and 1509) in accordance with the 8th edition of the National Research Council Guide for Care and Use of Laboratory animals

2.2. Breeding and Establishment of Timed Pregnancy for ASIC2a Mice

One to 3 female mice were paired with 1 male of the same age and genotype in the evening. The following day, male mice were returned to their home cages. As mice were paired for only 1 night, the exact date of pregnancy is known. The day of separation was considered as GD 0.5. Mice were monitored for changes in abdominal size, and pregnancy was confirmed on GD 13.5. Mice that did not become pregnant were subjected to one additional round of mating.

2.3. Mouse RUPP Model

On gestational day 13.5, mice were anesthetized using 3% isoflurane (2% maintenance). A modified RUPP procedure, which has been demonstrated elsewhere [19], was used to induce utero-placental ischemia. An abdominal incision was made after cleaning with bromide and 70% ethanol, and the uterus containing the placental–fetal unit was exteriorized. The uterine arteries between the ovaries and first pup were ligated on both uterine horns using braided silk sutures (5.0) in RUPP mice ($n = 7$). Uterine horns were exposed without occlusion in sham mice ($n = 6$). Following surgery, all mice were given an analgesic (5.0 mg/kg Carprofen, s.c.) on the day of surgery and Rimadyl tablets daily for three days following the surgery. Our a priori exclusion criteria included excluding dams with only 1 pup present at GD13.5, mice that were not pregnant, and mice with 100% fetal resorptions on GD18.5 ($n = 1$ from the RUPP group). Additionally, mice with pups on only 1 horn were assigned to the sham group. Mice were not excluded from analysis for not having a preeclampsia phenotype.

2.4. Euthanasia and Tissue Collection

At gestation day (GD) 18.5, mice were weighed and anesthetized using 3% isoflurane, and blood was collected via cardiac puncture. Organs were removed and weighed. The number of live and resorbed pups were counted, and pups and placentas were weighed. Blood samples were collected from the dams and pups and used for assessing hematocrit, which has been used in previous studies to assess whether an animal is in a hypoxic state [20]. Brains were removed and dissected to isolate the hippocampus, which was flash frozen in liquid nitrogen and stored at −80 °C until further processing. The mean pup weight, placenta weight, and pup hematocrit were calculated per dam.

2.5. ELISA Analysis

Blood samples were centrifuged at 2000 rpm for 20 min, and serum was collected and stored at −20 °C until processing. Quantikine®ELISA kits from R&D systems and Biotechne were used. Each kit was specific for vascular endothelial growth factor (VEGF, Catalog no. MMV00), VEGFR1 or Flt-1 (Catalog no. MVR100), and placental growth factor (PlGF2, Catalog no. MP200). According to the manufacturer's information, the mouse VEGFR1/Flt-1 ELISA kit measures natural and recombinant mouse sVEGFR1 (sFlt-1). Maternal serum was diluted with the dilution buffer included in the kit as follows: 1:20 for Flt-1, 1:4 for VEGF, and 1:4 for PlGF. Standards and samples were assayed in duplicate in 96-well plates coated with the appropriate polyclonal antibody following the manufacturer's instructions.

2.6. Western Blot Analysis

Placental and hippocampal tissues from sham and RUPP mice were homogenized using ChemCruz RIPA lysis buffer in a mini bead homogenizer at 4000 rpm. The supernatant for each sample was collected. A bicinchoninic acid (BCA) protein assay was used to determine the amount of protein in each sample. Samples were prepared for Western blot analysis using 1x PBS, 4X sample buffer which contained 2 beta-mercaptoethanol. Samples were electrophoresed through Criterion TGX 4–20% Stain-free gel at 120 V for 90 min. When completed, the gel was activated using the ChemiDoc MP Imaging System for 45 s. The proteins were transferred to nitrocellulose membranes using the Trans blot turbo transfer kit. Following 30 min of blocking in Odyssey blocking buffer, the primary antibodies for the proteins of interest (ASIC2a antibody, 1:1000, Thermo Fisher Scientific, catalog# OSR00097W) and alpha-tubulin antibody (1:15,000, Abcam, catalog# ab89984) were added and incubated overnight. After washing, the secondary antibodies, donkey anti rabbit (1:15,000, Li-Cor, Catalog# 131591), and donkey anti chicken (1:15,000, Li-Cor,

Catalog# 925-68028) were added. Membranes were imaged using the Chemidoc MP system. Protein expression was normalized to alpha tubulin expression. Blots were analyzed using the Bio-Rad Chemidoc Image Lab software.

2.7. Measurement of Hypoxia-Inducible Factor in Placental Samples

Placental homogenates were assayed in duplicate on a HIF1α ELISA kit (Human/Mouse total HIF1A ELISA, Cat #: DYC1935-2, R&D Systems) following the manufacturer's directions. Samples were run undiluted. Measured concentrations were normalized to total protein in each sample to obtain pg/mg protein.

2.8. Seizure Induction Procedure

On GD 18.5, pregnant ASIC2a+/+ (n = 7) and ASIC2a+/− (n = 14) mice were administered 40 mg/kg of pentylenetetrazol (PTZ) via intraperitoneal (i.p.) injection. After injection, animals were video monitored for 30 min. The observer was blind to the genotype. Behavior was scored using a revised version of the Racine seizure scoring scale [21] from 0 to 7, with 0 indicating no seizure or normal behavior and 7 indicating respiratory arrest. Seizure activity was analyzed using the Observer XT software (Noldus Information Technology, Leesburg, VA, USA). We measured latency to the first seizure behavior, the total duration of seizure activity, the total duration of and latency to severe seizures (score 4–7), and the highest seizure score of each animal throughout the behavior monitoring. Immediately after observation, mice were euthanized and organs harvested under isoflurane anesthesia.

2.9. ASIC2a Genotyping

Tail snips were collected and used to confirm ASIC2a genotype. The KAPA Mouse Genotyping Kit from KAPABIOSYSTEMS was used for DNA extraction and PCR genotyping. The protocol for using KAPPA Expression Extract required a 50 µL solution comprising 44 µL of di-deionized water, 5 µL of 10X KAPA Expression Extract Buffer, and 1 U/µL of KAPA E. E. Enzyme. The tail snip was added to the solution, and the tissue was lysed at 75 °C for 10 min. The enzyme was inactivated at 95 °C for 5 min. After lysis, the lysate was centrifuged for 5 min, and the supernatant was stored for genotyping. The following sequences were used (5′–3′): Reaction A: AGT CCT GCA CGG TGG GAG CTT CTA; GAA GAG GAA GGG AGC CAT GAT GAG; Reaction B: ATG GTT TCG GAG TGG TTT GGC ATT GTG and TGG ATG TGG AAT GTG TGC GA. (Heterozygous band: 365 and 450 bp; wild type: 365 bp.) Gels for running samples were 2% gels comprising 35 mL of di-deionized water, 0.7 g of agarose gel, and 3.5 µL of GelRed. An amount of 6 µL of 100 bp ruler was loaded, and 5 µL of sample was loaded before running the gel at 140 V for 45 min.

2.10. Data Analysis

Datasets were tested for normality using the Shapiro–Wilk test. We used Mann–Whitney U tests for non-normal data distributions and an unpaired t-test for normally distributed datasets. Data are presented as the mean ± SD. Differences were considered statistically significant at the $p < 0.05$ level. One mouse from the RUPP group was excluded from analysis due to having 100% resorptions at GD18.5 as a result of the surgery. Statistical analyses were performed, and graphs were generated using GraphPad Prism software (version 8.4.3).

3. Results

3.1. The Mouse RUPP Model Has Some Similar Clinical Characteristics to the Preeclampsia Patient and RUPP Rat

The maternal and fetal characteristics of the RUPP model at GD 18.5 are summarized in Table 1. There was no significant difference in maternal body weight or maternal hematocrit between groups at GD18.5. Additionally, at GD13.5, before the surgeries, there was no difference in body weight between the dams (28.0 ± 1.6 in dams designated to sham group vs. 29.5 ± 0.9 in the dams subjected to the RUPP; p = 0.218). In terms of fetal outcomes, there was a reduction in the number of live pups, and no difference in fetal resorptions, pup weight, or placental weight. Fetal hematocrit was increased in the RUPP group, indicating fetal hypoxia.

Table 1. General characteristics and pregnancy outcomes in RUPP mice.

Characteristic	Sham (n = 6)	RUPP (n = 6)	p-Value
Maternal			
Body weight (g)	39.2 ± 1.4	37.8 ± 1.9	0.289
Hematocrit (%)	35 ± 2	35 ± 1	‡ 0.446
Fetal Outcomes			
No. of live pups	8 ± 1	5 ± 1	0.0499 †
% resorptions	0 ± 0	21 ± 15	0.111
Pup weight (g)	1.13 ± 0.03	1.21 ± 0.06	§ 0.294
Placenta weight (g)	0.09 ± 0.00	0.10 ± 0.00	0.173
Pup Hematocrit (%)	30 ± 2	35 ± 1	0.037 †

† p = 0.0499 vs. sham; ‡ data analyzed using Welch's test; § data analyzed using Mann–Whitney U test.

To further characterize the mouse RUPP model, we measured maternal circulating factors using serum samples. The serum concentration of sFlt1 was significantly increased in RUPP mice when compared to sham mice (52.73 ± 9.5 vs. 32.39 ± 3.5 pg/mL; p = 0.0289; Figure 1A), while the VEGF concentration was not different between groups (71.04 ± 11.22 vs. 66.7 ± 11.35 pg/mL; p = 0.272; Figure 1B). RUPP mice had a higher serum concentration of PlGF (354.8 ± 18.02 vs. 259.2 ± 20.36 pg/mL; p = 0.0037; Figure 1C), a higher sFlt-1/VEGF ratio (31.87 ± 14.05 vs. 18.14 ± 4.840; p = 0.025; Figure 1D), and no difference in the sFlt-1/PlGF ratio (146.4 ± 46.77 vs. 128.3 ± 39.24; p = 0.250; Figure 1E). In response to RUPP, mice showed an increasing trend for circulating TNFα levels (2.56 ± 1.95 in RUPP vs. 0.98 ± 0.86 pg/mL in sham; p = 0.058; Figure 1F). Placental HIF1α was not different between the groups (3.90 ± 0.71 pg/mg protein in shams vs. 3.91 ± 0.60 pg/mg protein in RUPP; p = 0.497). Together, these data demonstrate that the mouse RUPP model displays few similar features of the rat RUPP model and clinical preeclampsia, with elevated sFlt-1 being the primary shared clinical characteristic.

3.2. The RUPP Model in the Mouse Induced Reduced ASIC2a Protein Expression in the Placenta and Hippocampus

Western blot analysis was used to determine if RUPP induces changes in the expression of ASIC2a in the hippocampus and placenta. In the hippocampus, there was a significant reduction in the expression of ASIC2a in RUPP mice compared to the sham control (0.49 ± 0.08 vs. 1.0 ± 0.22; p = 0.037; Figure 2A). In the placenta, RUPP mice had a reduced expression of ASIC2a compared to sham mice (0.78 ± 0.04 vs. 1.0 ± 0.09; p = 0.038; Figure 2B). Full images of the Western blots can be found in the Supplementary Materials Figures S1–S3. These data show that the RUPP procedure induced a reduction in ASIC2a expression in the hippocampus and placenta.

3.3. General Characteristics of Pregnant ASIC2a Wild-Type and Heterozygous Knockout Mice at GD18.5

Pregnant ASIC2a +/+ and +/− mice were used for assessing seizure susceptibility. A representative gel for each genotype can be found in Supplementary Materials Figure S4. Pregnancy characteristics for each genotype are shown in Table 2. There was no difference in body weight between +/+ and +/− dams, the number of live pups, fetal resorptions, pup weight, or placental weight at GD 18.5.

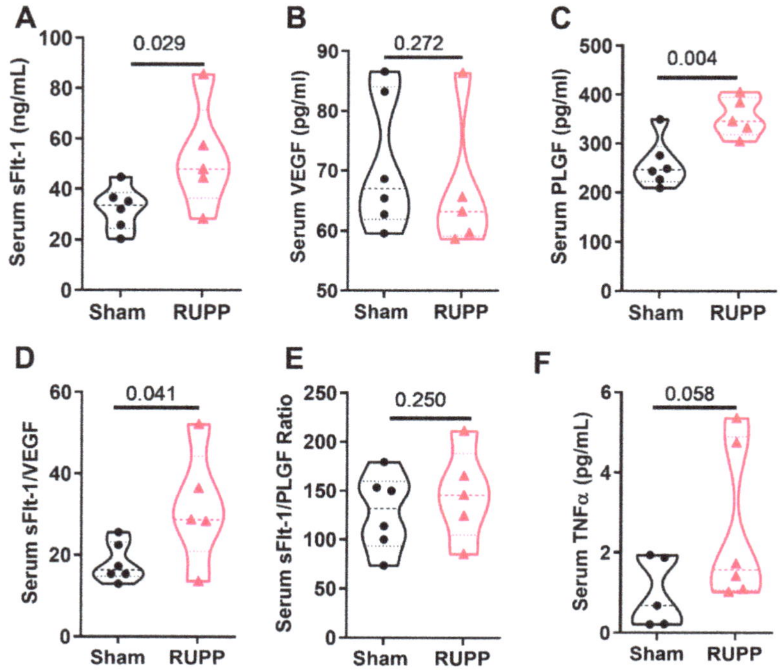

Figure 1. RUPP in the mouse induces some similar changes in angiogenic factors to the rat and preeclampsia patient. RUPP-induced (**A**) increased sFlt-1, (**B**) no change in VEGF, (**C**) increased PlGF, (**D**) increased sFlt-1/VEGF ratio, and (**E**) no change in sFlt-1/PlGF ratio in serum samples compared to sham controls. (**F**) Trend for increased TNFα levels in RUPP mice. Sham $n = 6$ mice and RUPP $n = 5$. Data analyzed using unpaired t-test. Points indicate values for individual dams. SFlt—soluble Fms-like tyrosine kinase 1; VEGF—vascular endothelial growth factor; PLGF—placental growth factor.

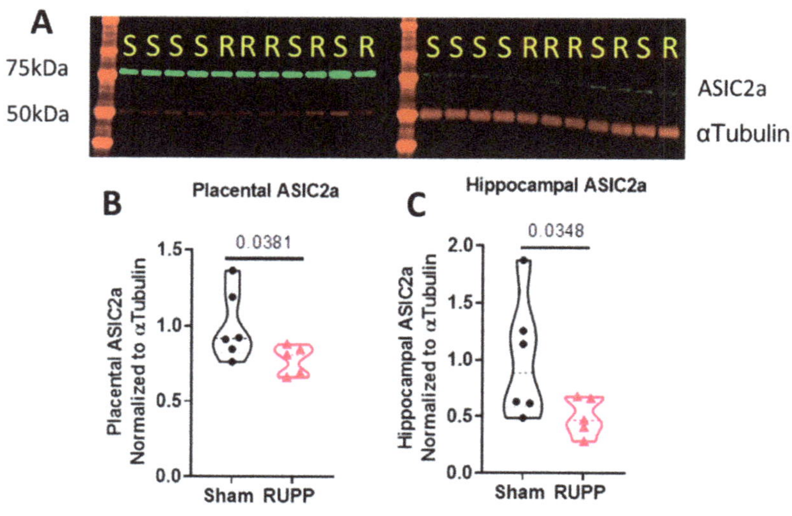

Figure 2. RUPP surgery reduces hippocampal and placental expression of ASIC2a. (**A**) Representative Western blot image from placental and hippocampal samples. Mice that underwent the RUPP procedure displayed reduced (**B**) placental and (**C**) hippocampal ASIC2a expression. Data were normalized to alpha tubulin followed by normalization to the sham levels. SHAM $n = 6$; RUPP $n = 5$. Data were analyzed using unpaired t-test (**B**) or Welch's t-test (**C**). Points indicate values for individual dams.

Table 2. General characteristics and pregnancy outcomes in ASIC2a mice at GD18.5.

Characteristic	ASIC2a+/+ (n = 7)	ASIC2a+/− (n = 14)	p-Value
Maternal			
Body weight (g)	35.5 ± 2.2	35.4±3.0	0.406
Fetal Outcomes			
No. of live pups	8 ± 1	7 ± 2	0.358
% resorptions	11 ± 13	5 ± 6	0.071
Pup weight (g)	0.87 ± 0.02	0.85 ± 0.01	0.096
Placenta weight (g)	0.10 ± 0.01	0.11 ± 0.01	0.066

Data analyzed using Mann–Whitney *U* test.

3.4. Knockdown of ASIC2a Results in Increased Seizure Severity and Longevity during Late Pregnancy

The Racine scale was used to analyze seizure behavior in ASIC2a mice. Pregnant ASIC2a +/− mice have significantly reduced latency to the first detected seizure behavior compared to pregnant +/+ mice (101.4 ± 17.61s vs. 247.7 ± 84.57s $p = 0.033$; Figure 3A). There was no differences in the latency to severe seizures ($p > 0.05$; Figure 3B); however, +/− mice had a higher overall seizure score compared to +/+ mice (4.9 ± 0.4 vs. 3.3 ± 0.5 $p = 0.014$; Figure 3C). ASIC2a+/− mice had an overall longer duration of seizure than +/+ mice (1253 ± 124.7s vs. 822.6 ± 136.6s $p = 0.019$; Figure 3D) as well as longer severe seizures (713.1 ± 187.7s vs. 18.29 ± 13.38s $p = 0.0036$; Figure 3E). These results indicate that reduced ASIC2a expression increases seizure susceptibility and severity during pregnancy.

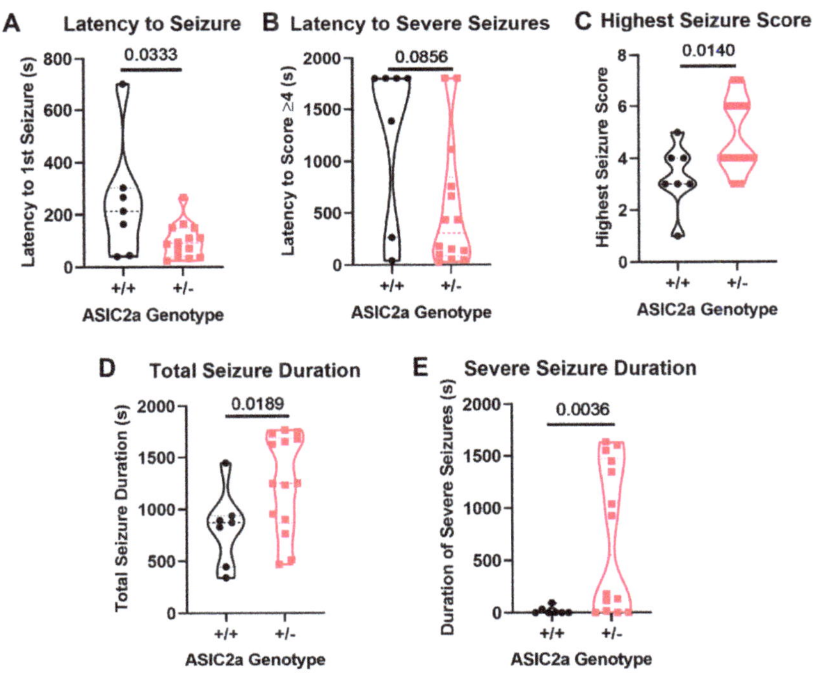

Figure 3. ASIC2a knockdown results in increased seizure severity and longevity in pregnancy. Pregnant ASIC2a+/− mice show (**A**) significantly reduced latency to the first seizure activity, (**B**) no difference in reduced latency to severe seizure activity, (**C**) higher seizure scores, (**D**) increased total duration of seizures, and (**E**) increased duration of severe seizures compared to pregnant ASIC2a+/+ mice. n = 7 (+/+) or n = 14 (+/−) mice. Data analyzed using Mann–Whitney U test. Points indicate values for individual dams.

4. Discussion

The mechanisms contributing to increased seizure sensitivity following RUPP are not fully known. The current study investigated whether inducing RUPP in mice would lead to a decrease in ASIC2a expression and whether the knockdown of ASIC2a would lead to increases in seizure susceptibility during normal pregnancy. We first measured the expression of ASIC2a in placental and hippocampal tissues from sham and RUPP mice and then determined whether reduced ASIC2a expression increased seizure severity and longevity in pregnant mice. The results indicate that RUPP mice have a reduction in ASIC2a protein expression in the hippocampus and the placenta at gestational day 18.5 of pregnancy. When seizures were induced in pregnant ASIC2a+/+ and +/− mice, ASIC2a+/− mice displayed seizure behavior suggesting increased seizure susceptibility and severity. These data support the hypothesis that the reduced hippocampal expression of ASIC2a in RUPP mice is a contributing factor to RUPP-induced increases in seizure susceptibility, although this hypothesis remains to be directly tested.

4.1. The Mouse RUPP Model

The current study used a version of the RUPP described by Fushima et al. where the authors reported several characteristics similar to the preeclampsia patient and the rat RUPP model [19]. While the majority of studies utilizing RUPP surgery are performed in rats [22–25], the use of a mouse RUPP model allows investigators to directly test the involvement of specific pathways by using genetic mouse models. Hallmarks of the rat RUPP model mimic features of clinical preeclampsia and include increased blood pressure; increased fetal demise; and an imbalance of maternal angiogenic factors PlGF, VEGF, and sFlt1 [26–29]. Here, we demonstrated that pregnant mice subjected to the RUPP procedure, as presented by Fushima et al. [19], display fewer similar features. We demonstrate fetal hypoxia (increased fetal hematocrit), reduced litter size at GD18.5, elevated sFlt-1, and a trend for increased TNFα in maternal serum. We do not know why we were not able to replicate all the findings reported in the Fushima study but think that the difference in the length of exposure to the RUPP may have contributed. In the study by Fushima and colleagues, the length of ischemia was 4 days (GD14–18), while our model was exposed for 5 days (GD13.5–18.5). Moreover, in our model, while hematocrit was increased in the fetuses, we did not find evidence of placental hypoxia, as placental HIF-1α was not different between the groups. We did not measure proteinuria or blood pressure from conscious mice and did not find evidence of intrauterine growth restriction in this study. It is possible that occluding only the uterine arteries near the ovaries resulted in compensatory flow from the abdominal aorta, supplying the inferior uterine horn. This compensatory flow could, therefore, contribute to preserving fetal weights. Indeed, in preliminary studies and published work, when the abdominal aorta is partially occluded, intrauterine growth restriction results [19,30]. We chose not to occlude the abdominal aorta in this study due to the complication of hind limb paralysis that occurred in a proportion of the mice used in pilot studies (data not shown).

The RUPP model is used in many studies investigating preeclampsia due to its ability to re-create symptoms of clinical preeclampsia, such as reduced concentrations of VEGF and PlGF in urine and maternal serum [31–34]. Although the modified RUPP model displayed similar symptoms to that of the rat RUPP model, such as an increase in serum sFlt-1 concentration, we found a significant increase in the concentration of PlGF, no change in the concentration of VEGF, and an increase in the sFlt1/VEGF ratio in maternal serum, whereas other models display a significant reduction in PlGF and VEGF [24,27]. It should be noted that the studies investigating changes in PlGF and VEGF were performed in rats and analyzed free or unbound VEGF and PlGF levels, whereas the ELISA kit in the current study measured total VEGF and PlGF. Importantly, when the ratio of sFLt-1/VEGF was calculated, we found a significant increase in the RUPP, suggesting that the level of unbound and bioavailable VEGF is reduced in response to the RUPP. The use of an ELISA kit that quantifies the unbound proangiogenic factors will aid in better characterizing the RUPP model.

4.2. Reduced ASIC2a Expression

ASICs are mechano- and chemoreceptors that have a significant role in neuroprotection [35,36]. The ability of ASICs to sense reductions in the extracellular pH, which occurs in adverse circumstances, such as seizures or hypertension [16,35,37], suggests that they may have a role in seizure susceptibility induced by RUPP. Several studies have shown that manipulating various ASIC isoforms via genetic and pharmacological approaches (ASIC1a reduction and ASIC3 elevation) increases seizure severity [38–40]. Studies investigating the role of ASIC2a show that ASIC2a overexpression increases seizure severity [16,17]. In spite of this phenomenon, there are no studies that have investigated the impact of reduced ASIC2a on seizure susceptibility in females. Additionally, no studies have investigated changes in ASIC2a expression in normal pregnancy or following the RUPP procedure. Due to this gap in knowledge, we assessed whether RUPP contributed to changes in ASIC2a expression.

It is currently not known what factors contribute to seizures in pregnancy and preeclampsia. This gap in knowledge is critical, given the life-threatening and long-term health risks of the mother following preeclampsia [41] and the dangers a seizure during pregnancy poses on the fetus [42]. Women with severe preeclampsia are given magnesium sulfate to prevent seizures; however, this method is not preventative in all cases of preeclampsia. Additionally, finding the optimal point at which treatments can prevent seizure as well as prevent the need to induce early delivery is imperative. Previous work from our laboratory demonstrated increased seizure susceptibility in a rat RUPP model [13]; however, the mechanism by which this susceptibility occurs had not been investigated. Here, we demonstrated, for the first time, that the pH-activated channel ASIC2a is significantly reduced in the placenta and hippocampus of the mouse RUPP model. Investigating seizure susceptibility in the mouse RUPP model is required and is being investigated in the lab. Additionally, whether manipulating ASIC2a expression in RUPP mice modulates seizure activity has not been investigated and is the next logical research question to answer.

Our finding of reduced ASIC2a expression in the placenta is interesting. Currently, there are no studies investigating the role of ASIC2a in the placenta. Nevertheless, work by Ali and colleagues showed that reductions in the extracellular pH led to marked vasodilation in placental arteries isolated from healthy pregnant women and that the response to reduced pH was markedly attenuated in vessels isolated from preeclampsia patients [43]. These findings, along with the finding of reduced ASIC2a in the RUPP placenta of our current study, suggest that pH-sensitive signaling is impaired in placentas from preeclampsia patients, and this may be due to a reduction in the expression of ASICs, such as ASIC2a.

4.3. Seizures in Pregnant ASIC2a Mice

The current findings demonstrate that pregnant mice with reduced ASIC2a have increased seizure susceptibility. Previous findings have suggested that the overexpression of ASIC2a increases seizure activity as well; however, the study used male rats in which ASIC2a was overexpressed in the hippocampus, whereas the current study used pregnant female mice with a global reduction in ASIC2a [16,17]. Whether there are sex- and species-dependent differences in the effect of ASIC2a manipulation on seizure susceptibility requires further investigation. It is also possible that there is a specific level of ASIC2a required for normal neuronal activity and that increasing or decreasing beyond that level would result in increased seizure susceptibility. The possibility of pregnancy having a significant impact on seizure susceptibility cannot be ruled out. Indeed, previous work by Johnson et al. found that normal pregnancy contributes to increased seizure susceptibility when compared to the non-pregnant control [14]. Although the previously mentioned study used rats, determining the contribution of pregnancy to seizure susceptibility in mice requires further investigation.

Another study analyzed the expression of ASIC2a in human hippocampal samples obtained from epilepsy patients and found elevated ASIC2a expression. Of note, only patients with a history of epilepsy were analyzed [17]; therefore, it is not known if these ASIC2a levels were gradually elevated as the epilepsy disorder progressed or whether ASIC2a levels were already different before the development of the seizure disorder. As the patients had a history of epilepsy, these human data may correspond to women who have epilepsy before becoming pregnant rather than de novo seizures during gestation or in the postpartum period [17]. We did not assess whether ASIC2a expression increases following seizures similarly to what is observed in epilepsy patients. This question will be investigated in future studies.

4.4. Limitations

The limitations of our study need to be considered. We did not measure blood pressure in our mice and cannot confidently say whether our findings were influenced by blood pressure changes in the RUPP. Moreover, we did not determine whether RUPP mice have evidence of proteinuria or kidney injury or even some of the cerebrovascular abnormalities reported in RUPP rats. Further characterization of the mouse RUPP model is required. In addition, to continue characterizing our RUPP mouse model, future work will investigate whether seizure induction using PTZ in RUPP mice will replicate the findings in the rat RUPP model. We will also need to investigate whether manipulating the expression of ASIC2a and inducing RUPP in these mice will lead to changes in seizure severity and longevity.

Another important consideration is that other isoforms of ASICs may also be affected by the RUPP procedure. ASIC2a forms complexes with other ASIC isoforms, such as ASIC2b and ASIC1a, and epithelial sodium channel (beta ENaC) proteins [15,44,45]. Ongoing studies are investigating whether RUPP leads to changes in the expression of these proteins. Another caveat to consider is that we used only one method to induce seizures and to analyze seizure behavior. Future studies will incorporate the use other pro-convulsive drugs and EEG recordings for further confirmation of seizure activity, similar to other studies [46].

5. Conclusions

In conclusion, our findings are the first, to our knowledge, to demonstrate that RUPP mice have a reduced expression of hippocampal and placental ASIC2a. This study is also the first to show that a reduction in ASIC2a expression leads to more severe seizure activity and increased duration of seizures during pregnancy. Thus, ASIC2a may be a therapeutic target to prevent the occurrence or to reduce the severity of seizures in pregnancy. Further investigation is required to determine if RUPP and/or ASIC2a reduction contributes to changes in the expression of excitatory and inhibitory neurotransmitter receptors or other ASIC isoforms in the context of epilepsy.

Supplementary Materials: The following are available online at https://www.mdpi.com/article/10.3390/cells10051135/s1, Figure S1: Full Western blot for ASIC2a (green) and alpha tubulin (red) staining. Membrane was probed for both ASIC2a and alpha tubulin. Placental samples are shown on the left, while hippocampal samples are shown on the right. The ASIC2a bands were observed at ~76 kDa, and alpha tubulin bands were visible at 50 kDa. Figure S2: Full Western blot (inverted greyscale image) for ASIC2a staining. On the left, placental samples were loaded. On the right, hippocampal samples are shown. The ASIC2a bands were observed at ~76 kDa. Figure S3: Full Western blot (inverted greyscale image) for alpha tubulin staining. On the left, placental samples were loaded. On the right, hippocampal samples are shown. The antibody recognized a band at 50 kDa. Extra bands were visible in the hippocampal samples. Figure S4: Representative bands from ASIC2a genotyping. ASIC2a+/+ mice have only the WT band (365bp), while ASIC2a+/− mice have the WT band plus the mutant band (450 bp).

Author Contributions: All authors had full access to all the data in the study and take responsibility for the integrity of the data and the accuracy of the data analysis. Conceptualization, J.P.W.; Methodology, M.J.-M., Q.S., J.P.S. and J.P.W.; Investigation, M.J.-M., Q.S. and J.P.W.; Formal Analysis, M.J.-M. and J.P.W.; Resources: J.P.S. and J.P.W.; Writing—Original Draft, M.J.-M. and J.P.W.; Writing—Review and Editing, M.J.-M., Q.S., L.C.-S., J.P.S. and J.P.W.; Visualization, M.J.-M. and J.P.W.; Supervision, J.P.W.; Funding Acquisition, J.P.W. All authors have read and agreed to the published version of the manuscript.

Funding: This research was funded by a COBRE pilot grant from the National Institutes of Health (P20GM104357), a K99/R00 award to JPW (R00HL129192), and a supplement R00HL129192-S1. Seizure behaviors were conducted in collaboration with the Animal Behavior Core, funded through an Institutional Development Award (IDeA) from the National Institute of General Medical Sciences of the NIH under grant number P30GM103328. MJM receives support from the T32 Training Grant T32HL105324. The APC was funded by R00HL129192.

Institutional Review Board Statement: The study was conducted according to the 8th edition of the National Research Council Guide for Care and Use of Laboratory animals, and approved by the Institutional Animal Care and Use Committee of the University of Mississippi Medical Center (protocol number: 1509, 1434, 1434A).

Informed Consent Statement: Not applicable.

Data Availability Statement: All data are available from the corresponding author upon reasonable request.

Acknowledgments: We would like to acknowledge the technical contribution of Rachael Thompson who helped with mouse colony maintenance and tissue harvest.

Conflicts of Interest: The authors declare no conflict of interest. The funders had no role in the design of the study; in the collection, analyses, or interpretation of data; in the writing of the manuscript, or in the decision to publish the results.

References

1. Fingar, K.R.; Mabry-Hernandez, I.; Ngo-Metzger, Q.; Wolff, T.; Steiner, C.A.; Elixhauser, A. Delivery Hospitalization Involving Preeclampsia and Eclampsia, 2005–2014. HCUP Statistical Brief #0222. April 2017. Agency for Healthcare Research and Quality, Rockville, MD. Available online: www.hcup-us.ahrq.gov/reports/statbriefs/sb222-Preeclampsia-Eclampsia-Delivery-Trends.pdf (accessed on 19 November 2019).
2. Vousden, N.; Lawley, E.; Seed, P.T.; Gidiri, M.F.; Goudar, S.; Sandall, J.; Chappell, L.C.; Shennan, A.H.; Group, C.T.C. Incidence of eclampsia and related complications across 10 low- and middle-resource geographical regions: Secondary analysis of a cluster randomised controlled trial. *PLoS Med.* **2019**, *16*, e1002775. [CrossRef] [PubMed]
3. Ghulmiyyah, L.; Sibai, B. Maternal mortality from preeclampsia/eclampsia. *Semin. Perinatol.* **2012**, *36*, 56–59. [CrossRef] [PubMed]
4. Sibai, B.M. Diagnosis, prevention, and management of eclampsia. *Obstet. Gynecol.* **2005**, *105*, 402–410. [CrossRef]
5. Ijomone, O.K.; Shallie, P.; Naicker, T. Changes in the structure and function of the brain years after Pre-eclampsia. *Ageing Res. Rev.* **2018**, *47*, 49–54. [CrossRef]
6. Goffin, S.M.; Derraik, J.G.B.; Groom, K.M.; Cutfield, W.S. Maternal pre-eclampsia and long-term offspring health: Is there a shadow cast? *Pregnancy Hypertens.* **2018**, *12*, 11–15. [CrossRef]
7. Dang, F.; Croy, B.A.; Stroman, P.W.; Figueiro-Filho, E.A. Impacts of Preeclampsia on the Brain of the Offspring. *Rev. Bras. Ginecol. Obstet.* **2016**, *38*, 416–422. [CrossRef]
8. Office of Communications. What Are the Treatments for Preeclampsia, Eclampsia, & HELLP Syndrome? Available online: https://www.nichd.nih.gov/health/topics/preeclampsia/conditioninfo/treatments (accessed on 29 April 2020).
9. Lubarsky, S.L.; Barton, J.R.; Friedman, S.A.; Nasreddine, S.; Ramadan, M.K.; Sibai, B.M. Late postpartum eclampsia revisited. *Obstet. Gynecol.* **1994**, *83*, 502–505. [CrossRef]
10. Duley, L.; Gulmezoglu, A.M.; Henderson-Smart, D.J.; Chou, D. Magnesium sulphate and other anticonvulsants for women with pre-eclampsia. *Cochrane Database Syst. Rev.* **2010**, *11*, CD000025. [CrossRef]
11. Huang, R.Q.; Bell-Horner, C.L.; Dibas, M.I.; Covey, D.F.; Drewe, J.A.; Dillon, G.H. Pentylenetetrazole-induced inhibition of recombinant gamma-aminobutyric acid type A (GABA(A)) receptors: Mechanism and site of action. *J. Pharmacol. Exp. Ther.* **2001**, *298*, 986–995.
12. Huang, Q.; Liu, L.; Hu, B.; Di, X.; Brennecke, S.P.; Liu, H. Decreased seizure threshold in an eclampsia-like model induced in pregnant rats with lipopolysaccharide and pentylenetetrazol treatments. *PLoS ONE* **2014**, *9*, e89333. [CrossRef] [PubMed]
13. Warrington, J.P. Placental ischemia increases seizure susceptibility and cerebrospinal fluid cytokines. *Physiol. Rep.* **2015**, *3*. [CrossRef]

14. Johnson, A.C.; Nagle, K.J.; Tremble, S.M.; Cipolla, M.J. The Contribution of Normal Pregnancy to Eclampsia. *PLoS ONE* **2015**, *10*, e0133953. [CrossRef]
15. Boscardin, E.; Alijevic, O.; Hummler, E.; Frateschi, S.; Kellenberger, S. The function and regulation of acid-sensing ion channels (ASICs) and the epithelial Na(+) channel (ENaC): IUPHAR Review 19. *Br. J. Pharmacol.* **2016**, *173*, 2671–2701. [CrossRef]
16. Wu, H.; Wang, C.; Liu, B.; Li, H.; Zhang, Y.; Dong, S.; Gao, G.; Zhang, H. Altered Expression Pattern of Acid-Sensing Ion Channel Isoforms in Piriform Cortex After Seizures. *Mol. Neurobiol.* **2016**, *53*, 1782–1793. [CrossRef]
17. Zhang, H.; Gao, G.; Zhang, Y.; Sun, Y.; Li, H.; Dong, S.; Ma, W.; Liu, B.; Wang, W.; Wu, H.; et al. Glucose Deficiency Elevates Acid-Sensing Ion Channel 2a Expression and Increases Seizure Susceptibility in Temporal Lobe Epilepsy. *Sci. Rep.* **2017**, *7*, 5870. [CrossRef]
18. Price, M.P.; Lewin, G.R.; McIlwrath, S.L.; Cheng, C.; Xie, J.; Heppenstall, P.A.; Stucky, C.L.; Mannsfeldt, A.G.; Brennan, T.J.; Drummond, H.A.; et al. The mammalian sodium channel BNC1 is required for normal touch sensation. *Nature* **2000**, *407*, 1007–1011. [CrossRef] [PubMed]
19. Fushima, T.; Sekimoto, A.; Minato, T.; Ito, T.; Oe, Y.; Kisu, K.; Sato, E.; Funamoto, K.; Hayase, T.; Kimura, Y.; et al. Reduced Uterine Perfusion Pressure (RUPP) Model of Preeclampsia in Mice. *PLoS ONE* **2016**, *11*, e0155426. [CrossRef] [PubMed]
20. Hunter, C.; Barer, G.R.; Shaw, J.W.; Clegg, E.J. Growth of the heart and lungs in hypoxic rodents: A model of human hypoxic disease. *Clin. Sci. Mol. Med.* **1974**, *46*, 375–391. [CrossRef] [PubMed]
21. Luttjohann, A.; Fabene, P.F.; van Luijtelaar, G. A revised Racine's scale for PTZ-induced seizures in rats. *Physiol. Behav.* **2009**, *98*, 579–586. [CrossRef]
22. Li, J.; LaMarca, B.; Reckelhoff, J.F. A model of preeclampsia in rats: The reduced uterine perfusion pressure (RUPP) model. *Am. J. Physiol. Heart C* **2012**, *303*, H1–H8. [CrossRef]
23. Morton, J.S.; Levasseur, J.; Ganguly, E.; Quon, A.; Kirschenman, R.; Dyck, J.R.B.; Fraser, G.M.; Davidge, S.T. Characterisation of the Selective Reduced Uteroplacental Perfusion (sRUPP) Model of Preeclampsia. *Sci. Rep.* **2019**, *9*, 9565. [CrossRef]
24. Gilbert, J.S.; Babcock, S.A.; Granger, J.P. Hypertension produced by reduced uterine perfusion in pregnant rats is associated with increased soluble fms-like tyrosine kinase-1 expression. *Hypertension* **2007**, *50*, 1142–1147. [CrossRef]
25. Granger, J.P.; LaMarca, B.B.; Cockrell, K.; Sedeek, M.; Balzi, C.; Chandler, D.; Bennett, W. Reduced uterine perfusion pressure (RUPP) model for studying cardiovascular-renal dysfunction in response to placental ischemia. *Methods Mol. Med.* **2006**, *122*, 383–392. [CrossRef]
26. Helske, S.; Vuorela, P.; Carpen, O.; Hornig, C.; Weich, H.; Halmesmaki, E. Expression of vascular endothelial growth factor receptors 1, 2 and 3 in placentas from normal and complicated pregnancies. *Mol. Hum. Reprod.* **2001**, *7*, 205–210. [CrossRef]
27. Maynard, S.E.; Min, J.Y.; Merchan, J.; Lim, K.H.; Li, J.; Mondal, S.; Libermann, T.A.; Morgan, J.P.; Sellke, F.W.; Stillman, I.E.; et al. Excess placental soluble fms-like tyrosine kinase 1 (sFlt1) may contribute to endothelial dysfunction, hypertension, and proteinuria in preeclampsia. *J. Clin. Investig.* **2003**, *111*, 649–658. [CrossRef] [PubMed]
28. Madazli, R.; Aydin, S.; Uludag, S.; Vildan, O.; Tolun, N. Maternal plasma levels of cytokines in normal and preeclamptic pregnancies and their relationship with diastolic blood pressure and fibronectin levels. *Acta Obstet. Gynecol. Scand.* **2003**, *82*, 797–802. [CrossRef]
29. Levine, R.J.; Maynard, S.E.; Qian, C.; Lim, K.H.; England, L.J.; Yu, K.F.; Schisterman, E.F.; Thadhani, R.; Sachs, B.P.; Epstein, F.H.; et al. Circulating angiogenic factors and the risk of preeclampsia. *N. Engl. J. Med.* **2004**, *350*, 672–683. [CrossRef] [PubMed]
30. Intapad, S.; Warrington, J.P.; Spradley, F.T.; Palei, A.C.; Drummond, H.A.; Ryan, M.J.; Granger, J.P.; Alexander, B.T. Reduced uterine perfusion pressure induces hypertension in the pregnant mouse. *Am. J. Physiol. Regul. Integr. Comp. Physiol.* **2014**, *307*, R1353–R1357. [CrossRef]
31. Aggarwal, P.K.; Jain, V.; Sakhuja, V.; Karumanchi, S.A.; Jha, V. Low urinary placental growth factor is a marker of pre-eclampsia. *Kidney Int* **2006**, *69*, 621–624. [CrossRef] [PubMed]
32. Livingston, J.C.; Chin, R.; Haddad, B.; McKinney, E.T.; Ahokas, R.; Sibai, B.M. Reductions of vascular endothelial growth factor and placental growth factor concentrations in severe preeclampsia. *Am. J. Obstet. Gynecol.* **2000**, *183*, 1554–1557. [CrossRef]
33. Levine, R.J.; Karumanchi, S.A. Circulating angiogenic factors in preeclampsia. *Clin. Obstet. Gynecol.* **2005**, *48*, 372–386. [CrossRef] [PubMed]
34. Taylor, R.N.; Grimwood, J.; Taylor, R.S.; McMaster, M.T.; Fisher, S.J.; North, R.A. Longitudinal serum concentrations of placental growth factor: Evidence for abnormal placental angiogenesis in pathologic pregnancies. *Am. J. Obstet. Gynecol.* **2003**, *188*, 177–182. [CrossRef] [PubMed]
35. Kweon, H.J.; Suh, B.C. Acid-sensing ion channels (ASICs): Therapeutic targets for neurological diseases and their regulation. *BMB Rep.* **2013**, *46*, 295–304. [CrossRef] [PubMed]
36. Liang, J.J.; Huang, L.F.; Chen, X.M.; Pan, S.Q.; Lu, Z.N.; Xiao, Z.M. Amiloride suppresses pilocarpine-induced seizures via ASICs other than NHE in rats. *Int. J. Clin. Exp. Pathol.* **2015**, *8*, 14507–14513. [PubMed]
37. Cristofori-Armstrong, B.; Rash, L.D. Acid-sensing ion channel (ASIC) structure and function: Insights from spider, snake and sea anemone venoms. *Neuropharmacology* **2017**, *127*, 173–184. [CrossRef] [PubMed]
38. Cao, Q.; Wang, W.; Gu, J.; Jiang, G.; Wang, K.; Xu, Z.; Li, J.; Chen, G.; Wang, X. Elevated Expression of Acid-Sensing Ion Channel 3 Inhibits Epilepsy via Activation of Interneurons. *Mol. Neurobiol.* **2016**, *53*, 485–498. [CrossRef]
39. Yang, F.; Sun, X.; Ding, Y.; Ma, H.; Yang, T.O.; Ma, Y.; Wei, D.; Li, W.; Xu, T.; Jiang, W. Astrocytic Acid-Sensing Ion Channel 1a Contributes to the Development of Chronic Epileptogenesis. *Sci. Rep.* **2016**, *6*, 31581. [CrossRef]

40. Yermolaieva, O.; Leonard, A.S.; Schnizler, M.K.; Abboud, F.M.; Welsh, M.J. Extracellular acidosis increases neuronal cell calcium by activating acid-sensing ion channel 1a. *Proc. Natl. Acad. Sci. USA* **2004**, *101*, 6752–6757. [CrossRef] [PubMed]
41. Dayan, N.; Kaur, A.; Elharram, M.; Rossi, A.M.; Pilote, L. Impact of Preeclampsia on Long-Term Cognitive Function. *Hypertension* **2018**, *72*, 1374–1380. [CrossRef]
42. Chen, Y.H.; Chiou, H.Y.; Lin, H.C.; Lin, H.L. Affect of seizures during gestation on pregnancy outcomes in women with epilepsy. *Arch. Neurol.* **2009**, *66*, 979–984. [CrossRef] [PubMed]
43. Ali, T.Y.; Broughton Pipkin, F.; Khan, R.N. The effect of pH and ion channel modulators on human placental arteries. *PLoS ONE* **2014**, *9*, e114405. [CrossRef] [PubMed]
44. Hanukoglu, I. ASIC and ENaC type sodium channels: Conformational states and the structures of the ion selectivity filters. *FEBS J.* **2017**, *284*, 525–545. [CrossRef] [PubMed]
45. Meltzer, R.H.; Kapoor, N.; Qadri, Y.J.; Anderson, S.J.; Fuller, C.M.; Benos, D.J. Heteromeric assembly of acid-sensitive ion channel and epithelial sodium channel subunits. *J. Biol. Chem.* **2007**, *282*, 25548–25559. [CrossRef]
46. Wang, L.; Long, X.; Arends, J.; Aarts, R.M. EEG analysis of seizure patterns using visibility graphs for detection of generalized seizures. *J. Neurosci. Methods* **2017**, *290*, 85–94. [CrossRef] [PubMed]

cells

Review

Vascular Dysfunction in Preeclampsia

Megan A. Opichka [1], Matthew W. Rappelt [2], David D. Gutterman [1,2,3], Justin L. Grobe [1,2,4,5,6] and Jennifer J. McIntosh [1,2,7,*]

1 Department of Physiology, Medical College of Wisconsin, Milwaukee, WI 53226, USA; mopichka@mcw.edu (M.A.O.); dgutt@mcw.edu (D.D.G.); jgrobe@mcw.edu (J.L.G.)
2 Cardiovascular Center, Medical College of Wisconsin, Milwaukee, WI 53226, USA; matthewrappelt@gmail.com
3 Department of Medicine, Medical College of Wisconsin, Milwaukee, WI 53226, USA
4 Neuroscience Research Center, Medical College of Wisconsin, Milwaukee, WI 53226, USA
5 Comprehensive Rodent Metabolic Phenotyping Core, Medical College of Wisconsin, Milwaukee, WI 53226, USA
6 Department of Biomedical Engineering, Medical College of Wisconsin, Milwaukee, WI 53226, USA
7 Department of Obstetrics and Gynecology, Medical College of Wisconsin, Milwaukee, WI 53226, USA
* Correspondence: jmcintosh@mcw.edu; Tel.: +1-(414)-805-9019

Abstract: Preeclampsia is a life-threatening pregnancy-associated cardiovascular disorder characterized by hypertension and proteinuria at 20 weeks of gestation. Though its exact underlying cause is not precisely defined and likely heterogenous, a plethora of research indicates that in some women with preeclampsia, both maternal and placental vascular dysfunction plays a role in the pathogenesis and can persist into the postpartum period. Potential abnormalities include impaired placentation, incomplete spiral artery remodeling, and endothelial damage, which are further propagated by immune factors, mitochondrial stress, and an imbalance of pro- and antiangiogenic substances. While the field has progressed, current gaps in knowledge include detailed initial molecular mechanisms and effective treatment options. Newfound evidence indicates that vasopressin is an early mediator and biomarker of the disorder, and promising future therapeutic avenues include mitigating mitochondrial dysfunction, excess oxidative stress, and the resulting inflammatory state. In this review, we provide a detailed overview of vascular defects present during preeclampsia and connect well-established notions to newer discoveries at the molecular, cellular, and whole-organism levels.

Keywords: preeclampsia; pregnancy; gestation; hypertension; vessel; blood pressure; placenta; trophoblast

Citation: Opichka, M.A.; Rappelt, M.W.; Gutterman, D.D.; Grobe, J.L.; McIntosh, J.J. Vascular Dysfunction in Preeclampsia. *Cells* **2021**, *10*, 3055. https://doi.org/10.3390/cells10113055

Academic Editors: Junie P. Warrington, Ana T. Palei, Mark W. Cunningham and Lorena M. Amaral

Received: 9 October 2021
Accepted: 4 November 2021
Published: 6 November 2021

1. Introduction

Preeclampsia is a pregnancy-related hypertensive disorder and a major cause of maternal and perinatal morbidity and mortality [1,2]. Despite its prevalence, well-cataloged risk factors, and clinical characteristics, the exact pathophysiology of this disorder remains unknown [2]. This deficit in knowledge has hampered the development of targeted therapies and limited treatment options for healthcare providers.

Clinically, preeclampsia is associated with a number of complications for both mother and fetus [3]. It is thought to lie on a spectrum of hypertensive diseases in pregnancy, with gestational hypertension at the mildest end of the spectrum, followed by preeclampsia, chronic hypertension with superimposed preeclampsia, hemolysis, elevated liver enzymes, low platelet count (HELLP) syndrome, and eclampsia at the most extreme end [1] (Figure 1). As preeclampsia is a true systemic disease, it may manifest in a number of different ways. Classically, it has been defined as maternal hypertension and renal dysfunction, specifically characterized by proteinuria [1]. However, more recent guidelines have noted thrombocytopenia, impaired liver function, pulmonary edema, and cerebral/visual symptoms as diagnostic features [1,2,4]. Additional maternal complications include seizures (eclampsia), cerebral hemorrhage, disseminated intravascular coagulation, and hepatic rupture [4]. Obstetric complications associated with preeclampsia consist of uteroplacental insufficiency, placental abruption, prematurity, and increased risk of cesarean delivery [2]. Additional fetal complications include intrapartum fetal distress, intrauterine growth restriction, oligohydramnios, and in severe cases, stillbirth [3]. Epidemiologic studies have linked approximately 15–20% of all fetal growth restriction and small for gestational age infants to preeclampsia, while 20% of all preterm births are associated with the disease [5,6]. Beyond the obstetric and neonatal consequences, preeclampsia confers long-term risk of complications, including cerebrovascular accident and hypertension [7,8].

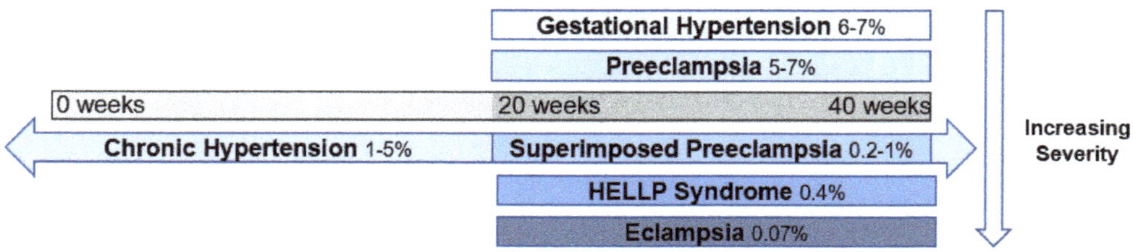

Figure 1. Spectrum of hypertensive disorders during pregnancy and their prevalence [9–11]. Gestational hypertension is defined by new-onset elevations in blood pressure (<140/90 mmHg) after 20 weeks of gestation, whereas preeclampsia is also accompanied by proteinuria and end-organ dysfunction. Chronic hypertension is present prior to 20 weeks of gestation or continues >12 weeks into the postnatal period and can occur in concert with preeclampsia. Hemolysis, elevated liver enzymes, and low platelets (HELLP) syndrome is classified as a subset of preeclampsia, and eclampsia is a complication of preeclampsia characterized by the addition of seizures.

An awareness of the syndrome dates back to Hippocrates in ~400 BC, yet preeclampsia remains a significant obstetric concern [12]. According to the World Health Organization, 16% of maternal deaths are attributable to preeclampsia and related gestational hypertensive disorders [13]. Furthermore, multiple studies have displayed an alarming increase in the incidence of gestational hypertension and preeclampsia over the past 3 decades [6,14–16]. Combining data from the Agency for Healthcare Research and Quality, the Center for Delivery, Organization, and Markets, Healthcare Cost and Utilization Project, and the National Inpatient Sample, the rate of overall preeclampsia and eclampsia increased by 21% from 2004 to 2014, with the incidence of severe preeclampsia rising by 50% during this period [11]. This imposes a severe clinical and economic burden, as the annual total cost associated with maternal and neonatal consequences of preeclampsia 12 months after delivery was USD 2.18 billion in the United States in 2012 [17].

As clinicians attempt to better manage this rising tide, a number of risk factors have been identified that reflect the complex nature of preeclampsia [15,18]. These include conditions such as chronic hypertension and other classical cardiovascular risk factors, along with chronic renal disease, antiphospholipid syndrome, collagen vascular diseases (e.g., lupus), and preexisting diabetes [18]. Additionally, factors such as nulliparity, a previous diagnosis of preeclampsia, abnormal placentation, multiple gestation, and maternal age at either end of the spectrum (<20 years or >35 years) also increase susceptibility [15]. The rate and severity of preeclampsia are higher among African Americans [19,20], and this likely reflects healthcare inequities as well as a higher incidence of determinants, including chronic hypertension, obesity, and type 2 diabetes, which are underdiagnosed in the African American community [21]. Finally, there appears to be a genetic component to preeclampsia, as a family history of preeclampsia, hypertension, and type II diabetes (either maternal or paternal) also portends higher risk [22,23].

Nonetheless, despite health disparities and the significant clinical impact of preeclampsia, the only "cure" is delivery, and even after childbirth, there remains an elevated risk of cardiovascular and metabolic disease later in life for these mothers and their children [24–28]. Thus, efforts to facilitate early detection, a better understanding of gestational mechanisms, and enhanced treatment modalities are imperative for improved management and health outcomes in patients with this complex condition. As the field advances, there is an increasing awareness that multiple subtypes of preeclampsia exist, and these subtypes may vary in their underlying cause, placental transcriptomic landscape, and disease severity [29–32]. Proposed classifications include early vs. late onset, in which early is more commonly associated with malplacentation, poor uterine perfusion, and fetal growth restriction. Alternatively, late-onset preeclampsia may be a consequence of placental overgrowth (resulting in compression of the chorionic villi), stress, or senescence towards the end of pregnancy [29–32]. Redman et al. recently highlighted the notion that regardless of the initiating factors, the placental syncytiotrophoblast layer is susceptible to cell stress throughout pregnancy (i.e., oxidative, mitochondrial, endoplasmic reticulum), and, ultimately, maternal responsiveness to these syncytial stress signals determines whether a woman will develop preeclampsia [29]. Though vascular dysfunction may elicit trophoblast stress or trophoblast stress may disrupt the vasculature, many of the risk factors, proposed mechanisms, and long-term implications of preeclampsia have a direct relationship with the maternal and placental vasculature. Hence, we aim to summarize key avenues of preeclampsia research and highlight the role of the vasculature within these areas.

2. Pathophysiology of Preeclampsia

2.1. Improper Decidualization and Placentation

It is widely accepted that placental development is disrupted in some pregnancies affected by preeclampsia, leading to cellular, molecular, immunological, and vascular changes [33–35], and the role of insufficient decidualization has also received increasing attention [36–39]. Early-onset preeclampsia is classically thought to be mediated by abnormal placentation and shallow trophoblast invasion within the uterus, thereby resulting in incomplete spiral artery remodeling [33–35]. This may lead to placental hypoxia, an aberrant angiogenic state, endothelial dysfunction, further decrements in placental formation, trophoblast stress, and ultimately the maternal presence of preeclampsia [2,35,40]. While much of the etiology remains unknown, research suggests that failed decidual differentiation prior to pregnancy can contribute to impaired trophoblast invasion and its sequela [36–39].

However, this is a complex syndrome, and the exact order of events in the pathogenesis is unclear. Predicting preeclampsia is imprecise, and determining whether physiological alterations cause preeclampsia or are a secondary result is quite challenging.

2.1.1. Cellular and Molecular Aspects

Decidualization occurs during the mid-secretory phase of the endometrial cycle to prepare the uterus for a potential incoming conceptus [41]. During this process, endometrial stromal cells differentiate into decidual stromal cells [38,39,41]. This process involves morphological alterations and genetic reprogramming to promote tolerance, optimal invasion, angiogenesis, and nutritional support to the embryo preceding placental development [38,39,41]. Research suggesting abnormal decidualization in preeclampsia indicates that human endometrial stromal cells derived from nonpregnant women following severe preeclampsia failed to differentiate, which was evident by a lack of structural changes and an absence of secretory markers [37]. Further, conditioned medium from severe preeclamptic decidual cells was unable to induce cytotrophoblast invasion, unlike culture medium from normal preterm birth decidual cells [37]. One molecular underpinning for this lack of decidualization and inability to stimulate invasion is an annexin A2 deficiency in preeclamptic human endometrial stromal cells [36], which may serve as a critical biomarker in assessing a woman's susceptibility to developing the disorder [36].

During normal placentation, trophoblasts, which are subdivided into cytotrophoblasts and syncytiotrophoblasts, descend from the blastocyst to form extraembryonic cells critical to placental formation [42,43]. Cytotrophoblasts comprise the inner villous placental layer closest to the fetal circulation, while syncytiotrophoblasts are derived from merging cytotrophoblasts and form the outer villous portion in contact with the maternal environment. Together, this arrangement of cells develops into a branch-like pattern referred to as the chorionic villi [42]. Though chorionic villi are essential for maternal–fetal exchange, the release of placenta-derived syncytiotrophoblast microvilli (STBM) into maternal circulation is elevated in preeclamptic women and impedes endothelial cell proliferation [44]. Furthermore, endothelium-dependent vasodilation is also impaired after infusion of placental STBM vesicles into adipose arterioles obtained during cesarean section. Electron micrographs following this perfusion confirm severe endothelial layer and intracellular organelle disruption, but the underlying smooth muscle remained intact [45]. Thus, the distribution of STBM into the bloodstream is a mechanism of vascular dysfunction in preeclampsia.

Cytotrophoblasts can also differentiate via the extravillous pathway to promote structural remodeling of maternal spiral arteries and sufficient oxygen exchange within the placenta [42,46,47]. In this process, extravillous trophoblasts infiltrate decidual and myometrial tissue during the first trimester to transform the resistance vessels within to those of greater capacitance [42,46–48]. Defects in this process during preeclampsia can allow spiral arteries to retain a muscular elastic phenotype, which leads to attenuated oxygen extraction between maternal and fetal circulation and instances of placental hypoxia [43,47].

Several recent findings indicate novel molecular mechanisms responsible for inadequate placentation [49–59] (Figure 2).

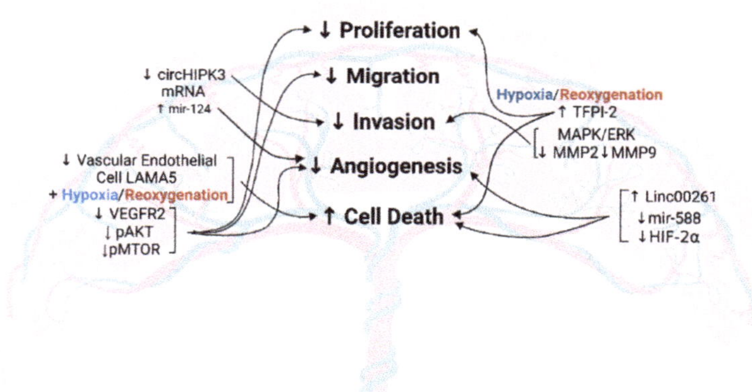

Figure 2. Identified and proposed mechanisms for abnormal placental development in preeclampsia [49–59] (circHIPK3, circular RNA homeodomain interacting protein kinase 3; LAMA5, laminin subunit alpha-5; VEGFR2, vascular endothelial growth factor receptor 2; pAKT, phosphorylated protein kinase B; pMTOR, phosphorylated mammalian target of rapamycin; TFPI-2, tissue factor pathway inhibitor-2; MAPK, mitogen-activated protein kinase; ERK, extracellular-signal-regulated kinase; MMP, matrix metalloproteinase; HIF-2α, hypoxia-induced factor-2α; ↑ refers to upregulation; ↓ refers to downregulation).

One proposed contributor involves decreased circular RNA homeodomain interacting protein kinase 3 (circHIPK3) mRNA in preeclamptic women [57]. CircHIPK3 is a circular RNA responsible for impeding microRNA (miRNA) activity [60]. Overexpression in cultured human first-trimester extravillous trophoblast cells facilitated migration and invasion, whereas small interfering RNA (siRNA) attenuated this effect [57]. CircHIPK3 has multiple known miRNA targets, including mir-124 and mir-558 [51,52], and dysregulation of miRNAs is a potential underlying cause for the progression of preeclampsia. It has been reported that mir-124 is upregulated in early-onset preeclampsia [53] and is considered a negative regulator of cardiac angiogenesis [58]. However, the exact role of mir-124 in placental tissue remains unknown. Similarly, the long non-coding RNA Linc00261 is upregulated in preeclampsia, leading to decreased mir-558 [50], which is another target of circHIPK3 [51]. In neuroblastoma cells, mir-558 has been shown to promote growth, invasion, and angiogenesis through the stimulation of hypoxia-induced factor-2α (HIF-2α) [55], so decreased mir-558 may be problematic in the context of preeclampsia if it has a similar role in placental tissue.

There is congruency between factors implicated in preeclampsia and those involved in other hypertensive disorders. In pulmonary hypertension, mir-124 is known to inhibit HIF-2α, initiating alterations in cell proliferation and the complications of this disease [61]. Members of the HIF transcription factor family, including HIF-1α and HIF-2α, have oxygen-sensing capabilities and thereby facilitate downstream regulation of angiogenic factors, including vascular endothelial growth factor (VEGF) and placental growth factor (PlGF) [62,63]. Two placental single-nucleotide polymorphisms in the coding region of *HIF1A* have been associated with preeclampsia as well as elevated HIF-1α activity [64]. HIF-1α also governs trophoblast differentiation during placentation. In a mouse model, transduction of a constitutively active version of HIF-1α into placental trophoblast cells established a preeclamptic-like phenotype, including irregular maternal spiral artery remodeling, altered labyrinth development, and elevated blood pressure [62]. While HIF-1α overactivation as a cause of preeclampsia may seem contradictory to the notion that HIFs are essential for angiogenesis, there is likely an optimal amount of activation, which may be dysregulated in preeclamptic pregnancies. Furthermore, previous studies indicate that different HIF1 isoforms have independent functions, despite overlapping gene targets [63].

Among the many molecular cascades implicated in preeclampsia, the phosphatidylinositol 3-kinase/protein kinase B (PI3K/AKT) and mitogen-activated protein kinase/extracellular-signal-regulated kinase (MAPK/ERK) signaling pathways are essential to trophoblast function and have downstream targets implicated in similar cellular processes, including migration and invasion [56,59,65]. Components of the extracellular matrix are essential for proper placental cell migration [66], and though some cases of preeclampsia were initially thought to be caused by reduced placental cell expansion [67], this deficiency has been linked to vascular dysfunction [56]. Laminin subunit alpha-5 (LAMA5) is a type of laminin, or an extracellular matrix glycoprotein, with decreased expression in vascular endothelial cells of preeclamptic placentas [56]. In human umbilical vein endothelial cells (HUVECs), LAMA5 knockdown combined with exposure to hypoxia/reoxygenation limited proliferation, migration, and angiogenic capacity while increasing cell death [56]. Independently, hypoxia/reperfusion and siLAMA5 caused reduced protein expression of vascular endothelial growth factor receptor 2 (VEGFR2), phosphorylated protein kinase B (pAKT), and phosphorylated mammalian target of rapamycin (pMTOR) (which are both downstream of PI3K), along with increased apoptosis and reduced proliferation [56]. However, the migratory function of trophoblasts relies not only on adhesion molecules such as laminins but also on proteolytic enzymes that degrade the surrounding matrix, called matrix metalloproteinases (MMPs) [68]. Tissue factor pathway inhibitor-2 (TFPI-2), an anticoagulant protein present in the plasma and on the endothelial cell surface, is upregulated in preeclamptic plasma and placental tissue [49,59]. Hypoxia also elevates TFPI-2 and MMP levels [54,59]. *TFPI2* knockdown in human trophoblast cells increased cell invasion and MMP2/9 protein levels via MAPK/ERK signaling [59], which highlights the importance of this pathway and its downstream effects in early processes that may contribute to preeclampsia.

2.1.2. Immunological and Vascular Aspects

The immune system has a profound effect on the maintenance of a healthy pregnancy, as an excessive inflammatory burden can result in the development of preeclampsia [69,70]. Normally, decidual natural killer cells are present in high abundance at the beginning of gestation to stimulate cytokines, MMPs, and angiogenic factors necessary for placental maturation [71,72]. Decidual macrophages have a similar role but also maintain fetal tolerance via immunosuppressive cytokines and phagocytosis of dead trophoblast cells [72].

Additionally, trophoblasts have a direct impact on the vascular arrangement and apoptosis by synthesizing tumor necrosis factor alpha (TNFα), fas-ligand, and other growth factors [73]. In preeclamptic cases, incomplete spiral artery remodeling, placental hypoxia, oxidative damage, and shear stress from uteroplacental blood flow impose a substantial immunological burden. The resulting proinflammatory environment can induce placental apoptosis or necrosis and the release of circulating inflammatory markers [72,74].

Preeclampsia is known to be a state of oxidative stress in which mitochondria are a major source of reactive oxygen species (ROS) [75,76]. Mitochondria-derived free radicals can induce apoptotic caspases and the circulation of cell-free mitochondrial DNA (cf-mtDNA) [76]. Cytosine phosphate guanine (CpG) dinucleotides present in cf-mtDNA are recognized by endosomal toll-like receptor 9 (TLR9) [77], which elicits downstream pro-inflammatory events, including interferon (IFN), MAPK/activator protein-1 (AP-1), and nuclear factor kappa B (NFκB) signaling [78]. Preeclampsia is marked by the release of cf-mtDNA and TLR9 activation [79], but other TLRs are attributed to this disease as well. For instance, administration of lipopolysaccharide, a bacterial endotoxin recognized by TLR4, in pregnant rats precipitated impaired endovascular spiral artery structure, along with the clinical symptoms of preeclampsia [80] (Table 1).

Table 1. Immune components altered in human preeclampsia and the effects of manipulation in cell culture and animal models.

Preeclampsia	Physiological Effects	Reference
↑ TLR3	Elevated systolic blood pressure	[81]
	Reduced aortic vasodilation	[81]
	Increased urinary protein concentrations	[81]
↑ TLR4	Impaired endovascular spiral artery structure	[80]
	Suppressed trophoblast migration	[82]
↑ TLR9	Suppressed trophoblast migration	[83]
	Decreased VEGFA, increased sFLT-1	[83]
↓ IL-10	Decreased buffering of antiangiogenic factors	[84]
	Increased endoplasmic reticulum stress	[85]
↓ Treg cells	Increased uterine artery vasoconstriction	[86]
	Elevated endothelin-1 production	[86]

TLR, toll-like receptor; VEGFA, vascular endothelial growth factor A; sFLT-1, soluble fms-like tyrosine kinase 1; IL-10, interleukin 10; Treg, regulatory T cell; ↑ refers to upregulation; ↓ refers to downregulation.

These inflammatory, blood pressure, and fetal growth effects were alleviated by treatment with a nitric oxide (NO) analog [80], highlighting the theory that vascular dysfunction is likely an etiological factor in preeclampsia. Both TLR4 and TLR9 have suppressive effects on trophoblast migration, suggesting that these receptors are attributed to placental aberrations [82,83]. Placental TLR3 is also upregulated in preeclampsia [87]. Confirming a more causative relationship, treatment with a viral mimetic in pregnant rats resulted in increased placental TLR3 expression, elevated systolic blood pressure, reduced aortic vasodilation, and higher urine protein excretion, and these effects were restricted to pregnant animals [81]. Uterine natural killer cells secrete the anti-inflammatory cytokine interleukin 10 (IL-10), which is highly involved in pregnancy [84,85,88–90]. Specific roles of IL-10 include preventing the maternal immune system from rejecting the fetal allograft, decreasing placental endoplasmic reticulum stress, and offsetting antiangiogenic factors [84,85,88,89]. In preeclampsia, there is decreased immunostaining of IL-10 and increased TNFα [91]. In mice, TLR3 activation and *Il10* knockout alone caused preeclamptic phenotypes, and together, these manipulations provoked a more severe state [92]. Exogenous recombinant IL-10 administration restored the impaired endothelium-dependent vasodilatory responses of these mice and may have useful therapeutic potential, considering the limited treatments available [92].

Maternal T cells have many subtypes and a diverse range of immunological functions in pregnancy [72,93]. The cluster of differentiation (CD)4+ class promotes fetal acceptance and consists of regulatory and helper subsets, whereas CD8+ T cells control trophoblast invasion [72]. An adequate ratio of T cell subtypes prevents immune overactivity and detrimental fetal or autoimmune attack [72,94,95]. Regulatory T cells (Tregs) control the defensive actions of T helper cells (Th) [94] and are thought to have imbalanced activity in preeclampsia [96]. More specifically, preeclamptic patients have a suppressed number of Treg cells with upregulated circulating and decidual activity of the proinflammatory Th1 and Th17 subsets [96]. Treg cell depletion in mice causes increased uterine artery vasoconstriction and endothelin-1 production [86], suggesting that altered vasoreactivity in preeclampsia may relate to the reduction in Treg cells.

A successful pregnancy begins with proper placentation, which involves the regulation of many cell types, including trophoblast, immune, and vascular subsets [42,43,46,72]. Placental stress via abnormal placental development or other gestational insults promotes a proinflammatory milieu, accompanied by TLR activation, decreased IL-10, increased TNFα, changes in T cell ratios, and disrupted vasodilatory function [79,80,82,83,91,92,96]. It is probable that immunological and vascular anomalies are present before the onset of maternal preeclamptic symptoms and are further exacerbated throughout its progression.

3. Circulating and Placenta-Derived Vascular Substances Associated with Preeclampsia (VEGF, PlGF, sFLT-1, ENG, and sENG)

Preeclampsia is characterized by an imbalance of pro- and antiangiogenic factors, which directly impacts endothelial function [97–100]. VEGFA stimulates angiogenesis, vascular permeability, and cell migration by binding to its tyrosine kinase receptors VEGFR1 and VEGFR2 [101,102]. VEGFA binding to VEGFR2 elicits stronger signaling than VEGFR1 [103] via activation of the phospholipase C gamma (PLCy)/protein kinase C (PKC)/MAPK pathway involved in endothelial cell proliferation [101]. During placental villous development, VEGFA is present in trophoblasts and perivascular cells to support de novo vascular development (i.e., vasculogenesis), as well as vessel expansion via endothelial sprouting (i.e., angiogenesis) [104]. In pregnancy, VEGF induces a more robust activation of endothelial nitric oxide synthase (eNOS), with NO production primarily occurring through VEGFR2-induced PI3K/AKT signaling [105,106]. PlGF, its proangiogenic counterpart, binds to VEGFR1 to increase the likelihood of VEGF binding to VEGFR-2 [107]. PlGF's interaction with VEGFR1 also promotes other critical events, such as transphosphorylation of VEGFR2, augmenting its downstream signaling cascade [107]. Similar to VEGF, the actions of PlGF facilitate the growth and migration of endothelial and trophoblast cells [107–109]. In healthy pregnancies, PlGF increases until 32 weeks and then subsequently declines [110,111]. However, in preeclampsia, there is a marked reduction in venous levels as early as 13–16 weeks, occurring before the onset of other clinical symptoms [108,112]. Not only does this have adverse cardiovascular consequences during pregnancy, but these vascular pathologies and unfavorable cardiac remodeling can persist for years beyond pregnancy [113]. This suggests that though overt symptoms are often resolved after delivery, preeclamptic mothers are at risk for cardiovascular disease years or decades later.

While the VEGFR is membrane bound, a truncated, soluble version known as soluble fms-like tyrosine kinase 1 (sFLT-1) sequesters ligands that bind to the VEGFR, specifically VEGF and PlGF [114]. sFLT-1 is similar to VEGFR1 but lacks the membrane-spanning domain [115]. sFLT-1 plays a role in the pathogenesis of preeclampsia and can be detected in the serum and placenta before other clinical manifestations, including proteinuria and hypertension [99,116]. The sFLT-1/PlGF ratio from 24 weeks to 36 weeks 6 days has been deemed a useful tool for predicting the absence of preeclampsia within 1 week of measurement, with a negative predictive value of 99.3% at a cutoff of 38 [117]. However, the positive predictive value for the diagnosis of preeclampsia within the next 4 weeks at this threshold was only 36.7% [117]. Thus, while this test is informative, earlier detection methods and tools with greater positive predictive value would be more powerful for improving the care of women prior to preeclampsia.

Adding to the complexity of this disorder, it has been discovered that a unique isoform of sFLT-1 predominates after the first trimester in both healthy and preeclamptic pregnancies [115,118]. At term, preeclamptic placentas express much higher *sFLT1* mRNA, specifically in syncytial knots, which are a source of circulating sFLT [118]. Furthermore, hypoxic conditions stimulate mRNA expression of the novel *sFLT1* variant and sFLT-1 peptide release in cultured primary cytotrophoblast and syncytiotrophoblast cells [115,119], but these effects are cell-specific, such that low oxygen tension does not increase the release of sFLT-1 protein in the culture medium of HUVECS or villous fibroblasts [119]. In cytotrophoblasts, no changes in free VEGF protein were detected under conditions of reduced oxygen tension despite an increase in total VEGF protein production, suggesting sequestration by sFLT-1 [119]. This imbalance leads to elevated oxidative stress, evident by increased lipid peroxidase relative to superoxide dismutase [120], a key antioxidant for scavenging oxygen free radicals [121].

Endoglin (ENG) is yet another hypoxia-induced protein implicated in preeclampsia [54,122,123]. This protein is located on endothelial cells, syncytiotrophoblasts, and columnar cytotrophoblasts prior to uterine invasion and exists in either a proangiogenic membrane-bound form or an antiangiogenic soluble form [54,122,124]. ENG serves as a receptor for the cytokines transforming growth factor-beta 1 and 3 (TGF-β1 and TGF-β3), which perform functions related to cell proliferation and apoptosis [122,125,126]. There are contradictory findings regarding the role of TGF-β in preeclampsia, which may be a product of gestational-age-related variations. Ayatollahi et al. reported no differences in serum TGF-β1 between healthy and preeclamptic pregnancies at 36–40 weeks and suggested that the anti-inflammatory properties of TGF-β may be instrumental for fetal-allograft survival [127]. Alam et al. found elevated TGF-β1 levels in preeclamptic patients from 30 to 33 weeks with a slight drop after 33 weeks [125], which may explain the lack of difference in the aforementioned study. One proposed explanation for the increase in TGF-β1 involves the shedding of necrotic placental trophoblast cells into the circulation, which are thought to undergo apoptosis in healthy pregnancies. When necrotic trophoblasts are phagocytosed by endothelial cells, the endothelial cells secrete TGF-β1, leading to IL-6 release. This causes placental soluble endoglin (sENG) secretion, sENG occupancy of ENG receptors, and impaired NO synthesis. Because TGF-β cannot bind to its receptor, its circulatory presence is increased, explaining the rise in TGF-β in preeclamptic women [125].

Both membrane and sENG are elevated in preeclampsia and have been considered as a diagnostic tool for detecting the syndrome [123,128]. Excess sENG has detrimental effects on endothelial tube development. In animal models, injection of both sENG and sFLT-1 results in a severe preeclamptic phenotype, accompanied by liver defects, fetal weight restriction, and neurological impairments [129]. More favorably, transmembrane ENG has been associated with NO-mediated vasodilation via eNOS expression [130,131] as well as trophoblast differentiation prior to invasion [126,130] (Figure 3). However, fluctuations in membrane ENG concentration and localization throughout pregnancy may be necessary for the maintenance of placental function because *ENG* knockdown has been shown to support extravillous trophoblast invasion [132].

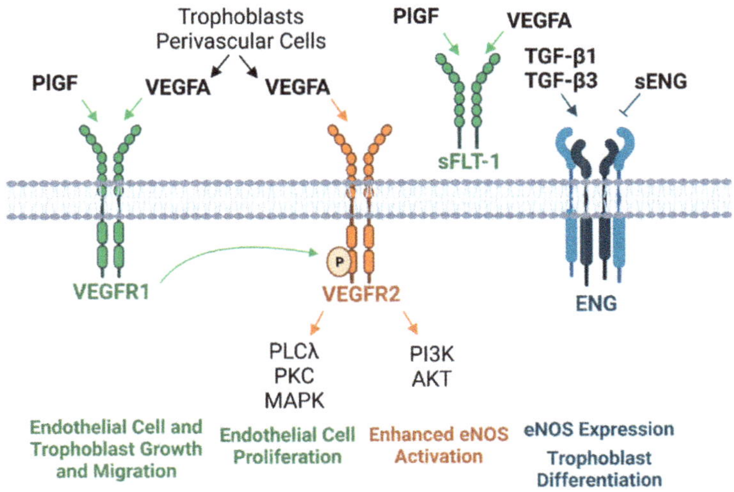

Figure 3. Circulating and placenta-derived vascular substances associated with preeclampsia and their downstream cellular effects [54,101,102,104–107,114,115,122,123,125,126,130,131,133]. PlGF, VEGFA, and ENG are considered "proangiogenic" factors, whereas sFLT-1 and ENG are "antiangiogenic." sFLT-1 sequesters PlGF and VEGFA, and sENG blocks TGF-β binding to the ENG receptor (PlGF, placental growth factor; VEGFA, vascular endothelial growth factor A; sFLT-1, soluble fms-like tyrosine kinase 1; ENG, endoglin; sENG, soluble endoglin; VEGFR1 and VEGFR2, vascular endothelial growth factor receptor 1 and 2; TGF-β, transforming growth factor-beta; PLCλ, phospholipase C gamma; PKC, protein kinase C; MAPK, mitogen-activated protein kinase; PI3K, phosphatidylinositol 3-kinase; AKT, protein kinase B; eNOS, endothelial nitric oxide synthase).

Several hypotheses exist regarding the mechanism of elevated sENG in preeclampsia, many of which are thought to independently contribute to the disorder [54,130]. Proposed notions include insufficient endogenous oxidative protective factors, NFκB activation, the MAPK stress response, MMP-regulated cleavage from endothelial plasma membranes, and angiotensin II type 1 receptor autoantibodies (AT1-AA) [54,130]. Experimental AT1A receptor activation elicits many preeclamptic-like phenotypes, including cell proliferation, vasoconstriction, renal fluid reabsorption, and vascular smooth muscle cell (VSMC) hypertrophy [134]. Mechanistically, rat models suggest that reduced uterine perfusion pressure stimulates increases in TNFα, which is coupled to AT1AA production [135,136]. Further, AT1AA infusion induced elevations in sENG as well as endothelin-1, and sFLT-1 [137,138].

While there are no reliable preventive or treatment therapeutics for preeclampsia [139–141], numerous studies support the notion that proangiogenic factors may be a viable future target against the high sFLT-1 levels found in preeclampsia [99,142–144]. Administration of VEGF and PlGF reestablished endothelial tube formation in HUVECs after it was impeded by preeclamptic serum [99]. In vitro, VEGF and PlGF produce dose-dependent arteriole vasodilation, and this response is blocked by sFLT1 [99]. Along with compromised vasomotor function, sFLT-1 increases placental and vascular superoxide production, and vasodilatory impairment can be reversed by free-radical scavenging [145]. In vivo, animal models indicate that PlGF infusion eliminates hypertension caused by both sFLT-1 and reduced uteroplacental perfusion [142,143]. Removal of sFLT-1 from women prolonged pregnancy [144]. Therefore, reversing the ratio of circulating vascular substances in preeclampsia has the potential to restore the dysregulated homeostatic state of preeclamptic women or prevent the progression of this syndrome in those at high risk.

4. Endothelial Damage in Preeclampsia

The endothelium, or inner layer of blood vessels, serves a wide array of functions that encompass but are not constrained to hemostasis, fibrinolysis, regulation of vascular tone, mediation of inflammatory cascades, and permeability [146–149]. Endothelial dysfunction, specifically in the form of barrier disruption and impaired vasodilatory capacity, is prevalent in preeclampsia and implicated in many stages of the disease [150–154]. Hemodynamic shifts accompanying compromised endothelial junction integrity, specifically those related to vasopressin, may precede early malfunctions in placental development [155], or in late-onset preeclampsia, endothelial damage may render a mother unable to buffer trophoblast-derived stress signals that accumulate throughout pregnancy [29,32]. The vascular defects of late-stage preeclampsia appear to be targeted to the endothelium in certain vascular beds, which is evident by the incubation of uterine myometrial resistance vessels with preeclamptic plasma [152]. Preeclamptic plasma restricted endothelium-dependent relaxation, but the effect only occurred with an intact endothelium, demonstrating that the vascular smooth muscle was unimpaired [152]. Similarly, placental vessels obtained from preeclamptic pregnancies show attenuated endothelial function with unaltered smooth muscle-mediated dilation [153]. These endothelial-specific findings have also been confirmed in vivo, as endothelium-dependent flow-mediated dilation was impaired in women with a history of preeclampsia compared to those without [154]. Together, the imbalance between constriction and relaxation and hemodynamic modifications that alter body fluid homeostasis are prominent features of preeclampsia [156].

4.1. Volemic Changes Associated with Endothelial Barrier Integrity

Under normal physiologic conditions, pregnancy is accompanied by systemic arterial vasodilation that is countered by an increase in cardiac output, sympathetic activation, stimulation of the renin–angiotensin–aldosterone system, and non-osmotic vasopressin release [157,158]. As a result, maternal plasma volume expansion begins in the first trimester, resulting in a subsequent decrease in plasma osmolality, and this expansion continues until around 32 weeks [159]. The maintenance of maternal blood volume is essential for fetal development [158], and overall blood and plasma volumes are reduced in those with preeclampsia despite elevations in blood pressure [160,161]. In these hypovolemic women, body fluid is distributed more to the interstitial space rather than the intravascular space, which is indicative of capillary leak [162–164] and paralleled in the deoxycorticosterone acid salt rat model of preeclampsia [165]. Proposed mechanisms of disrupted endothelial barrier integrity include the release of placenta-derived HtrA serine peptidase 4 (HTRA4), which is increased in serum from patients with early-onset preeclampsia and responsible for cleaving VE-cadherin, an endothelial junctional protein [150,166]. In vitro incubation of HUVECs with similar levels of HTRA4 altered cell morphology and the integrity of cell junctions [150]. Likewise, the expression of chymotrypsin-like protease is increased in the vascular endothelium of women diagnosed with preeclampsia and promotes VE-cadherin disruption and endothelial permeability via its activation of protease-activated receptor 2 (PAR-2) [167,168].

The presence of hypovolemia within the vessels due to endothelial leak shifts the plasma osmolality vs. vasopressin relationship such that plasma vasopressin levels are higher at a given osmolality [169,170]. Santillan et al. demonstrated that plasma copeptin, the pro-segment of vasopressin, is elevated as early as the 6th week of pregnancy in women prior to the onset of preeclampsia, independent of confounding variables such as maternal age and body mass index [171]. With a cutoff value of 811 pg/mL in the first trimester, this test has a sensitivity of 88% and a specificity of 81%. Even within the third trimester, a threshold of 758 pg/mL maintains a sensitivity and specificity of 78% and 71%, respectively [171]. Further, a causative relationship between vasopressin and preeclampsia was established in a mouse model, where infusion of vasopressin had no effect prior to pregnancy but then led to hypertension, renal glomerular endotheliosis, spiral artery malformation, decreased PlGF, and elevated placental oxidative markers during gestation [155]. In conclusion, increased vasopressin during preeclampsia may be a result of hypovolemia or another mechanism, such as a hypothalamic disturbance; regardless, the secretion of this hormone elicits many maternal and placental phenotypes typical of this syndrome and may be used as an early biomarker [155,171].

4.2. Altered Vasomotor Tone

To offset pregnancy-induced increases in blood volume, compensatory alterations include decreased sensitivity to vasoconstrictive hormones, along with increased production and responsiveness to vasodilators such as NO and prostaglandins [156,172].

However, studies have demonstrated that NO availability is diminished in preeclamptic women [173,174]. NO is a prominent endothelium-derived vasodilator catalyzed by the conversion of L-arginine to L-citrulline via the enzyme eNOS [148,175]. Its release is mediated by acetylcholine, bradykinin, mechanical shear stress, and a variety of other stimuli [148]. When NO is synthesized, it diffuses from the endothelium to the surrounding vascular smooth muscle cells. This facilitates vasodilation through a cascade of events involving the formation of cyclic guanosine monophosphate, activation of protein kinase G, increased potassium channel conductance, hyperpolarization of the smooth muscle sarcolemma, and decreased cytosolic calcium content [148,175,176]. eNOS is a calcium calmodulin-dependent enzyme; hence, its activation is heavily reliant on calcium [177]. Additionally, other eNOS modifications, including phosphorylation, render it more sensitive to calcium signaling [177], but phosphorylation alone is not adequate for eNOS activation in pregnant uterine artery endothelial cells [178].

Research indicates that augmented eNOS activation during pregnancy is mediated by adaptive signaling mechanisms to permit calcium influx, which are deficient in preeclampsia [179]. This calcium entry occurs through transient receptor potential channels and is maintained only with proper connexin 43 (Cx43) gap junction communication [179,180]. Interestingly, the most common variant of VEGFA, referred to as VEGF165, has been shown to initially facilitate NO generation in endothelial cells via elevated calcium responses [176]. However, longer exposure to VEGF165 impairs calcium-provoked vasodilation, presumably by promoting Cx43 phosphorylation, which attenuates calcium signaling in these cells [176,180]. VEGF has also been shown to facilitate the phosphorylation of eNOS on serine 1177 in glomerular endothelial cells, which is necessary for eNOS activation [181]. In summary, VEGF, which is sequestered by high levels of sFLT-1 in preeclampsia, promotes both calcium-dependent and -independent (i.e., phosphorylation) stimulation of eNOS, whereas vasodilators such as acetylcholine signal through the Gαq pathway to increase endothelial calcium and activate eNOS [182].

NO is also positively regulated by the endogenous vasodilator hydrogen sulfide [183–187], and both of these gaseous molecules work in concert [185]. Though not specific to preeclampsia, hydrogen sulfide enhances NO function by impeding cGMP degradation [185], activating eNOS via a PI3K/Akt-induced phosphorylation event [185], facilitating mir-455-3p and subsequent eNOS expression [186], and arbitrating the reduction of nitrite to NO through xanthine oxidase [187]. Conversely, NO is also necessary for the maximal vasodilatory and angiogenic effects of hydrogen sulfide [185]. Instances of both lowered and elevated plasma hydrogen sulfide levels have been reported in preeclampsia [188–190]. This molecule can be separately synthesized by cystathionine β synthase (CBS), cystathionine γ lyase (CSE), and 3-mercaptopyruvate sulfurtransferase (MPST) [191]. Within the placenta, CBS and CSE proteins were localized to fetal endothelial cells [192], and decreases in placental mRNA expression of CBS [192] and CSE [189] have been documented in preeclampsia. Administration of a CSE inhibitor beginning at gestational day 8.5 in mice led to hypertension, altered placental labyrinth vascular patterning, and fetal growth restriction, which was restored by hydrogen sulfide, establishing the direct significance of hydrogen sulfide in preeclampsia [189]. However, it should be noted that just as there is crosstalk among angiogenic, NO, and migration pathways [105–109], a similar relationship is paralleled by hydrogen sulfide [193]. Cell culture models indicate that CSE deficiency has a negative impact on trophoblast invasion and propagates antiangiogenic factor release [189], which, at least in part, may be responsible for the physiological effects exhibited by CSE inhibition in pregnant mice [189].

Oxidative stress in preeclampsia may be linked to the nitric oxide pathway [194,195]. For instance, arginase is upregulated in preeclamptic plasma and competes for L-arginine, the substrate used by NOS [194]. Similarly, circulating maternal asymmetric dimethylarginine (ADMA), a competitive inhibitor of arginine binding to NOS, is elevated mid-gestation and at the time of delivery in women with a diagnosis of preeclampsia [195,196]. Both arginase and ADMA result in the formation of superoxide [194,195], and exposing HUVECs to preeclamptic plasma also elevates oxidative markers, including superoxide [197]. Human data confirm that scavenging free radicals using ascorbic acid restores endothelium-dependent vasodilation [154].

In addition to vasodilatory substances such as nitric oxide, the vasoconstrictive peptides angiotensin II and endothelin-1 also play a role in the symptomology of preeclampsia [198,199]. While angiotensin II exerts contractive effects on vascular smooth muscle cells, it can also act on endothelial cells, prompting prostacyclin synthesis and endothelin mRNA activation [200]. During pregnancy, women with preeclampsia have increased angiotensin II sensitivity and decreased circulating levels [201]. The increased sensitivity to angiotensin II persists postpartum, along with microvascular dysfunction, and angiotensin II receptor blockers facilitate vasodilation in this cohort [202].

Endothelin-1 is secreted from endothelial cells and is a strong regulator of vasomotor tone through its interactions with endothelin-A (ETA) and endothelin-B (ETB) receptors [203]. ETA receptors are localized to smooth muscle cells and evoke contraction, whereas ETB receptors are present on both the endothelial and smooth muscle vascular layers and elicit differential responses in each: contraction in smooth muscle and vasodilation in the endothelium [204]. In the context of preeclampsia, endothelin-1 is a powerful regulator of uterine artery resistance, which has a direct impact on uteroplacental perfusion [205,206]. Pharmacological antagonism of the ETA receptor blunted the rise in mean arterial pressure in a rat model of uterine ischemia, with no effect in nonpregnant animals [207]. After testing the effects of different classes of antihypertensive medications, dihydropyridines (a type of calcium channel blocker) were the most effective in blocking and reversing endothelin-1-mediated constriction in human uterine arteries obtained after hysterectomy [205]. More specifically, ex vivo pretreatment with these calcium channel antagonists in endothelium-denuded vessels reduced tension and maximal contraction to endothelin-1, even at high doses [205]. While a combination of selective ETA and ETB receptor antagonists did decrease contractile responses to endothelin-1 at lower concentrations, these compounds had no effect under conditions of high endothehlin-1, suggesting that dihydropyridines may be the most useful for enhancing blood supply to the placenta [205]. However, the function of endothelin-1 extends far beyond vasoconstriction and can also induce inflammation, angiogenesis, VSMC proliferation, and vasodilation, depending on the receptor subtype being activated [208].

Angiotensin II, endothelin-1, and other vasoconstrictive hormones, including vasopressin, predominantly signal through G protein-coupled receptors (GPCRs) [209,210], and excess input from these substances has been attributed to the symptomology of preeclampsia [155,171,198,199]. Regulator of G protein signaling (RGS) proteins dampen the strength and duration of GPCR activation by hydrolyzing the guanosine-5′-triphosphate bound to an active Gα subunit [211–214]. Of the large family of RGS proteins, RGS2 is of particular importance in buffering this GPCR signaling associated with preeclampsia [213,215] and is present in vascular cells, including endothelial cells [216], as well as many other cell types of the placenta, including trophoblasts, immune cells, and fibroblasts [213,217]. Recently, Perschbacher et al. found that *RGS2* was decreased in the placenta during human preeclampsia, and heterozygous knockout in only the fetoplacental unit of mice was sufficient to cause key hallmarks of the disorder, such as diastolic hypertension and proteinuria [213]. Further, transcriptomic analyses revealed that there was overlap in molecular pathways enriched in human preeclamptic placenta and mice with disrupted *Rgs2*, including those related to mitochondrial dysfunction, the unfolded protein response, and oxidative stress [213]. Regarding the maternal vasculature and complementing this work, Koch et al. found that elimination of *Rgs2* in mice attenuated uterine artery blood flow and increased the resistive index during mid-gestation [215]. Hence, these studies indicate that preeclampsia may also be a result of disinhibition of GPCR signaling to various hormones, which contribute to placental and vascular dysfunction [213,215].

Overall, the pathogenic vascular state of preeclampsia can be attributed to the decreased synthesis of relaxing substances and elevated vasoconstrictive signaling [218,219]. This signaling may be a consequence of increased abundance or sensitivity to vasoconstrictive hormones as well as disinhibition of these pathways [213,215,218,219]. Vascular resistance has profound effects, both systemically and on specific vascular beds, and has been demonstrated experimentally in vivo, ex vivo, and in vitro [152,154,197,205,207]. Systemically, previously preeclamptic women display impaired endothelium-dependent vasodilation, which can be ameliorated by angiotensin II receptor blockers [154,202]. Additionally, diminished uteroplacental perfusion secondary to elevations in mean arterial pressure is prevented by ETA antagonism in rats [207]. Both uterine and placental arteries displayed attenuated endothelium-dependent vasodilatory function in the context of preeclampsia [152,153], which may be partially caused by oxidative stress [154,197]. In combination, uteroplacental resistance negatively impacts fetoplacental exchange, while systemic resistance can contribute to glomerular endotheliosis, liver failure, and central nervous system damage. Together, this creates an array of multiorgan dysfunctions [99,220,221].

5. Preeclampsia-Associated Platelet Alterations

Trophoblast stress is a common feature of preeclampsia and results in a compensatory surge of inflammatory mediators [29,75,222–224]. Of these, prostacyclin is a product of arachidonic acid metabolism [224,225]. Its role as a vasodilatory factor opposes the action of thromboxane, which elicits vascular smooth muscle constriction and platelet aggregation [226] (Figure 4).

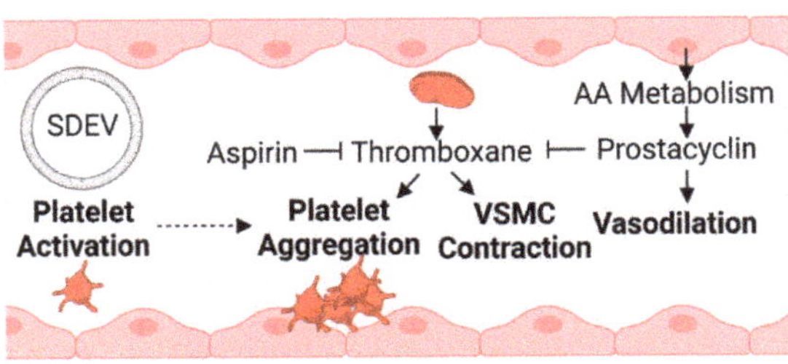

Figure 4. An imbalance in coagulation factors promotes a more prothrombotic environment during preeclampsia [226–228]. Prostacyclin is a vasodilatory product of arachidonic acid metabolism derived from endothelial cells [225]. Prostacyclin and aspirin oppose the actions of thromboxane, and the ratio of prostacyclin to thromboxane decreases in preeclampsia [226,229]. The accumulation of placental stress can lead to the release of SDEVs into maternal circulation. SDEVs can promote platelet activation, a precursory step to platelet aggregation and the formation of blood clots [29,230] (SDEV, syncytiotrophoblast-derived extracellular vesicle; VSMC, vascular smooth muscle cell; AA, arachidonic acid).

These two vasoactive substances are produced by the endothelium, platelets, and reproductive tissues, and the balance between them is disrupted in preeclampsia, with the level of placental and plasma thromboxane exceeding that of prostacyclin [226–228]. The pharmacological effects of aspirin diminish platelet thromboxane production, altering the prostaglandin thromboxane ratio [227]. A meta-analysis revealed that starting low-dose aspirin early in pregnancy as a preventative measure for preeclampsia has moderately favorable results [231]. The American College of Obstetricians and Gynecologists, the Society for Maternal-Fetal Medicine, and the U.S. Preventative Services Task Force now recommend low-dose aspirin for women with a high risk of preeclampsia [232].

Platelet activation, aggregation, and blood coagulation (clotting) are interrelated processes [233]. Briefly, platelets have adhesive properties and, upon binding to an injured endothelium, release substances such as thromboxane to promote aggregation. Platelet aggregation encourages the formation of a platelet plug and thrombin-mediated generation of a cross-linked fibrin clot [233]. A recent systemic review and meta-analysis suggests that preeclamptic patients have higher mean platelet volume (indicating platelet activation) and a higher likelihood of adhesion and aggregation [234]. In this paper, Jakobsen et al. reported inconsistent findings regarding aggregation, with more studies suggesting no difference or decreased aggregation, but these particular studies did not assess adhesion [234]. One study that did assess platelet adhesion reported decreased immunohistochemical expression of platelet endothelial cell adhesion molecule-1 and increased intercellular adhesion molecule-1 in the human placenta of preeclamptic individuals, which has been proposed to play a role in trophoblast invasion and vascular dysfunction [234,235]. Converging the idea that syncytiotrophoblast stress is a final common factor that leads to the maternal elements of preeclampsia [29] with the importance of platelet function in this syndrome, syncytiotrophoblast-derived extracellular vesicles (SDEVs) have been shown to activate platelets ex vivo [230]. SDEVs obtained from preeclamptic placentas evoke greater platelet activation than those from normal pregnancies, but platelet aggregation is prevented by aspirin treatment [230].

During pregnancy, there is a natural decline in platelet count throughout gestation [236], part of which may be attributed to sequestration of blood cells in the intervillous space [237], increases in plasma volume [238], and heightened aggregation from thromboxane A2 [239]. Thrombocytopenia beyond normal pregnancy-induced platelet decreases is commonly seen in preeclampsia and may specifically be accompanied by reduced platelet numbers [239–241] and activated coagulation [240,242]. Together, these hemostatic effects contribute to bleeding and microthrombi risk in preeclamptic mothers [239,243].

6. Oxidative Stress, Mitochondrial DNA Damage, and TLR9 Activation

The generation of reactive oxygen species via placental hypoxia, immune activation, and other cellular insults has numerous consequences, including mitochondrial DNA (mtDNA) damage [244–247]. The DNA repair capabilities of the mitochondria are less extensive than those for nuclear DNA, which renders cells with mtDNA mutations more susceptible to death by apoptosis or necrosis [248,249]. This causes the release of DNA into the maternal circulation, which is considered a damage-associated molecular pattern (DAMP), recognized by pattern recognition receptors such as TLR9 [250]. TLR9 is a pro-inflammatory innate immune component activated by hypomethylated CpG dinucleotides, which are prevalent in mtDNA and bacteria [251,252]. Although surface receptors are also present, the recognition of DNA by TLR9 primarily occurs in endolysosomes because the acidic environment allows TLR9 to more easily bind to negative DNA [251]. Thus, DNA enters the cell through endocytosis [253] and, upon TLR9 binding, causes a downstream proinflammatory cascade of events, including signaling for IFNs, NFκB, and AP-1 [83].

Supporting these notions, there is an increased abundance of serum mtDNA in preeclamptic plasma [254], and TLR9 activity is elevated upon the presentation of preeclamptic symptoms [247] (Figure 5). Recent findings by He et al. link the TLR9 inflammatory response to other aspects of preeclampsia, including angiogenesis and trophoblast function [83]. In this study, human placental VEGFA was decreased, but TLR9 and sFLT-1 were increased in preeclamptic samples [83]. Applying these findings to a mouse model, a TLR9 agonist induced the traditional hallmarks of preeclampsia and also reproduced the downregulated VEGFA and elevated sFLT-1 observed in human tissue [83]. siRNA knockdown of *TLR9* in human trophoblast cells facilitated migration and invasion, which highlights the importance of TLR9 in early phases of placentation as well [83]. The dendritic cells of preeclamptic women appear to be hyperresponsive to immune-evoking substances, suggesting a potential source for this excess TLR9 engagement. In preeclampsia, dendritic cell TLR9 expression levels were higher, and upon stimulation, these receptors evoked more proinflammatory cytokines compared to healthy pregnant controls [255]. These data reveal an interconnected relationship among inflammation, oxidative stress, TLR9 activation, the release of antiangiogenic factors, and trophoblast dysfunction [83,244,245,248–250] and reiterate the complex interplay between many molecular mediators in women with preeclampsia. Understanding each of these components and how they interact is essential to mitigating the disorder but has proven difficult considering its heterogeneity and the lack of a single, well-understood initiating mechanism [29,31].

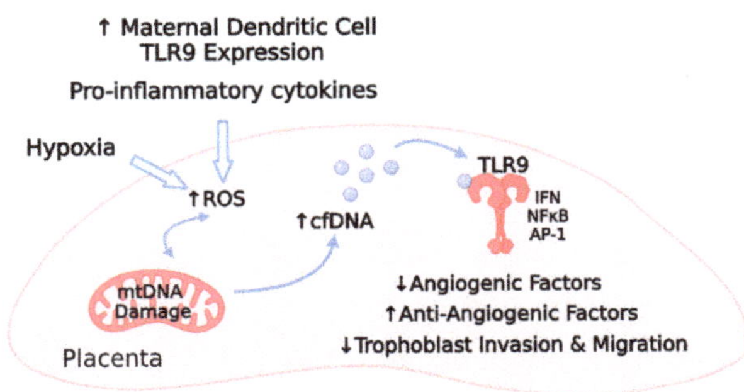

Figure 5. Maternal factors, including TLR9-induced proinflammatory cytokine release and placental hypoxia, promote a cascade of oxidative stress, mitochondrial DNA damage, cell-free DNA release, TLR9 activation, and subsequent TLR9 activation within the placenta [83,244–252,254,255]. Animal and cell culture models indicate that TLR9-related signaling results in decreased angiogenic factors, increased antiangiogenic factors, and impaired trophoblast function [83] (TLR9, toll-like receptor 9; ROS, reactive oxygen species; mtDNA, mitochondrial DNA; cfDNA, cell-free DNA; IFN, interferon; NFκB, nuclear factor kappa B; AP-1, activator protein-1; ↑ refers to upregulation; ↓ refers to downregulation).

7. Conclusions

Despite the profound detrimental impact that preeclampsia has upon maternal and fetal health, its pathogenesis has yet to be definitively determined and likely varies [30,256], which has limited the development of treatment options [256]. Thus, management of preeclampsia has been primarily symptomatic, focusing on maintaining acceptable blood pressure ranges, neuroprotection, and seizure prophylaxis, with prompt delivery at term or 34 weeks for cases with severe features [1]. However, it has been established that instances of vascular impairments are evident in all stages of preeclampsia, beginning with placentation and extending well beyond delivery [7,202], and are likely a product of some combination of insufficient trophoblast invasion, poor placental oxygen extraction, a proinflammatory immune environment, antiangiogenic factors, endothelial dysfunction, and oxidative stress [34,100,135,257–259].

Due to a lack of robust research assessing vascular indices prior to pregnancy and preceding the onset of preeclampsia, it is unclear whether women who develop this syndrome have an underlying vascular pathology [260,261] or if the detrimental vascular effects are solely a byproduct of exacerbated trophoblast stress signals to the mother [262]. It is likely that both aspects play a role, and the dysregulated physiological state of preeclampsia begins well before the clinical diagnosis. Therefore, advancing early detection methods and screening tools is of critical importance. Though not yet utilized in the clinic, routinely measuring vasopressin during pregnancy is a promising avenue for predicting the future development of preeclampsia and providing more proactive care in these patients [171]. In terms of molecular targets, less explored areas include the modulation of RGS proteins to mitigate the negative effects of excessive GPCR induction via hormones such as angiotensin II, endothelin-1, and vasopressin [213,263] or the alleviation of cellular stress that leads to mitochondrial dysfunction, cell death, circulating DNA, and subsequent TLR9 activation [79,83,244,247,252,264–266]. Though much remains undiscovered, translational research [152–154], basic animal models [155,213,215], and mechanistic cell work [56,57,150,213] have made a profound impact in the field thus far, and emerging technologies such as trophoblast organoid cultures [267] provide great potential for new insight. Thus, collaborating across the spectrum, from bench to bedside, will allow the most rapid acceleration in our understanding of preeclampsia and may foster the development of novel targeted therapeutics.

Author Contributions: Conceptualization, J.J.M.; writing—original draft preparation, M.A.O. and M.W.R.; writing—review and editing, J.J.M., J.L.G., D.D.G. and M.W.R.; visualization, M.A.O. and J.J.M.; supervision, J.J.M. All authors have read and agreed to the published version of the manuscript.

Funding: National Center for Advancing Translational Sciences, National Institutes of Health (UL1TR001436).Jennifer McIntosh receives funding through the National Heart, Lung, and Blood Institute (HL150340-01). Justin Grobe receives funding through the National Heart, Lung, and Blood Institute (HL134850, HL084207) and the American Heart Association (18EIA33890055). David Gutterman receives funding through the National Heart, Lung, and Blood Institute (HL135901). Megan Opichka has a predoctoral fellowship from the American Heart Association (AHA826132).

Institutional Review Board Statement: Not applicable.

Informed Consent Statement: Not applicable.

Data Availability Statement: Not applicable.

Acknowledgments: Figures 2–5 were created using Biorender.com. Graphical abstract was adapted from "Gut-Brain Axis Regulators" by Biorender.com (2021). Retrieved from https://app.biorender.com/biorender-templates (accessed on 2 October 2021).

Conflicts of Interest: The authors declare no conflict of interest.

References

1. Hypertension in pregnancy. Report of the American College of Obstetricians and Gynecologists' Task Force on Hypertension in Pregnancy. *Obs. Gynecol* **2013**, *122*, 1122–1131. [CrossRef]
2. Rana, S.; Lemoine, E.; Granger, J.; Karumanchi, S.A. Preeclampsia: Pathophysiology, challenges, and perspectives. *Circ. Res.* **2019**, *124*, 1094–1112. [CrossRef] [PubMed]
3. Fox, R.; Kitt, J.; Leeson, P.; Aye, C.Y.L.; Lewandowski, A.J. Preeclampsia: Risk Factors, Diagnosis, Management, and the Cardiovascular Impact on the Offspring. *J. Clin. Med.* **2019**, *8*, 1625. [CrossRef] [PubMed]
4. Pankiewicz, K.; Szczerba, E.; Maciejewski, T.; Fijałkowska, A. Non-obstetric complications in preeclampsia. *Prz. Menopauzalny* **2019**, *18*, 99–109. [CrossRef] [PubMed]
5. Duley, L. The global impact of pre-eclampsia and eclampsia. *Semin. Perinatol.* **2009**, *33*, 130–137. [CrossRef]
6. Jeyabalan, A. Epidemiology of preeclampsia: Impact of obesity. *Nutr. Rev.* **2013**, *71*. [CrossRef] [PubMed]
7. Bellamy, L.; Casas, J.-P.; Hingorani, A.D.; Williams, D.J. Pre-eclampsia and risk of cardiovascular disease and cancer in later life: Systematic review and meta-analysis. *BMJ Clin. Res. Ed.* **2007**, *335*, 974. [CrossRef]
8. Aykas, F.; Solak, Y.; Erden, A.; Bulut, K.; Dogan, S.; Sarli, B.; Acmaz, G.; Afsar, B.; Siriopol, D.; Covic, A.; et al. Persistence of cardiovascular risk factors in women with previous preeclampsia: A long-term follow-up study. *J. Investig. Med.* **2015**, *63*, 641–645. [CrossRef]
9. Ying, W.; Catov, J.M.; Ouyang, P. Hypertensive Disorders of Pregnancy and Future Maternal Cardiovascular Risk. *J. Am. Heart Assoc.* **2018**, *7*, e009382. [CrossRef]
10. Perez Botero, J.; Reese, J.A.; George, J.N.; McIntosh, J.J. Severe thrombocytopenia and microangiopathic hemolytic anemia in pregnancy: A guide for the consulting hematologist. *Am. J. Hematol.* **2021**. [CrossRef]
11. Fingar, K.R.; Mabry-Hernandez, I.; Ngo-Metzger, Q.; Wolff, T.; Steiner, C.A.; Elixhauser, A. Delivery Hospitalizations Involving Preeclampsia and Eclampsia, 2005–2014: Statistical Brief #222. In *Healthcare Cost and Utilization Project (HCUP) Statistical Briefs*; Agency for Healthcare Research and Quality (US): Rockville, MD, USA, 2006.
12. Bell, M.J. A historical overview of preeclampsia-eclampsia. *J. Obstet. Gynecol. Neonatal Nurs. JOGNN* **2010**, *39*, 510–518. [CrossRef] [PubMed]
13. Högberg, U. The World Health Report 2005: "make every mother and child count"—including Africans. *Scand. J. Public Health* **2005**, *33*, 409–411. [CrossRef] [PubMed]

14. Ananth, C.V.; Keyes, K.M.; Wapner, R.J. Pre-eclampsia rates in the United States, 1980–2010: Age-period-cohort analysis. *BMJ* **2013**, *347*, f6564. [CrossRef] [PubMed]
15. Shih, T.; Peneva, D.; Xu, X.; Sutton, A.; Triche, E.; Ehrenkranz, R.A.; Paidas, M.; Stevens, W. The Rising Burden of Preeclampsia in the United States Impacts Both Maternal and Child Health. *Am. J. Perinatol.* **2016**, *33*, 329–338. [CrossRef]
16. Wallis, A.B.; Saftlas, A.F.; Hsia, J.; Atrash, H.K. Secular trends in the rates of preeclampsia, eclampsia, and gestational hypertension, United States, 1987–2004. *Am. J. Hypertens* **2008**, *21*, 521–526. [CrossRef] [PubMed]
17. Stevens, W.; Shih, T.; Incerti, D.; Ton, T.G.N.; Lee, H.C.; Peneva, D.; Macones, G.A.; Sibai, B.M.; Jena, A.B. Short-term costs of preeclampsia to the United States health care system. *Am. J. Obs. Gynecol.* **2017**, *217*, 237–248. [CrossRef]
18. Duckitt, K.; Harrington, D. Risk factors for pre-eclampsia at antenatal booking: Systematic review of controlled studies. *BMJ* **2005**, *330*, 565. [CrossRef]
19. Ghosh, G.; Grewal, J.; Männistö, T.; Mendola, P.; Chen, Z.; Xie, Y.; Laughon, S.K. Racial/ethnic differences in pregnancy-related hypertensive disease in nulliparous women. *Ethn. Dis.* **2014**, *24*, 283–289.
20. Zhang, M.; Wan, P.; Ng, K.; Singh, K.; Cheng, T.H.; Velickovic, I.; Dalloul, M.; Wlody, D. Preeclampsia Among African American Pregnant Women: An Update on Prevalence, Complications, Etiology, and Biomarkers. *Obs. Gynecol. Surv.* **2020**, *75*, 111–120. [CrossRef]
21. Shahul, S.; Tung, A.; Minhaj, M.; Nizamuddin, J.; Wenger, J.; Mahmood, E.; Mueller, A.; Shaefi, S.; Scavone, B.; Kociol, R.D.; et al. Racial Disparities in Comorbidities, Complications, and Maternal and Fetal Outcomes in Women With Preeclampsia/eclampsia. *Hypertens. Pregnancy* **2015**, *34*, 506–515. [CrossRef]
22. Bezerra, P.C.; Leão, M.D.; Queiroz, J.W.; Melo, E.M.; Pereira, F.V.; Nóbrega, M.H.; Jeronimo, A.K.; Ferreira, L.C.; Jerônimo, S.M.; de Araújo, A.C. Family history of hypertension as an important risk factor for the development of severe preeclampsia. *Acta Obs. Gynecol. Scand.* **2010**, *89*, 612–617. [CrossRef] [PubMed]
23. Qiu, C.; Williams, M.A.; Leisenring, W.M.; Sorensen, T.K.; Frederick, I.O.; Dempsey, J.C.; Luthy, D.A. Family history of hypertension and type 2 diabetes in relation to preeclampsia risk. *Hypertension* **2003**, *41*, 408–413. [CrossRef] [PubMed]
24. Cho, G.J.; Jung, U.S.; Sim, J.Y.; Lee, Y.J.; Bae, N.Y.; Choi, H.J.; Park, J.H.; Kim, H.-J.; Oh, M.-J. Is preeclampsia itself a risk factor for the development of metabolic syndrome after delivery? *Obs. Gynecol. Sci.* **2019**, *62*, 233–241. [CrossRef] [PubMed]
25. Ryckman, K.K.; Borowski, K.S.; Parikh, N.I.; Saftlas, A.F. Pregnancy Complications and the Risk of Metabolic Syndrome for the Offspring. *Curr. Cardiovasc. Risk Rep.* **2013**, *7*, 217–223. [CrossRef] [PubMed]
26. Muijsers, H.E.C.; Roeleveld, N.; van der Heijden, O.W.H.; Maas, A.H.E.M. Consider Preeclampsia as a First Cardiovascular Event. *Curr. Cardiovasc. Risk Rep.* **2019**, *13*, 21. [CrossRef]
27. Lu, H.Q.; Hu, R. Lasting Effects of Intrauterine Exposure to Preeclampsia on Offspring and the Underlying Mechanism. *AJP Rep.* **2019**, *9*, e275–e291. [CrossRef]
28. Anderson, C.M. Preeclampsia: Exposing future cardiovascular risk in mothers and their children. *J. Obs. Gynecol. Neonatal. Nurs.* **2007**, *36*, 3–8. [CrossRef]
29. Redman, C.W.G.; Staff, A.C.; Roberts, J.M. Syncytiotrophoblast stress in preeclampsia: The convergence point for multiple pathways. *Am. J. Obstet. Gynecol.* **2021**. [CrossRef]
30. Roberts, J.M.; Rich-Edwards, J.W.; McElrath, T.F.; Garmire, L.; Myatt, L. Subtypes of Preeclampsia: Recognition and Determining Clinical Usefulness. *Hypertension* **2021**, *77*, 1430–1441. [CrossRef]
31. Leavey, K.; Benton, S.J.; Grynspan, D.; Kingdom, J.C.; Bainbridge, S.A.; Cox, B.J. Unsupervised Placental Gene Expression Profiling Identifies Clinically Relevant Subclasses of Human Preeclampsia. *Hypertension* **2016**, *68*, 137–147. [CrossRef]
32. Phipps, E.; Prasanna, D.; Brima, W.; Jim, B. Preeclampsia: Updates in Pathogenesis, Definitions, and Guidelines. *Clin. J. Am. Soc. Nephrol.* **2016**, *11*, 1102–1113. [CrossRef]
33. Staff, A.C. The two-stage placental model of preeclampsia: An update. *J. Reprod. Immunol.* **2019**, *134–135*, 1–10. [CrossRef] [PubMed]
34. Roberts, J.M.; Escudero, C. The placenta in preeclampsia. *Pregnancy Hypertens.* **2012**, *2*, 72–83. [CrossRef]
35. Fisher, S.J. Why is placentation abnormal in preeclampsia? *Am. J. Obstet. Gynecol.* **2015**, *213*, S115–S122. [CrossRef] [PubMed]
36. Garrido-Gomez, T.; Quiñonero, A.; Dominguez, F.; Rubert, L.; Perales, A.; Hajjar, K.A.; Simon, C. Preeclampsia: A defect in decidualization is associated with deficiency of Annexin A2. *Am. J. Obs. Gynecol.* **2020**, *222*, 376–376.e371. [CrossRef] [PubMed]
37. Garrido-Gomez, T.; Dominguez, F.; Quiñonero, A.; Diaz-Gimeno, P.; Kapidzic, M.; Gormley, M.; Ona, K.; Padilla-Iserte, P.; McMaster, M.; Genbacev, O.; et al. Defective decidualization during and after severe preeclampsia reveals a possible maternal contribution to the etiology. *Proc. Natl. Acad. Sci. USA* **2017**, *114*, E8468–E8477. [CrossRef]
38. Ng, S.-W.; Norwitz, G.A.; Pavlicev, M.; Tilburgs, T.; Simón, C.; Norwitz, E.R. Endometrial Decidualization: The Primary Driver of Pregnancy Health. *Int. J. Mol. Sci.* **2020**, *21*, 4092. [CrossRef]
39. Garrido-Gómez, T.; Castillo-Marco, N.; Cordero, T.; Simón, C. Decidualization resistance in the origin of preeclampsia. *Am. J. Obs. Gynecol.* **2020**. [CrossRef] [PubMed]
40. Sánchez-Aranguren, L.C.; Prada, C.E.; Riaño-Medina, C.E.; Lopez, M. Endothelial dysfunction and preeclampsia: Role of oxidative stress. *Front. Physiol.* **2014**, *5*, 372. [CrossRef]
41. Okada, H.; Tsuzuki, T.; Murata, H. Decidualization of the human endometrium. *Reprod. Med. Biol.* **2018**, *17*, 220–227. [CrossRef] [PubMed]
42. Baergen, R.N. *Manual of Pathology of the Human Placenta*; Springer Science & Business Media: Berlin/Heidelberg, Germany, 2011.

43. Wang, Y.; Zhao, S. Cell Types of the Placenta. In *Vascular Biology of the Placenta*; Morgan & Claypool Life Sciences: San Rafael, CA, USA, 2010.
44. Knight, M.; Redman, C.W.G.; Linton, E.A.; Sargent, I.L. Shedding of syncytiotrophoblast microvilli into the maternal circulation in pre-eclamptic pregnancies. *BJOG Int. J. Obstet. Gynaecol.* **1998**, *105*, 632–640. [CrossRef] [PubMed]
45. Cockell, A.P.; Learmont, J.G.; Smárason, A.K.; Redman, C.W.G.; Sargent, I.L.; Poston, L. Human placental syncytiotrophoblast microvillous membranes impair maternal vascular endothelial function. *BJOG Int. J. Obstet. Gynaecol.* **1997**, *104*, 235–240. [CrossRef] [PubMed]
46. Ji, L.; Brkić, J.; Liu, M.; Fu, G.; Peng, C.; Wang, Y.L. Placental trophoblast cell differentiation: Physiological regulation and pathological relevance to preeclampsia. *Mol. Asp. Med.* **2013**, *34*, 981–1023. [CrossRef] [PubMed]
47. Pollheimer, J.; Vondra, S.; Baltayeva, J.; Beristain, A.G.; Knöfler, M. Regulation of Placental Extravillous Trophoblasts by the Maternal Uterine Environment. *Front. Immunol.* **2018**, *9*, 2597. [CrossRef]
48. Lyall, F.; Bulmer, J.N.; Duffie, E.; Cousins, F.; Theriault, A.; Robson, S.C. Human trophoblast invasion and spiral artery transformation: The role of PECAM-1 in normal pregnancy, preeclampsia, and fetal growth restriction. *Am. J. Pathol.* **2001**, *158*, 1713–1721. [CrossRef]
49. Briones, M.A. *Chapter 131—General Overview of the Hypercoaguable State In Transfusion Medicine and Hemostasis*; Hillyer, C.D., Shaz, B.H., Zimring, J.C., Abshire, T.C., Eds.; Academic Press: San Diego, CA, USA, 2009; pp. 677–679.
50. Cheng, D.; Jiang, S.; Chen, J.; Li, J.; Ao, L.; Zhang, Y. Upregulated long noncoding RNA Linc00261 in pre-eclampsia and its effect on trophoblast invasion and migration via regulating miR-558/TIMP4 signaling pathway. *J. Cell. Biochem.* **2019**, *120*, 13243–13253. [CrossRef]
51. Li, Y.; Zheng, F.; Xiao, X.; Xie, F.; Tao, D.; Huang, C.; Liu, D.; Wang, M.; Wang, L.; Zeng, F.; et al. CircHIPK3 sponges miR-558 to suppress heparanase expression in bladder cancer cells. *EMBO Rep.* **2017**, *18*, 1646–1659. [CrossRef] [PubMed]
52. Liu, Y.; Xia, L.; Dong, L.; Wang, J.; Xiao, Q.; Yu, X.; Zhu, H. CircHIPK3 Promotes Gemcitabine (GEM) Resistance in Pancreatic Cancer Cells by Sponging miR-330-5p and Targets RASSF1. *Cancer Manag. Res.* **2020**, *12*, 921–929. [CrossRef]
53. Lykoudi, A.; Kolialexi, A.; Lambrou, G.I.; Braoudaki, M.; Siristatidis, C.; Papaioanou, G.K.; Tzetis, M.; Mavrou, A.; Papantoniou, N. Dysregulated placental microRNAs in Early and Late onset Preeclampsia. *Placenta* **2018**, *61*, 24–32. [CrossRef]
54. Oujo, B.; Perez-Barriocanal, F.; Bernabeu, C.; Lopez-Novoa, J. Membrane and Soluble Forms of Endoglin in Preeclampsia. *Curr. Mol. Med.* **2013**, *13*, 1345–1357. [CrossRef] [PubMed]
55. Qu, H.; Zheng, L.; Song, H.; Jiao, W.; Li, D.; Fang, E.; Wang, X.; Mei, H.; Pu, J.; Huang, K.; et al. microRNA-558 facilitates the expression of hypoxia-inducible factor 2 alpha through binding to 5′-untranslated region in neuroblastoma. *Oncotarget* **2016**, *7*, 40657–40673. [CrossRef] [PubMed]
56. Zhang, X.M.; Li, Q.; Jiang, W.; Xiong, X.; Li, H.Y.; Zhao, J.L.; Qi, H.B. LAMA5 promotes human umbilical vein endothelial cells migration, proliferation, and angiogenesis and is decreased in preeclampsia. *J. Matern. Fetal Neonatal Med.* **2020**, *33*, 1114–1124. [CrossRef] [PubMed]
57. Zhang, Y.; Cao, L.; Jia, J.; Ye, L.; Wang, Y.; Zhou, B.; Zhou, R. CircHIPK3 is decreased in preeclampsia and affects migration, invasion, proliferation, and tube formation of human trophoblast cells. *Placenta* **2019**, *85*, 1–8. [CrossRef]
58. Zhao, Y.; Yan, M.; Chen, C.; Gong, W.; Yin, Z.; Li, H.; Fan, J.; Zhang, X.A.; Wang, D.W.; Zuo, H. MiR-124 aggravates failing hearts by suppressing CD151-facilitated angiogenesis in heart. *Oncotarget* **2018**, *9*, 14382–14396. [CrossRef] [PubMed]
59. Zheng, L.; Huang, J.; Su, Y.; Wang, F.; Kong, H.; Xin, H. Overexpression of tissue factor pathway inhibitor 2 attenuates trophoblast proliferation and invasion in preeclampsia. *Hum. Cell* **2020**. [CrossRef]
60. Han, C.; Seebacher, N.A.; Hornicek, F.J.; Kan, Q.; Duan, Z. Regulation of microRNAs function by circular RNAs in human cancer. *Oncotarget* **2017**, *8*, 64622–64637. [CrossRef]
61. Babicheva, A.; McDermott, K.M.; Williams, S.C.; Yee, A.M.; Dash, S.; Rodriquez, M.; Ingabire, N.; Makino, A.; Yuan, J.X.-J. Pathogenic and Therapeutic Role of MicroRNA in Pulmonary Arterial Hypertension. In *Diagnosis and Treatment of Pulmonary Hypertension: From Bench to Bedside*; Fukumoto, Y., Ed.; Springer: Singapore, 2017; pp. 31–54.
62. Albers, R.E.; Kaufman, M.R.; Natale, B.V.; Keoni, C.; Kulkarni-Datar, K.; Min, S.; Williams, C.R.; Natale, D.R.C.; Brown, T.L. Trophoblast-specific expression of hif-1α results in preeclampsia-like symptoms and fetal growth restriction. *Sci. Rep.* **2019**, *9*, 1–13. [CrossRef]
63. Rathmell, W.K.; Chen, S. VHL inactivation in renal cell carcinoma: Implications for diagnosis, prognosis and treatment. *Expert Rev. Anticancer Ther.* **2008**, *8*, 63–73. [CrossRef]
64. Harati-Sadegh, M.; Kohan, L.; Teimoori, B.; Mehrabani, M.; Salimi, S. The association of the placental Hypoxia-inducible factor1-α polymorphisms and HIF1-α mRNA expression with preeclampsia. *Placenta* **2018**, *67*, 31–37. [CrossRef]
65. Gupta, S.K.; Malhotra, S.S.; Malik, A.; Verma, S.; Chaudhary, P. Cell Signaling Pathways Involved During Invasion and Syncytialization of Trophoblast Cells. *Am. J. Reprod. Immunol.* **2016**, *75*, 361–371. [CrossRef]
66. O'Connor, B.B.; Pope, B.D.; Peters, M.M.; Ris-Stalpers, C.; Parker, K.K. The role of extracellular matrix in normal and pathological pregnancy: Future applications of microphysiological systems in reproductive medicine. *Exp. Biol. Med.* **2020**, *245*, 1163–1174. [CrossRef] [PubMed]
67. Kaufmann, P.; Black, S.; Huppertz, B. Endovascular trophoblast invasion: Implications for the pathogenesis of intrauterine growth retardation and preeclampsia. *Biol. Reprod.* **2003**, *69*, 1–7. [CrossRef] [PubMed]

68. Matthiesen, L.; Berg, G.; Ernerudh, J.; Ekerfelt, C.; Jonsson, Y.; Sharma, S. Immunology of preeclampsia. *Chem. Immunol. Allergy* **2005**, *89*, 49–61. [CrossRef] [PubMed]
69. Harmon, A.C.; Cornelius, D.C.; Amaral, L.M.; Faulkner, J.L.; Cunningham, M.W., Jr.; Wallace, K.; LaMarca, B. The role of inflammation in the pathology of preeclampsia. *Clin. Sci.* **2016**, *130*, 409–419. [CrossRef] [PubMed]
70. Aneman, I.; Pienaar, D.; Suvakov, S.; Simic, T.P.; Garovic, V.D.; McClements, L. Mechanisms of Key Innate Immune Cells in Early- and Late-Onset Preeclampsia. *Front. Immunol.* **2020**, *11*, 1864. [CrossRef]
71. Zhang, X.; Wei, H. Role of Decidual Natural Killer Cells in Human Pregnancy and Related Pregnancy Complications. *Front. Immunol.* **2021**, *12*, 728291. [CrossRef]
72. Geldenhuys, J.; Rossouw, T.M.; Lombaard, H.A.; Ehlers, M.M.; Kock, M.M. Disruption in the regulation of immune responses in the placental subtype of preeclampsia. *Front. Immunol.* **2018**, *9*. [CrossRef]
73. Whitley, G.S.; Cartwright, J.E. Trophoblast-mediated spiral artery remodelling: A role for apoptosis. *J. Anat.* **2009**, *215*, 21–26. [CrossRef] [PubMed]
74. Furuya, M.; Ishida, J.; Aoki, I.; Fukamizu, A. Pathophysiology of placentation abnormalities in pregnancy-induced hypertension. *Vasc. Health Risk Manag.* **2008**, *4*, 1301–1313. [CrossRef]
75. Aouache, R.; Biquard, L.; Vaiman, D.; Miralles, F. Oxidative Stress in Preeclampsia and Placental Diseases. *Int. J. Mol. Sci.* **2018**, *19*, 1496. [CrossRef]
76. Chen, Y.; Zhou, Z.; Min, W. Mitochondria, Oxidative Stress and Innate Immunity. *Front. Physiol.* **2018**, *9*, 1487. [CrossRef] [PubMed]
77. Goulopoulou, S.; Davidge, S.T. Molecular mechanisms of maternal vascular dysfunction in preeclampsia. *Trends Mol. Med.* **2015**, *21*, 88–97. [CrossRef] [PubMed]
78. Prearo Moço, N.; Ribeiro de Andrade Ramos, B.; de Castro Silva, M.; Polettini, J.; Menon, R.; Guimarães da Silva, M. Spontaneous Prematurity, Innate Immune System, and Oxidative Stress at the Maternal-Fetal Interface: An Overview. In *Translational Studies on Inflammation*; IntechOpen: London, UK, 2020.
79. Ozeki, A.; Tani, K.; Takahashi, H.; Suzuki, H.; Nagayama, S.; Hirashima, C.; Iwata, H.; Kuwayama, T.; Ohkuchi, A.; Shirasuna, K. Preeclamptic patient-derived circulating cell-free DNA activates the production of inflammatory cytokines via toll-like receptor 9 signalling in the human placenta. *J. Hypertens.* **2019**, *37*, 2452–2460. [CrossRef]
80. Cotechini, T.; Komisarenko, M.; Sperou, A.; Macdonald-Goodfellow, S.; Adams, M.A.; Graham, C.H. Inflammation in rat pregnancy inhibits spiral artery remodeling leading to fetal growth restriction and features of preeclampsia. *J. Exp. Med.* **2014**, *211*, 165–179. [CrossRef] [PubMed]
81. Tinsley, J.H.; Chiasson, V.L.; Mahajan, A.; Young, K.J.; Mitchell, B.M. Toll-like receptor 3 activation during pregnancy elicits preeclampsia-like symptoms in rats. *Am. J. Hypertens* **2009**, *22*, 1314–1319. [CrossRef]
82. Xue, P.; Fan, W.; Diao, Z.; Li, Y.; Kong, C.; Dai, X.; Peng, Y.; Chen, L.; Wang, H.; Hu, Y.; et al. Up-regulation of PTEN via LPS/AP-1/NF-κB pathway inhibits trophoblast invasion contributing to preeclampsia. *Mol. Immunol.* **2020**, *118*, 182–190. [CrossRef] [PubMed]
83. He, B.; Yang, X.; Li, Y.; Huang, D.; Xu, X.; Yang, W.; Dai, Y.; Zhang, H.; Chen, Z.; Cheng, W. TLR9 (toll-like receptor 9) agonist suppresses angiogenesis by differentially regulating VEGFA (Vascular Endothelial Growth Factor A) and sFLT1 (Soluble Vascular Endothelial Growth Factor Receptor 1) in Preeclampsia. *Hypertension* **2018**, *71*, 671–680. [CrossRef]
84. Kalkunte, S.; Boij, R.; Norris, W.; Friedman, J.; Lai, Z.; Kurtis, J.; Lim, K.-H.; Padbury, J.F.; Matthiesen, L.; Sharma, S. Sera from preeclampsia patients elicit symptoms of human disease in mice and provide a basis for an in vitro predictive assay. *Am. J. Pathol.* **2010**, *177*, 2387–2398. [CrossRef]
85. Cheng, S.-B.; Sharma, S. Interleukin-10: A pleiotropic regulator in pregnancy. *Am. J. Reprod. Immunol.* **2015**, *73*, 487–500. [CrossRef]
86. Care, A.S.; Bourque, S.L.; Morton, J.S.; Hjartarson, E.P.; Robertson, S.A.; Davidge, S.T. Reduction in regulatory t cells in early pregnancy causes uterine artery dysfunction in mice. *Hypertension* **2018**, *72*, 177–187. [CrossRef]
87. Gierman, L.M.; Silva, G.B.; Pervaiz, Z.; Rakner, J.J.; Mundal, S.B.; Thaning, A.J.; Nervik, I.; Elschot, M.; Mathew, S.; Thomsen, L.C.V.; et al. TLR3 expression by maternal and fetal cells at the maternal-fetal interface in normal and preeclamptic pregnancies. *J. Leukoc. Biol.* **2021**, *109*, 173–183. [CrossRef]
88. Kalkunte, S.; Nevers, T.; Norris, W.E.; Sharma, S. Vascular IL-10: A protective role in preeclampsia. *J. Reprod. Immunol.* **2011**, *88*, 165–169. [CrossRef] [PubMed]
89. Thaxton, J.E.; Sharma, S. Interleukin-10: A multi-faceted agent of pregnancy. *Am. J. Reprod. Immunol.* **2010**, *63*, 482–491. [CrossRef]
90. Blois, S.M.; Freitag, N.; Tirado-González, I.; Cheng, S.B.; Heimesaat, M.M.; Bereswill, S.; Rose, M.; Conrad, M.L.; Barrientos, G.; Sharma, S. NK cell-derived IL-10 is critical for DC-NK cell dialogue at the maternal-fetal interface. *Sci. Rep.* **2017**, *7*, 2189. [CrossRef] [PubMed]
91. Hennessy, P.A.; Pilmore, H.L.; Simmons, L.A.; Painter, D.M. A Deficiency of Placental IL-10 in preeclampsia. *J. Immunol.* **1999**, *163*, 3491–3495.
92. Chatterjee, P.; Chiasson, V.L.; Kopriva, S.E.; Young, K.J.; Chatterjee, V.; Jones, K.A.; Mitchell, B.M. Interleukin 10 deficiency exacerbates toll-like receptor 3-induced preeclampsia-like symptoms in mice. *Hypertension* **2011**, *58*, 489–496. [CrossRef] [PubMed]
93. Abu-Raya, B.; Michalski, C.; Sadarangani, M.; Lavoie, P.M. Maternal Immunological Adaptation During Normal Pregnancy. *Front. Immunol.* **2020**, *11*, 575197. [CrossRef]

94. Corthay, A. How do regulatory t cells work? *Scand. J. Immunol.* **2009**, *70*, 326–336. [CrossRef]
95. Zhao, H.; Liao, X.; Kang, Y. Tregs: Where we are and what comes next? *Front. Immunol.* **2017**, *8*, 1578. [CrossRef]
96. Robertson, S.A.; Green, E.S.; Care, A.S.; Moldenhauer, L.M.; Prins, J.R.; Louise Hull, M.; Barry, S.C.; Dekker, G. Therapeutic potential of regulatory T cells in preeclampsia-opportunities and challenges. *Front. Immunol.* **2019**, *10*, 478. [CrossRef]
97. Wang, A.; Rana, S.; Karumanchi, S.A. Preeclampsia: The role of angiogenic factors in its pathogenesis. *Physiology* **2009**, *24*, 147–158. [CrossRef] [PubMed]
98. Lee, E.S.; Oh, M.-J.; Jung, J.W.; Lim, J.-E.; Seol, H.-J.; Lee, K.-J.; Kim, H.-J. The levels of circulating vascular endothelial growth factor and soluble Flt-1 in pregnancies complicated by preeclampsia. *J. Korean Med. Sci.* **2007**, *22*, 94–98. [CrossRef] [PubMed]
99. Maynard, S.E.; Min, J.Y.; Merchan, J.; Lim, K.H.; Li, J.; Mondal, S.; Libermann, T.A.; Morgan, J.P.; Sellke, F.W.; Stillman, I.E.; et al. Excess placental soluble fms-like tyrosine kinase 1 (sFlt1) may contribute to endothelial dysfunction hypertension, and proteinuria in preeclampsia. *J. Clin. Investig.* **2003**, *111*, 649–658. [CrossRef] [PubMed]
100. Maynard, S.E.; Karumanchi, S.A. Angiogenic factors and preeclampsia. *Semin. Nephrol.* **2011**, *31*, 33–46. [CrossRef] [PubMed]
101. Shibuya, M. Vascular Endothelial Growth Factor (VEGF) and Its Receptor (VEGFR) Signaling in Angiogenesis: A Crucial Target for Anti- and Pro-Angiogenic Therapies. *Genes Cancer* **2011**, *2*, 1097–1105. [CrossRef]
102. Smani, T.; Gómez, L.J.; Regodon, S.; Woodard, G.E.; Siegfried, G.; Khatib, A.M.; Rosado, J.A. Trp channels in angiogenesis and other endothelial functions. *Front. Physiol.* **2018**, *9*, 1731. [CrossRef] [PubMed]
103. Peach, C.J.; Mignone, V.W.; Arruda, M.A.; Alcobia, D.C.; Hill, S.J.; Kilpatrick, L.E.; Woolard, J. Molecular Pharmacology of VEGF-A Isoforms: Binding and Signalling at VEGFR2. *Int. J. Mol. Sci.* **2018**, *19*, 1264. [CrossRef]
104. Geva, E.; Ginzinger, D.G.; Zaloudek, C.J.; Moore, D.H.; Byrne, A.; Jaffe, R.B. Human placental vascular development: Vasculogenic and angiogenic (branching and nonbranching) transformation is regulated by vascular endothelial growth factor-a, angiopoietin-1, and angiopoietin-2. *J. Clin. Endocrinol. Metab.* **2002**, *87*, 4213–4224. [CrossRef]
105. Grummer, M.A.; Sullivan, J.A.; Magness, R.R.; Bird, I.M. Vascular endothelial growth factor acts through novel, pregnancy-enhanced receptor signalling pathways to stimulate endothelial nitric oxide synthase activity in uterine artery endothelial cells. *Biochem. J.* **2009**, *417*, 501–511. [CrossRef]
106. Pandey, A.K.; Singhi, E.K.; Arroyo, J.P.; Ikizler, T.A.; Gould, E.R.; Brown, J.; Beckman, J.A.; Harrison, D.G.; Moslehi, J. Mechanisms of VEGF (Vascular Endothelial Growth Factor) Inhibitor–Associated Hypertension and Vascular Disease. *Hypertension* **2018**, *71*, E1–E8. [CrossRef]
107. Chau, K.; Hennessy, A.; Makris, A. Placental growth factor and pre-eclampsia. *J. Hum. Hypertens.* **2017**, *31*, 782–786. [CrossRef] [PubMed]
108. Williams, D.; Kenyon, A.; Adamson, D. Physiology. In *Basic Science in Obstetrics and Gynaecology*; Elsevier: Amsterdam, The Netherlands, 2010; pp. 173–230.
109. Dewerchin, M.; Carmeliet, P. PlGF: A multitasking cytokine with disease-restricted activity. *Cold Spring Harb. Perspect. Med.* **2012**, *2*, a011056. [CrossRef]
110. Lobmaier, S.M.; Figueras, F.; Mercade, I.; Crovetto, F.; Peguero, A.; Parra-Saavedra, M.; Ortiz, J.U.; Crispi, F.; Gratacós, E. Levels of maternal serum angiogenic factors in third-trimester normal pregnancies: Reference ranges, influence of maternal and pregnancy factors and fetoplacental Doppler indices. *Fetal. Diagn.* **2014**, *36*, 38–43. [CrossRef] [PubMed]
111. Ng, Q.J.; Han, J.Y.; Saffari, S.E.; Yeo, G.S.-H.; Chern, B.S.M.; Tan, K.H. Longitudinal circulating placental growth factor (PlGF) and soluble FMS-like tyrosine kinase-1 (sFlt-1) concentrations during pregnancy in Asian women: A prospective cohort study. *BMJ Open* **2019**, *9*, e028321. [CrossRef]
112. Levine, R.J.; Maynard, S.E.; Qian, C.; Lim, K.H.; England, L.J.; Yu, K.F.; Schisterman, E.F.; Thadhani, R.; Sachs, B.P.; Epstein, F.H.; et al. Circulating Angiogenic Factors and the Risk of Preeclampsia. *N. Engl. J. Med.* **2004**, *350*, 672–683. [CrossRef]
113. Benschop, L.; Schalekamp-Timmermans, S.; Broere-Brown, Z.A.; Roeters van Lennep, J.E.; Jaddoe, V.W.V.; Roos-Hesselink, J.W.; Ikram, M.K.; Steegers, E.A.P.; Roberts, J.M.; Gandley, R.E. Placental Growth Factor as an Indicator of Maternal Cardiovascular Risk After Pregnancy. *Circulation* **2019**, *139*, 1698–1709. [CrossRef] [PubMed]
114. Roberts, J.M.; Rajakumar, A. Preeclampsia and soluble fms-like tyrosine kinase 1. *J. Clin. Endocrinol. Metab.* **2009**, *94*, 2252–2254. [CrossRef]
115. Thomas, C.P.; Andrews, J.I.; Raikwar, N.S.; Kelley, E.A.; Herse, F.; Dechend, R.; Golos, T.G.; Liu, K.Z. A recently evolved novel trophoblast-enriched secreted form of fms-like tyrosine kinase-1 variant is up-regulated in hypoxia and preeclampsia. *J. Clin. Endocrinol. Metab.* **2009**, *94*, 2524–2530. [CrossRef] [PubMed]
116. Adamson, S.L. sFLT1 in preeclampsia: Trophoblast defense against a decidual VEGFA barrage? *J. Clin. Investig.* **2014**, *124*, 4690–4692. [CrossRef]
117. Zeisler, H.; Llurba, E.; Chantraine, F.; Vatish, M.; Staff, A.C.; Sennström, M.; Olovsson, M.; Brennecke, S.P.; Stepan, H.; Allegranza, D.; et al. Predictive Value of the sFlt-1:PlGF Ratio in Women with Suspected Preeclampsia. *N. Engl. J. Med.* **2016**, *374*, 13–22. [CrossRef]
118. Sela, S.; Itin, A.; Natanson-Yaron, S.; Greenfield, C.; Goldman-Wohl, D.; Yagel, S.; Keshet, E. A novel human-specific soluble vascular endothelial growth factor receptor 1: Cell type-specific splicing and implications to vascular endothelial growth factor homeostasis and preeclampsia. *Circ. Res.* **2008**, *102*, 1566–1574. [CrossRef] [PubMed]

119. Nagamatsu, T.; Fujii, T.; Kusumi, M.; Zou, L.; Yamashita, T.; Osuga, Y.; Momoeda, M.; Kozuma, S.; Taketani, Y. Cytotrophoblasts up-regulate soluble fms-like tyrosine kinase-1 expression under reduced oxygen: An implication for the placental vascular development and the pathophysiology of preeclampsia. *Endocrinology* **2004**, *145*, 4838–4845. [CrossRef] [PubMed]
120. Li, H.; Gu, B.; Zhang, Y.; Lewis, D.F.; Wang, Y. Hypoxia-induced increase in soluble Flt-1 production correlates with enhanced oxidative stress in trophoblast cells from the human placenta. *Placenta* **2005**, *26*, 210–217. [CrossRef] [PubMed]
121. Fukai, T.; Ushio-Fukai, M. Superoxide dismutases: Role in redox signaling, vascular function, and diseases. *Antioxid. Redox Signal.* **2011**, *15*, 1583–1606. [CrossRef] [PubMed]
122. Nassiri, F.; Cusimano, M.D.; Scheithauer, B.W.; Rotondo, F.; Fazio, A.; Yousef, G.M.; Syro, L.V.; Kovacs, K.; Lloyd, R.V. Endoglin (CD105): A review of its role in angiogenesis and tumor diagnosis, progression and therapy. *Anticancer Res.* **2011**, *31*, 2283–2290.
123. Venkatesha, S.; Toporsian, M.; Lam, C.; Hanai, J.I.; Mammoto, T.; Kim, Y.M.; Bdolah, Y.; Lim, K.H.; Yuan, H.T.; Libermann, T.A.; et al. Soluble endoglin contributes to the pathogenesis of preeclampsia. *Nat. Med.* **2006**, *12*, 642–649. [CrossRef]
124. St-Jacques, S.; Forte, M.; Lye, S.J.; Letarte, M. Localization of endoglin, a transforming growth factor-β binding protein, and of CD44 and integrins in placenta during the first trimester of pregnancy. *Biol. Reprod.* **1994**, *51*, 405–413. [CrossRef]
125. Alam, S.; Ahmad, S.; Zafeer, F.; Rizvi, A.A.; Gulati, R.; Rashid, S. Role of TGF-B1 in The Pathogenesis of Pre-Eclampsia. *Ann. Int. Med. Dent. Res.* **2017**, *3*, 1. [CrossRef]
126. Caniggia, I.; Taylor, C.V.; Ritchie, J.W.K.; Lye, S.J.; Letarte, M. Endoglin regulates trophoblast differentiation along the invasive pathway in human placental villous explants. *Endocrinology* **1997**, *138*, 4977–4988. [CrossRef]
127. Ayatollahi, M.; Geramizadeh, B.; Samsami, A. Transforming growth factor beta-1 influence on fetal allografts during pregnancy. *Transplant. Proc.* **2005**, *37*, 4603–4604. [CrossRef]
128. Nikuei, P.; Rajaei, M.; Malekzadeh, K.; Nejatizadeh, A.; Mohseni, F.; Atashabparvar, A. Accuracy of soluble endoglin for diagnosis of preeclampsia and its severity. *Iran. Biomed. J.* **2017**, *21*, 312–320. [CrossRef] [PubMed]
129. Powe, C.E.; Levine, R.J.; Karumanchi, S.A. Preeclampsia, a disease of the maternal endothelium: The role of antiangiogenic factors and implications for later cardiovascular disease. *Circulation* **2011**, *123*, 2856–2869. [CrossRef]
130. Bell, M.J.; Roberts, J.M.; Founds, S.A.; Jeyabalan, A.; Terhorst, L.; Conley, Y.P. Variation in endoglin pathway genes is associated with preeclampsia: A case-control candidate gene association study. *BMC Pregnancy Childbirth* **2013**, *13*. [CrossRef]
131. Jerkic, M.; Rivas-Elena, J.V.; Prieto, M.; Carrón, R.; Sanz-Rodríguez, F.; Pérez-Barriocanal, F.; Rodríguez-Barbero, A.; Bernabéu, C.; López-Novoa, J.M. Endoglin regulates nitric oxide-dependent vasodilatation. *FASEB J.* **2004**, *18*, 609–611. [CrossRef] [PubMed]
132. Mano, Y.; Kotani, T.; Shibata, K.; Matsumura, H.; Tsuda, H.; Sumigama, S.; Yamamoto, E.; Iwase, A.; Senga, T.; Kikkawa, F. The loss of endoglin promotes the invasion of extravillous trophoblasts. *Endocrinology* **2011**, *152*, 4386–4394. [CrossRef]
133. Williams, P.J.; Mistry, H.D.; Innes, B.A.; Bulmer, J.N.; Broughton Pipkin, F. Expression of AT1R, AT2R and AT4R and Their Roles in Extravillous Trophoblast Invasion in the Human. *Placenta* **2010**, *31*, 448–455. [CrossRef] [PubMed]
134. Guo, D.F.; Sun, Y.L.; Hamet, P.; Inagami, T. The angiotensin II type 1 receptor and receptor-associated proteins. *Cell Res.* **2001**, *11*, 165–180. [CrossRef]
135. LaMarca, B. Endothelial dysfunction. An important mediator in the pathophysiology of hypertension during pre-eclampsia. *Minerva Ginecol.* **2012**, *64*, 309–320.
136. LaMarca, B.; Wallukat, G.; Llinas, M.; Herse, F.; Dechend, R.; Granger, J.P. Autoantibodies to the angiotensin type I receptor in response to placental ischemia and tumor necrosis factor alpha in pregnant rats. *Hypertension* **2008**, *52*, 1168–1172. [CrossRef]
137. LaMarca, B.; Parrish, M.; Ray, L.F.; Murphy, S.R.; Roberts, L.; Glover, P.; Wallukat, G.; Wenzel, K.; Cockrell, K.; Martin, J.N., Jr.; et al. Hypertension in response to autoantibodies to the angiotensin II type I receptor (AT1-AA) in pregnant rats: Role of endothelin-1. *Hypertension* **2009**, *54*, 905–909. [CrossRef]
138. Parrish, M.R.; Murphy, S.R.; Rutland, S.; Wallace, K.; Wenzel, K.; Wallukat, G.; Keiser, S.; Ray, L.F.; Dechend, R.; Martin, J.N.; et al. The effect of immune factors, tumor necrosis factor-alpha, and agonistic autoantibodies to the angiotensin II type I receptor on soluble fms-like tyrosine-1 and soluble endoglin production in response to hypertension during pregnancy. *Am. J. Hypertens.* **2010**, *23*, 911–916. [CrossRef]
139. Bezerra Maia E Holanda Moura, S.; Marques Lopes, L.; Murthi, P.; da Silva Costa, F. Prevention of preeclampsia. *J. Pregnancy* **2012**, *2012*, 435090. [CrossRef] [PubMed]
140. Brown, C.M.; Garovic, V.D. Drug treatment of hypertension in pregnancy. *Drugs* **2014**, *74*, 283–296. [CrossRef]
141. Duhig, K.; Vandermolen, B.; Shennan, A. Recent advances in the diagnosis and management of pre-eclampsia. *F1000Res* **2018**, *7*, 242. [CrossRef]
142. Spradley, F.T.; Tan, A.Y.; Joo, W.S.; Daniels, G.; Kussie, P.; Karumanchi, S.A.; Granger, J.P. Placental growth factor administration abolishes placental ischemia-induced hypertension. *Hypertension* **2016**, *67*, 740–747. [CrossRef] [PubMed]
143. Zhu, M.; Ren, Z.; Possomato-Vieira, J.S.; Khalil, R.A. Restoring placental growth factor-soluble fms-like tyrosine kinase-1 balance reverses vascular hyper-reactivity and hypertension in pregnancy. *Am. J. Physiol. Regul. Integr. Comp. Physiol.* **2016**, *311*, R505–R521. [CrossRef] [PubMed]
144. Thadhani, R.; Hagmann, H.; Schaarschmidt, W.; Roth, B.; Cingoez, T.; Karumanchi, S.A.; Wenger, J.; Lucchesi, K.J.; Tamez, H.; Lindner, T.; et al. Removal of soluble fms-like tyrosine kinase-1 by dextran sulfate apheresis in preeclampsia. *J. Am. Soc. Nephrol.* **2016**, *27*, 903–913. [CrossRef] [PubMed]

145. Bridges, J.P.; Gilbert, J.S.; Colson, D.; Gilbert, S.A.; Dukes, M.P.; Ryan, M.J.; Granger, J.P. Oxidative stress contributes to soluble fms-like tyrosine kinase-1 induced vascular dysfunction in pregnant rats. *Am. J. Hypertens.* **2009**, *22*, 564–568. [CrossRef] [PubMed]
146. Félétou, M. The Endothelium: Part 1: Multiple Functions of the Endothelial Cells- Focus on Endothelium-Derived Vasoactive Mediators. In *Colloquium Series on Integrated Systems Physiology*; Morgan & Claypool Life Sciences: San Rafael, CA, USA, 2011.
147. Galley, H.F.; Webster, N.R. Physiology of the endothelium. *Br. J. Anaesth.* **2004**, *93*, 105–113. [CrossRef]
148. Marti, C.N.; Gheorghiade, M.; Kalogeropoulos, A.P.; Georgiopoulou, V.V.; Quyyumi, A.A.; Butler, J. Endothelial dysfunction, arterial stiffness, and heart failure. *J. Am. Coll. Cardiol.* **2012**, *60*, 1455–1469. [CrossRef]
149. Sukriti, S.; Tauseef, M.; Yazbeck, P.; Mehta, D. Mechanisms regulating endothelial permeability. *Pulm. Circ.* **2014**, *4*, 535–551. [CrossRef]
150. Tseng, E.; Yee Teoh, S.S.; Wang, Y.; Nie, G. Elevated protease HtrA4 in the maternal circulation of preeclampsia may contribute to endothelial barrier disruption by cleaving key junctional protein VE-cadherin. *Placenta* **2019**, *76*, 51–53. [CrossRef]
151. Wang, Y.; Lewis, D.F.; Alexander, J.S.; Granger, D.N. Endothelial barrier function in preeclampsia. *Front. Biosci.* **2007**, *12*, 2412–2424. [CrossRef] [PubMed]
152. Hayman, R.; Warren, A.; Brockelsby, J.; Johnson, I.; Baker, P. Plasma from women with pre-eclampsia induces an in vitro alteration in the endothelium-dependent behaviour of myometrial resistance arteries. *BJOG Int. J. Obstet. Gynaecol.* **2000**, *107*, 108–115. [CrossRef] [PubMed]
153. McIntosh, J.J.; Derayunan, A.; Hader, S.; Lohr, N.; Beyer, A.; Gutterman, D. Impaired Microvascular Endothelial Function in Preeclampsia. *FASEB J.* **2020**, *34*, 1. [CrossRef]
154. Chambers, J.C.; Fusi, L.; Malik, I.S.; Haskard, D.O.; De Swiet, M.; Kooner, J.S. Association of maternal endothelial dysfunction with preeclampsia. *J. Am. Med. Assoc.* **2001**, *285*, 1607–1612. [CrossRef]
155. Sandgren, J.A.; Deng, G.; Linggonegoro, D.W.; Scroggins, S.M.; Perschbacher, K.J.; Nair, A.R.; Nishimura, T.E.; Zhang, S.Y.; Agbor, L.N.; Wu, J.; et al. Arginine vasopressin infusion is sufficient to model clinical features of preeclampsia in mice. *JCI Insight* **2018**, *3*, e99403. [CrossRef] [PubMed]
156. Vanwijk, M.J.; Kublickiene, K.; Boer, K.; VanBavel, E. Vascular function in preeclampsia. *Cardiovasc. Res.* **2000**, *47*, 38–48. [CrossRef]
157. Schrier, R.W. Pathogenesis of sodium and water retention in high-output and low-output cardiac failure, nephrotic syndrome, cirrhosis, and pregnancy (1). *N. Engl. J. Med.* **1988**, *319*, 1065–1072. [CrossRef]
158. Tkachenko, O.; Shchekochikhin, D.; Schrier, R.W. Hormones and hemodynamics in pregnancy. *Int. J. Endocrinol. Metab.* **2014**, *12*, e14098. [CrossRef]
159. Vricella, L.K. Emerging understanding and measurement of plasma volume expansion in pregnancy. *Am. J. Clin. Nutr.* **2017**, *106*, 1620S–1625S. [CrossRef] [PubMed]
160. Silver, H.M.; Seebeck, M.; Carlson, R. Comparison of total blood volume in normal, preeclamptic, and nonproteinuric gestational hypertensive pregnancy by simultaneous measurement of red blood cell and plasma volumes. *Am. J. Obs. Gynecol.* **1998**, *179*, 87–93. [CrossRef]
161. Salas, S.P.; Marshall, G.; Gutiérrez, B.L.; Rosso, P. Time Course of Maternal Plasma Volume and Hormonal Changes in Women With Preeclampsia or Fetal Growth Restriction. *Hypertension* **2006**, *47*, 203–208. [CrossRef]
162. Soffronoff, E.C.; Kaufmann, B.M.; Connaughton, J.F. Intravascular volume determinations and fetal outcome in hypertensive diseases of pregnancy. *Am. J. Obs. Gynecol.* **1977**, *127*, 4–9. [CrossRef]
163. Arias, F. Expansion of intravascular volume and fetal outcome in patients with chronic hypertension and pregnancy. *Am. J. Obs. Gynecol.* **1975**, *123*, 610–616. [CrossRef]
164. Anthony, J.; Schoeman, L.K. Fluid management in pre-eclampsia. *Obs. Med.* **2013**, *6*, 100–104. [CrossRef]
165. Uddin, M.N.; McLean, L.B.; Hunter, F.A.; Horvat, D.; Severson, J.; Tharakan, B.; Childs, E.W.; Puschett, J.B. Vascular leak in a rat model of preeclampsia. *Am. J. Nephrol.* **2009**, *30*, 26–33. [CrossRef] [PubMed]
166. Singh, H.; Zhao, M.; Chen, Q.; Wang, Y.; Li, Y.; Kaitu'u-Lino, T.J.; Tong, S.; Nie, G. Human HtrA4 Expression Is Restricted to the Placenta, Is Significantly Up-Regulated in Early-Onset Preeclampsia, and High Levels of HtrA4 Cause Endothelial Dysfunction. *J. Clin. Endocrinol. Metab.* **2015**, *100*, E936–E945. [CrossRef]
167. Wang, Y.; Gu, Y.; Lewis, D.F.; Alexander, J.S.; Granger, D.N. Elevated plasma chymotrypsin-like protease (chymase) activity in women with preeclampsia. *Hypertens Pregnancy* **2010**, *29*, 253–261. [CrossRef]
168. Gu, Y.; Groome, L.J.; Alexander, J.S.; Wang, Y. PAR-2 triggers placenta-derived protease-induced altered VE-cadherin reorganization at endothelial junctions in preeclampsia. *Placenta* **2012**, *33*, 803–809. [CrossRef] [PubMed]
169. Robertson, G.L.; Shelton, R.L.; Athar, S. The osmoregulation of vasopressin. *Kidney Int.* **1976**, *10*, 25–37. [CrossRef] [PubMed]
170. Dunn, F.L.; Brennan, T.J.; Nelson, A.E.; Robertson, G.L. The role of blood osmolality and volume in regulating vasopressin secretion in the rat. *J. Clin. Investig.* **1973**, *52*, 3212–3219. [CrossRef]
171. Santillan, M.K.; Santillan, D.A.; Scroggins, S.M.; Min, J.Y.; Sandgren, J.A.; Pearson, N.A.; Leslie, K.K.; Hunter, S.K.; Zamba, G.K.D.; Gibson-Corley, K.N.; et al. Vasopressin in preeclampsia: A novel very early human pregnancy biomarker and clinically relevant mouse model. *Hypertension* **2014**, *64*, 852–859. [CrossRef] [PubMed]

172. Lopes van Balen, V.A.; van Gansewinkel, T.A.G.; de Haas, S.; van Kuijk, S.M.J.; van Drongelen, J.; Ghossein-Doha, C.; Spaanderman, M.E.A. Physiological adaptation of endothelial function to pregnancy: Systematic review and meta-analysis. *Ultrasound Obstet. Gynecol.* **2017**, *50*, 697–708. [CrossRef]
173. Choi, J.W.; Im, M.W.; Pai, S.H. Nitric oxide production increases during normal pregnancy and decreases in preeclampsia. *Ann. Clin. Lab. Sci.* **2002**, *32*, 257–263.
174. Seligman, S.P.; Buyon, J.P.; Clancy, R.M.; Young, B.K.; Abramson, S.B. The role of nitric oxide in the pathogenesis of preeclampsia. *Am. J. Obstet. Gynecol.* **1994**, *171*, 944–948. [CrossRef]
175. Lind, L. Endothelium-dependent Vasodilation in Hypertension: A Review. *Blood Press.* **2000**, *9*, 4–15. [CrossRef] [PubMed]
176. Boeldt, D.S.; Krupp, J.; Yi, F.-X.; Khurshid, N.; Shah, D.M.; Bird, I.M. Positive versus negative effects of VEGF165 on Ca^{2+} signaling and NO production in human endothelial cells. *Am. J. Physiol. -Heart Circ. Physiol.* **2016**, *312*, H173–H181. [CrossRef]
177. Fleming, I.; Busse, R. Signal transduction of eNOS activation. *Cardiovasc. Res.* **1999**, *43*, 532–541. [CrossRef]
178. Cale, J.M.; Bird, I.M. Dissociation of endothelial nitric oxide synthase phosphorylation and activity in uterine artery endothelial cells. *Am. J. Physiol. Heart Circ. Physiol.* **2006**, *290*. [CrossRef] [PubMed]
179. Boeldt, D.S.; Yi, F.X.; Bird, I.M. eNOS activation and NO function: Pregnancy adaptive programming of capacitative entry responses alters nitric oxide (NO) output in vascular endothelium-new insights into eNOS regulation through adaptive cell signaling. *J. Endocrinol.* **2011**, *210*, 243–258. [CrossRef] [PubMed]
180. Boeldt, D.S.; Grummer, M.A.; Yi, F.X.; Magness, R.R.; Bird, I.M. Phosphorylation of Ser-279/282 and Tyr-265 positions on Cx43 as possible mediators of VEGF-165 inhibition of pregnancy-adapted Ca2+ burst function in ovine uterine artery endothelial cells. *Mol. Cell. Endocrinol.* **2015**, *412*, 73–84. [CrossRef]
181. Feliers, D.; Chen, X.; Akis, N.; Choudhury, G.G.; Madaio, M.; Kasinath, B.S. VEGF regulation of endothelial nitric oxide synthase in glomerular endothelial cells. *Kidney Int.* **2005**, *68*, 1648–1659. [CrossRef]
182. Craici, I.M.; Wagner, S.J.; Weissgerber, T.L.; Grande, J.P.; Garovic, V.D. Advances in the pathophysiology of pre-eclampsia and related podocyte injury. *Kidney Int.* **2014**, *86*, 275–285. [CrossRef]
183. Kolluru, G.K.; Shen, X.; Kevil, C.G. A tale of two gases: NO and H2S, foes or friends for life? *Redox Biol.* **2013**, *1*, 313–318. [CrossRef]
184. Naik, J.S.; Osmond, J.M.; Walker, B.R.; Kanagy, N.L. Hydrogen sulfide-induced vasodilation mediated by endothelial TRPV4 channels. *Am. J. Physiol. Heart Circ. Physiol.* **2016**, *311*, H1437–H1444. [CrossRef]
185. Coletta, C.; Papapetropoulos, A.; Erdelyi, K.; Olah, G.; Módis, K.; Panopoulos, P.; Asimakopoulou, A.; Gerö, D.; Sharina, I.; Martin, E.; et al. Hydrogen sulfide and nitric oxide are mutually dependent in the regulation of angiogenesis and endothelium-dependent vasorelaxation. *Proc. Natl. Acad. Sci. USA* **2012**, *109*, 9161–9166. [CrossRef] [PubMed]
186. Li, X.-H.; Xue, W.-L.; Wang, M.-J.; Zhou, Y.; Zhang, C.-C.; Sun, C.; Zhu, L.; Liang, K.; Chen, Y.; Tao, B.-B.; et al. H2S regulates endothelial nitric oxide synthase protein stability by promoting microRNA-455-3p expression. *Sci. Rep.* **2017**, *7*, 44807. [CrossRef]
187. Bir, S.C.; Kolluru, G.K.; McCarthy, P.; Shen, X.; Pardue, S.; Pattillo, C.B.; Kevil, C.G. Hydrogen sulfide stimulates ischemic vascular remodeling through nitric oxide synthase and nitrite reduction activity regulating hypoxia-inducible factor-1α and vascular endothelial growth factor-dependent angiogenesis. *J. Am. Heart Assoc.* **2012**, *1*, e004093. [CrossRef] [PubMed]
188. Ahmed, A. Molecular mechanisms and therapeutic implications of the carbon monoxide/hmox1 and the hydrogen sulfide/CSE pathways in the prevention of pre-eclampsia and fetal growth restriction. *Pregnancy Hypertens* **2014**, *4*, 243–244. [CrossRef]
189. Wang, K.; Ahmad, S.; Cai, M.; Rennie, J.; Fujisawa, T.; Crispi, F.; Baily, J.; Miller, M.R.; Cudmore, M.; Hadoke, P.W.; et al. Dysregulation of hydrogen sulfide producing enzyme cystathionine γ-lyase contributes to maternal hypertension and placental abnormalities in preeclampsia. *Circulation* **2013**, *127*, 2514–2522. [CrossRef]
190. Possomato-Vieira, J.S.; Palei, A.C.; Pinto-Souza, C.C.; Cavalli, R.; Dias-Junior, C.A.; Sandrim, V. Circulating levels of hydrogen sulphide negatively correlate to nitrite levels in gestational hypertensive and preeclamptic pregnant women. *Clin. Exp. Pharm. Physiol.* **2021**, *48*, 1224–1230. [CrossRef] [PubMed]
191. Rajendran, S.; Shen, X.; Glawe, J.; Kolluru, G.K.; Kevil, C.G. Nitric Oxide and Hydrogen Sulfide Regulation of Ischemic Vascular Growth and Remodeling. *Compr. Physiol.* **2019**, *9*, 1213–1247. [CrossRef]
192. Holwerda, K.M.; Bos, E.M.; Rajakumar, A.; Ris-Stalpers, C.; van Pampus, M.G.; Timmer, A.; Erwich, J.J.; Faas, M.M.; van Goor, H.; Lely, A.T. Hydrogen sulfide producing enzymes in pregnancy and preeclampsia. *Placenta* **2012**, *33*, 518–521. [CrossRef] [PubMed]
193. Wang, R.; Szabo, C.; Ichinose, F.; Ahmed, A.; Whiteman, M.; Papapetropoulos, A. The role of H2S bioavailability in endothelial dysfunction. *Trends Pharm. Sci* **2015**, *36*, 568–578. [CrossRef] [PubMed]
194. Sankaralingam, S.; Xu, H.; Davidge, S.T. Arginase contributes to endothelial cell oxidative stress in response to plasma from women with preeclampsia. *Cardiovasc. Res.* **2010**, *85*, 194–203. [CrossRef]
195. Speer, P.D.; Powers, R.W.; Frank, M.P.; Harger, G.; Markovic, N.; Roberts, J.M. Elevated asymmetric dimethylarginine concentrations precede clinical preeclampsia, but not pregnancies with small-for-gestational-age infants. *Am. J. Obstet. Gynecol.* **2008**, *198*, 112.e111–112.e127. [CrossRef]
196. Braekke, K.; Ueland, P.M.; Harsem, N.K.; Staff, A.C. Asymmetric Dimethylarginine in the Maternal and Fetal Circulation in Preeclampsia. *Pediatric Res.* **2009**, *66*, 411–415. [CrossRef]
197. Davidge, S.T.; Baker, P.N.; Roberts, J.M. NOS expression is increased in endothelial cells exposed to plasma from women with preeclampsia. *Am. J. Physiol. Heart Circ. Physiol.* **1995**, *269*. [CrossRef]

198. Aggarwal, P.K.; Chandel, N.; Jain, V.; Jha, V. The relationship between circulating endothelin-1, soluble fms-like tyrosine kinase-1 and soluble endoglin in preeclampsia. *J. Hum. Hypertens.* **2012**, *26*, 236–241. [CrossRef]
199. Gant, N.F.; Daley, G.L.; Chand, S.; Whalley, P.J.; MacDonald, P.C. A study of angiotensin II pressor response throughout primigravid pregnancy. *J. Clin. Investig.* **1973**, *52*, 2682–2689. [CrossRef] [PubMed]
200. Lüscher, T.F. Angiotensin, ACE-inhibitors and endothelial control of vasomotor tone. *Basic Res. Cardiol.* **1993**, *88* (Suppl 1), 15–24. [CrossRef]
201. Leaños-Miranda, A.; Inova, C.-G.; Méndez-Aguilar, F.; Molina-Pérez, C.J.; Ramírez-Valenzuela, K.L.; Sillas-Pardo, L.J.; Uraga-Camacho, N.C.; Isordia-Salas, I.; Berumen-Lechuga, M.G. Lower circulating angiotensin II levels are related to the severity of preeclampsia and its risk as disclosed by a specific bioassay. *Medicine* **2018**, *97*, e12498. [CrossRef] [PubMed]
202. Stanhewicz, A.E.; Jandu, S.; Santhanam, L.; Alexander, L.M. Increased angiotensin II sensitivity contributes to microvascular dysfunction in women who have had preeclampsia. *Hypertension* **2017**, *70*, 382–389. [CrossRef]
203. Jain, A. Endothelin-1: A key pathological factor in pre-eclampsia? *Reprod. Biomed. Online* **2012**, *25*, 443–449. [CrossRef]
204. Taddei, S.; Virdis, A.; Ghiadoni, L.; Sudano, I.; Magagna, A.; Salvetti, A. Role of Endothelin in the Control of Peripheral Vascular Tone in Human Hypertension. *Heart Fail. Rev.* **2001**, *6*, 277–285. [CrossRef]
205. Dechanet, C.; Fort, A.; Barbero-Camps, E.; Dechaud, H.; Richard, S.; Virsolvy, A. Endothelin-Dependent Vasoconstriction in Human Uterine Artery: Application to Preeclampsia. *PLoS ONE* **2011**, *6*, e16540. [CrossRef]
206. Saleh, L.; Verdonk, K.; Visser, W.; van den Meiracker, A.H.; Danser, A.H.J. The emerging role of endothelin-1 in the pathogenesis of pre-eclampsia. *Adv. Cardiovasc. Dis.* **2016**, *10*, 282–293. [CrossRef]
207. Alexander, B.T.; Rinewalt, A.N.; Cockrell, K.L.; Massey, M.B.; Bennett, W.A.; Granger, J.P. Endothelin type a receptor blockade attenuates the hypertension in response to chronic reductions in uterine perfusion pressure. *Hypertension* **2001**, *37*, 485–489. [CrossRef] [PubMed]
208. Thorin, E.; Webb, D.J. Endothelium-derived endothelin-1. *Pflug. Arch.* **2010**, *459*, 951–958. [CrossRef]
209. Wynne, B.M.; Chiao, C.-W.; Webb, R.C. Vascular smooth muscle cell signaling mechanisms for contraction to angiotensin II and endothelin-1. *J. Am. Soc. Hypertens* **2009**, *3*, 84–95. [CrossRef]
210. Hoffert, J.D.; Pisitkun, T.; Saeed, F.; Song, J.H.; Chou, C.-L.; Knepper, M.A. Dynamics of the G protein-coupled vasopressin V2 receptor signaling network revealed by quantitative phosphoproteomics. *Mol. Cell Proteom.* **2012**, *11*, M111.014613. [CrossRef]
211. Kimple, A.J.; Bosch, D.E.; Giguère, P.M.; Siderovski, D.P. Regulators of G-protein signaling and their Gα substrates: Promises and challenges in their use as drug discovery targets. *Pharm. Rev.* **2011**, *63*, 728–749. [CrossRef]
212. Osei-Owusu, P.; Blumer, K.J. Regulator of G protein signaling 2: A versatile regulator of vascular function. *Prog. Mol. Biol. Transl. Sci.* **2015**, *133*, 77–92.
213. Perschbacher, K.J.; Deng, G.; Sandgren, J.A.; Walsh, J.W.; Witcher, P.C.; Sapouckey, S.A.; Owens, C.E.; Zhang, S.Y.; Scroggins, S.M.; Pearson, N.A. Reduced mRNA Expression of RGS2 (Regulator of G Protein Signaling-2) in the Placenta Is Associated With Human Preeclampsia and Sufficient to Cause Features of the Disorder in Mice. *Hypertension* **2020**, *75*, 569–579. [CrossRef]
214. Perschbacher, K.J.; Deng, G.; Fisher, R.A.; Gibson-Corley, K.N.; Santillan, M.K.; Grobe, J.L. Regulators of G protein signaling in cardiovascular function during pregnancy. *Physiol. Genom.* **2018**, *50*, 590–604. [CrossRef] [PubMed]
215. Koch, J.N.; Dahlen, S.A.; Owens, E.A.; Osei-Owusu, P.; Osei-Owusu, P.; N, K.J.; A, D.S.; A, O.E.; Patrick, O.O. Regulator of G protein signaling 2 facilitates uterine artery adaptation during pregnancy in mice. *J. Am. Heart Assoc.* **2019**, *8*, e010917. [CrossRef]
216. Osei-Owusu, P.; Sabharwal, R.; Kaltenbronn, K.M.; Rhee, M.H.; Chapleau, M.W.; Dietrich, H.H.; Blumer, K.J. Regulator of G protein signaling 2 deficiency causes endothelial dysfunction and impaired endothelium-derived hyperpolarizing factor-mediated relaxation by dysregulating G i/o signaling. *J. Biol. Chem.* **2012**, *287*, 12541–12549. [CrossRef] [PubMed]
217. Vento-Tormo, R.; Efremova, M.; Botting, R.A.; Turco, M.Y.; Vento-Tormo, M.; Meyer, K.B.; Park, J.E.; Stephenson, E.; Polański, K.; Goncalves, A.; et al. Single-cell reconstruction of the early maternal–fetal interface in humans. *Nature* **2018**, *563*, 347–353. [CrossRef] [PubMed]
218. Lu, Y.-P.; Hasan, A.A.; Zeng, S.; Hocher, B. Plasma ET-1 concentrations are elevated in pregnant women with hypertension -meta-analysis of clinical studies. *Kidney Blood Press. Res.* **2017**, *42*, 654–663. [CrossRef]
219. Shah, D.M. The role of RAS in the pathogenesis of preeclampsia. *Curr. Hypertens. Rep.* **2006**, *8*, 144–152. [CrossRef] [PubMed]
220. Eastabrook, G.; Brown, M.; Sargent, I. The origins and end-organ consequence of pre-eclampsia. *Best Pr. Res. Clin. Obs. Gynaecol.* **2011**, *25*, 435–447. [CrossRef] [PubMed]
221. Lambert, G.; Brichant, J.F.; Hartstein, G.; Bonhomme, V.; Dewandre, P.Y. Preeclampsia: An update. *Acta Anaesthesiol. Belg.* **2014**, *65*, 137–149.
222. Can, M.; Guven, B.; Bektas, S.; Arikan, I. Oxidative stress and apoptosis in preeclampsia. *Tissue Cell* **2014**, *46*, 477–481. [CrossRef]
223. Cornelius, D.C. Preeclampsia: From inflammation to immunoregulation. *Clin. Med. Insights Blood Disord.* **2018**, *11*. [CrossRef]
224. Toniolo, A.; Buccellati, C.; Pinna, C.; Gaion, R.M.; Sala, A.; Bolego, C. Cyclooxygenase-1 and prostacyclin production by endothelial cells in the presence of mild oxidative stress. *PLoS ONE* **2013**, *8*, e56683. [CrossRef]
225. Brock, T.G.; McNish, R.W.; Peters-Golden, M. Arachidonic acid is preferentially metabolized by cyclooxygenase-2 to prostacyclin and prostaglandin E2. *J. Biol. Chem.* **1999**, *274*, 11660–11666. [CrossRef]
226. Walsh, S.W. Preeclampsia: An imbalance in placental prostacyclin and thromboxane Production. *Am. J. Obstet. Gynecol.* **1985**, *152*, 335–340. [CrossRef]

227. Atallah, A.; Lecarpentier, E.; Goffinet, F.; Doret-Dion, M.; Gaucherand, P.; Tsatsaris, V. Aspirin for Prevention of Preeclampsia. *Drugs* **2017**, *77*, 1819–1831. [CrossRef] [PubMed]
228. Lewis, D.F.; Canzoneri, B.J.; Gu, Y.; Zhao, S.; Wang, Y. Maternal Levels of Prostacyclin, Thromboxane, ICAM, and VCAM in Normal and Preeclamptic Pregnancies. *Am. J. Reprod. Immunol.* **2010**, *64*, 376–383. [CrossRef]
229. Walsh, S.W. Low-Dose Aspirin: Treatment for the Imbalance of Increased Thromboxane and Decreased Prostacyclin in Preeclampsia. *Am. J. Perinatol.* **1989**, *6*, 124–132. [CrossRef] [PubMed]
230. Tannetta, D.S.; Hunt, K.; Jones, C.I.; Davidson, N.; Coxon, C.H.; Ferguson, D.; Redman, C.W.; Gibbins, J.M.; Sargent, I.L.; Tucker, K.L. Syncytiotrophoblast Extracellular Vesicles from Pre-Eclampsia Placentas Differentially Affect Platelet Function. *PLoS ONE* **2015**, *10*, e0142538. [CrossRef]
231. Duley, L.; Meher, S.; Hunter, K.E.; Seidler, A.L.; Askie, L.M. Antiplatelet agents for preventing pre-eclampsia and its complications. *Cochrane Database Syst. Rev.* **2007**, CD004659. [CrossRef] [PubMed]
232. ACOG Committee Opinion No. 743: Low-Dose Aspirin Use During Pregnancy. *Obs. Gynecol* **2018**, *132*, e44–e52. [CrossRef]
233. Periayah, M.H.; Halim, A.S.; Saad, A.Z.M. Mechanism action of platelets and crucial blood coagulation pathways in Hemostasis. *Int. J. Hematol. Oncol. Stem Cell Res.* **2017**, *11*, 319–327.
234. Jakobsen, C.; Larsen, J.B.; Fuglsang, J.; Hvas, A.M. Platelet function in preeclampsia–a systematic review and meta-analysis. *Platelets* **2019**, *30*, 549–562. [CrossRef]
235. Goksu Erol, A.Y.; Nazli, M.; Yildiz, S.E. Significance of platelet endothelial cell adhesion molecule-1 (PECAM-1) and intercellular adhesion molecule-1 (ICAM-1) expressions in preeclamptic placentae. *Endocrine* **2012**, *42*, 125–131. [CrossRef] [PubMed]
236. Reese, J.A.; Peck, J.D.; Deschamps, D.R.; McIntosh, J.J.; Knudtson, E.J.; Terrell, D.R.; Vesely, S.K.; George, J.N. Platelet counts during pregnancy. *N. Engl. J. Med.* **2018**, *379*, 32–43. [CrossRef]
237. Reese, J.A.; Peck, J.D.; Yu, Z.; Scordino, T.A.; Deschamps, D.R.; McIntosh, J.J.; Terrell, D.R.; Vesely, S.K.; George, J.N. Platelet sequestration and consumption in the placental intervillous space contribute to lower platelet counts during pregnancy. *Am. J. Hematol.* **2019**, *94*, E8–E11. [CrossRef] [PubMed]
238. Cines, D.B.; Levine, L.D. Thrombocytopenia in pregnancy. *Blood* **2017**, *130*, 2271–2277. [CrossRef]
239. Ciobanu, A.M.; Colibaba, S.; Cimpoca, B.; Peltecu, G.; Panaitescu, A.M. Thrombocytopenia in Pregnancy. *Maedica* **2016**, *11*, 55–60.
240. Heilmann, L.; Rath, W.; Pollow, K. Hemostatic abnormalities in patients with severe preeclampsia. *Clin. Appl. Thromb. Hemost.* **2007**, *13*, 285–291. [CrossRef] [PubMed]
241. Han, L.; Liu, X.; Li, H.; Zou, J.; Yang, Z.; Han, J.; Huang, W.; Yu, L.; Zheng, Y.; Li, L. Blood coagulation parameters and platelet indices: Changes in normal and preeclamptic pregnancies and predictive values for preeclampsia. *PLoS ONE* **2014**, *9*, e114488. [CrossRef]
242. Youssef, L.; Miranda, J.; Blasco, M.; Paules, C.; Crovetto, F.; Palomo, M.; Torramade-Moix, S.; García-Calderó, H.; Tura-Ceide, O.; Dantas, A.P.; et al. Complement and coagulation cascades activation is the main pathophysiological pathway in early-onset severe preeclampsia revealed by maternal proteomics. *Sci. Rep.* **2021**, *11*, 3048. [CrossRef] [PubMed]
243. Higgins, J.R.; Bonnar, J.; Norris, L.A.; Darling, M.R.; Walshe, J.J. The effect of pre-eclampsia on coagulation and fibrinolytic activation in the neonate. *Thromb. Res.* **2000**, *99*, 567–570. [CrossRef]
244. Shokolenko, I.; Venediktova, N.; Bochkareva, A.; Wilson, G.L.; Alexeyev, M.F. Oxidative stress induces degradation of mitochondrial DNA. *Nucleic Acids Res.* **2009**, *37*, 2539–2548. [CrossRef]
245. Tenório, M.B.; Ferreira, R.C.; Moura, F.A.; Bueno, N.B.; de Oliveira, A.C.M.; Goulart, M.O.F. Cross-Talk between Oxidative Stress and Inflammation in Preeclampsia. *Oxidative Med. Cell. Longev.* **2019**, *2019*, 8238727. [CrossRef]
246. Walsh, S.W. Plasma from preeclamptic women stimulates transendothelial migration of neutrophils. *Reprod. Sci.* **2009**, *16*, 320–325. [CrossRef] [PubMed]
247. Williamson, R.D.; McCarthy, F.P.; Kenny, L.C.; McCarthy, C.M. Activation of a TLR9 mediated innate immune response in preeclampsia. *Sci. Rep.* **2019**, *9*, 1–8. [CrossRef]
248. Lemasters, J.J.; Nieminen, A.L.; Qian, T.; Trost, L.C.; Elmore, S.P.; Nishimura, Y.; Crowe, R.A.; Cascio, W.E.; Bradham, C.A.; Brenner, D.A.; et al. The mitochondrial permeability transition in cell death: A common mechanism in necrosis, apoptosis and autophagy. *Biochim. Et Biophys. Acta Bioenerg.* **1998**, *1366*, 177–196. [CrossRef]
249. Van Houten, B.; Hunter, S.E.; Meyer, J.N. Mitochondrial DNA damage induced autophagy, cell death, and disease. *Front. Biosci.* **2016**, *21*, 42–54. [CrossRef]
250. Thurairajah, K.; Briggs, G.D.; Balogh, Z.J. The source of cell-free mitochondrial DNA in trauma and potential therapeutic strategies. *Eur. J. Trauma Emerg. Surg.* **2018**, *44*, 325–334. [CrossRef]
251. Chockalingam, A.; Brooks, J.C.; Cameron, J.L.; Blum, L.K.; Leifer, C.A. TLR9 traffics through the Golgi complex to localize to endolysosomes and respond to CpG DNA. *Immunol. Cell Biol.* **2009**, *87*, 209–217. [CrossRef]
252. Goulopoulou, S.; Matsumoto, T.; Bomfim, G.F.; Webb, R.C. Toll-like receptor 9 activation: A novel mechanism linking placenta-derived mitochondrial DNA and vascular dysfunction in pre-eclampsia. *Clin. Sci.* **2012**, *123*, 429–435. [CrossRef]
253. Bao, W.; Xia, H.; Liang, Y.; Ye, Y.; Lu, Y.; Xu, X.; Duan, A.; He, J.; Chen, Z.; Wu, Y.; et al. Toll-like Receptor 9 Can be Activated by Endogenous Mitochondrial DNA to Induce Podocyte Apoptosis. *Sci. Rep.* **2016**, *6*, 22579. [CrossRef] [PubMed]
254. Marschalek, J.; Wohlrab, P.; Ott, J.; Wojta, J.; Speidl, W.; Klein, K.U.; Kiss, H.; Pateisky, P.; Zeisler, H.; Kuessel, L. Maternal serum mitochondrial DNA (mtDNA) levels are elevated in preeclampsia–a matched case-control study. *Pregnancy Hypertens.* **2018**, *14*, 195–199. [CrossRef] [PubMed]

255. Panda, B.; Panda, A.; Ueda, I.; Abrahams, V.M.; Norwitz, E.R.; Stanic, A.K.; Young, B.C.; Ecker, J.L.; Altfeld, M.; Shaw, A.C.; et al. Dendritic cells in the circulation of women with preeclampsia demonstrate a pro-inflammatory bias secondary to dysregulation of TLR receptors. *J. Reprod. Immunol.* **2012**, *94*, 210–215. [CrossRef]
256. Redman, C. Pre-eclampsia: A complex and variable disease. *Pregnancy Hypertens* **2014**, *4*, 241–242. [CrossRef] [PubMed]
257. Barrientos, G.; Pussetto, M.; Rose, M.; Staff, A.C.; Blois, S.M.; Toblli, J.E. Defective trophoblast invasion underlies fetal growth restriction and preeclampsia-like symptoms in the stroke-prone spontaneously hypertensive rat. *Mol. Hum. Reprod.* **2017**, *23*, 509–519. [CrossRef]
258. Ma, Y.; Ye, Y.; Zhang, J.; Ruan, C.C.; Gao, P.J. Immune imbalance is associated with the development of preeclampsia. *Medicine* **2019**, *98*, e15080. [CrossRef]
259. Raijmakers, M.T.; Dechend, R.; Poston, L. Oxidative stress and preeclampsia: Rationale for antioxidant clinical trials. *Hypertension* **2004**, *44*, 374–380. [CrossRef]
260. Enkhmaa, D.; Wall, D.; Mehta, P.K.; Stuart, J.J.; Rich-Edwards, J.W.; Merz, C.N.B.; Shufelt, C. Preeclampsia and Vascular Function: A Window to Future Cardiovascular Disease Risk. *J. Womens Health* **2016**, *25*, 284–291. [CrossRef] [PubMed]
261. Romundstad, P.R.; Magnussen, E.B.; Smith, G.D.; Vatten, L.J. Hypertension in pregnancy and later cardiovascular risk: Common antecedents? *Circulation* **2010**, *122*, 579–584. [CrossRef]
262. Redman, C.W. Stress responses and pre-eclampsia. *Pregnancy Hypertens* **2013**, *3*, 57. [CrossRef] [PubMed]
263. Roman, D.L.; Traynor, J.R. Regulators of G protein signaling (RGS) proteins as drug targets: Modulating G-protein-coupled receptor (GPCR) signal transduction. *J. Med. Chem.* **2011**, *54*, 7433–7440. [CrossRef]
264. Vangrieken, P.; Al-Nasiry, S.; Bast, A.; Leermakers, P.A.; Tulen, C.B.M.; Schiffers, P.M.H.; van Schooten, F.J.; Remels, A.H.V. Placental Mitochondrial Abnormalities in Preeclampsia. *Reprod. Sci.* **2021**, *28*, 2186–2199. [CrossRef] [PubMed]
265. Cushen, S.C.; Osikoya, O.; Blessing, A.; Phillips, N.R.; Goulopoulou, S. Placental Exposure to Hypoxia and Oxidative Stress Causes Mitochondrial DNA Release into the Extracellular Space. *FASEB J.* **2019**, *33*, 703. [CrossRef]
266. Marín, R.; Chiarello, D.I.; Abad, C.; Rojas, D.; Toledo, F.; Sobrevia, L. Oxidative stress and mitochondrial dysfunction in early-onset and late-onset preeclampsia. *Biochim. Biophys. Acta Mol. Basis Dis.* **2020**, *1866*, 165961. [CrossRef] [PubMed]
267. Turco, M.Y.; Gardner, L.; Kay, R.G.; Hamilton, R.S.; Prater, M.; Hollinshead, M.S.; McWhinnie, A.; Esposito, L.; Fernando, R.; Skelton, H.; et al. Trophoblast organoids as a model for maternal-fetal interactions during human placentation. *Nature* **2018**, *564*, 263–267. [CrossRef] [PubMed]

cells

Review

NLRP3 Activation and Its Relationship to Endothelial Dysfunction and Oxidative Stress: Implications for Preeclampsia and Pharmacological Interventions

Priscila Rezeck Nunes *, Sarah Viana Mattioli and Valeria Cristina Sandrim

Department of Biophysics and Pharmacology, Institute of Biosciences, Sao Paulo State University, Sao Paulo 01049-010, Brazil; sarah.mattioli@unesp.br (S.V.M.); valeria.sandrim@unesp.br (V.C.S.)
* Correspondence: priscila.nunes@unesp.br

Abstract: Preeclampsia (PE) is a specific syndrome of human pregnancy, being one of the main causes of maternal death. Persistent inflammation in the endothelium stimulates the secretion of several inflammatory mediators, activating different signaling patterns. One of these mechanisms is related to NLRP3 activation, initiated by high levels of danger signals such as cholesterol, urate, and glucose, producing IL-1, IL-18, and cell death by pyroptosis. Furthermore, reactive oxygen species (ROS), act as an intermediate to activate NLRP3, contributing to subsequent inflammatory cascades and cell damage. Moreover, increased production of ROS may elevate nitric oxide (NO) catabolism and consequently decrease NO bioavailability. NO has many roles in immune responses, including the regulation of signaling cascades. At the site of inflammation, vascular endothelium is crucial in the regulation of systemic inflammation with important implications for homeostasis. In this review, we present the important role of NLRP3 activation in exacerbating oxidative stress and endothelial dysfunction. Considering that the causes related to these processes and inflammation in PE remain a challenge for clinical practice, the use of drugs related to inhibition of the NLRP3 may be a good option for future solutions for this disease.

Keywords: NLRP3; preeclampsia; nitric oxide; endothelial dysfunction; oxidative stress; inflammation

Citation: Nunes, P.R.; Mattioli, S.V.; Sandrim, V.C. NLRP3 Activation and Its Relationship to Endothelial Dysfunction and Oxidative Stress: Implications for Preeclampsia and Pharmacological Interventions. *Cells* **2021**, *10*, 2828. https://doi.org/10.3390/cells10112828

Academic Editors: Junie P Warrington, Ana T. Palei, Mark W. Cunningham and Lorena M. Amaral

Received: 10 August 2021
Accepted: 15 October 2021
Published: 21 October 2021

1. Introduction

Preeclampsia (PE) is a specific syndrome of human pregnancy, considered the main cause of morbidity and mortality in 2 to 8% of pregnancies worldwide [1], and one of the main causes of maternal death. The clinical parameters that characterize this pathology are arterial hypertension and proteinuria from the twentieth week of pregnancy or in the first days after delivery. However, other maternal dysfunctions may also be related to PE, such as renal failure, liver involvement, neurological or hematological complications, uteroplacental dysfunction, or fetal growth restriction [2,3]. This pathology increases the risk of maternal and fetal mortality, through placental abruption, cerebrovascular events, organ failure, and disseminated intravascular coagulation [4].

In a healthy pregnancy, during the second trimester of gestation, maternal spiral arteries are invaded by the trophoblast that phenotypically differentiates into endothelial cells, resulting in remodeling of the spiral arteries [5]. Vasculogenesis ensures adequate blood supply to the placenta and fetal growth, however, it is observed that in placentas of pregnant women with PE, trophoblastic invasion is inadequate, occurring in only 30–50% of the arteries [6]. This failure in vascular remodeling can lead to poor placental perfusion and ischemia [7]. Ischemia occurs since the arteries are not sufficiently remodeled, causing disorderly perfusion of blood flow to the intervillous space. Added to an inadequate supply of nutrients and oxygen, there is a reduction in the surface area available for exchange between mother and fetus, which can contribute to unfavorable pregnancy-related outcomes [8]. The pathophysiology of PE is not fully understood, but it is currently known that placental ischemia is of fundamental importance in this process, since the release of products resulting from poor perfusion in the maternal circulation can lead to systemic endothelial dysfunction [9].

Endothelial cells have different functions during non-inflammatory conditions, such as maintaining blood fluidity, regulating blood flow, and maintaining leukocytes in a basal state circulating [10]. In cases of infection or inflammation, these cells recognize danger signals and they act as active regulators of the inflammatory response [10], and receptors in these cells help the response to a range of external signals [11]. This meeting between endothelial cells and danger signals, such as ATP and high mobility group box 1 protein (HMGB1), can activate the NOD-like receptor family, pyrin domain-containing protein 3 (NLRP3) [12,13]. Activation of NLRP3 inflammasome in endothelial cells was already observed in animal models, and production of IL-1β by these cells has been shown to contribute to diverse pathological conditions [14,15].

Several recent studies in the literature have demonstrated that women with PE present a significantly higher expression of NLRP3, and related mediators such as caspase-1, IL-1, and IL-18 compared to normotensive healthy pregnant women [16–18].

This article aims to summarize the role of NLRP3 in PE related to endothelial dysfunction and oxidative stress, proposing different approaches for future therapies.

2. Preeclampsia and Endothelial Dysfunction

Endothelial cells form a monolayer that covers the interior of blood vessels, creating a barrier between blood and the extravascular matrix. These cells have a fundamental role in maintaining a dynamic modulation of homeostasis, angiogenesis, and vascular tone, besides maintaining an antioxidant, anti-inflammatory, and antithrombotic profile in healthy individuals [19]. Endothelial dysfunction is the term used to describe an imbalance in these endothelial functions affecting vasoprotective homeostasis [20].

In normal pregnancies, the typical increase in blood volume is commonly compensated by a slight decrease in blood pressure. For a long time, this rise in blood pressure has been associated with reduced maternal vascular resistance [21,22]. However, in PE, the compensatory maternal vascular adaptations are insufficient, and it has been associated with systemic endothelial dysfunction [23–26]. In this syndrome, this characteristic is associated with the PE poorly perfused placenta, which releases proinflammatory and antiangiogenetic factors into maternal circulation [27,28]. This hypothesis has been reinforced by many studies so far. For example, Myers and colleagues demonstrated that healthy myometrium vessels incubated with plasma from PE pregnant women had reduced endothelium mediated vascular relaxation compared to those incubated with plasma from healthy pregnant women [29]. Plasma from women with PE can modify the endothelial function, altering the balance between vasoactive substances. Despite evidence showing the release of placental factors into maternal circulation could alter endothelial function, the exact mechanism is not fully understood. [30].

Until today, many studies demonstrated several alterations, both locally and in circulation, in multiple bioactive factors in PE. For example, the angiogenic balance disturbance has been described by decreases in pro-angiogenic vascular endothelial growth factor (VEGF) and placental growth factor (PlGF) by the action of the placental soluble fms-like tyrosine kinase-1 (sFlt-1) and soluble endoglin (sEng) [29,31–33]. Moreover, proinflammatory molecules such as tumor necrosis factor-α (TNF-α), endocan, interleukin-6 (IL-6), and IL-1β have also been reported to be altered in PE [34–37]. Altogether, these alterations lead to systemic endothelial dysfunction in PE, and it also is possible that there may be more mechanisms involved that were not discovered yet.

These alterations seem to contribute to the development of the significant symptoms of the maternal syndrome in PE, such as hypertension, edema, proteinuria, and platelet aggregation [38]. For example, the decreased levels of NO production and increased levels of ET-1 and sFlt-1 leads towards a vasoconstrictive and hypertensive maternal profile [39]. Regarding endothelial monolayer barrier integrity, the reduced levels of pro-angiogenic and increased levels of proinflammatory molecules lead to a more permeable profile of the vascular endothelium, which may lead to edema, proteinuria, and even cerebral endotheliosis, that leads to seizures in severe cases [40].

The different mechanisms overlapping each other lead to a common end, endothelial dysfunction, and this condition represent a major hallmark of PE, contributing to the clinical consequences of the disease. Therefore, fully understanding and identifying the factors that lead to endothelial damage is the key to further understand the pathogenesis of PE and provide early diagnosis and effective therapies.

3. NLRP3 Inflammasome Activation and Regulation in Preeclampsia

3.1. *Inflammasome Formation and the Role of NLRP3 in the Pathogenesis of PE*

The immune response is divided into innate and adaptive immunity. Contact with pathogens or any danger signal activates the innate immune system, as the first line of defense. This process starts quickly as possible to protect the organism and to maintain homeostasis. The immune cells detect the signals from invaders, expressing molecules known as pathogen-associated molecular patterns (PAMPs). Besides that, these cells also identify molecules associated with inflammation and cell death, in cases of sterile inflammation, without any external microbial sign. These molecules associated with inflammation are named damage-associated molecular patterns (DAMPs). PAMPs and DAMPs are recognized by pattern recognition receptors (PRRs). Two of the most studied PRRs are Toll-like receptors (TLRs) and nucleotide-binding domain leucine-rich repeat-containing receptors (NLRs) [41].

There are 22 recognized members of the NLR family, between them, NLRP3 (NOD-like receptor family, pyrin domain-containing protein 3) is the most studied and investigated, because this NLR forms complexes with other proteins, forming multimeric complexes, called inflammasomes [42].

NLRP3 inflammasome complex is constituted by NLRP3, ASC (apoptosis-associated speck-like protein containing a caspase recruitment domain), and the cysteine protease precursor procaspase-1 (Figure 1).

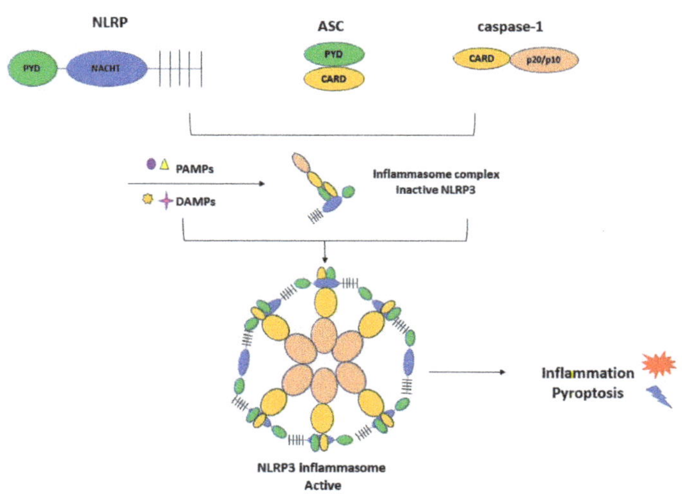

Figure 1. The NLRP3 inflammasome consists of NLRP3, ASC, and caspase-1. NLRP3 is composed of C-terminal leucine-rich repeats (LRRs), a central nucleotide-binding and oligomerization domain (NACHT), and an N-terminal pyrin domain (PYD). ASC is also termed Pycard, containing an N-terminal PYD and a C-terminal caspase recruitment domain (CARD). The last element of the CARD and caspase domains. PAMPs and DAMPs can activate the inflammasome complex and triggers inflammation and pyroptosis.

Recently, reports regarding the NLRP3 inflammasome activation in PE have been increased. The literature shows higher expression of the NLRP3 inflammasome components in blood cells and placenta from PE women compared with normotensive healthy pregnant women [16,43,44]. Furthermore, trophoblastic cells also express NLRP3, ASC, and caspase-1 [45–47], and IL-1β secretion occurs in human trophoblast cells in response to activators of the NLRP3 inflammasome [46,47]. The interaction between alarmin-induced activation of placental NLRP3 inflammasome and the resulting placental inflammation presented in pregnancy complications such as preeclampsia has been shown by in vivo studies [46,48–50].

These recent contributions suggest that NLRP3 inflammasome activation is implicated in the inflammatory processes associated with the pathophysiology of preeclampsia. Moreover, in vitro and in vivo studies have shown that inflammatory stimuli induce the activation of the NLRP3 inflammasome in the placenta, also contributing to other pregnancy-related disorders [51].

3.2. Activation of NLRP3 and Pyroptosis: The Cell Death Related to Inflammatory Processes

The literature data showed significantly higher expression of the NLRP3 and related mediators such as caspase-1, IL-1, and IL-18 in samples from women with PE compared to controls [16,44]. Other groups highlighted the NLRP3 gene polymorphisms associated with a significantly higher risk of disease development [17,18].

Inflammasome activation starts with two signals, both initiated by DAMPs or PAMPs [52,53]. Figure 2 shows these two different signals in the activation of NLRP3. The first one is the priming signal, leading nuclear factor kappa B (NF-κB) activation through membrane receptors. NF-κB is important in the activation of the transcription and regulators of several genes, inducing the expression of pro-IL-1 and NLRP3. In the second signal, PAMPs and DAMPs appear to bind directly to NLRP3 [53]. Once activated, NLRP3 interacts with ASC, recruiting and activating procaspase-1. The interaction between NLRP3 and ASC activates caspase-1, as well as pro-IL-1 and IL-18, releasing these cytokines in their active forms.

Pyroptosis, a programmed necrosis type, involves recruitment of its executor gasdermin D, (GSDMD) leading to inflammatory cascades, releasing alarmins or DAMPs. Besides cleavage of pro-IL-18/pro-IL-1β in inflammasome activation, Cheng et al., 2019 also demonstrated that pro-GSDMD is also cleaved into N-terminal which are translocated to the plasma membrane and form pores, which leads to pyroptosis and subsequent release of cell particulates, including DAMPs. GSDMD is significantly expressed in the placenta from early-onset PE and in cellular models of PE pathophysiology. They concluded that placental pyroptosis is a major sterile inflammatory pathway in PE that may lead to excessive production of IL-1β and IL-18, contributing to the systemic manifestation of this disease [54].

This type of cell death-related with NLRP3 activation is caspase-1-dependent because it depends on plasma membrane rupture. This process releases DAMPs and cytokines into the extracellular milieu, leading to sterile inflammation, as it occurs in PE. This type of programmed cell death generates highly inflammatory species [55]. This process releases IL-1β, IL-18, and HMGB1 (high mobility group box 1), which distinguishes this type of cell death from others. Pyroptosis has been identified as a potent cause of endothelial cell death [56].

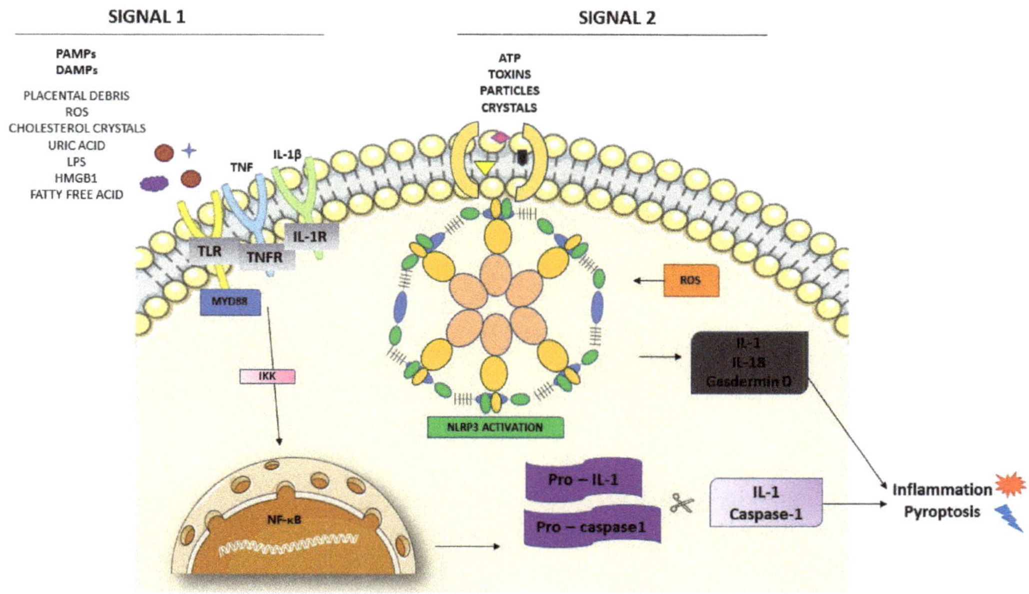

Figure 2. NLRP3 inflammasome activation. The priming signal (signal 1) occurs in the presence of danger signals (PAMPs and DAMPs), leading to the activation of the NF-κB and subsequent upregulation of NLRP3 and pro-IL-1 and pro-caspase1. The activation signal (signal 2) starts with the direct activation of the NLRP3 inflammasome with ROS recruitment. The process leads to inflammation and pyroptosis.

4. NLRP3 and its Relation with Endothelial Dysfunction and Oxidative Stress

According to Burton et al., 2019, the release of products resulting from poor perfusion in the maternal circulation can lead to systemic endothelial dysfunction [9]. The mechanism by which these products are released into the maternal circulation, how they modify endothelial function in pregnant women with PE, how they change the balance between vasoactive substances, such as NO, prostacyclin, and endothelin, is not yet fully understood [30]. The literature data suggest that the generalized endothelial dysfunction seen in PE is the main cause of the clinical abnormalities observed in this disease [38,57]. Vascular endothelial cells cover the inner layer of blood vessels, forming a barrier between blood and the extravascular matrix. This barrier maintains the transport of solutes, fluids, and cells [58]. Endothelial barrier dysfunction is characterized by loss of contact between endothelial cells and the extravasation of plasma, proteins, cells, and solutes [59].

The products resulting from endothelial dysfunction can act as inflammatory mediators, activating the innate immune system, the first mechanism by which the body responds immediately to infections and injuries [60]. Cells from the innate immune system play an important role in the inflammatory response initiated by PRRs, but cells outside the immune system, such as endothelial cells, still need to be better studied in this process [61].

Generally, DAMPs can trigger NLRP3 inflammasome activation, producing mature forms of IL-1β and IL-18 from cells to promote further inflammatory processes and oxidative stress in the endothelium [62]. Endothelial cells (ECs) are a target of IL-1β, and it also produces IL-1β during inflammation [63], activating other inflammatory mediators, contributing to secreting adhesion molecules and chemokines in ECs, inducing a potent pro-inflammatory response [64]. Endothelial inflammation may initiate the occurrence and progression of endothelial dysfunction.

Oxidative stress and inflammation are inseparable events in inflammatory diseases and both play an essential role in the pathogenesis of PE (Figure 3). The NLRP3 activation initiates from various stimuli, including the production of reactive oxygen species

(ROS) [65]. They are the first intermediate reactive products generated during inflammasome activation, being responsible for the release of inflammatory agents in the immune response [66].

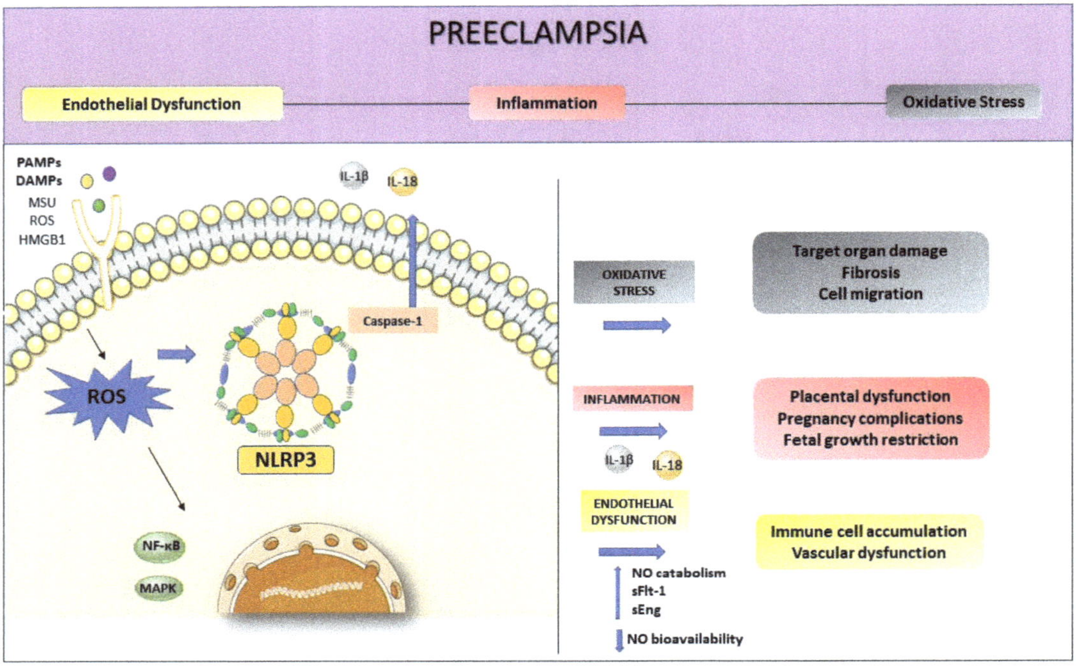

Figure 3. Preeclampsia is characterized by intense oxidative stress, inflammation, and endothelial dysfunction. The activation of NLRP3 may start with the production of ROS. Inflammasome activation is responsible for the release of inflammatory agents during the immune response, such as IL-1β and IL-18. High levels of ROS increase NO catabolism and decrease NO bioavailability as well as increasing factors such as sFlt-1 and sEng. This process enhanced inflammation-related genes expression, contributing to endothelial dysfunction.

In this way, ROS mediate the interaction between NLRP3 inflammasome and endothelial dysfunction, being the first participant in the NLRP3 activation, promoting inflammation, and activating immune responses [66]. Three important proteins, thioredoxin-interacting protein (TXNIP), nuclear factor kappaB (NF-κB), and the transcription factor nuclear factor erythroid 2-related factor 2 (Nrf2) are involved in the oxidative stress, connecting ROS to NLRP3 activation [67]. In a state of increased oxidative stress, as occurs in preeclampsia, the imbalance between pro and antioxidants, coupled with higher ROS production may increase NO catabolism, and decrease NO bioavailability. The oxidative stress enhanced inflammation-related genes expression and increased inflammatory proteins, impairing endothelial function [68].

5. Pharmacological Interventions: Selective and Non-Selective Drugs

The association of NLRP3 inflammasome activation with various inflammatory diseases involves interest in the scientific community to explore the actions of the effective NLRP3 inflammasome inhibitors. Several inhibitors of NLRP3 inflammasome have been reported. Here, we summarize recent pharmacological inhibitors in Table 1. A diverse range of targets can be used for its inhibition due to the fact of its complex signaling cascade.

Table 1. Potential inhibitors of NLRP3 inflammasome. NLRP3-specificity and targets (including the mode of action) are also represented.

Drug	NLRP3-Specific	Direct Inhibition	Action
Glyburide [17,69–75]	Yes	No	Induces the closure of ATP-sensitive K+ channels; Raises the intracellular K+ concentration
16673-34-0 [71,72]	Yes	No	Interferes with downstream events involved in NLRP3 conformational changes secondary to activation or binding to ASC
JC124 [76,77]	Yes	No	Blocks ASC aggregation, caspase-1 activation, and IL-1β secretion
FC11A-2 [78]	Yes	No	Repress IL-1β/18 release; induces autocleavage of procaspase-1, resulting in a reduced amount of activated caspase-1
MCC950 [79–84]	Yes	Yes	Blocks the release of IL-1β induced by NLRP3 activators
CY-09 [85–87]	Yes	Yes	Blocks the ATP, monosodium urate (MSU), and nigericin-induced activation of caspase-1 and resultant release of IL-1β
Tranilast [80,88–90]	Yes	Yes	Impairs the endogenous NLRP3-ASC interaction
OLT1177 [91,92]	Yes	Yes	Binds with NLRP3 to block its ATPase activity
Oridonin [93–97]	Yes	Yes	Inhibits the NF-κB or MAPK activation and repress the release of inflammasome-independent proinflammatory cytokines release
Parthenolide [98–101]	No	No	Inhibits caspase-1 activation; Targets ATPase activity of NLRP3
VX-740/VX-765 [102–108]	No	No	Block caspase-1 and resultant cleavage of pro-IL-1β/18
Bay 11-7082 [87,99]	No	No	Prevents the organization of ASC pyroptosome
BHB [109]	No	No	Lowered the production of IL-1ß and IL-1; reduces the oligomerization and speck formation of ASC

Different strategies may be used for inflammasome inhibition, such as suppression of activation signals, blockade of inflammasome complex formation, inhibition of caspase-1 activation, blockade of pore-forming protein gasdermin D, avoid inflammatory cytokines production and release. Here, we describe some drugs that are summarized in Table 1.

Glyburide, also known as glibenclamide, is a drug from the sulfonylurea family widely prescribed to treat type 2 diabetes mellitus (T2D), and it is suggested to effectively inhibit the migration of inflammatory cells, as it prevents the assembly of the inflammasome. Specifically, glibenclamide inhibits NLRP3 activation by inducing the closure of ATP-sensitive potassium channels, increasing the intracellular potassium concentration [69].

Thus, there is a reduction in inflammatory cell infiltration, preventing further organ damage in ischemic tissue [70,71]. This drug works by improving endothelial dysfunction and has also been described as an inhibitor of NLRP3 in endothelial cells in the blood–brain barrier [17]. Furthermore, some authors have shown that this drug was able to cause vessel relaxation in vascular reactivity studies in rats [72]. Studies with glyburide administration in vitro or in vivo [69,73–75] showed inhibitory activity of NLRP3 activation. However, the necessary dose in vivo is high to exert an inhibitory effect and can cause hypoglycemia. Because of that, the use of glyburide is still restricted to T2D [75]. 16673-34-0 is a glyburide intermediate substrate produced during its synthesis and appears not to affect glucose metabolism. A study conducted by Marchetti et al. showed that 16673-34-0 inhibits NLRP3 inflammasome formation in murine macrophages and rat cardiomyocytes. In vivo tests showed positive results in mouse models of acute myocardial infarction. This substrate was tested in the presence of diverse stimuli of the NLRP3 inflammasome, and independent of the stimuli, the inhibitory effects of 16673-34-0 remained the same. This information suggests that this molecule interferes with downstream events involved in both NLRP3 activation and binding to ASC [71,72].

A novel small molecule developed by Kuwar et al., named JC124, mimetics the structure of glyburide and attends to minimize the hypoglycemic effects of glyburide. It was tested in traumatic brain injury (TBI) therapy and exerted a significant anti-inflammatory effect to protect the injured brain. Treatment with this molecule reduced the expression of NLRP3, ASC, caspase-1, pro-IL-1β, TNFα, and inducible nitric oxide synthase (iNOS) [76]. Besides that, JC124 also showed protective effects in a mouse model of acute myocardial infarction [77].

Another synthetic molecule created by Liu et al., 1-ethyl-5-methyl-2-phenyl-1H-benzodimidazole, also known as FC11A-2, has inhibitory potential for NLRP3 inflammasome. This molecule was tested in THP-1 cells and a mouse model of experimental colitis, showing a blockage in IL-1β/18 release and reduced activated caspase-1, in an NF-κB independent pathway [78].

A selective NLRP3 inhibitor, MCC950, specifically acts to inhibit this inflammasome. This drug blocks the oligomerization of ASC and the hydrolysis of ATP [79,80] and has been studied in several human diseases, proving to be effective in the treatment of vascular dysfunction in diabetes [81], and sepsis [82,83]. This inhibitor was reported to decreased inflammation in skin and lungs in mice [84] and some other in vivo experiments showed that MCC950 alleviates the severity of experimental autoimmune encephalomyelitis (EAE) [79].

CY-09 is described as an analog of CFTR (inh)-172 (C172), inhibiting the cystic fibrosis transmembrane conductance regulator (CFTR) channel [85]. Jiang et al. identified an effect of this molecule in NLRP3 activation with significant inhibition in vivo in mice models and ex vivo in human cells [86]. CY-09 acts blocking the ATP, MSU, and nigericin-induced activation of caspase-1 and consequently IL-1β release. CY-09 demonstrated preventive and therapeutic actions in the mice models of gout, T2D, and cryopyrin-associated periodic syndromes (CAPS). A great achievement about this molecule is that CY-09 showed good oral bioavailability, safety, and stability [87].

Tranilast (N-[3′,4′-dimethoxycinnamoyl]-anthranilic acid, also known as TR) is a tryptophan metabolite analog [88]. TR prevents the interaction between NLRP3 and ASC, proving that it affects the NLRP3 activation directly. TR has already been demonstrated as a significant therapy for the prevention of poor outcomes in gout, CAPS, and T2D mice models [80]. This drug has an important aspect that is the safety in high doses with appropriate tolerance levels in patients [89,90] This tolerance is important because it allows the use for tests in NLRP3-related diseases treatments.

OLT1177 is an active β-sulfonyl nitrile compound used in experimental clinical tests for the treatment of degenerative arthritis [91]. OLT1177 blocked in vitro activation of NLRP3 and direct binding with NLRP3 to block its ATPase activity. This drug was already given orally in clinical trials, showing safety and tolerability. OLT1177 presents a long half-life and did not show organ or hematological toxicity [92]. Thus, it seems to show significant potential for the treatment of NLRP3-related disorders.

Oridonin is a bioactive compound of Rabdosia rubescens, which is extensively utilized in traditional Chinese medicine [93,94]. This drug acts by inhibiting the NF-κB/MAPK activation and the release of inflammasome-independent proinflammatory cytokines [95,96]. This drug was tested in mice models of T2D, peritonitis, and gouty arthritis, exhibiting significant preventive, and therapeutic effects [97]. It could be used in future studies as a clinically applicable inhibitor of NLRP3 inflammasome.

A plant sesquiterpene lactone named Parthenolide has numerous anti-inflammatory effects and has been utilized in herbal medicine for the treatment of various inflammatory diseases [98]. It acts by inhibiting caspase-1 activation in response to NLRP1, NLRC4, and NLRP3 activation via caspase-1. This drug can also target the ATPase activity of NLRP3 [99]. The main concern for the use of Parthenolide is the poor solubility and bioavailability [100,101].

VX-740 (also known as Pralnacasan) and its analog VX-765 are inhibitors of caspase-1 [102,103], blocking this protein and resultant cleavage of pro-IL-1β/18 [104]. In rheumatoid arthritis (RA) clinical trials, these pro-drugs exhibited significant anti-inflammatory effects with a good pharmacokinetic profile [105,106]. Moreover, it had positive outcomes for the treatment of epilepsy and psoriasis in mouse models nonetheless, hepatic toxicity in animals after long-term exposure remains a concern [103,107,108].

Bay 11-7082 is a phenyl vinyl sulfone and acts inhibiting the NF-κB pathway [87]. Tests with NG5 cells and mouse primary bone marrow-derived macrophages (BMDMs) showed that this drug prevents the organization of ASC pyroptosome and NLRP3 inflammasome. Initial clinical trials showed that these compounds are well-tolerated, non-mutagenic, with suitable pharmacokinetic profiles, as well as also having the ability to permeate the cell membrane easily [99].

The last drug listed here is β-hydroxybutyrate (BHB), a ketone metabolite, which was tested for NLRP3 inflammasome blockade by Youm et al. BHB was able to decrease the production of IL-1ß and IL-18 in human monocytes in response to activated NLRP3 inflammasome. It blocks the activation of NLRP3 inflammasome independent of ROS, AMP-activated protein kinase, glycolytic inhibition, or autophagy [109]. Thus, BHB could be used in trials to reduce the severity of NLRP3-mediated chronic inflammatory diseases.

Many inhibitors for NLRP3 inflammasome have been reported in the literature and some of them have shown remarkable therapeutic potential. More research is needed to develop specific and safe molecules to inhibit NLRP3 inflammasome. The use of drugs with clinical positive results may be the ideal choice for the treatment of endothelial dysfunction, providing a new strategy to treat related illnesses. Considering that the causes related to endothelial dysfunction, oxidative stress, and inflammation in PE remain a challenge for clinical practice, the use of pharmacological substances related to the inhibition of the NLRP3 inflammasome may be a good choice to propose future treatments and strategies for PE. In addition, expanding research into the role of NLRP3 in endothelial dysfunction may enrich the understanding of several inflammatory diseases.

6. Conclusions

NLRP3 activation plays an important role in the development of PE. Although NLRP3 has been the most intensively investigated type of inflammasome, a total mechanism for its activation has not yet been elucidated. Therefore, inhibitors of NLRP3 could be a very effective treatment for PE. With new research, the mechanisms regarding endothelial function and its relation to the NLRP3 inflammasome activation pathway can be better elucidated. Meanwhile, the interactions between endothelial dysfunction, oxidative stress, and the NLRP3 inflammasome-regulated pathways may improve the treatments of inflammation-related disorders, such as PE.

Author Contributions: Conceptualization, P.R.N. and V.C.S.; writing—original draft preparation, P.R.N., V.C.S. and S.V.M.; writing—review and editing, P.R.N. and V.C.S.; funding acquisition, V.C.S. All authors have read and agreed to the published version of the manuscript.

Funding: This research was funded by Fundação de Amparo à Pesquisa do Estado de São Paulo (FAPESP), grant numbers 2019/07230-8; 2020/14610-9 and Coordenação de Aperfeiçoamento de Pessoal de Nível Superior grant number 88887.604855/2021-00.

Institutional Review Board Statement: Not applicable.

Informed Consent Statement: Not applicable.

Data Availability Statement: Not applicable.

Conflicts of Interest: The authors declare no conflict of interest.

References

1. American College of Obstetricians and Gynecologists. ACOG practice bulletin number 222. Gestational Hypertension and Preeclampsia. *Obstet. Gynecol.* **2020**, *135*, 237–260. [CrossRef]
2. Tranquilli, A.L.; Dekker, G.; Magee, L.; Roberts, J.; Sibai, B.M.; Steyn, W.; Zeeman, G.G.; Brown, M.A. The classification, diagnosis, and management of the hypertensive disorders of pregnancy: A revised statement from the ISSHP. *Pregnancy Hypertens.* **2014**, *4*, 97–104. [CrossRef]
3. Mol, B.W.; Roberts, C.T.; Thangaratinam, S.; Magee, L.A.; de Groot, C.J.; Hofmeyr, G.J. Pre-eclampsia. *Lancet* **2016**, *387*, 999–1011. [CrossRef]
4. Kintiraki, E.; Papakatsika, S.; Kotronis, G.; Goulis, D.G.; Kotsis, V. Pregnancy-Induced hypertension. *Horm* **2015**, *14*, 211–223. [CrossRef]
5. Collins, S.L.; Birks, J.S.; Stevenson, G.N.; Papageorghiou, A.T.; Noble, J.A.; Impey, L. Measurement of spiral artery jets: General principles and differences observed in small-for-gestational-age pregnancies. *Ultrasound Obstet. Gynecol.* **2012**, *40*, 171–178. [CrossRef]
6. Von Dadelszen, P.; Magee, L.A.; Roberts, J.M. Subclassification of Preeclampsia. *Hypertens. Pregnancy* **2003**, *22*, 143–148. [CrossRef]
7. George, E.M.; Granger, J.P. Mechanisms and potential therapies for preeclampsia. *Curr. Hypertens. Rep.* **2011**, *13*, 269–275. [CrossRef]
8. Turco, M.Y.; Moffett, A. Development of the human placenta. *Development* **2019**, *146*, dev163428. [CrossRef]
9. Burton, G.J.; Redman, C.W.; Roberts, J.M.; Moffett, A. Pre-eclampsia: Pathophysiology and clinical implications. *BMJ* **2019**, *366*, l2381. [CrossRef] [PubMed]
10. Pober, J.S.; Sessa, W.C. Evolving functions of endothelial cells in inflammation. *Nat. Rev. Immunol.* **2007**, *7*, 803–815. [CrossRef] [PubMed]
11. Opitz, B.; Hippenstiel, S.; Eitel, J.; Suttorp, N. Extra- and intracellular innate immune recognition in endothelial cells. *Thromb. Haemost.* **2007**, *98*, 319–326.
12. Li, Y.X.; Wang, P.; Yang, X.; Wang, W.; Zhang, J.; He, Y.; Zhang, W.; Jing, T.; Wang, B.; Lin, R. SIRT1 inhibits inflammatory response partly through regulation of NLRP3 inflammasome in vascular endothelial cells. *Mol. Immunol.* **2016**, *77*, 148–156. [CrossRef]
13. Chen, Y.; Pitzer, A.L.; Li, X.; Li, P.L.; Wang, L.; Zhang, Y. Instigation of endothelial Nlrp3 inflammasome by adipokine visfatin promotes inter-endothelial junction disruption: Role of HMGB1. *J. Cell Mol. Med.* **2015**, *19*, 2715–2727. [CrossRef] [PubMed]
14. Chen, Y.; Wang, L.; Pitzer, A.L.; Li, X.; Li, P.L.; Zhang, Y. Contribution of redox-dependent activation of endothelial Nlrp3 inflammasomes to hyperglycemia-induced endothelial dysfunction. *J. Mol. Med.* **2016**, *94*, 1335–1347. [CrossRef] [PubMed]
15. Zhang, Y.; Li, X.; Pitzer, A.L.; Chen, Y.; Wang, L.; Li, P.L. Coronary endothelial dysfunction induced by nucleotide oligomerization domain-like receptor protein with pyrin domain containing 3 inflammasome activation during hypercholesterolemia: Beyond inflammation. *Antioxid. Redox Signal* **2015**, *22*, 1084–1096. [CrossRef]
16. Weel, I.C.; Romão-Veiga, M.; Matias, M.L.; Fioratti, E.G.; Peraçoli, J.C.; Borges, V.T.; Araujo Jr, J.P.; Peraçoli, M.T. Increased expression of NLRP3 inflammasome in placentas from pregnant women with severe preeclampsia. *J. Reprod. Immunol.* **2017**, *123*, 40–47. [CrossRef] [PubMed]
17. Xu, L.; Li, S.; Liu, Z.; Jiang, S.; Wang, J.; Guo, M.; Zhao, X.; Song, W.; Liu, S. The NLRP3 rs10754558 polymorphism is a risk factor for preeclampsia in a Chinese Han population. *J. Matern. Fetal Neonatal Med.* **2019**, *32*, 1792–1799. [CrossRef] [PubMed]
18. Pontillo, A.; Reis, E.C.; Bricher, P.N.; Vianna, P.; Diniz, S.; Fernandes, K.S.; Chies, J.A.; Sandrim, V. NLRP1 L155H Polymorphism is a Risk Factor for Preeclampsia Development. *Am. J. Reprod. Immunol.* **2015**, *73*, 577–581. [CrossRef]
19. Gimbrone, M.A.; García-cardeña, G. Endothelial Cell Dysfunction and the Pathobiology of Atherosclerosis. *Circ. Res.* **2016**, *118*, 620–636. [CrossRef]
20. Xu, S.; Ilyas, I.; Little, P.J.; Li, H.; Kamato, D.; Zheng, X.; Luo, S.; Li, Z.; Liu, P.; Han, J.; et al. Endothelial dysfunction in atherosclerotic cardiovascular diseases and beyond: From mechanism to pharmacotherapies. *Pharmacol. Rev.* **2021**, *73*, 924–967. [CrossRef] [PubMed]
21. Chesley, L.C.; Talledo, E.; Bohler, C.S.; Zuspan, F.P. Vascular reactivity to angiotensin II and norepinephrine in pregnant and nonpregnant women. *Am. J. Obstet. Gynecol.* **1965**, *91*, 837–842. [CrossRef]
22. Duvekot, J.J.; Cheriex, E.C.; Pieters, F.A.; Menheere, P.P.; Peeters, L.H. Early pregnancy changes in hemodynamics and volume homeostasis are consecutive adjustments triggered by a primary fall in systemic vascular tone. *Am. J. Obstet. Gynecol.* **1993**, *169*, 382–392. [CrossRef]
23. Amaral, L.M.; Wallace, K.; Owens, M.; LaMarca, B. Pathophysiology and Current Clinical Management of Preeclampsia. *Curr. Hypertens. Rep.* **2017**, *19*, 19–21. [CrossRef] [PubMed]
24. Rana, S.; Lemoine, E.; Granger, J.P.; Karumanchi, A. Preeclampsia: Pathophysiology, Challenges, and Perspectives. *Circ. Res.* **2019**, *124*, 1094–1112. [CrossRef]
25. Redman, C.W.G. Pre-eclampsia and the placenta. *Placenta* **1991**, *12*, 301–308. [CrossRef]
26. Yoshida, A.; Nakao, S.; Kobayashi, M.; Kobayashi, H. Flow-mediated vasodilation and plasma fibronectin levels in preeclampsia. *Hypertension* **2000**, *36*, 400–404. [CrossRef]
27. Maynard, S.E.; Min, J.; Merchan, J.; Lim, K.; Li, J.; Mondal, S. Excess placental soluble fms-like tyrosine kinase 1 (sFlt1) may contribute to endothelial dysfunction hypertension and proteinuria in preeclampsia. *J. Clin. Investig.* **2003**, *111*, 649–658. [CrossRef] [PubMed]

28. Rinehart, B.K.; Terrone, D.A.; Lagoo-Deenadayalan, S.; Barber, W.H.; Hale, E.A.; Martin, J.N., Jr.; Bennett, W.A. Expression of the placental cytokines tumor necrosis factor-alpha, interleukin 1beta, and interleukin 10 is increased in preeclampsia. *Am. J. Obstet. Gynecol.* **1999**, *181*, 915–920. [CrossRef]
29. Myers, J.; Mires, G.; Macleod, M.; Baker, P. In preeclampsia, the circulating factors capable of altering in vitro endothelial function precede clinical disease. *Hypertension* **2005**, *45*, 258–263. [CrossRef]
30. Roberts, J. Endothelial Dysfunction in Preeclampsia. *Semin. Reprod. Endocrinol.* **1998**, *16*, 5–15. [CrossRef]
31. Akhilesh, M.; Mahalingam, V.; Nalliah, S.; Ali, R.M.; Ganesalingam, M.; Haleagrahara, N. Hypoxia-inducible factor-1α as a predictive marker in pre-eclampsia. *Biomed. Rep.* **2013**, *1*, 257–258. [CrossRef] [PubMed]
32. Rana, S.; Karumanchi, S.A.; Levine, R.J.; Venkatesha, S.; Rauh-Hain, J.A.; Tamez, H.; Thadhani, R. Sequential changes in antiangiogenic factors in early pregnancy and risk of developing preeclampsia. *Hypertension* **2007**, *50*, 137–142. [CrossRef] [PubMed]
33. Romero, R.; Nien, J.K.; Espinoza, J.; Todem, D.; Fu, W.; Chung, W.; Kusanovic, J.P.; Gotsch, F.; Erez, O.; Mazaki-Tovi, S.; et al. A longitudinal study of angiogenic (placental growth factor) and anti-angiogenic (soluble endoglin and soluble vascular endothelial growth factor receptor-1) factors in normal pregnancy and patients destined to develop preeclampsia and deliver a small for gestational age neonate. *J. Matern. Fetal Neonatal Med.* **2008**, *21*, 9–23. [PubMed]
34. Cakmak, M.; Yilmaz, H.; Bağlar, E.; Darcin, T.; Inan, O.; Aktas, A.; Celik, H.T.; Ozdemir, O.; Atalay, C.R.; Akcay, A. Serum levels of endocan correlate with the presence and severity of pre-eclampsia. *Clin. Exp. Hypertens.* **2016**, *38*, 137–142. [CrossRef]
35. Pinheiro, M.B.; Gomes, K.B.; Ronda, C.R.S.C.; Guimarães, G.G.; Freitas, L.G.; Teixeira-Carvalho, A.; Martins-Filho, O.A.; Dusse, L.M. Severe preeclampsia: Association of genes polymorphisms and maternal cytokines production in Brazilian population. *Cytokine* **2014**, *71*, 232–237. [CrossRef]
36. Ramma, W.; Buhimschi, I.A.; Zhao, G.; Dulay, A.T.; Nayeri, U.A.; Buhimschi, C.S.; Ahmed, A. The elevation in circulating anti-angiogenic factors is independent of markers of neutrophil activation in preeclampsia. *Angiogenesis* **2012**, *15*, 333–340. [CrossRef]
37. Siljee, J.E.; Wortelboer, E.J.; Koster, M.P.H.; Imholz, S.; Rodenburg, W.; Visser, G.H.A.; de Vries, A.; Schielen, P.C.J.I.; Pennings, J.L.A. Identification of interleukin-1 beta, but no other inflammatory proteins, as an early onset pre-eclampsia biomarker in first-trimester serum by bead-based multiplexed immunoassays. *Prenat. Diagn.* **2013**, *33*, 1183–1188. [CrossRef]
38. Brennan, L.J.; Morton, J.S.; Davidge, S.T. Vascular Dysfunction in Preeclampsia. *Microcirculation* **2014**, *21*, 4–14. [CrossRef]
39. Ives, C.W.; Sinkey, R.; Rajapreyar, I.; Tita, A.T.N.; Oparil, S. Preeclampsia-Pathophysiology, and Clinical Presentations: JACC State-of-the-Art Review. *J. Am. Coll. Cardiol.* **2020**, *76*, 1690–1702. [CrossRef]
40. Shah, D.A.; Khalil, R.A. Bioactive factors in uteroplacental and systemic circulation link placental ischemia to generalized vascular dysfunction in hypertensive pregnancy and preeclampsia. *Biochem. Pharmacol.* **2015**, *95*, 211–226. [CrossRef] [PubMed]
41. Socha, M.W.; Malinowski, B.; Puk, O.; Dubiel, M.; Wiciński, M. The NLRP3 Inflammasome Role in the Pathogenesis of Pregnancy Induced Hypertension and Preeclampsia. *Cells* **2020**, *9*, 1642. [CrossRef]
42. Kelley, N.; Jeltema, D.; Duan, Y.; He, Y. The NLRP3 Inflammasome: An Overview of Mechanisms of Activation and Regulation. *Int. J. Mol. Sci.* **2019**, *20*, 3328. [CrossRef]
43. Xie, F.; Hu, Y.; Turvey, S.E.; Magee, L.A.; Brunham, R.M.; Choi, K.C.; Krajden, M.; Leung, P.C.K.; Money, D.M.; Patrick, D.M.; et al. Toll-like receptors 2 and 4 and the cryopyrin inflammasome in normal pregnancy and pre-eclampsia. *BJOG* **2010**, *117*, 99–108. [CrossRef] [PubMed]
44. Matias, M.L.; Romão, M.; Weel, I.C.; Ribeiro, V.R.; Nunes, P.R.; Borges, V.T.M.; Araújo, J.P., Jr.; Peraçoli., J.C.; de Oliveira, L.; Peraçoli, M.T. Endogenous and uric acid-induced activation of NLRP3 inflammasome in pregnant women with preeclampsia. *PLoS ONE* **2015**, *10*, e0129095. [CrossRef] [PubMed]
45. Mulla, M.J.; Myrtolli, K.; Potter, J.; Boeras, C.; Kavathas, P.B.; Sfakianaki, A.K.; Tadesse, S.; Norwitz, E.R.; Guller, S.; Abrahams, V.M. Uric acid induces trophoblast IL-1beta production via the inflammasome: Implications for the pathogenesis of preeclampsia. *Am. J. Reprod. Immunol.* **2011**, *65*, 542–548. [CrossRef] [PubMed]
46. Shirasuna, K.; Usui, F.; Karasawa, T.; Kimura, H.; Kawashima, A.; Mizukami, H.; Ohkuchi, A.; Nishimura, S.; Sagara, J.; Noda, T.; et al. Nanosilica-induced placental inflammation and pregnancy complications: Different roles of the inflammasome components NLRP3 and ASC. *Nanotoxicology* **2015**, *9*, 554–567. [CrossRef]
47. Tamura, K.; Ishikawa, G.; Yoshie, M.; Ohneda, W.; Nakai, A.; Takeshita, T.; Tachikawa, E. Glibenclamide inhibits NLRP3 inflammasome-mediated IL-1beta secretion in human trophoblasts. *J. Pharmacol. Sci.* **2017**, *135*, 89–95. [CrossRef] [PubMed]
48. Brien, M.E.; Duval, J.C.; Palacios, J.; Boufaied, I.; Hudon-Thibeault, A.-A.; Nadeau-Valle'e, M.; Vaillancourt, C.; Sibley, C.P.; Abrahams, V.M.; Jones, R.L.; et al. Uric acid crystals induce placental inflammation and alter trophoblast function via an IL-1-dependent pathway: Implications for fetal growth restriction. *J. Immunol.* **2017**, *198*, 443–451. [CrossRef]
49. Shirasuna, K.; Karasawa, F.T.; Usui, M.; Kobayashi, T.; Komada, H.; Kimura, A.; Kawashima, A.; Ohkuchi, A.; Taniguchi, S.; Takahashi, M. NLRP3 deficiency improves angiotensin II-induced hypertension but not fetal growth restriction during pregnancy. *Endocrinology* **2015**, *156*, 4281–4292. [CrossRef]
50. Seno, K.; Sase, S.; Ozeki, A.; Takahashi, H.; Ohkuchi, A.; Suzuki, H.; Matsubara, S.; Iwata, H.; Kuwayama, T.; Shirasuna, K. Advanced glycation end products regulate interleukin-1b production in human placenta. *J. Reprod. Dev.* **2017**, *63*, 401–408. [CrossRef]

51. Gomez-Lopez, N.; Motomura, K.; Miller, D.; Garcia-Flores, V.; Galaz, J.; Romero, R. Inflammasomes: Their Role in Normal and Complicated Pregnancies. *J. Immunol.* **2019**, *203*, 2757–2769. [CrossRef]
52. Sutterwala, F.S.; Haasken, S.; Cassel, S.L. Mechanism of NLRP3 inflammasome activation. *Ann. N. Y. Acad. Sci.* **2014**, *1319*, 82–95. [CrossRef] [PubMed]
53. Liu, D.; Zeng, X.; Li, X.; Mehta, J.L.; Wang, X. Role of NLRP3 inflammasome in the pathogenesis of cardiovascular diseases. *Basic Res. Cardiol.* **2017**, *113*, 5. [CrossRef] [PubMed]
54. Cheng, S.; Nakashima, A.; Huber, W.J.; Davis, S.; Banerjee, S.; Huang, Z.; Saito, S.; Sadovsky, Y.; Sharma, S. Pyroptosis is a critical inflammatory pathway in the placenta from early-onset preeclampsia and in human trophoblasts exposed to hypoxia and endoplasmic reticulum stressors. *Cell Death Dis.* **2019**, *10*, 927. [CrossRef] [PubMed]
55. Keyel, P.A. How is inflammation initiated? Individual influences of IL-1, IL-18, and HMGB1. *Cytokine* **2014**, *69*, 136–145. [CrossRef]
56. Shi, J.; Gao, W.; Shao, F. Pyroptosis: Gasdermin-mediated programmed necrotic cell death. *Trends Biochem. Sci.* **2017**, *42*, 245–254. [CrossRef]
57. Goulopoulou, S.; Davidge, S.T. Molecular mechanisms of maternal vascular dysfunction in preeclampsia. *Trends Mol. Med.* **2015**, *21*, 88–97. [CrossRef]
58. Komarova, Y.A.; Kruse, K.; Mehta, D.; Malik, A.B. Protein interactions at endothelial junctions and signaling mechanisms regulating endothelial permeability. *Circ. Res.* **2017**, *120*, 179–206. [CrossRef]
59. Radeva, M.Y.; Waschke, J. Mind the gap: Mechanisms regulating the endothelial barrier. *Acta Physiol.* **2018**, *222*, e12860. [CrossRef]
60. Gong, T.; Liu, L.; Jiang, W.; Zhou, R. DAMP-sensing receptors in sterile inflammation and inflammatory diseases. *Nat. Rev. Immunol.* **2020**, *20*, 95–112. [CrossRef]
61. Takeuchi, O.; Akira, S. Pattern recognition receptors and inflammation. *Cell* **2010**, *140*, 805–820. [CrossRef]
62. Chen, Z.; Martin, M.; Li, Z.; Shyy, J.Y. Endothelial dysfunction: The role of sterol regulatory element-binding protein-induced NOD-like receptor Family pyrin domain-containing protein 3 inflammasome in atherosclerosis. *Curr. Opin. Lipido.* **2014**, *25*, 339–349. [CrossRef]
63. Wang, J.; Williams, J.C.; Davis, B.K.; Jacobson, K.; Doerschuk, C.M.; Ting, J.P.; Mackman, N. Monocytic microparticles activate endothelial cells in an IL- 1beta-dependent manner. *Blood* **2011**, *118*, 2366–2374. [CrossRef]
64. Grebe, A.; Hoss, F.; Latz, E. NLRP3 inflammasome and the IL-1 pathway in atherosclerosis. *Circ. Res.* **2018**, *122*, 1722–1740. [CrossRef]
65. He, Y.; Hara, H.; Nunez, G. Mechanism and regulation of NLRP3 inflammasome activation. *Trends Biochem. Sci.* **2016**, *41*, 1012–1021. [CrossRef]
66. Abais, J.M.; Xia, M.; Zhang, Y.; Boini, K.M.; Li, P.L. Redox regulation of NLRP3 inflammasomes: ROS as trigger or effector? *Antioxid. Redox Signal* **2015**, *22*, 1111–1129. [CrossRef]
67. Dunn, D.; Alvarez, L.A.; Zhang, X.; Soldati, T. Reactive oxygen species, and mitochondria: A nexus of cellular homeostasis. *Redox Biol.* **2015**, *6*, 472–485. [CrossRef] [PubMed]
68. Shihata, W.A.; Michell, D.L.; Andrews, K.L.; Chin-Dusting, J.P. Caveolae: A role in endothelial inflammation and mechanotransduction? *Front. Physiol.* **2016**, *7*, 628. [CrossRef] [PubMed]
69. Lamkanfi, M.; Mueller, J.L.; Vitari, A.C.; Misaghi, S.; Fedorova, A.; Deshayes, K.; Lee, W.P.; Hoffman, H.M.; Dixit, V.M. Glyburide inhibits the Cryopyrin/Nalp3 inflammasome. *J. Cell Biol.* **2009**, *187*, 61–70. [CrossRef] [PubMed]
70. Perregaux, D.G.; McNiff, P.; Laliberte, R. Hawryluk, N.; Peurano, H.; Stam Eggler, J.; Griffiths, R.; Dombroski, M.A.; Gabel, C.A. Identification and characterization of a novel class of interleukin-1 post-translational processing inhibitors. *J. Pharmacol. Exp. Ther.* **2001**, *299*, 187–197.
71. Marchetti, C.; Toldo, S.; Chojnacki, J.; Mezzaroma, E.; Liu, K.; Salloum, F.N.; Nordio, A.; Carbone, S.; Mauro, A.G.; Das, A.; et al. Pharmacologic inhibition of the NLRP3 inflammasome preserves cardiac function after ischemic and non-ischemic injury in the mouse. *J. Cardiovasc. Pharmacol.* **2015**, *66*, 1–8. [CrossRef] [PubMed]
72. Marchetti, C.; Chojnacki, J.; Toldo, S.; Mezzaroma, E.; Tranchida, N.; Rose, S.W.; Federici, M.; Van Tassell, B.W.; Zhang, S.; Abbate, A. A novel pharmacologic inhibitor of the NLRP3 inflammasome limits myocardial injury following ischemia-reperfusion in the mouse. *J. Cardiovasc. Pharmacol.* **2014**, *63*, 316–322. [CrossRef] [PubMed]
73. Hamon, Y.; Luciani, M.F.; Becq, F.; Verrier, B.; Rubartelli, A.; Chimini, G. Interleukin-1beta secretion is impaired by inhibitors of the Atp binding cassette transporter, ABC1. *Blood* **1997**, *90*, 2911–2915. [CrossRef] [PubMed]
74. Lottaz, D.; Beleznay, Z.; Bickel, M. Inhibition of ATP-binding cassette transporter downregulates interleukin-1beta-mediated autocrine activation of human dermal fibroblasts. *J. Investig. Dermatol.* **2001**, *117*, 871–876. [CrossRef] [PubMed]
75. Mangan, M.S.; Olhava, E.J.; Roush, W.R.; Seidel, H.M.; Glick, G.D.; Latz, E. Targeting the NLRP3 inflammasome in inflammatory diseases. *Nat. Rev. Drug Discov.* **2018**, *17*, 588–606. [CrossRef] [PubMed]
76. Kuwar, R.; Rolfe, A.; Di, L.; Xu, H.; He, L.; Jiang, Y.; Zhang, S.; Sun, D. A novel small molecular NLRP3 inflammasome inhibitor alleviates neuroinflammatory response following traumatic brain injury. *J. Neuroinflammation* **2019**, *16*, 81. [CrossRef] [PubMed]
77. Fulp, J.; He, L.; Toldo, S.; Jiang, Y.; Boice, A.; Guo, C.; Li, X.; Rolfe, A.; Sun, D.; Abbate, A.; et al. Structural insights of benzenesulfonamide analogues as NLRP3 inflammasome inhibitors: Design, synthesis, and biological characterization. *J. Med. Chem.* **2018**, *61*, 5412–5423. [CrossRef] [PubMed]

78. Liu, W.; Guo, W.; Wu, J.; Luo, Q.; Tao, F.; Gu, Y.; Shen, Y.; Li, J.; Tan, R.; Xu, Q.; et al. A novel benzo [d] imidazole derivate prevents the development of dextran sulfate sodium-induced murine experimental colitis via inhibition of NLRP3 inflammasome. *Biochem. Pharmacol.* **2013**, *85*, 1504–1512. [CrossRef]
79. Coll, R.C.; Robertson, A.A.; Chae, J.J.; Higgins, S.C.; Muñoz-Planillo, R.; Inserra, M.C.; Vetter, I.; Dungan, L.S.; Monks, B.G.; Stutz, A.; et al. A small-molecule inhibitor of the NLRP3 inflammasome for the treatment of inflammatory diseases. *Nat. Med.* **2015**, *21*, 248–255. [CrossRef]
80. Coll, R.C.; Hill, J.R.; Day, C.J.; Zamoshnikova, A.; Boucher, D.; Massey, N.L.; Chitty, J.L.; Fraser, J.A.; Jennings, M.P.; Robertson, A.A.B.; et al. MCC950 directly targets the NLRP3 ATP-hydrolysis motif for inflammasome inhibition. *Nat. Chem. Biol.* **2019**, *15*, 556–559. [CrossRef]
81. Ferreira, N.S.; Bruder-Nascimento, T.; Pereira, C.A.; Zanotto, C.Z.; Prado, D.S.; Silva, J.F.; Rassi, D.M.; Foss-Freitas, M.C.; Alves-Filho, J.C.; Carlos, D.; et al. NLRP3 Inflammasome and Mineralocorticoid Receptors Are Associated with Vascular Dysfunction in Type 2 Diabetes Mellitus. *Cells* **2019**, *8*, 1595. [CrossRef]
82. Cornelius, D.C.; Travis, O.K.; Tramel, R.W.; Borges-Rodriguez, M.; Baik, C.H.; Greer, M.; Giachelli, C.A.; Tardo, G.A.; Williams, J.M. NLRP3 inflammasome inhibition attenuates sepsis induced platelet activation and prevents multi-organ injury in cecal-ligation puncture. *PLoS ONE* **2020**, *15*, e0234039. [CrossRef] [PubMed]
83. Luo, M.; Meng, J.; Yan, J.; Shang, F.; Zhang, T.; Lv, D.; Li, C.; Yang, X.; Luo, S. Role of the nucleotide-binding domain-like receptor protein 3 inflammasome in the endothelial dysfunction of early sepsis. *Inflammation* **2020**, *43*, 1561–1571. [CrossRef]
84. Primiano, M.J.; Lefker, B.A.; Bowman, M.R.; Bree, A.G.; Hubeau, C.; Bonin, P.D.; Mangan, M.; Dower, K.; Monks, B.G.; Cushing, L.; et al. Efficacy and pharmacology of the NLRP3 inflammasome inhibitor CP-456,773 (CRID3) in murine models of dermal and pulmonary inflammation. *J. Immunol.* **2016**, *197*, 2421–2433. [CrossRef]
85. Ma, T.; Thiagarajah, J.R.; Yang, H.; Sonawane, N.D.; Folli, C.; Galietta, L.J.; Verkman, A.S. Thiazolidinone CFTR inhibitor identified by high-throughput screening blocks cholera toxin–induced intestinal fluid secretion. *J. Clin. Investig.* **2002**, *110*, 1651–1658. [CrossRef]
86. Jiang, H.; He, H.; Chen, Y.; Huang, W.; Cheng, J.; Ye, J.; Wang, A.; Tao, J.; Wang, C.; Liu, Q.; et al. Identification of a selective and direct NLRP3 inhibitor to treat inflammatory disorders. *J. Exp. Med.* **2017**, *214*, 3219–3238. [CrossRef]
87. Zahid, A.; Li, B.; Kombe, A.J.; Jin, T.; Tao, J. Pharmacological Inhibitors of the NLRP3 Inflammasome. *Front. Immunol.* **2019**, *10*, 2538. [CrossRef]
88. Darakhshan, S.; Pour, A.B. Tranilast: A review of its therapeutic applications. *Pharmacol. Res.* **2015**, *91*, 15–28. [CrossRef]
89. Konneh, M. Tranilast Kissei pharmaceutical. *Idrugs* **1998**, *1*, 141–146. [PubMed]
90. Platten, M.; Ho, P.P.; Youssef, S.; Fontoura, P.; Garren, H.; Hur, E.M.; Gupta, R.; Lee, L.Y.; Kidd, B.A.; Robinson, W.H.; et al. Treatment of autoimmune neuroinflammation with a synthetic tryptophan metabolite. *Science* **2005**, *310*, 850–855. [CrossRef] [PubMed]
91. Toldo, S.; Abbate, A. The NLRP3 inflammasome in acute myocardial infarction. *Nat. Rev. Cardiol.* **2018**, *15*, 203–214. [CrossRef] [PubMed]
92. Marchetti, C.; Swartzwelter, B.; Gamboni, F.; Neff, C.P.; Richter, K.; Azam, T.; Carta, S.; Tengesdal, I.; Nemkov, T.; D'Alessandro, A.; et al. OLT1177, a β-sulfonyl nitrile compound, safe in humans, inhibits the NLRP3 inflammasome and reverses the metabolic cost of inflammation. *Proc. Natl. Acad. Sci. USA* **2018**, *115*, E1530–E1539. [CrossRef] [PubMed]
93. Kadota, S.; Basnet, P.; Ishii, E.; Tamura, T.; Namba, T. Antibacterial activity of trichorabdal A from Rabdosia trichocarpa against Helicobacter pylori. *Zent. Für Bakteriol.* **1997**, *286*, 63–67. [CrossRef]
94. Kuo, L.-M.; Kuo, C.-Y.; Lin, C.-Y.; Hung, M.-F.; Shen, J.-J.; Hwang, T.-L. Intracellular glutathione depletion by oridonin leads to apoptosis in hepatic stellate cells. *Molecules* **2014**, *19*, 3327–3344. [CrossRef] [PubMed]
95. Huang, J.; Wu, L.; Tashiro, S.; Onodera, S.; Ikejima, T. A comparison of the signal pathways between the TNFα-and oridonin-induced Murine L929 fibrosarcoma cell death. *Acta Med. Okayama* **2005**, *59*, 261–270.
96. Zhao, G.; Zhang, T.; Ma, X.; Jiang, K.; Wu, H.; Qiu, C.; Guo, M.; Deng, G. Oridonin attenuates the release of pro-inflammatory cytokines in lipopolysaccharide-induced RAW264. 7 cells and acute lung injury. *Oncotarget* **2017**, *8*, 68153. [CrossRef]
97. He, H.; Jiang, H.; Chen, Y.; Ye, J.; Wang, A.; Wang, C.; Liu, Q.; Liang, G.; Deng, X.; Jiang, W.; et al. Oridonin is a covalent NLRP3 inhibitor with strong anti-inflammasome activity. *Nat. Commun.* **2018**, *9*, 2550. [CrossRef]
98. Heinrich, M.; Robles, M.; West, J.E.; Ortiz de Montellano, B.R.; Rodriguez, E. Ethnopharmacology of Mexican asteraceae (compositae). *Ann. Rev. Pharmacol. Toxicol.* **1998**, *38*, 539–565. [CrossRef]
99. Juliana, C.; Fernandes-Alnemri, T.; Wu, J.; Datta, P.; Solorzano, L.; Yu, J.-W.; Meng, R.; Quong, A.A.; Latz, E.; Scott, C.P.; et al. Anti-inflammatory compounds parthenolide and Bay 11–7082 are direct inhibitors of the inflammasome. *J. Biol. Chem.* **2010**, *285*, 9792–9802. [CrossRef]
100. Guzman, M.L.; Rossi, R.M.; Neelakantan, S.; Li, X.; Corbett, C.A.; Hassane, D.C.; Becker, M.W.; Bennett, J.M.; Sullivan, E.; Lachowicz, J.L.; et al. An orally bioavailable parthenolide analog selectively eradicates acute myelogenous leukemia stem and progenitor cells. *Blood* **2007**, *110*, 4427–4435. [CrossRef]
101. D'anneo, A.; Carlisi, D.; Lauricella, M.; Puleio, R.; Martinez, R.; Di Bella, S.; Di Marco, P.; Emanuele, S.; Di Fiore, R.; Guercio, A.; et al. Parthenolide generates reactive oxygen species and autophagy in MDA-MB231 cells. A soluble parthenolide analogue inhibits tumour growth and metastasis in a xenograft model of breast cancer. *Cell Death Dis.* **2013**, *4*, e891. [CrossRef] [PubMed]

102. Wannamaker, W.; Davies, R.; Namchuk, M.; Pollard, J.; Ford, P.; Ku, G.; Decker, C.; Charifson, P.; Weber, P.; Germann, U.A.; et al. (S)-1-((S)-2-{[1-(4-Amino-3-chloro-phenyl)-methanoyl]-amino}-3,3-dimethyl-butanoyl)-pyrrolidine-2-carboxylic acid ((2R,3S)-2-ethoxy-5-oxo-tetrahydro-furan-3-yl)-amide (VX-765), an Orally Available Selective Interleukin (IL)-Converting Enzyme/Caspase-1 Inhi. *J. Pharmacol. Exp. Ther.* **2007**, *321*, 509–516. [CrossRef] [PubMed]
103. Mackenzie, S.H.; Schipper, J.L.; Clark, A.C. The potential for caspases in drug discovery. *Curr. Opin. Drug Discov. Devel.* **2010**, *13*, 568–576.
104. Boxer, M.B.; Shen, M.; Auld, D.S.; Wells, J.A.; Thomas, C.J. *A Small Molecule Inhibitor of Caspase 1. Probe Reports from the NIH Molecular Libraries Program [Internet]*; National Center for Biotechnology Information: Bethesda, MD, USA, 2010.
105. Siegmund, B.; Zeitz, M. Pralnacasan (vertex pharmaceuticals). *Idrugs* **2003**, *6*, 154–158.
106. Strand, V.; Sokolove, J. Randomized controlled trial design in rheumatoid arthritis: The past decade. *Arthritis Res. Ther.* **2009**, *11*, 1–11. [CrossRef]
107. Fischer, U.; Schulzeosthoff, K. Apoptosis-based therapies and drug targets. *Cell Death Differ.* **2005**, *12*, 942–961. [CrossRef]
108. Maroso, M.; Balosso, S.; Ravizza, T.; Liu, J.; Bianchi, M.E.; Vezzani, A. Interleukin-1 type 1 receptor/Toll-like receptor signalling in epilepsy: The importance of IL-1beta and high-mobility group box 1. *J. Int. Med.* **2011**, *270*, 319–326. [CrossRef] [PubMed]
109. Youm, Y.-H.; Nguyen, K.Y.; Grant, R.W.; Goldberg, E.L.; Bodogai, M.; Kim, D.; D'Agostino, D.; Planavsky, N.; Lupfer, C.; Kanneganti, T.D.; et al. The ketone metabolite β-hydroxybutyrate blocks NLRP3 inflammasome–mediated inflammatory disease. *Nat. Med.* **2015**, *21*, 263–269. [CrossRef] [PubMed]

cells

MDPI

Article

Sustained Elevated Circulating Activin A Impairs Global Longitudinal Strain in Pregnant Rats: A Potential Mechanism for Preeclampsia-Related Cardiac Dysfunction

Bhavisha A. Bakrania [1,2,*], Ana C. Palei [3], Umesh Bhattarai [2], Yingjie Chen [2], Joey P. Granger [2] and Sajid Shahul [4]

1 UQ Centre for Clinical Research, Faculty of Medicine, University of Queensland, Brisbane 4029, Australia
2 Department of Physiology and Biophysics, University of Mississippi Medical Center, Jackson, MS 39216, USA; ubhattarai@umc.edu (U.B.); ychen2@umc.edu (Y.C.); jgranger@umc.edu (J.P.G.)
3 Department of Surgery, University of Mississippi Medical Center, Jackson, MS 39216, USA; apalei@umc.edu
4 Department of Anesthesia and Critical Care, University of Chicago, Chicago, IL 60637, USA; sshahul1@dacc.uchicago.edu
* Correspondence: b.bakrania@uq.edu.au

Abstract: Mediators of cardiac injury in preeclampsia are not well understood. Preeclamptic women have decreased cardiac global longitudinal strain (GLS), a sensitive measure of systolic function that indicates fibrosis and tissue injury. GLS is worse in preeclampsia compared to gestational hypertension, despite comparable blood pressure, suggesting that placental factors may be involved. We previously showed that Activin A, a pro-fibrotic factor produced in excess by the placenta in preeclampsia, predicts impaired GLS postpartum. Here, we hypothesized that chronic excess levels of Activin A during pregnancy induces cardiac dysfunction. Rats were assigned to sham or activin A infusion (1.25–6 μg/day) on a gestational day (GD) 14 (n = 6–10/group). All animals underwent blood pressure measurement and comprehensive echocardiography followed by euthanasia and the collection of tissue samples on GD 19. Increased circulating activin A (sham: 0.59 ± 0.05 ng/mL, 6 μg/day: 2.8 ± 0.41 ng/mL, $p < 0.01$) was associated with impaired GLS (Sham: −22.1 ± 0.8%, 6 μg/day: −14.7 ± 1.14%, $p < 0.01$). Activin A infusion (6 μg/day) increased beta-myosin heavy chain expression in heart tissue, indicating cardiac injury. In summary, our findings indicate that increasing levels of activin A during pregnancy induces cardiac dysfunction and supports the concept that activin A may serve as a possible mediator of PE-induced cardiac dysfunction.

Keywords: preeclampsia; cardiac dysfunction; placental factors; activin A

Citation: Bakrania, B.A.; Palei, A.C.; Bhattarai, U.; Chen, Y.; Granger, J.P.; Shahul, S. Sustained Elevated Circulating Activin A Impairs Global Longitudinal Strain in Pregnant Rats: A Potential Mechanism for Preeclampsia-Related Cardiac Dysfunction. *Cells* **2022**, *11*, 742. https://doi.org/10.3390/cells11040742

Academic Editor: Daniel Surbek

Received: 21 January 2022
Accepted: 17 February 2022
Published: 21 February 2022

1. Introduction

During normal gestation, the woman's cardiovascular system endures major structural and hemodynamic alterations to meet the demands of the developing fetus. The pregnancy-induced hemodynamic shift usually initiates prior to placentation, peaks in the second trimester of gestation, and persists until delivery [1]. However, disturbed cardiovascular adaptations may lead to adverse pregnancy outcomes, including peripartum cardiomyopathy and preeclampsia (PE). PE is a syndrome characterized by new-onset hypertension and significant end-organ damage in the last half of gestation or immediate postpartum period [2]. It affects 3–5% of pregnancies worldwide, with incidence increasing over the recent decades due to the higher prevalence of risk factors such as advanced maternal age, obesity, and other chronic health conditions [3–5]. In addition to low plasma volume, high peripheral vascular resistance, and reduced cardiac output [6–8], the heart of PE women often exhibits impaired global longitudinal strain (GLS) [9–15]–a sensitive measure of systolic function that indicates cardiac injury and fibrosis. Although hypertension and cardiac dysfunction generally resolve after delivery, PE is associated with a significant risk of mortality as well as short- and long-term morbidity for both mother mothers and offspring [16]. Despite the negative clinical, social, and public health impact of PE, the only definitive treatment is early delivery [17]. Therefore, it is imperative to determine the mechanisms underlying cardiovascular dysfunction during pregnancies complicated by PE in order to advance the discovery of novel therapies to improve maternal and fetal outcomes.

The cardiovascular adaptations of pregnancy are mainly regulated by hormonal and neural mechanisms, with multiple pathways interacting to control blood pressure [18]. However, as an initial step in the pathophysiology of PE, abnormal placental morphogenesis and perfusion stimulate the release of placental factors into the maternal circulation, and compelling evidence indicates that vascular and cardiac dysfunction in PE are also induced by many of these placental factors [17]. Indeed, it was found that GLS gradually deteriorates from normotensive pregnancy to gestational hypertension (GH, allegedly without placental compromise) to PE [9,14,15], despite comparable blood pressure in these hypertensive groups [14]. Additionally, GLS remains impaired postpartum even after adjustment for blood pressure and other clinically and biologically relevant variables [19–21], further supporting a role for placental factors in the development of cardiac dysfunction in PE.

One potential placental factor that may contribute to detrimental changes to the heart in PE is activin A [22]. Activin A is a glycoprotein member of the transforming growth factor β superfamily involved in multiple biological functions, including reproduction, embryogenesis, inflammation, and fibrosis. While the placenta is the major source of activin A in the maternal circulation, the pituitary gland, ovaries, uterus, and inflammatory cells are also able to produce activin A [23]. Increased expression of activin A was reported in patients with heart failure as well as in experimental animal models of myocardium infarction and dilated cardiomyopathy [24–26]. Furthermore, activin A was implicated in cardiac remodeling and fibrosis by promoting the release of atrial natriuretic peptide (ANP), brain natriuretic peptide (BNP), reactive oxygen species, and cytokines from cardiac myocytes in vitro [27,28].

Dysregulation of the activin A signaling pathway is linked to the manifestation and prediction of PE. Activin A is increased in the blood and placental tissue of PE patients compared with both GH and normal pregnant women [29–43]. Moreover, circulating levels of activin A are higher from 10 to 26 weeks of gestation in women who subsequently developed PE [33,35,44–50]. We previously showed that serum activin A levels during the third trimester in normotensive pregnancy, gestational or chronic hypertension, and PE correlate positively with abnormal GLS at one year postpartum [21] and this relationship persists 10 years after pregnancy [20]. Earlier studies demonstrated sustained infusion of activin A-induced PE-like features in pregnant mice, including hypertension, proteinuria, preterm birth, and fetal growth restriction [51]. However, the importance of activin A in mediating cardiac dysfunction related to PE is not known. Thus, the present study aimed to test the hypothesis that chronic excess levels of activin A induce cardiac dysfunction in pregnant rats.

2. Materials and Methods

2.1. Experimental Animals and Protocols

All experiments were approved by the Institutional Animal Care and Use Committee at the University of Mississippi Medical Center (UMMC) and conducted in accordance with the National Institutes of Health (NIH) Guide for the Care and Use of Laboratory Animals. Female Sprague Dawley rats were received from Charles River Laboratories (Wilmington, MA, USA) on gestational day (GD) 10 and housed in the Center for Comparative Research facility at the UMMC. On GD 14, animals were randomly assigned to either the sham ($n = 10$) or the recombinant activin A (AnshLabs, Webster, TX, USA) group. For activin A infusion, rats were anesthetized (~3% isoflurane in 2 L/min O_2), and mini-osmotic pumps (Alzet, Cupertino, CA, USA) containing varying doses were placed intraperitoneally. Pumps were prepared to infuse 1.25 μg/day, 1.9 μg/day, 3 μg/day, or 6 μg/day of activin A continuously from GD 14 to 19 ($n = 6$ per dose). These doses were selected to achieve a five-fold increase of Activin A in pregnant rats, to mimic levels seen in PE, compared to healthy pregnancies, clinically [21,40]. All animals were fed nutritionally complete diets (Envigo Teklad 8640, Indianapolis, IN, USA) and water, ad libitum.

On GD 19, blood pressure was measured, followed by an echocardiogram and sacrifice. These procedures are detailed below. During sacrifice, blood was collected from the abdominal aorta into EDTA-coated vacutainers (BD, Franklin Lakes, NJ, USA). The heart was excised, arrested in ice-cold cardioplegic solution, and weighed. The coronary circulation was cleared with a cardioplegic solution by retrograde perfusion, snap-frozen, and stored at −80 °C. To calculate fetal outcomes, pup and placental weights were averaged per animal to constitute a single data point.

2.2. Maternal Blood Pressure Measurement

Blood pressure was assessed as we previously described [52]. Briefly, on GD 18, indwelling carotid catheters were placed to measure blood pressure via a pressure transducer. On GD 19, conscious blood pressure was measured in all animals for a period of two hours. The final 30 min was used to average mean arterial pressure (MAP).

2.3. Echocardiography with Speckle Tracking Technology

Comprehensive echocardiography was performed using the Vevo3100 (VisualSonics, Toronto, ON, Canada) and an MX250 scan head for small animals on GD 19. Rats were anesthetized with constant temperature and heart rate monitoring. Cardiac output and ejection fraction were determined in B-mode, whereas mass was determined in M-mode. Speckle tracking analysis to determine GLS was performed using the accompanying VisualSonics VevoStrain software.

2.4. Western Blotting

Quantification of the β-myosin heavy chain (β-MHC) was performed by Western blotting as we previously described [53,54]. Briefly, rat heart samples were homogenized in RIPA lysis buffer (Thermo Scientific, Rockford, IL, USA) containing protease and phosphatase inhibitors. Protein concentration in lysed samples was estimated using the Pierce BCA protein assay kit (Thermo Scientific). Samples were loaded and separated by SDS-PAGE gel and transferred to nitrocellulose membrane. The membrane was then blocked with 5% non-fat milk in TBST for 1 h and incubated with primary antibodies overnight at 4 °C. The antibodies used were mouse anti-β-MHC (cat# ab50967, Abcam, Waltham, MA, USA) or mouse anti-GAPDH (cat# 97166S, Cell Signaling Technology, Inc., Danvers, MA, USA). The membrane was washed with TBST three times, 10 min each, and then incubated with hrp-conjugated donkey anti-mouse secondary antibody for 1 h. The protein bands were detected using the iBright FL1500 instrument (Thermo Scientific), and the bands were quantified using NIH ImageJ analysis software (Version: 1.52k, Bethesda, MD, USA).

2.5. Histology for Detection of Cardiac Fibrosis

Fibrosis was determined by cardiac histology as we previously described [53,55]. Briefly, heart samples were collected, fixed in 10% formalin, and embedded in paraffin. The paraffin blocks were sectioned into 10 μm sections and stained with Sirius Red and the Fast Green Collagen Staining Kit (Chondrex Inc., Woodinville, WA, USA), according to the manufacturer's instructions. The slides were then washed and mounted with a mounting medium. Images were taken with a bright field microscope and analyzed using NIH ImageJ software.

2.6. Biochemical Analyses

Following sacrifice on GD 19, blood was spun at 2500 RPM for 12 min at 4 °C and stored at −20 °C. Activin A (AnshLabs), ANP, and BNP (ab108797 and ab108815, respectively, Abcam) were measured in serum using commercially available sandwich ELISA kits, as per the manufacturers' instructions. ANP and BNP levels were also quantified in heart homogenate samples by ELISA (Abcam) and normalized by total protein concentration determined by BCA assay (Thermo Scientific). A Tecan GENios plate reader (Mannedorf, Switzerland) with Magellan version 4.1 software (Mannedorf, Switzerland) was used to read the plates.

2.7. Statistical Analysis

All graphs and statistical analysis were performed using Graphpad Prism (Version 9.0.0 GraphPad Software, La Jolla, CA, USA). The specific n for each data set is detailed in the figure legend. Data followed a normal distribution and were analyzed using either a two-tailed t-test or one-way ANOVA followed by Tukey's multiple comparison test depending on the number of groups being compared. Pearson's correlation was performed to determine the relationship between circulating activin A and markers of cardiac injury. Data are presented as mean ± SEM. A p-value of 0.05 was considered statistically significant.

3. Results

Activin A-infused pregnant rats had circulating activin-A levels on GD 19 that increased in a dose-dependent manner. We trialed four different doses, and at 6 µg/day, activin A concentration in serum was approximately five times that of the sham group (Figure 1), which is comparable to the fold-change increase seen in PE women [21,40]. GLS, which is a sensitive measure of cardiac systolic function, was significantly decreased in pregnant rats treated with 1.25, 1.9, 3, or 6 µg per day of activin A compared to sham (Figure 2). Interestingly, MAP was not different among groups regardless of the activin A dose infused (Figure 3), suggesting that the effects of activin A on GLS were independent of blood pressure.

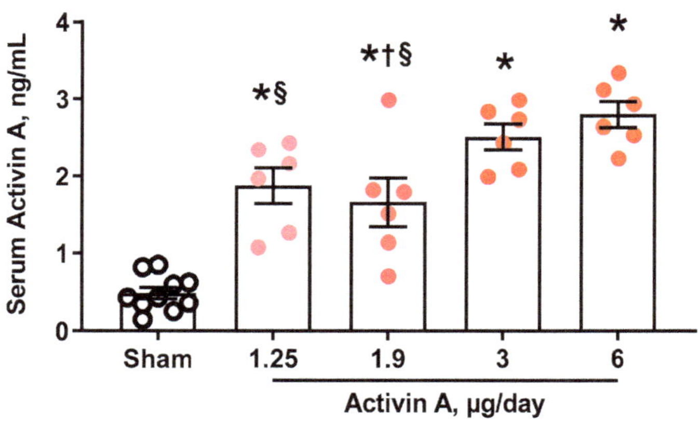

Figure 1. Circulating levels of activin A on gestational day 19 in sham (black open circles) and pregnant rats infused with 1.25 µg/day, 1.9 µg/day, 3 µg/day, and 6 µg/day of human recombinant activin A (solid red circles). Data are mean ± SEM. * $p < 0.05$ vs. sham, † $p < 0.05$ vs. 3 µg/day, § $p < 0.05$ vs. 6 µg/day.

Pregnant rats infused with the highest dose of activin A (6 µg/day) did not have any significant differences in fetal parameters on GD 19, including fetal weight and placental weight (Table 1), compared to sham rats. Similarly, maternal body weight, heart weight, heart rate, cardiac output, ejection fraction, and fractional shortening were not significantly altered in activin A-infused pregnant rats (Table 1, Figure 4).

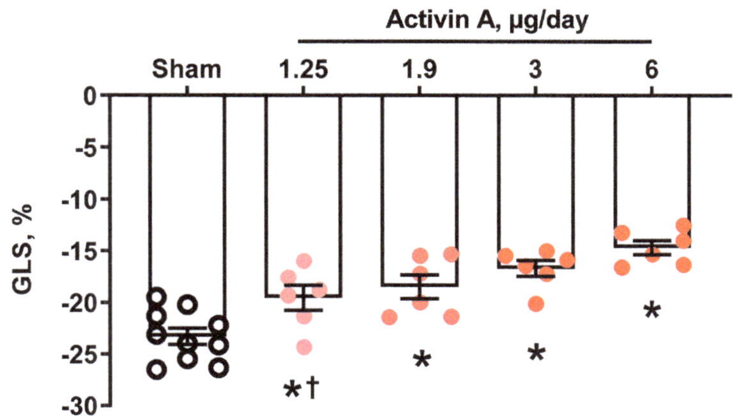

Figure 2. Cardiac global longitudinal strain (GLS) on gestational day 19 in sham (open black circles) and pregnant rats infused with 1.25 μg/day, 1.9 μg/day, 3 μg/day, and 6 μg/day of human recombinant activin A (solid red circles). Data are mean ± SEM. * $p < 0.05$ vs. sham, † $p < 0.05$ vs. 6 μg/day.

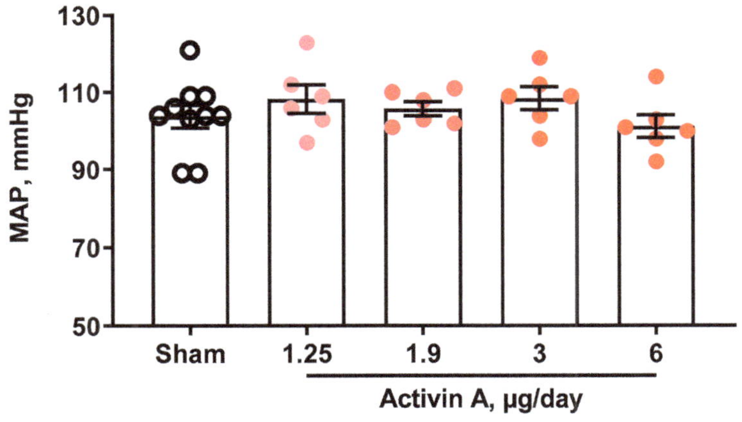

Figure 3. Mean arterial pressure (MAP) on gestational day 19 in sham (open black circles) and pregnant rats infused with 1.25 μg/day, 1.9 μg/day, 3 μg/day, and 6 μg/day of human recombinant activin A (solid red circles). Data are mean ± SEM.

Table 1. Characteristics of sham vs. activin A (6 μg/day) infused rats on gestational day 19.

	Sham	Activin A	*p*-Value
n	10	6	
Body weight, g	302 ± 10	305 ± 8	0.85
Fetal weight, g	2.66 ± 0.124	2.74 ± 0.072	0.65
Placental weight, g	0.65 ± 0.0213	0.64 ± 0.026	0.89
Heart weight, g	0.84 ± 0.011	0.87 ± 0.064	0.24
Heart rate, bpm	405 ± 14	408 ± 11	0.91
Cardiac output, mL/min	87 ± 5	101 ± 5	0.08

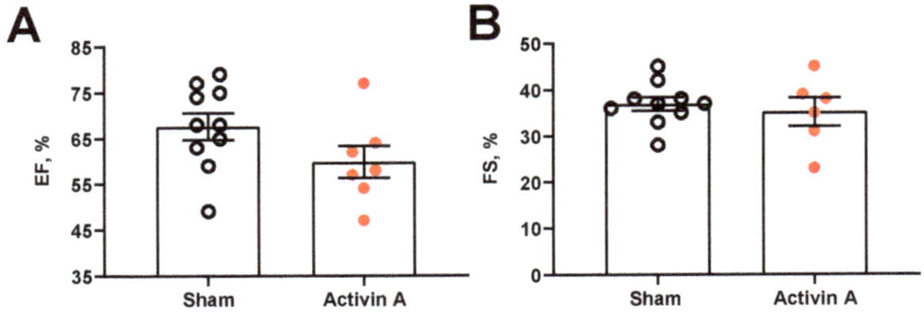

Figure 4. Cardiac parameters. Ejection fraction (EF; (**A**)); and fractional shortening (FS; (**B**)); measured on gestational day 19 in sham (open black circles) and activin A-infused (6 μg/day; solid red circles) pregnant rats. Data are mean ± SEM.

Previous studies demonstrated that pathological left ventricular hypertrophy is often associated with abnormal accumulation of collagen within the heart extracellular space, with the resultant fibrosis leading to increased ventricular stiffness. Conversely, perivascular fibrosis may cause myocardial ischemia. Both ventricular stiffness and myocardial ischemia were demonstrated to play important roles in the development of cardiac dysfunction [56]. Furthermore, in a variety of pathophysiologic conditions, including hypertrophy and ischemia, the postnatal heart undergoes adaptive mechanisms to support cardiac structure and function, such as the switch from α-MHC to β-MHC. However, at a certain point, this fetal-like reprogramming no longer suffices, with increased β-MHC linked to heart failure [57]. To characterize whether circulating activin A in excess induces cardiac fibrosis during pregnancy, collagen I and III fibers were visualized and quantified in histological heart slides using Sirus red staining. We also determined the fetal gene program by quantifying β-MHC in heart homogenates using Western blotting. While there were no significant differences between sham and activin A (6 μg/day)-infused pregnant rats in left ventricular fibrosis (Figure 5A,B), β-MHC protein expression was significantly increased in the left ventricle of activin A-infused rats on GD 19 (Figure 5C,D).

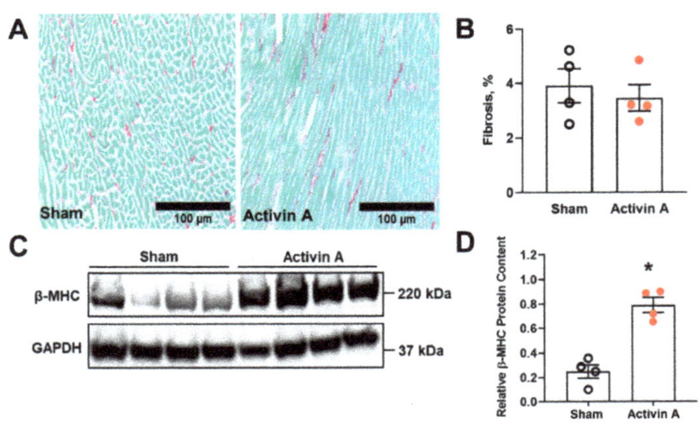

Figure 5. Markers of cardiac injury. Left ventricular fibrosis (**A**,**B**) indicated by Sirius Red and Fast Green staining, and beta-cardiac myosin heavy chain (β-MHC) protein content (**C**,**D**) on gestational day 19 in sham (open black circles) and activin A-infused (6 μg/day; solid red circles) pregnant rats. Data are mean ± SEM, * $p < 0.05$ vs. sham.

Finally, we measured ANP and BNP in the circulation and heart tissue of sham rats and rats that were infused with 6 μg/day of activin A on GD 19 as indicators of presence and severity of hemodynamic cardiac stress and heart failure [58]. There were no statistically significant differences in either serum or left ventricular ANP and BNP levels between groups (Figure 6A–D). Due to limitations on the standard curve of the ELISA kit, six (out of 10) serum ANP values in the sham group were undetectable compared with one (out of six) value in the activin A-infused group (Figure 6A). Nonetheless, circulating BNP levels were measurable in all serum samples from both groups (Figure 6B). Unfortunately, we had no remaining heart samples from two of the activin A-infused pregnant rats to run the ANP and BNP assays (Figure 6C,D). Interestingly, circulating activin A was positively correlated to circulating BNP, but a significant relationship was not detected for circulating ANP, suggesting that activin A induced some degree of cardiac injury (Figure 7A,B).

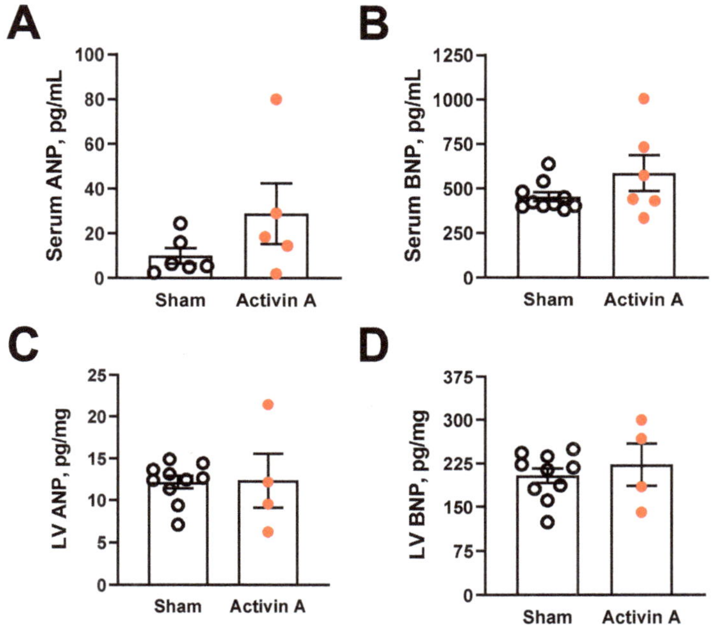

Figure 6. Circulating markers of cardiac injury. Serum (**A**,**B**) and left ventricular (LV; (**C**,**D**)); concentration of atrial natriuretic peptide (ANP) and brain natriuretic peptide (BNP) on gestational day 19 in sham (open black circles) and activin A-infused (6 µg/day; solid red circles) pregnant rats. Data are mean ± SEM.

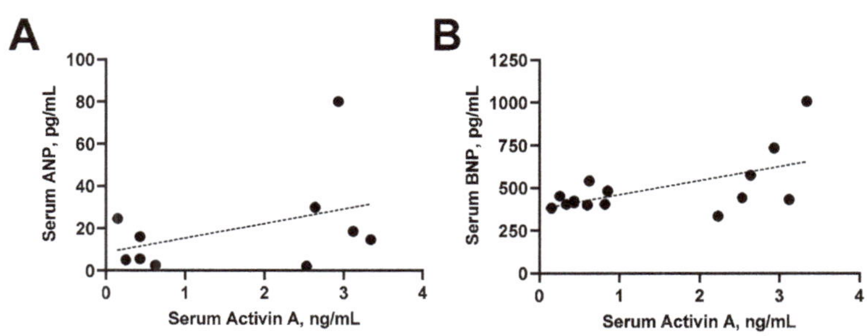

Figure 7. Relationship between activin A and circulating markers of cardiac injury. Correlation of serum activin A with serum atrial natriuretic peptide (ANP; r = 0.40, p = 0.25; (**A**)) or serum brain natriuretic peptide (BNP; r = 0.58, p = 0.02; (**B**)) in sham and activin A-infused (6 µg/day) pregnant rats.

4. Discussion

This study sought to determine whether increases in circulating activin A induces cardiac dysfunction during late pregnancy and whether this was associated with fibrosis and markers of cardiac injury. We previously showed that a five-fold increase in plasma activin A levels of PE patients, compared to normal pregnant women, predicts cardiac dysfunction at 1 year postpartum [21]. Similarly, we demonstrated here that a five-fold increase in circulating activin A levels induces cardiac dysfunction, as assessed by GLS, in pregnant rats. While fibrosis was not detected in the left ventricle of activin A-infused rats, we report a pronounced activation of the fetal gene program, which is indicative of cardiac injury. These data suggest that elevated circulating activin A in PE may play a role in mediating cardiac dysfunction in this cohort.

4.1. Cardiac Dysfunction in Preeclampsia Is Associated with Elevated Activin A

Circulating placental factors play a central role in promoting end-organ damage in PE. For instance, maternal angiogenesis imbalance, including increased soluble fms-like tyrosine kinase (sFlt)-1 and soluble endoglin (sEng) as well as decreased placental growth factor (PlGF), were associated with cardiac abnormalities during gestation and postpartum [59,60]. More recently, elevated antepartum activin A was associated with impaired GLS during gestation and also with worsened GLS one year postpartum. At the plasma activin A levels above 23.74 ng/mL, 84.7% of patients developed impaired GLS postpartum compared to 24.5% of those below the cutoff value [21]. These associations between circulating placental-derived factors and cardiac dysfunction remain significant after multivariable adjustment for clinically relevant confounders, including blood pressure [21,59,60]. Importantly, while circulating levels of other placental factors are similar between normal pregnancy and PE [59], activin A levels are comparable at one year postpartum but significantly elevated at approximately 10 years after a pregnancy complicated by PE [20,21]. These clinical data provide strong evidence of a relationship between plasma activin A and a lifelong risk of cardiac dysfunction. However, the direct effect of excess circulating activin A on heart function during pregnancy has yet to be examined.

The primary source of activin A during preeclampsia is thought to be the placenta; indeed, placental activin A expression positively correlates with circulating activin A levels in PE patients [29,31,37,41]. However, in vitro studies show that trophoblasts as well as endothelial and immune cells also secret activin A when exposed to pro-inflammatory and stress oxidative conditions, both hallmarks of PE [23]. Although the source of elevated circulating activin A levels in the years following the index pregnancy is not known, it is plausible that these other cell types may contribute.

Prior studies showed that activin A-infused pregnant mice exhibit increased blood pressure levels compared with saline-infused counterparts, along with preterm birth and fetal growth restriction [51]. However, we noted that impaired GLS following activin A infusion in pregnant rats was independent of alterations in blood pressure. Moreover, there were no differences in gestational length and fetal weight in our study. Differences in the effect of activin A on maternal-fetal outcomes between studies might be related to the activin A administration regimen, methods employed for blood pressure measurement, and/or even inter-species variability. Lim et al. infused activin A subcutaneously at a dose of 360 μg/kg/day from GD 10 to 16, resulting in more than an eight-fold increase in serum activin A levels (216 ± 102 vs. 1845 ± 286 ng/mL). Additionally, they assessed maternal blood pressure by noninvasive tail-cuff [51]. In contrast, we invasively determined a method considered more accurate and reliable than plethysmography via a catheter implanted in the carotid [61]. Indeed, the results of our study isolate the role of elevated circulating activin A levels in the cardiac function of pregnant rats, independent of a PE-like phenotype or hypertension itself.

4.2. Mechanisms of Activin A-Induced Cardiac Dysfunction

The mechanisms by which activin A results in cardiac dysfunction were not previously studied in the context of pregnancy. In mouse models of aging and left ventricular pressure overload, circulating activin A levels and cardiac activin type II receptor (ActIIR) signaling is significantly increased [62]. Interestingly, cardiac activin A expression was similar in aged (28 months old) animals compared to young animals (4 months old), suggesting that elevated activin A in the circulation and consequent activation of the ActIIR signaling in heart tissue largely originates from outside the heart. However, cardiac activin A expression also increased after pressure overload in young animals, indicating that both local and systemic activin A production were up-regulated [62]. Furthermore, overexpression of activin A with the administration of an adenoviral construct into young mice was sufficient to impair radial systolic and early diastolic strain rates without significantly altering blood pressure in only 96 h after injection [62]. Similarly, we report here that five days of continuous infusion of activin A at 1.25, 1.9, 3, or 6 μg/day result in significant decreases in GLS with no effect on blood pressure. These data are consistent with clinical data and suggest that the activin A-ActIIR pathway plays a causal role in cardiac dysfunction.

A number of studies have implicated activin A overexpression in the development of fibrosis [63,64]. While the impaired systolic strain is linked to increased fibrosis [65], few studies investigated the direct relationship between activin A overexpression and the development of cardiac fibrosis, and whether this effect could be ameliorated by inhibition of activin A receptors. Hu et al. showed that activin A stimulates cardiac fibroblast proliferation and differentiation [66]. Roh et al. demonstrated that a monoclonal antibody that blocks ActIIR named CDD866 (a murinized version of bimagrumab) ameliorates subclinical left ventricular systolic function of old mice. Moreover, CDD866 was not only able to mitigate substantially the decline in systolic function induced by pressure overload but also to attenuate the expression of fibrosis-related genes in heart tissue, both in prevention and treatment protocol [62]. The same research group found that cardiac myocyte-specific ActIIR knockout mice display normal cardiac structure and function at baseline but were protected from pressure overload-induced systolic dysfunction [62]. Castillero et al. also showed that inhibition of the ActIIR signaling either with decoy myostatin to prevent ligands from binding to ActIIR or with follistatin was associated with preserved cardiac function and fibroblast-driven decrease in cardiac fibrosis in mice following coronary ligation to induce myocardial infarction [63]. Importantly, treatment with these ActIIR inhibitors was first administered two weeks after the initial insult [62,63]. As mentioned previously, we continuously infused activin A for only 5 days. Thus, the infusion window may have been too short to detect significant differences in cardiac fibrosis between groups of pregnant rats.

4.3. The Relationship between Activin A and Recognized Markers of Cardiac Dysfunction

Elevated β-MHC protein expression indicates the activation of the fetal gene program, which occurs in response to stress. The predominant sarcomeric proteins in the fetal heart switch from β-MHC to α-MHC in rodents to support the mechanical performance and efficiency of the heart ex utero [57]. Abnormal expression of these MHC isoforms was reported in cardiac hypertrophy and heart failure. Although β-MHC presents lower adenosine triphosphatase activity and lower filament sliding velocity, it can generate cross-bridge force with a higher economy of energy consumption than α-MHC, suggesting that a shift from α- to β-MHC may be an adaptative response in order to preserve energy. Furthermore, culminating evidence indicates that increased β-MHC expression decreases contractile function, eventually leading to cardiac dysfunction and dictating clinical outcomes in cardiac hypertrophy and heart failure [67]. These proteins were not measured in other models of increased activin A expression; however, we previously found that the reduced uterine perfusion pressure (RUPP) rat model of PE, which develops cardiac dysfunction following 5 days of placental ischemia-induced hypertension [68], also exhibits decreased cardiac α-MHC/β-MHC mRNA ratios [52]. The activin A-infused rats in the current study did not have elevated blood pressure, suggesting that activation of the fetal gene program is a direct effect of excess activin A on the heart.

Cardiac dysfunction was proposed as a state of reduced effectiveness of the natriuretic peptide system [69]. ANP and BNP, which are synthetized and secreted by cardiac myocytes, promote cardiovascular protection by antagonizing the actions of the renin–angiotensin–aldosterone system concerning blood pressure regulation and salt-water balance. The identification of these natriuretic peptides as sensitive markers of the cardiac load has profoundly impacted research into cardiovascular homeostasis and disease [70]. In the case of cardiac dysfunction, particularly in heart failure, despite dramatic increases in circulating levels of ANP and BNP, their effects become blunted [69]. Still, they serve as accurate diagnosis and prognosis markers for various cardiac disorders [58]. Increased circulating levels of ANP and BNP were described in PE patients, including those with reported impaired GLS [15,71–73]. Likewise, we and others found elevated ANP and BNP levels in the circulation and/or heart tissue of RUPP rats [52,74,75]. Previous studies demonstrated that activin A overexpression increases, whereas inhibition of ActIIR signaling following pressure overload or myocardial infarction decreases cardiac gene expression of both ANP and BNP in mice [62,63]. We were unable to detect a significant rise in either serum or heart levels of ANP and BNP. However, we showed that serum activin A is positively correlated to serum BNP. Since these earlier studies have measured ANP and BNP mRNA levels after 8–10 weeks of insult/treatment, the lack of effect of activin A on triggering marked natriuretic peptide release in the current study might be a consequence of the short duration of activin A infusion.

4.4. Study Limitations

Our study has some limitations that should be noted. For instance, continuous activin A infusion induced subclinical cardiac dysfunction in pregnant rats, similarly to what was reported in PE patients [9,10,12,14]. Whether this insult is sufficient to increase the risk for long-term cardiac disorders after PE is still unknown. As placental anti-angiogenic factors were also associated with cardiac dysfunction in PE, follow-up studies should examine whether an infusion of activin A with or without sFlt-1 and/or sEng into pregnant animals results in worsened GLS during and after gestation. Furthermore, future studies should determine whether placental and heart expression of activin A and ActIIR is increased in models of PE-induced cardiac dysfunction, such as in RUPP rats [52,68,74,75] and whether treatment with inhibitors of the activin A-ActIIR pathway improves GLS.

5. Conclusions

In conclusion, we found that circulating activin A in excess induces cardiac stress and subclinical dysfunction in pregnant rats, represented by increased heart β-MHC expression and decreased GLS. As the placenta is the major source of activin A during pregnancy and placental activin expression was positively correlated with circulating activin A levels in PE patients [29,31,37,41], our data provide evidence that activin A may serve as a causal link between placental abnormalities and cardiac dysfunction in PE.

Author Contributions: Conceptualization, J.P.G., S.S. and B.A.B.; methodology, J.P.G. and B.A.B.; validation, B.A.B., A.C.P. and U.B.; formal analysis, B.A.B., A.C.P., U.B. and Y.C.; investigation, B.A.B., A.C.P., U.B. and Y.C.; resources, J.P.G. and Y.C.; data curation, B.A.B. and A.C.P.; writing—original draft preparation, B.A.B. and A.C.P.; writing—review and editing, B.A.B. and A.C.P.; visualization, B.A.B., A.C.P. and U.B.; supervision, J.P.G.; project administration, B.A.B. and A.C.P.; funding acquisition, S.S., J.P.G. and B.A.B. All authors have read and agreed to the published version of the manuscript.

Funding: This research was funded by the National Institutes of Health (grant numbers, 5R01HL148191-02, P01HL051971, P20GM104357, R01HL139797, U54GM115428) and the American Heart Association (grant number, 18POST33990293).

Institutional Review Board Statement: The animal study protocol was approved by the Institutional Animal Use and Care Committee of the University of Mississippi Medical Centre (protocol code 1557 on 3 April 2019).

Informed Consent Statement: Not applicable.

Data Availability Statement: The authors confirm that data supporting the findings of this study are presented within the article. Raw data are available from the corresponding author upon reasonable request.

Conflicts of Interest: The authors declare no conflict of interest.

References

1. Morton, A. Physiological Changes and Cardiovascular Investigations in Pregnancy. *Heart Lung Circ.* **2021**, *30*, e6–e15. [CrossRef] [PubMed]
2. ACOG. Gestational Hypertension and Preeclampsia: ACOG Practice Bulletin Summary, Number 222. *Obstet. Gynecol.* **2020**, *135*, 1492–1495. [CrossRef] [PubMed]
3. Ananth, C.V.; Keyes, K.M.; Wapner, R.J. Pre-eclampsia rates in the United States, 1980–2010: Age-period-cohort analysis. *BMJ* **2013**, *347*, f6564. [CrossRef] [PubMed]
4. Hutcheon, J.A.; Lisonkova, S.; Joseph, K.S. Epidemiology of pre-eclampsia and the other hypertensive disorders of pregnancy. *Best Pract. Res. Clin. Obstet. Gynaecol.* **2011**, *25*, 391–403. [CrossRef]
5. Shraga, Y.; Pariente, G.; Rotem, R.; Baumfeld, Y.; Miodownik, S.; Weintraub, A.Y. Changes in trends over time for the specific contribution of different risk factors for pre-eclampsia. *Arch. Gynecol. Obstet.* **2020**, *302*, 977–982. [CrossRef] [PubMed]
6. Bosio, P.M.; McKenna, P.J.; Conroy, R.; O'Herlihy, C. Maternal central hemodynamics in hypertensive disorders of pregnancy. *Obstet. Gynecol.* **1999**, *94*, 978–984. [PubMed]
7. Rang, S.; Wolf, H.; van Montfrans, G.A.; Karemaker, J.M. Serial assessment of cardiovascular control shows early signs of developing pre-eclampsia. *J. Hypertens.* **2004**, *22*, 369–376. [CrossRef] [PubMed]
8. Salas, S.P.; Marshall, G.; Gutierrez, B.L.; Rosso, P. Time course of maternal plasma volume and hormonal changes in women with preeclampsia or fetal growth restriction. *Hypertension* **2006**, *47*, 203–208. [CrossRef]
9. Ajmi, H.; Abid, D.; Milouchi, S.; Louati, D.; Sghaier, A.; Choura, D.; Chaaben, K.; Abid, L.; Kammoun, S. Interest of speckle tracking in the detection of cardiac involvement in pregnant women with hypertensive disorder. *Pregnancy Hypertens.* **2018**, *11*, 136–141. [CrossRef]
10. Buddeberg, B.S.; Sharma, R.; O'Driscoll, J.M.; Kaelin Agten, A.; Khalil, A.; Thilaganathan, B. Cardiac maladaptation in term pregnancies with preeclampsia. *Pregnancy Hypertens.* **2018**, *13*, 198–203. [CrossRef]
11. Cong, J.; Fan, T.; Yang, X.; Shen, J.; Cheng, G.; Zhang, Z. Maternal cardiac remodeling and dysfunction in preeclampsia: A three-dimensional speckle-tracking echocardiography study. *Int. J. Cardiovasc. Imaging* **2015**, *31*, 1361–1368. [CrossRef] [PubMed]
12. Kraker, K.; Schütte, T.; O'Driscoll, J.; Birukov, A.; Patey, O.; Herse, F.; Müller, D.N.; Thilaganathan, B.; Haase, N.; Dechend, R. Speckle Tracking Echocardiography: New Ways of Translational Approaches in Preeclampsia to Detect Cardiovascular Dysfunction. *Int. J. Mol. Sci.* **2020**, *21*, 1162. [CrossRef] [PubMed]
13. Mostafavi, A.; Tase Zar, Y.; Nikdoust, F.; Tabatabaei, S.A. Comparison of left ventricular systolic function by 2D speckle-tracking echocardiography between normal pregnant women and pregnant women with preeclampsia. *J. Cardiovasc. Thorac. Res.* **2019**, *11*, 309–313. [CrossRef] [PubMed]
14. Shahul, S.; Rhee, J.; Hacker, M.R.; Gulati, G.; Mitchell, J.D.; Hess, P.; Mahmood, F.; Arany, Z.; Rana, S.; Talmor, D. Subclinical left ventricular dysfunction in preeclamptic women with preserved left ventricular ejection fraction: A 2D speckle-tracking imaging study. *Circ. Cardiovasc. Imaging* **2012**, *5*, 734–739. [CrossRef]
15. Yu, L.; Zhou, Q.; Peng, Q.; Yang, Z. Left ventricular function of patients with pregnancy-induced hypertension evaluated using velocity vector imaging echocardiography and N-terminal pro-brain natriuretic peptide. *Echocardiography* **2018**, *35*, 459–466. [CrossRef]
16. Bakrania, B.A.; Spradley, F.T.; Drummond, H.A.; LaMarca, B.; Ryan, M.J.; Granger, J.P. Preeclampsia: Linking Placental Ischemia with Maternal Endothelial and Vascular Dysfunction. *Compr. Physiol.* **2020**, *11*, 1315–1349.
17. Rana, S.; Lemoine, E.; Granger, J.P.; Karumanchi, S.A. Preeclampsia: Pathophysiology, Challenges, and Perspectives. *Circ. Res.* **2019**, *124*, 1094–1112. [CrossRef]
18. Ganzevoort, W.; Rep, A.; Bonsel, G.J.; de Vries, J.I.; Wolf, H. Plasma volume and blood pressure regulation in hypertensive pregnancy. *J. Hypertens.* **2004**, *22*, 1235–1242. [CrossRef]
19. Boardman, H.; Lamata, P.; Lazdam, M.; Verburg, A.; Siepmann, T.; Upton, R.; Bilderbeck, A.; Dore, R.; Smedley, C.; Kenworthy, Y.; et al. Variations in Cardiovascular Structure, Function, and Geometry in Midlife Associated With a History of Hypertensive Pregnancy. *Hypertension* **2020**, *75*, 1542–1550. [CrossRef]
20. deMartelly, V.A.; Dreixler, J.; Tung, A.; Mueller, A.; Heimberger, S.; Fazal, A.A.; Naseem, H.; Lang, R.; Kruse, E.; Yamat, M.; et al. Long-Term Postpartum Cardiac Function and Its Association With Preeclampsia. *J. Am. Heart Assoc.* **2021**, *10*, e018526. [CrossRef]

21. Shahul, S.; Ramadan, H.; Nizamuddin, J.; Mueller, A.; Paterl, V.; Dreixler, J.; Tung, A.; Lang, R.M.; Weinert, L.; Nasim, R.; et al. Activin A and Late Postpartum Cardiac Dysfunction Among Women With Hypertensive Disorders of Pregnancy. *Hypertension* **2018**, *72*, 188–193. [CrossRef] [PubMed]
22. Bovee, E.M.; Gulati, M.; Maas, A.H. Novel Cardiovascular Biomarkers Associated with Increased Cardiovascular Risk in Women With Prior Preeclampsia/HELLP Syndrome: A Narrative Review. *Eur. Cardiol.* **2021**, *16*, e36. [CrossRef] [PubMed]
23. Muttukrishna, S.; Tannetta, D.; Groome, N.; Sargent, I. Activin and follistatin in female reproduction. *Mol. Cell. Endocrinol.* **2004**, *225*, 45–56. [CrossRef]
24. Fukushima, N.; Matsuura, K.; Akazawa, H.; Honda, A.; Nagai, T.; Takahashi, T.; Seki, A.; Murasaki, K.M.; Shimizu, T.; Okano, T.; et al. A crucial role of activin A-mediated growth hormone suppression in mouse and human heart failure. *PLoS ONE* **2011**, *6*, e27901. [CrossRef] [PubMed]
25. Mahmoudabady, M.; Mathieu, M.; Dewachter, L.; Hadad, I.; Ray, L.; Jespers, P.; Brimioulle, S.; Naeije, R.; McEntee, K. Activin-A, transforming growth factor-beta, and myostatin signaling pathway in experimental dilated cardiomyopathy. *J. Card. Fail.* **2008**, *14*, 703–709. [CrossRef]
26. Wei, Q.; Wang, Y.; Liu, H.; Yang, J.; Yang, C.; Liu, M.; Liu, Y.; Yang, P.; Liu, Z. The expression and role of activin A and follistatin in heart failure rats after myocardial infarction. *Int. J. Cardiol.* **2013**, *168*, 2994–2997. [CrossRef] [PubMed]
27. Li, S.; Li, Z.; Yin, R.; Nie, J.; Fu, Y.; Ying, R. Knockdown of dual oxidase 1 suppresses activin A-induced fibrosis in cardiomyocytes via the reactive oxygen species-dependent pyroptotic pathway. *Int. J. Biochem. Cell Biol.* **2021**, *131*, 105902. [CrossRef]
28. Yndestad, A.; Ueland, T.; Øie, E.; Florholmen, G.; Halvorsen, B.; Attramadal, H.; Simonsen, S.; Frøland, S.S.; Gullestad, L.; Christensen, G.; et al. Elevated levels of activin A in heart failure: Potential role in myocardial remodeling. *Circulation* **2004**, *109*, 1379–1385. [CrossRef]
29. Bersinger, N.A.; Groome, N.; Muttukrishna, S. Pregnancy-associated and placental proteins in the placental tissue of normal pregnant women and patients with pre-eclampsia at term. *Eur. J. Endocrinol.* **2002**, *147*, 785–793. [CrossRef]
30. Bersinger, N.A.; Smarason, A.K.; Muttukrishna, S.; Groome, N.P.; Redman, C.W. Women with preeclampsia have increased serum levels of pregnancy-associated plasma protein A (PAPP-A), inhibin A, activin A and soluble E-selectin. *Hypertens. Pregnancy* **2003**, *22*, 45–55. [CrossRef]
31. Casagrandi, D.; Bearfield, C.; Geary, J.; Redman, C.W.; Muttukrishna, S. Inhibin, activin, follistatin, activin receptors and beta-glycan gene expression in the placental tissue of patients with pre-eclampsia. *Mol. Hum. Reprod.* **2003**, *9*, 199–203. [CrossRef] [PubMed]
32. D'Antona, D.; Reis, F.M.; Benedetto, C.; Evans, L.W.; Groome, N.P.; de Kretser, D.M.; Wallace, E.M.; Petraglia, F. Increased maternal serum activin A but not follistatin levels in pregnant women with hypertensive disorders. *J. Endocrinol.* **2000**, *165*, 157–162. [CrossRef] [PubMed]
33. Diesch, C.H.; Holzgreve, W.; Hahn, S.; Zhong, X.Y. Comparison of activin A and cell-free fetal DNA levels in maternal plasma from patients at high risk for preeclampsia. *Prenat. Diagn.* **2006**, *26*, 1267–1270. [CrossRef] [PubMed]
34. Florio, P.; Ciarmela, P.; Luisi, S.; Palumbo, M.A.; Lambert-Messerlian, G.; Severi, F.M.; Petraglia, F. Pre-eclampsia with fetal growth restriction: Placental and serum activin A and inhibin A levels. *Gynecol. Endocrinol.* **2002**, *16*, 365–372. [CrossRef]
35. Keelan, J.A.; Taylor, R.; Schellenberg, J.C.; Groome, N.P.; Mitchell, M.D.; North, R.A. Serum activin A, inhibin A, and follistatin concentrations in preeclampsia or small for gestational age pregnancies. *Obstet. Gynecol.* **2002**, *99*, 267–274. [PubMed]
36. Laivuori, H.; Kaaja, R.; Turpeinen, U.; Stenman, U.H.; Ylikorkala, O. Serum activin A and inhibin A elevated in pre-eclampsia: No relation to insulin sensitivity. *Br. J. Obstet. Gynaecol.* **1999**, *106*, 1298–1303. [CrossRef] [PubMed]
37. Manuelpillai, U.; Schneider-Kolsky, M.; Dole, A.; Wallace, E.M. Activin A and activin receptors in gestational tissue from preeclamptic pregnancies. *J. Endocrinol.* **2001**, *171*, 57–64. [CrossRef]
38. Muttukrishna, S.; Hyett, J.; Paine, M.; Moodley, J.; Groome, N.; Rodeck, C. Uterine vein and maternal urinary levels of activin A and inhibin A in pre-eclampsia patients. *Clin. Endocrinol.* **2006**, *64*, 469–473. [CrossRef]
39. Muttukrishna, S.; Knight, P.G.; Groome, N.P.; Redman, C.W.; Ledger, W.L. Activin A and inhibin A as possible endocrine markers for pre-eclampsia. *Lancet* **1997**, *349*, 1285–1288. [CrossRef]
40. Petraglia, F.; Aguzzoli, L.; Gallinelli, A.; Florio, P.; Zonca, M.; Benedetto, C.; Woodruff, K. Hypertension in pregnancy: Changes in activin A maternal serum concentration. *Placenta* **1995**, *16*, 447–454. [CrossRef]
41. Silver, H.M.; Lambert-Messerlian, G.M.; Reis, F.M.; Diblasio, A.M.; Petraglia, F.; Canick, J.A. Mechanism of increased maternal serum total activin a and inhibin a in preeclampsia. *J. Soc. Gynecol. Investig.* **2002**, *9*, 308–312. [CrossRef]
42. Silver, H.M.; Lambert-Messerlian, G.M.; Star, J.A.; Hogan, J.; Canick, J.A. Comparison of maternal serum total activin A and inhibin A in normal, preeclamptic, and nonproteinuric gestationally hypertensive pregnancies. *Am. J. Obstet. Gynecol.* **1999**, *180*, 1131–1137. [CrossRef]
43. Yair, D.; Eshed-Englender, T.; Kupferminc, M.J.; Geva, E.; Frenkel, J.; Sherman, D. Serum levels of inhibin B, unlike inhibin A and activin A, are not altered in women with preeclampsia. *Am. J. Reprod. Immunol.* **2001**, *45*, 180–187. [CrossRef]
44. Ay, E.; Kavak, Z.N.; Elter, K.; Gokaslan, H.; Pekin, T. Screening for pre-eclampsia by using maternal serum inhibin A, activin A, human chorionic gonadotropin, unconjugated estriol, and alpha-fetoprotein levels and uterine artery Doppler in the second trimester of pregnancy. *Aust. N. Z. J. Obstet. Gynaecol.* **2005**, *45*, 283–288. [CrossRef]

45. Florio, P.; Reis, F.M.; Pezzani, I.; Luisi, S.; Severi, F.M.; Petraglia, F. The addition of activin A and inhibin A measurement to uterine artery Doppler velocimetry to improve the early prediction of pre-eclampsia. *Ultrasound Obstet. Gynecol.* **2003**, *21*, 165–169. [CrossRef]
46. Madazli, R.; Kuseyrioglu, B.; Uzun, H.; Uludag, S.; Ocak, V. Prediction of preeclampsia with maternal mid-trimester placental growth factor, activin A, fibronectin and uterine artery Doppler velocimetry. *Int. J. Gynaecol. Obstet.* **2005**, *89*, 251–257. [CrossRef]
47. Muttukrishna, S.; North, R.A.; Morris, J.; Schellenberg, J.C.; Taylor, R.S.; Asselin, J.; Ledger, W.; Groome, N.; Redman, C.W. Serum inhibin A and activin A are elevated prior to the onset of pre-eclampsia. *Hum. Reprod.* **2000**, *15*, 1640–1645. [CrossRef]
48. Ong, C.Y.; Liao, A.W.; Munim, S.; Spencer, K.; Nicolaides, K.H. First-trimester maternal serum activin A in pre-eclampsia and fetal growth restriction. *J. Matern. Fetal Neonatal Med.* **2004**, *15*, 176–180. [CrossRef]
49. Spencer, K.; Yu, C.K.; Savvidou, M.; Papageorghiou, A.T.; Nicolaides, K.H. Prediction of pre-eclampsia by uterine artery Doppler ultrasonography and maternal serum pregnancy-associated plasma protein-A, free beta-human chorionic gonadotropin, activin A and inhibin A at 22 + 0 to 24 + 6 weeks' gestation. *Ultrasound Obstet. Gynecol.* **2006**, *27*, 658–663. [CrossRef]
50. Yu, J.; Shixia, C.Z.; Wu, Y.; Duan, T. Inhibin A, activin A, placental growth factor and uterine artery Doppler pulsatility index in the prediction of pre-eclampsia. *Ultrasound Obstet. Gynecol.* **2011**, *37*, 528–533. [CrossRef]
51. Lim, R.; Acharya, R.; Delpachitra, P.; Hobson, S.; Sobey, C.G.; Drummond, G.R.; Wallace, E.M. Activin and NADPH-oxidase in preeclampsia: Insights from in vitro and murine studies. *Am. J. Obstet. Gynecol.* **2015**, *212*, 86.e1–86.e12. [CrossRef] [PubMed]
52. Gutkowska, J.; Granger, J.P.; Lamarca, B.B.; Danalache, B.A.; Wang, D.; Jankowski, M. Changes in cardiac structure in hypertension produced by placental ischemia in pregnant rats: Effect of tumor necrosis factor blockade. *J. Hypertens.* **2011**, *29*, 1203–1212. [CrossRef] [PubMed]
53. Shang, L.; Yue, W.; Wang, D.; Weng, X.; Hall, M.E.; Xu, Y.; Hou, M.; Chen, Y. Systolic overload-induced pulmonary inflammation, fibrosis, oxidative stress and heart failure progression through interleukin-1beta. *J. Mol. Cell. Cardiol.* **2020**, *146*, 84–94. [CrossRef]
54. Wang, H.; Xu, X.; Fassett, J.; Kwak, D.; Liu, X.; Hu, X.; Falls, T.J.; Bell, J.B.; Li, H.; Bitterman, P.; et al. Double-stranded RNA-dependent protein kinase deficiency protects the heart from systolic overload-induced congestive heart failure. *Circulation* **2014**, *129*, 1397–1406. [CrossRef]
55. Wang, H.; Hou, L.; Kwak, D.; Fassett, J.; Xu, X.; Chen, A.; Chen, W.; Blazar, B.R.; Xu, Y.; Hall, J.L.; et al. Increasing Regulatory T Cells With Interleukin-2 and Interleukin-2 Antibody Complexes Attenuates Lung Inflammation and Heart Failure Progression. *Hypertension* **2016**, *68*, 114–122. [CrossRef] [PubMed]
56. Yamazaki, T.; Komuro, I.; Shiojima, I.; Yazaki, Y. The molecular mechanism of cardiac hypertrophy and failure. *Ann. N. Y. Acad. Sci.* **1999**, *874*, 38–48. [CrossRef]
57. Taegtmeyer, H.; Sen, S.; Vela, D. Return to the fetal gene program: A suggested metabolic link to gene expression in the heart. *Ann. N. Y. Acad. Sci.* **2010**, *1188*, 191–198. [CrossRef]
58. Mueller, C.; McDonald, K.; de Boer, R.A.; Maisel, A.; Cleland, J.G.F.; Kozhuharov, N.; Coats, A.J.S.; Metra, M.; Mebazaa, A.; Ruschitzka, F.; et al. Heart Failure Association of the European Society of Cardiology practical guidance on the use of natriuretic peptide concentrations. *Eur. J. Heart Fail.* **2019**, *21*, 715–731. [CrossRef]
59. Garrido-Gimenez, C.; Mendoza, M.; Cruz-Lemini, M.; Galian-Gay, L.; Sanchez-Garcia, O.; Granato, C.; Rodriguez-Sureda, V.; Rodriguez-Palomares, J.; Carreras-Moratonas, E.; Cabero-Roura, L.; et al. Angiogenic Factors and Long-Term Cardiovascular Risk in Women That Developed Preeclampsia During Pregnancy. *Hypertension* **2020**, *76*, 1808–1816. [CrossRef]
60. Shahul, S.; Medvedofsky, D.; Wenger, J.B.; Nizamuddin, J.; Brown, S.M.; Bajracharya, S.; Salahuddin, S.; Thadhani, R.; Mueller, A.; Tung, A.; et al. Circulating Antiangiogenic Factors and Myocardial Dysfunction in Hypertensive Disorders of Pregnancy. *Hypertension* **2016**, *67*, 1273–1280. [CrossRef]
61. Kurtz, T.W.; Griffin, K.A.; Bidani, A.K.; Davisson, R.L.; Hall, J.E. Subcommittee of Professional and Public Education of the American Heart Association. Recommendations for blood pressure measurement in humans and experimental animals. Part 2: Blood pressure measurement in experimental animals: A statement for professionals from the subcommittee of professional and public education of the American Heart Association council on high blood pressure research. *Hypertension* **2005**, *45*, 299–310. [PubMed]
62. Roh, J.D.; Hobson, R.; Chaudhari, V.; Quintero, P.; Yeri, A.; Benson, M.; Xiao, C.; Zlotoff, D.; Bezzerides, V.; Houstis, N.; et al. Activin type II receptor signaling in cardiac aging and heart failure. *Sci. Transl. Med.* **2019**, *11*, eaau8680. [CrossRef] [PubMed]
63. Castillero, E.; Akashi, H.; Najjar, M.; Ji, R.; Brandstetter, L.M.; Wang, C.; Liao, X.; Zhang, X.; Sperry, A.; Gailes, M.; et al. Activin type II receptor ligand signaling inhibition after experimental ischemic heart failure attenuates cardiac remodeling and prevents fibrosis. *Am. J. Physiol. Heart Circ. Physiol.* **2020**, *318*, H378–H390. [CrossRef] [PubMed]
64. Haland, T.F.; Almaas, V.M.; Hasselberg, N.E.; Saberniak, J.; Leren, I.S.; Hopp, E.; Edvardsen, T.; Haugaa, K.H. Strain echocardiography is related to fibrosis and ventricular arrhythmias in hypertrophic cardiomyopathy. *Eur. Heart J. Cardiovasc. Imaging* **2016**, *17*, 613–621. [CrossRef] [PubMed]
65. Xie, T.; Han, L.; Chen, Y.; Wu, H. Progranulin and Activin A Concentrations are Elevated in Serum from Patients with Acute Exacerbations of Idiopathic Pulmonary Fibrosis. *Lung* **2021**, *199*, 467–473. [CrossRef] [PubMed]
66. Hu, J.; Wang, X.; Wei, S.M.; Tang, Y.H.; Zhou, Q.; Huang, C.X. Activin A stimulates the proliferation and differentiation of cardiac fibroblasts via the ERK1/2 and p38-MAPK pathways. *Eur. J. Pharmacol.* **2016**, *789*, 319–327. [CrossRef]
67. Krenz, M.; Robbins, J. Impact of beta-myosin heavy chain expression on cardiac function during stress. *J. Am. Coll. Cardiol.* **2004**, *44*, 2390–2397. [CrossRef]

68. Bakrania, B.A.; Hall, M.E.; Shahul, S.; Granger, J.P. The Reduced Uterine Perfusion Pressure (RUPP) rat model of preeclampsia exhibits impaired systolic function and global longitudinal strain during pregnancy. *Pregnancy Hypertens.* **2019**, *18*, 169–172. [CrossRef]
69. Diez, J. Chronic heart failure as a state of reduced effectiveness of the natriuretic peptide system: Implications for therapy. *Eur. J. Heart Fail.* **2017**, *19*, 167–176. [CrossRef]
70. Kuwahara, K. The natriuretic peptide system in heart failure: Diagnostic and therapeutic implications. *Pharmacol. Ther.* **2021**, *227*, 107863. [CrossRef]
71. Grunewald, C.; Nisell, H.; Carlstrom, K.; Kublickas, M.; Randmaa, I.; Nylund, L. Acute volume expansion in normal pregnancy and preeclampsia. Effects on plasma atrial natriuretic peptide (ANP) and cyclic guanosine monophosphate (cGMP) concentrations and feto-maternal circulation. *Acta Obstet. Gynecol. Scand.* **1994**, *73*, 294–299. [CrossRef] [PubMed]
72. Itoh, H.; Sagawa, N.; Nanno, H.; Mori, T.; Mukoyama, M.; Itoh, H.; Nakao, K. Impaired guanosine 3′,5′-cyclic phosphate production in severe pregnancy-induced hypertension with high plasma levels of atrial and brain natriuretic peptides. *Endocr. J.* **1997**, *44*, 389–393. [CrossRef]
73. Sandrim, V.C.; Palei, A.C.; Sertorio, J.T.; Amaral, L.M.; Cavalli, R.C.; Tanus-Santos, J.E. Alterations in cyclic GMP levels in preeclampsia may reflect increased B-type natriuretic peptide levels and not impaired nitric oxide activity. *Clin. Biochem.* **2011**, *44*, 1012–1014. [CrossRef] [PubMed]
74. Ding, L.; Bai, C.; Liu, Y. Interleukin-6 contributes to myocardial damage in pregnant rats with reduced uterine perfusion pressure. *Braz. J. Med. Biol. Res.* **2018**, *51*, e6921. [CrossRef] [PubMed]
75. Richards, C.; Sesperez, K.; Chhor, M.; Ghorbanpour, S.; Rennie, C.; Ming, C.L.C.; Evenhuis, C.; Nikolic, V.; Orlic, N.K.; Mikovic, K.; et al. Characterisation of cardiac health in the reduced uterine perfusion pressure model and a 3D cardiac spheroid model, of preeclampsia. *Biol. Sex Differ.* **2021**, *12*, 31. [CrossRef]

cells MDPI

Review

The Relationship between Obesity and Pre-Eclampsia: Incidental Risks and Identification of Potential Biomarkers for Pre-Eclampsia

Talitha Abraham and Andrea M. P. Romani *

Department of Physiology and Biophysics, School of Medicine, Case Western Reserve University, 10900 Euclid Avenue, Cleveland, OH 44106-4970, USA; talitha.abraham@case.edu
* Correspondence: amr5@po.cwru.edu; Tel.: +1-216-3681625

Abstract: Obesity has been steadily increasing over the past decade in the US and worldwide. Since 1975, the prevalence of obesity has increased by 2% per decade, unabated despite new and more stringent guidelines set by WHO, CDC, and other public health organizations. Likewise, maternal obesity has also increased worldwide over the past several years. In the United States, pre-pregnancy rates have increased proportionally across all racial groups. Obesity during pregnancy has been directly linked to obstetric complications including gestational diabetes, HTN, hematomas, pre-eclampsia, and congenital defects. In the particular case of pre-eclampsia, the incidence rate across the globe is 2.16%, but the condition accounts for 30% of maternal deaths, and a robust body of evidence underscored the relationship between obesity and pre-eclampsia. More recently, attention has focused on the identification of reliable biomarkers predictive of an elevated risk for pre-eclampsia. The aim of this literature review is to elucidate the relationship between obesity and these predictive biomarkers for future prediction and prevention of pre-eclampsia condition in women at risk.

Keywords: pre-eclampsia; obesity; biomarkers; adipokines; adiponectin; leptin; ROS; angiogenic factors

Citation: Abraham, T.; Romani, A.M.P. The Relationship between Obesity and Pre-Eclampsia: Incidental Risks and Identification of Potential Biomarkers for Pre-Eclampsia. *Cells* **2022**, *11*, 1548. https://doi.org/10.3390/cells11091548

Academic Editor: Ana T. Palei

Received: 20 December 2021
Accepted: 3 May 2022
Published: 5 May 2022

1. Introduction

Obesity has been steadily increasing over the past decade in the U.S. and worldwide [1]. Starting in 1975, the prevalence of obesity has increased by 2% per decade [1] but has remained largely dismissed or ignored until a decade ago, when progressively more attention has been paid to this public health condition owing to the large number of individuals affected and the associated complications [1,2]. First reported as an epidemic in the U.S., obesity quickly earned the label of pandemic due to its rapid progression and spreading to other developed and developing nations [3]. A rapid increase in upward trend was first noted as early as 1988, and since then prevalence has increased from 22.9% to 30.5% by the year 2000 [1,2]. Concomitantly, morbid obesity has also increased by 2%. Recent reports from the WHO and the CDC indicate that approximately 1 billion people worldwide present with obesity at various stages, regardless of age, gender, and ethnicity [4–6]. According to the National Institutes of Health (NIH), obesity is the second leading cause of preventable death in the U.S. causing 300,000 deaths annually [7].

Maternal obesity has also increased significantly over the past decade, in line with the general uptrend of obesity. In 1980, 29.8% of women were overweight or obese, and by 2013 the incidence rate had increased to 38% in both high- and middle-income countries [8]. In the U.S., pre-pregnancy rates have increased by 11% between 2016 and 2019, and this increase has occurred proportionally across all racial groups [2]. Obesity during pregnancy has been linked primarily to obstetric complications including gestational diabetes, HTN, hematomas, pre-eclampsia, and congenital defects [9]. Further, a strong association has been observed between increasing BMI and risk of still-birth, with obesity being responsible for 25% of stillbirths between 37- and 42-week gestation [10].

A significant body of studies has investigated the systemic effects of obesity, its role in multiple comorbidities, and the effects of high BMI on mortality [11]. Research has also highlighted the relationship between obesity and pre-eclampsia [12,13]. At the same time, a large body of research has attempted to identify possible biomarkers predictive of an elevated risk for pre-eclampsia. Incidence of several adverse pregnancy outcomes are associated with increased weight gain. This is particularly true for pre-eclampsia. Pre-eclampsia is observed in 2–8% of pregnancies globally. Multiple studies have confirmed that maternal obesity increases the risk of pre-eclampsia by three-four times when compared to normo-weight mothers [14–16]. This increased risk is especially concerning if we consider that at front of an average incidence rate of 2.16% across the globe as an obstetric complication, pre-eclampsia accounts for 30% of maternal deaths [17]. By monitoring more than 39 million births over 10 years, the National Hospital Discharge Survey has estimated to ~6% the incidence of hypertensive disorders in pregnancy, which include pre-eclampsia, eclampsia, gestational hypertension and chronic hypertension, and between 3- and 25-fold the increased risk of severe pregnancy complications in women with pre-eclampsia and eclampsia [18].

The goal of this literature review is to elucidate the current notion in the field that obesity is a major risk factor for pre-eclampsia, and discuss the potential biomarkers that can predict pre-eclampsia onset and progression in over-weight mothers as compared to normo-weight ones. In particular, this review aims at determining whether abnormalities in the levels of the currently most accepted biomarkers for pre-eclampsia are related, directly or indirectly, to obesity and/or the underlying dysmetabolic conditions associated with weight gain, overweight, or frank obesity in pregnant women.

For this review, recent literature was collected using online databases (MEDLINE and BIOMED) searches. Used keywords included: gestational weight gain, sugary dietary consumption, pre-eclampsia risk, and biomarkers, such as adiponectin, leptin, resistin, vascular endothelial growth factor (VEGF), placental growth factor (PlGF), soluble fms-like tyrosine kinase (sFlt1), soluble endoglin (sEng), oxidative stress, and HLA-antigens. A total of 632 articles were initially obtained. Articles were excluded based on the following: publication date prior to 2000 (to be more in tune with the current biomedical perception of the disease); studies addressing only obesity or pre-eclampsia but not both conditions; assessment of only long term implications for mother and child; studies with limited or too small subject groups, or with subject groups selected based on criteria that strictly applied to particular geographical areas (e.g., exposure to local diseases, limited access to health care, reliance on traditional medicine, cultural beliefs); studies using exclusively age or socioeconomic status of the mothers without mention or consideration of biological risk-factors; studies conducted on non-human subjects only; absence or incomplete reporting of results. These exclusion criteria (Figure 1) reduced the number of utilizable articles to 63 original articles plus 8 review articles. Of the 63 articles, 8 addressed adipokines, 6 use of sugary dietary products, 14 serum factors, 2 histocompatibility antigens, 23 obesity and 10 pre-eclampsia or eclampsia. The references listed include populations from Northern Sweden, USA, Tanzania, China, Northwest Ethiopia, Saudi Arabia, Northeast Brazil, Norway, Denmark, and Australia, with sample size varying from 60 to 33,000. The major findings of each study reported in Table 1 will be discussed in this review.

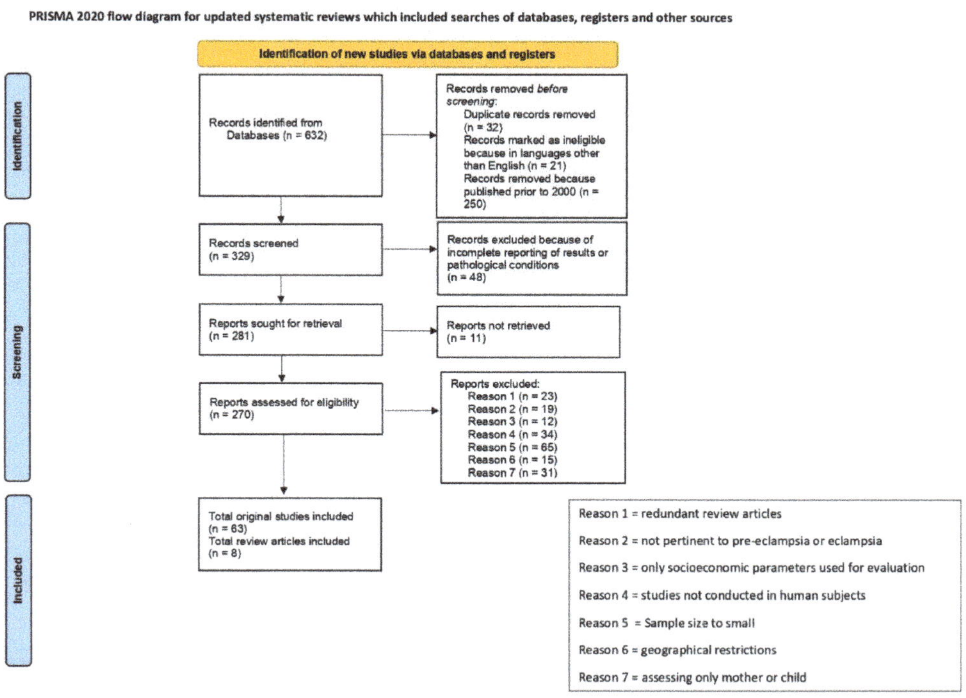

Figure 1. The flow chart illustrates the number of articles identified through the literature search and the inclusion/exclusion screening criteria to select the pertinent articles to be included in the review.

Table 1. The articles used for this review are ordered by the reference number by which the articles were introduced and discussed in the manuscript (see list of abbreviations at the end of the review for clarification).

Researchers (Reference)	Publication Date	Country	Sample	Study Type	Goal	Findings
Heslehurst, N.; Simpson, H.; Ells, L.J.; Rankin, J.; Wilkinson, J.; Lang, R.; Brown, T.J.; Summerbell, C.D. [9]	2008	UK	Obese women with proper comparison group	Meta-analysis of cohort studies of pregnant women from 16 weeks of pregnancy to delivery	To investigate relationships between obesity and impact on obstetric care	The study shows a significant relationship between obesity and increased odds of C-section and other delivery complications including longer duration of hospital stay and increased neonatal intensive care requirement. Maternal obesity significantly contributes to a poorer prognosis for mother and baby
Yao, R.; Ananth, C.V.; Park, B.Y.; Pereira, L.; Plante, L.A. [10]	2014	US	Almost 3 million singleton births were evaluated	Population-based retrospective cohort study	To examine the association of maternal pre-pregnancy body mass index and risk of stillbirth	The results of this study indicate an increase in the risk of stillbirth with increasing BMI. The association is strongest at early- and late-term gestation periods. Extreme maternal obesity is a significant risk factor for stillbirth
Afshin, A.; Forouzanfar, M.H.; Reitsma, M.B.; Sur, P.; Estep, K.; Lee, A.; Marczak, L.; Mokdad, A.H.; Moradi-Lakeh, M.; Naghavi, M. et al. [11]	2017	195 countries	Data from 68.5 million persons in 195 countries	Data analysis: 1990 to 2015	To assess the trends in the prevalence of overweight and obesity among children and adults between 1980 and 2015	The study shows that more than two thirds of deaths related to high BMI were due to cardiovascular disease
Lisonkova, S.; Joseph, K.S. [12]	2013	US	456,668	Cox and logistic regression models	To examine the gestational age–specific incidence of pre-eclampsia onset and identify the associated risk factors and birth outcomes	The study indicates that early-and late-onset pre-eclampsia shares some etiological features, differ with regard to several risk factors, and lead to different outcomes. The 2 pre-eclampsia types should be treated as distinct entities from an etiological and prognostic standpoint

Table 1. *Cont.*

Researchers (Reference)	Publication Date	Country	Sample	Study Type	Goal	Findings
Rasmussen, S.; Irgens, L.M.; Espinoza, J. [13]	2014	Norway	77,294 singleton pregnancies	Observational study comparing birthweight percentiles and z-scores	To assess whether excess of LGA neonates in pre-eclamptic women delivering at term is attributable to maternal obesity	The study indicates that accelerated fetal growth is observed in a subset of pre-eclamptic women delivering at term. The excess of LGA neonates is attributable to maternal obesity
Mbah, A.; Kornosky, J.; Kristensen, S.; August, E.; Alio, A.; Marty, P.; Belogolovkin, V.; Bruder, K.; Salihu, H. [14]	2010	US	All singleton live births in the state of Missouri from 1989 to 2005	Logistic regression models with adjustment for intra-cluster correlation	To determine the association between obesity and the risk of pre-eclampsia	The study shows that the rate of pre-eclampsia increases with increasing BMI. Obese women ($BMI \geq 30\ kg/m^2$) have a higher risk for pre-eclampsia. Super-obese women have the highest incidence (13.4%)
Bodnar, L.M.; Catov, J.M.; Klebanoff, M.A.; Ness, R.B.; Roberts, J.M. [15]	2007	US	38,188 pregnant women	Prospective cohort study	To assess the association of pre-pregnancy BMI with severe and mild pre-eclampsia and transient hypertension of pregnancy	The study identifies a monotonic, dose-response relation between pre-pregnancy BMI and risk of both severe and mild pre-eclampsia as well as the risk of severe and mild transient hypertension of pregnancy
Bodnar, L.M.; Ness, R.B.; Markovic, N.; Roberts, J.M. [16]	2005	US	1179 primiparous women	Prospective cohort study	To explore the relation between pre-pregnancy BMI and the risk of pre-eclampsia	The study indicates that the risk of pre-eclampsia rises with increasing pre-pregnancy body mass index (from 15 to 30)
Abalos, E.; Cuesta, C.; Carroli, G.; Qureshi, Z.; Widmer, M.; Vogel, J.P.; Souza, J.P. [17]	2014	Multi-country: 29 countries from Africa, Asia, Latin America and the Middle East.	357 health facilities conducting 1000 or more deliveries annually	Secondary analysis of the WHOMCS database	To assess the incidence of hypertensive disorders of pregnancy and related severe complications	This WHOMCS on maternal and newborn health research network identifies hypertensive disorders of pregnancy as pre-eclampsia, eclampsia as adverse, life-threatening maternal and perinatal outcomes

Table 1. *Cont.*

Researchers (Reference)	Publication Date	Country	Sample	Study Type	Goal	Findings
Zhang, J.; Meikle, S.; Trumble, A. [19]	2003	US	300,000 deliveries assessed	Data Analysis	To study the incidence of severe maternal morbidity associated with hypertensive disorders of pregnancy in the US	The study shows that pre-eclampsia and eclampsia carry a high risk for severe maternal morbidity. Compared to Caucasians, African Americans have higher incidence of hypertensive disorders in pregnancy and suffer from more severe complications
Funai, E.F.; Friedlander, Y.; Paltiel, O.; Tiram, E.; Xue, X.; Deutsch, L.; Harlap, S. [20]	2005	Israel	37,061 women	Cox-proportional model	To investigate the long-term risk of mortality in women with pre-eclampsia	The study indicates that among women with pre-eclampsia who have subsequent births without pre-eclampsia, the excess risk of mortality became manifest only after 20 years
Wahabi, H.A.; Fayed, A.A.; Alzeidan, R.A.; Mandil, A.A. [21]	2014	Saudi Arabia	2701 women	Retrospective Study	To investigate the independent effect of GDM and obesity on the adverse pregnancy outcomes at term	The study shows a significant increase in the percentage of macrosomia, high birth weight, and pre-eclampsia in women with GDM and obesity. The study also shows a two- fold increase in C-section delivery in obese women
Mrema, D.; Lie, R.T.; Østbye, T.; Mahande, M.J.; Daltveit, A.K. [22]	2018	Tanzania	17,738 singleton birth women	Multi-variable analysis of registry based data	To examine the association between pre pregnancy BMI and the risk of pre-eclampsia in Tanzania	The study indicates that pre-pregnancy maternal overweight and obesity are associated with an increased risk of pre-eclampsia

Table 1. *Cont.*

Researchers (Reference)	Publication Date	Country	Sample	Study Type	Goal	Findings
Sohlberg, S.; Stephansson, O.; Cnattingius, S.; Wikström, A.K. [23]	2012	Sweden	503,179 nulliparous women	Population-based Cohort study	To determine whether BMI has an effect on pre-eclampsia of all severities	The study shows that short maternal stature and high BMI increase risks of pre-eclampsia of all severities. The associations is especially strong between short stature and severe types of pre-eclampsia, and high BMI and mild types of pre-eclampsia
Dantas, E.M.D.M.; Pereira, F.V.M.; Queiroz, J.W.; Dantas, D.L.D.M.; Monteiro, G.R.G.; Duggal, P.; Azevedo, M.D.F.; Jeronimo, S.M.B.; Araujo, A.C.P.F. [24]	2013	Brazil	242 women	Prospective case control study	To determine the frequency of and risk factors for pre-eclampsia in a low income population	The study indicates that women with pre-eclampsia develope chronic hypertension more often than normotensive controls
Shi, P.; Liu, A.; Yin, X. [25]	2021	China	1606 with GDM	Retrospective Cohort	To examine association between gestational weight gain in women with GDM and adverse pregnancy outcomes	The study indicates higher risk for pre-eclampsia and pregnancies complicated by hypertension in women with higher BMIs and high rates of gestational weight gain (above IOM guidelines)
Shao, Y.; Qiu, J.; Huang, H.; Mao, B.; Dai, W.; He, X.; Cui, H.; Lin, X.; Ly, L.; Wang, D.; Tang, Z.; Xu, S.; Zhao, N.; Zhou, M.; Xu, X.; Qiu, W.; Liu, Q.; Zhang, Y. [26]	2017	Lanzhou, China	9516	Cohort Study	To evaluate independent and joint effects of pre-pregnancy BMI and GWG on pre-eclampsia and its subtypes	The study shows that women overweight or obese have an increased risk for pre-eclampsia. Women with higher GWG also present with increased risk for pre-eclampsia. Similar increased risk was reported for all subtypes of pre-eclampsia. The highest risk for pre-eclampsia was observed to be directly proportional to the level of weight gain during gestation

Table 1. *Cont.*

Researchers (Reference)	Publication Date	Country	Sample	Study Type	Goal	Findings
Endershaw, M.; Abebe, F.; Worku, S.; Menber, L.; Assress, M.; Assefa, M. [27]	2016	Northwest Ethiopia	151 Pregnant women; 302 controls	Case-control study	To estimate the effect of obesity and dietary habits on pre-eclampsia	The study indicates that the risk of pre-eclampsia is higher among obese women compared to leaner women. The effect of obesity on pre-eclampsia is significant in women younger than 35 y.o. Folate supplementation is associated with a reduced risk of pre-eclampsia.
Meander, L.; Lindqvist, M.; Mogren, I.; Sandlund, J.; West, C.E.; Domellöf, M. [28]	2021	North Sweden	2203	Epidemiological study	To examine the level of physical activity and sedentary time in the Sweden population and explore effects of gestation weight gain, mode of delivery, birth weight of the child, and blood loss	The study shows that higher levels of physical activity are associated with reduced risk of emergency C-section and low gestational weight gain. Only 27.3% of the women considered in the sample achieve recommended level of physical activity, which is associated with more favorable pregnancy outcomes
Borgen, I.; Aamodt, G.; Harsem, H.; Haugen, M.; Meltzer, H.M.; Brantsaeter, A.L. [29]	2012	Norway	32,933 nulliparous women	Mother and Child Cohort Study	To determine whether maternal sugar consumption increases the risk of pre-eclampsia in nulliparous Norwegian women	The study indicates that sugar-sweetened carbonated and non-carbonated beverages are significantly associated with increased risk of pre-eclampsia, both independently and combined
Schoenaker, D.A.J.M.; Soedamah-Muthu, S.S.; Callaway, L.K.; Mishra, G.D. [30]	2015	Australia	292 GDM	population-based cohort study	To examine the associations between pre-pregnancy dietary patterns and the incidence of GDM	The study shows that the 'Meats, snacks and sweets' pattern is associated with higher GDM risk after adjustment for socioeconomic, reproductive and lifestyle factors

Table 1. *Cont.*

Researchers (Reference)	Publication Date	Country	Sample	Study Type	Goal	Findings
Clausen, T.; Slott, M.; Solvoll, K.; Drevon, C.A.; Vollset, S.E.; Henriksen, T. [31]	2001	Norway	3133 women	prospective, population-based, cohort study of pregnant women	To investigate prospectively whether diet in the first half of pregnancy is associated with risk for pre-eclampsia	The study indicates that high intake of energy, sucrose, and polyunsaturated fatty acids is associated with increased risk of pre-eclampsia
Kibret, K.T.; Chojenta, C.; Gresham, E.; Tegegne, T.K.; Loxton, D. [32]	2018	Australia	21 studies were assessed	A systematic review and meta-analysis	To assess the association between dietary patterns and the risk of adverse pregnancy and birth outcomes	The study shows that dietary patterns with a higher intake of fruits, vegetables, legumes, whole grains and fish are associated with a decreased likelihood of adverse pregnancy and birth outcomes
Schoenaker, D.A.J.M.; Soedamah-Muthu, S.S.; Mishra, G.D. [33]	2014	Australia	In total, 23 cohort and 15 case-control studies were identified	systematic review and meta-analyses	To determine whether dietary factors play a role in the prevention of HDP	The study indicates that higher total energy and lower magnesium and calcium intake measured during pregnancy are identified as related to HDP
Ikem, E.; Halldorsson, T.; Birgisdottir, B.; Rasmussen, M.; Olsen, S.; Maslova, E. [34]	2019	Denmark	55,139 Danish women	Prospective Longitudinal Study	To examine the association between mid-pregnancy dietary patterns and PAH	The study shows a protective association of seafood diet and a harmful association of Western diet with PAH
Wang, D.; Darling, A.M.; McDonald, C.R.; Perumal, N.; Liu, E.; Wang, M.; Aboud, S.; Urassa, W.; Conroy, A.L.; Hayford, K.T.; Liles, W.C.; Kain, K.C.; Fawzi, W.W. [35]	2021	Tanzania	1002 women	Prospective cohort study	To evaluate associations between a panel of inflammatory, angiogenic, and metabolic proteins measured in mid-pregnancy and gestational weight gain	The study shows that plasma concentrations of leptin at mid-pregnancy are associated with gestational weight gain among pregnant women in Tanzania

Table 1. *Cont.*

Researchers (Reference)	Publication Date	Country	Sample	Study Type	Goal	Findings
Molvarec, A.; Szarka, A.; Walentin, S.; Beko, G.; Karadi, I.; Prohazska, Z. Rigo, J., Jr [36]	2011	Hungary	Sixty pre-eclamptic patients, 60 healthy pregnant women and 59 healthy non-pregnant women	Case Control study	To investigate whether serum leptin levels are related to the clinical characteristics of healthy non-pregnant and pregnant women and pre-eclamptic patients	The study shows that serum leptin levels correlate inversely with fetal birth weight in healthy pregnant women. Elevated serum leptin concentrations directly correlate with adipose tissue mass, systemic inflammation, and systolic blood pressure, and negatively correlate with birth weight in normal pregnancies. In both normal and pre-eclampsia pregnancies, increased leptin levels correlate with interferon-y-inducible protein (IP-10) levels. Elevated serum leptin levels and sFlt-1/PlGF ratio have an additive effect on the risk of pre-eclampsia
Bawah, A.T.; Yeboah, F.A.; Nanga, S.; Alidu, H.; Ngala, R.A. [37]	2020	Ghana	90 PE	Case-control study	To determine the levels of serum adiponectin, leptin, resistin, visfatin, and lipids during the first trimester of pregnancy and to evaluate the relation between these markers and pre-eclampsia	The study shows the presence of significant differences in adipokines levels between the pre-eclampsia group and the group without pre-eclampsia. Adiponectin, leptin, resistin, and visfatin are identified as significant predictors of pre-eclampsia, with resistin being the best predictor after controlling for BMI

Table 1. *Cont.*

Researchers (Reference)	Publication Date	Country	Sample	Study Type	Goal	Findings
Hu, W.; Wang, Z.; Wang, H.; Huang, H.; Dong, M. [38]	2008	China	27 women with pre-eclampsia, 28 women in the third trimester of normal pregnancy, and 28 normal non-pregnant women	Case Control study	To characterize the changes in serum visfatin levels in late normal pregnancy and pre-eclampsia	The study indicates a decrease in visfatin level in pre-eclampsia, suggesting that visfatin and adipokine-associated metabolic abnormalities are involved in the pathogenesis of the disease
Kapustin, R.V.; Tcybuk, E.M.; Chepanov, S.V.; Alekseenkova, E.N.; Kopteeva, E.V.; Arzhanova, O.N. [39]	2021	Russia	140 pregnant women	Case Control study	To evaluate sFlt-1 and PlGF levels in the blood of pregnant women	The study shows that blood level alterations of PlGF and sFlt-1 are characteristic of patients with diabetes mellitus in the first and third trimesters of pregnancy. Determination of the sFlt-1/PlGF ratio is a valid method for predicting the development or absence of pre-eclampsia in women with diabetes mellitus
Nikuei, P.; Rajaei, M.; Roozbeh, N.; Mohsenu, F.; Poordarvishi, F.; Azas, M.; Haidari, S. [40]	2020	Iran	23 mild, 15 severe pre-eclamptic patients, and 20 normal term pregnant women	ROC curve analysis	To evaluate the diagnostic accuracy of sFlt-1 to PlGF ratio for diagnosis of pre-eclampsia in an Iranian population	The study shows that sFlt-1/PlGF ratio has higher accuracy than each individual parameter in differentiating pre-eclampsia patients from non-pre-eclampsia patients
Andraweera, P.; Dekker, G.; Roberts, C. [41]	2012	Australia	18 women with pre-eclampsia; 15 women with gestational hypertension; 13 normo-tensive women with SGA; 10 women with spontaneous pre-term birth, and 30 women with uncomplicated pregnancy	Retrospective analysis	To elucidate the role of angiogenic factors in placentation and to evaluate the predictive value of their protein concentrations and genetic variations in pregnancy complications	The study concludes that the current predictive value of the VEGF family as biomarkers appears to be limited to early-onset pre-eclampsia

Table 1. *Cont.*

Researchers (Reference)	Publication Date	Country	Sample	Study Type	Goal	Findings
León-Reyes, G.; Maida-Claros, R.F.; Urrutia-Medina, A.X.; Jorge-Galarza, E.; Guzman-Grenfell, A.M.; Fuentes-Garcia, S.; Medina-Navarro, R.; Moreno-Eutimio, M.A.; Muñoz-Sánchez, J.L.; Hicks, J.J.; Torres-Ramos, Y.D. [42]	2017	Mexico	Thirty women diagnosed with pre-eclampsia and thirty women without pre-eclampsia were included in the study	Transversal and Observational	To evaluate the oxidative profile of lipoproteins isolated from women with pre-eclampsia	The study demonstrates evident oxidative changes in the lipids and proteins in HDL-c and LDL-c particles in PE women
Sharabi-Nov, A.; Srsen, T.P.; Kumer, K.; Vodusek, V.F.; Fabjan, T.; Tul, N.; Meiri, H.; Nicolaides, K.H.; Osredkar, J. [43]	2021	Slovenia	31 cases of pre-eclampsia, 16 of FGR, 42 of pre-eclampsia + FGR, 15 preterm delivery, and 21 unaffected controls	Secondary Analysis	To examine the potential additive value of maternal serum Inhibin-A	The study shows that maternal serum Inhibin-A augments the value of maternal serum PIGF and sFlt-1/PIGF ratio to predict pre-eclampsia near delivery
Kumer, K.; Sharabi-Nov, A.; Vodusek, V.F.; Srsen, T.P.; Tul, N.; Fabian, T.; Meiri, H.; Nicolaides, K.H.; Osredkar, J. [44]	2021	Slovenia	31 cases of pre-eclampsia, 16 of FGR, 42 of pre-eclampsia + FGR, 15 cases who developed with unrelated complications before 37 weeks, and 21 unaffected controls	Secondary Analysis	To assess the accuracy of PlGF, sFlt-1, and sEng in the diagnosis of suspected pre-eclampsia with and without FGR near delivery	The study shows that pro- and anti-angiogenic markers are important clinical tools to identify pre-eclampsia near delivery even in the absence of changes in FGR
Hawkins, M.; Braun, B.; Marcus, B.H.; Stanek, E.; Markenson, G.; Chasan-Taber, L. [45]	2015	US	171 women divided into 84 in exercise protocol and 87 wellness protocol	Randomized control trial	To evaluate the impact of an individually-tailored motivationally-matched exercise intervention on CRP in pregnant women	The study shows that CRP decreases from pre-to post-intervention in the exercise arm and increases in the health and wellness arm; however, the between group difference is not statistically significant ($p = 0.14$). Findings do not differ according to ethnic group or pre-pregnancy body mass index

Table 1. *Cont.*

Researchers (Reference)	Publication Date	Country	Sample	Study Type	Goal	Findings
Johnsen, G.M.; Fjeldstad, H.E.S.; Drabbels, J.J.M.; Haasnoot, G.W.; Eikmans, M.; Størvold, G.L.; Alnaes-Katjavivi, P.; Jacobsen, D.P.; Scherjon, S.A.; Redman, C.W.G.; Claas, F.H.J.; Staff, A.C. [46]	2021	Norway	83 normo-tensive and 83 pre-eclamptic pregnancies	Case Control study	To investigate whether variants of the 3′UTR of the HLA-G gene in mother and fetus are associated with acute atherosis, a pregnancy specific arterial lesion of the decidua basalis that is prevalent in pre-eclampsia	The study shows that HLA-G polymorphisms in the fetus are associated with acute atherosis. These polymorphisms lead to altered HLA-G expression in the decidua basalis, affecting local feto-maternal immune tolerance and development of acute atherosis in pre-eclampsia

2. Pathophysiology of Obesity and Pre-Eclampsia and Discussion of the Findings

To better frame the findings reported in Table 1, the main pathophysiological and clinical observations relative to obesity and pre-eclampsia will be briefly summarized here, before discussing the main findings reported in the cited literature.

3. Obesity

Obesity is defined as a medical condition in which excess body fat has accumulated in an individual to an extent that it may have a negative effect on the individual's health. Metabolic imbalance as observed in the metabolically active fat surrounding the abdominal organs, i.e., visceral fat, is implicated in metabolic dysregulation [47], and the greater is the amount of adiposity accumulated, the higher and more complex are the associated metabolic issues [47]. The main criterion used to determine obesity in an individual is the BMI or body mass index (kg/height in m^2). According to NIH guidelines [https://www.nhlbi.nih.gov/health/educational/lose_wt/BMI/bmi_tbl.pdf (accessed on 10 April 2022)], an individual is defined as normo-weight when the BMI is between 19 and 24.9, overweight when the BMI is between 25 and 29.9, obese when the BMI is between 30 and 39.9, and morbidly obese when the BMI is 40 or higher. Additionally, central obesity is used as a criterion to assess cardiovascular risk. Abdominal circumference above 94 cm in men and 80 cm in women indicate increased cardiovascular risk [19,48], although this concept has recently been revised to take into account race and ethnicity [47]. Several risk factors are considered in the propensity to develop obesity, including genetics and dietary regimen, but primarily lifestyle habits. The most common genetic markers are responsible for <1.5% cases of obesity. High caloric diet, reduced physical activity, extreme amounts of sleep, and socio-economic status play a larger role as risk factors [47].

Metabolic syndrome is a particular health condition associated with obesity. This syndrome is characterized by a set of pathologies that co-exist in a certain patient. These pathologies include central obesity, insulin resistance with higher than normal fasting glycemia, HTN, liver steatosis, and dislipidemia, which together with chronic inflammation can lead to aggressive atherosclerosis [49]. Diabetes and insulin resistance are often present and are considered to be the result of lipotoxicity [50]. Triacylglycerols deposition within adipocytes provide a protective function in low BMI individuals by preventing the circulation of free fatty acids and their biochemical conversion by oxidative stress. At higher BMIs, where there is excess energy storage, the sympathetic nervous system is activated [51], leading to lipolysis and increased free fatty acid release into circulation. Accumulation of poorly utilized lipids within tissues like liver, heart, and skeletal muscles, among others, promotes lipotoxicity and tissue damage. Insulin-receptor in these tissues is among the most common sites of damage, and the ensuing dysfunction of the receptor-together with β-islet cells fatigue as an attempt to increase insulin output to counteract elevated blood glucose levels [50]-result in an insulin-resistant state that further exacerbates the existing hyperglycemia.

On the other hand, HTN and chronic inflammation are the result of the release of cytokines from adipocytes [50]. The adipokines released include interleukin-1, interleukin-6, resistin and TNF-α (tumor necrotic factor-alpha), the latter two being directly involved in promoting insulin resistance. Other adipokines involved are leptin, whose release is increased, and adiponectin, the release of which is reduced [52]. As adiponectin antagonizes Angiotensin-II, its decrease can explain, at least in part, the hypertension observed in the metabolic syndrome but also in pre-eclampsia in obese mothers [50].

The increased circulation of inflammatory adipokines, in particular interleukin-1 and interleukin-6, give rise to multiple pathologies that share an inflammatory base. Among these pathologies, polycystic ovary syndrome, depression, infertility, and pre-eclampsia are the main clinical conditions associated with pregnancy [50].

4. Pre-Eclampsia

Pre-eclampsia is an obstetric complication affecting 2–8% of pregnancies globally [11]. This multisystem progressive disorder is characterized by " ... the new onset of hypertension and proteinuria, or the new onset of hypertension and significant end-organ dysfunction with or without proteinuria in the last half of pregnancy or postpartum ... " (https://www.uptodate.com/contents/preeclampsia-clinical-features-and-diagnosis; accessed 10 April 2022). Hypertension and proteinuria resolve following child delivery or by the 6th week postpartum [53]. Pre-eclampsia and its immediate complications, including eclampsia, are responsible for 10 to 15% of maternal deaths worldwide [54]. Because of this high death toll, the pathology continues to be extensively investigated through the lens of obesity being an important, modifiable risk factor. In fact, according to a prospective cohort study carried out in 2005, pre-eclampsia risk increases at least three folds in women with a BMI of 30 as compared to women with normal (i.e., <24.9) BMI [14].

While the etiology of pre-eclampsia is still not fully elucidated, abnormal placentation due to defective invasion of cytotrophoblast by the spiral arteries has been pinpointed to play a causal role [53]. It has been suggested that inhibition of nitric oxide synthesis is involved in the abnormal placentation through increased arterial resistance [55]. The resulting oxidative stress induces a release of cytokines, oxidized lipids, and free radicals that directly affect the functionality of the vascular endothelium [56]. The ensuing endothelial dysfunction and the abnormalities in Angiotensin-II regulation mentioned previously [50] have been invoked to cause the elevated blood pressure observed in pre-eclampsia. The effects of elevated systolic pressure on endothelial cells extends to the glomerular filtration barrier in the kidneys. The associated depletion of vascular endothelial growth factors in the podocytes further compromises the glomerular filtration process, giving rise to proteinuria [53]. The progressive imbalance between increased systolic pressure and decreased oncotic pressure, and the hyper-permeability of the vascular endothelium can then explain the insurgence of edemas in the lower extremities and the lungs, typical symptoms and complications of pre-eclampsia.

The maternal immune system is also highly involved in pre-eclampsia pathophysiology. Due to lack of recognition of the feto-placental unit, immune cells are overproduced, leading to elevated TNF-α levels and inducing apoptosis of the cytotrophoblast [53,57]. The histocompatibility antigen HLA-G is important for correct invasion of the cytotrophoblast, and its expression is reduced in all forms of pre-eclampsia [58]. The interactions between HLA-G and cytotrophoblasts are possibly mediated by VEGF and placental growth factors, and it has been proposed that the levels of all these growth factors are strong predictors of pre-eclampsia [53]. Because these biological, vascular, and immunological components appear to contribute differently to the etiology of pre-eclampsia, recent literature divides pre-eclampsia in different types/subtypes, each characterized by distinct pathophysiological processes, risk factors, clinical outcomes, and long-term prognosis [10]. One of the most commonly proposed distinctions is between early-onset (before the 34th week of gestation) versus late-onset (34th week of gestation, or later) based on the spiral artery remodeling, which is present in the early-onset but not in the late-onset [10].

Aside for risk of pre-eclampsia progressing to eclampsia in the short-term, the long-term effects of pre-eclampsia on both mother and child are also significant. The mortality risk for mothers who experience pre-eclampsia is elevated primarily as a result of cardiovascular complications [20]. Some of these complications can persist for a very long time, as pre-eclampsia has been reported to cause cardiovascular complications and cardiac diseases in the mother 20 years after the pre-eclampsia event. As for the children delivered by pre-eclamptic mothers, they are usually small in size and with low birth weight, conditions that can lead to increased risk for coronary artery disease and other cardiovascular etiologies later in life [53,59].

5. The Effect of Obesity on Risk for Pre-Eclampsia

By now, it is a well-accepted notion that an elevated BMI as observed in over-weight and obese women represents a major risk factor for pre-eclampsia (see ref. [9] as an example). It is less clear, however, to which extent obesity promotes the onset of other biological parameters identified as potentially clinically relevant risk-factors to predict pre-eclampsia and its impact on maternal and newborn health.

Our analysis of the literature cited here reaffirms the notion that obesity imposes major significant negative effects on pregnancy, directly, and indirectly through the associated metabolic dysfunctions and the increase in basal inflammatory state. Both these conditions can lead to various obstetric complications such as gestational diabetes and hypertension, thus setting a baseline of increased pre-eclampsia risk.

Consistent with this assertion, the majority of the studies reported in Table 1 substantiate that overweight and obese women have a higher risk for pre-eclampsia as compare to normo-weight mothers. A retrospective study carried out in 2014 at King Khalid University Hospital grouped mothers by BMI levels and presence of gestational diabetes (GDM) to investigate the independent effects of GDM and obesity on adverse pregnancy outcomes. The results obtained from about 2700 women assessed clearly showed the independent negative effects of obesity and GDM on pregnancy outcomes including pre-eclampsia. Interestingly, the risk for adverse outcomes increased synergistically when both obesity and GDM were present as compared to when only one of these two conditions was present. In agreement with the trend observed in many Arab countries in the last decade, the study reported an incidence of maternal obesity of 44%, which the investigators attributed to the increased sedentary lifestyle in 75% of the participants [21].

Similar results were reported by the registry-based 2020 study carried out in Tanzania in more than 17,000 women and in a similar study carried out in Sweden. The Tanzanian study confirmed the strong association between pre-eclampsia and obesity as overweight and obese women presented higher risk of pre-eclampsia [22]. This association was independent of the socio-economic status of the women as it was observed in both wealthy and low-income communities [22]. The Swedish birth-registry study, which actually preceded the Tanzanian study, corroborated the importance of obesity as a risk factor in that elevated BMI increased the risk for all types of pre-eclampsia and its complications including delivery at term [23]. Similar support is provided by the prospective study performed in Brazil in 2013 [24]. Out of the initial 212 women recruited for the study, 30 suffered pre-eclampsia (approximately 14%). On average, these women had higher BMI values than the normo-weight, normotensive controls. By including a socioeconomic stratification, this study evidenced that women who developed pre-eclampsia had lower levels of education than normotensive women, as only 20% of them had completed high school. This discrepancy in outcome was attributed to reduced access to quality health care and poor compliance to treatment and nutrition [24].

Aside from pre-pregnancy obesity, weight gain during pregnancy (gestational weight gain or GWG) has been implicated in adverse pregnancy outcomes. Weight gain during pregnancy should occur within the normal range based on the pre-pregnancy BMI of the mother and the expected development and growth of the placenta and the fetus. Gestation weight gain in excess of the predicted range appears to lead to detrimental pregnancy outcomes. The results of a recent retrospective cohort study involving 1606 women with GDM [25] are in agreement with this statement. In this study, the mothers were divided in 2 groups: one group with normal GWG, and a second group with above normal GWG. The results indicated an increased risk of C-section, pre-eclampsia, and pre-term labor in women with above normal GWG and with GDM. One detracting weakness of this study is that while the study focused on GWG, it did not provide weight parameters for the women involved in the study prior to pregnancy, thus preventing from obtaining meaningful insights on the possible link adverse pregnancy outcomes in overweight or obese women [25].

The relevance of weight gain as a risk factor for pre-eclampsia, is also supported by the studies carried out in China and in Northwestern Ethiopia. The cohort study from Lanzou, China [26], supports the notion that gestational weight gain increases the risk of obesity, and pre-pregnancy obesity and gestational weight gain increase the risk for pre-eclampsia, both independently and in conjunction, as the combined effects of maternal obesity and gestational weight gain further heightens the risk for pre-eclampsia. The case-control study from Northwestern Ethiopia [27] included socio-demographic characteristics and diet to the variables tested for pre-eclampsia risk. After controlling for confounding variables such as education, residence, anemia, and alcohol/meat consumption, five variables remained to have a significant impact on pre-eclampsia risk, including coffee consumption, which doubled the risk of pre-eclampsia by elevating systolic blood pressure, but primarily obesity, which showed the strongest association with pre-eclampsia risk especially in younger women (less than 35 years of age) [27].

Congruent with the role of obesity as a risk factor for pre-eclampsia, sedentary lifestyle and sugary food consumption also represents major risk factors of pre-eclampsia in that both contribute to an increase in gestational weight gain. The epidemiology study conducted in North Sweden in 2021 on a sample population of more than 2000 women [28] validates the observation that higher levels of physical activity can reduce the risk of adverse pregnancy outcomes including emergency C-section and gestational weight gain [28]. This notion is supported by the study out of King Khalid University Hospital [21] that identified sedentary lifestyle as the most direct reason behind the high incidence of maternal obesity in the sample population assessed. Consumption of sugary foods and products has also been associated with pre-eclampsia. However, the number of studies addressing this particular aspect is rather limited [29–33]. A prospective study carried out on almost 33,000 normo-weight and overweight pregnant Norwegian women indicated that intake of more than 125 mL of sugary beverages per day was associated with a higher risk of pre-eclampsia as compared to the consumption of sugars as fruit (both dried and fresh) [29]. The results of this study corroborate the observation by Clausen et al. [31] in 3133 pregnant Norwegian women that the risk of pre-eclampsia was increased in women whose sucrose intake covered more than 25% of the total energy intake per day as compared to women whose sucrose intake contributed less than 8.5% to total energy intake. Similarly, a prospective longitudinal cohort study on 55,139 Danish women reported that the incidences of gestational hypertension and pre-eclampsia were strongly associated with a higher adherence to a typical Western diet [34]. At the same time, Schoenaker et al. [30] observed that a Mediterranean-style diet was inversely related to the risk of pregnancy-associated hypertension, including pre-eclampsia, in 3582 Australian women who participated in the 9-year Australian Longitudinal Study on Women's health.

6. Potential Biochemical Biomarkers Linking Obesity to Pre-Eclampsia

Significant attention has been paid to investigating and identifying biomarkers that can consistently predict the development of pre-eclampsia. It is only in the last decade that the clinical practice of measuring multiple markers has started and developed [53,60]. Since 2011, the clinical assessment of different growth factors including placental growth factor, VEGF, and anti-angiogenic markers as endoglin and sFlt1, has become more of a routine in women with pre-eclampsia [53]. It has to be noted, however, that the clinical use of these biomarkers as predictive or diagnostic tools still needs proper validation. Based on our analysis of the literature, it is not surprising that the biomarkers that appear to more reliably predict pre-eclampsia are adipokines associated with obesity and gestational weight gain. A secondary analysis of a randomized control trial in Tanzania [35] assessed leptin levels during pregnancy and its relation to gestational weight gain, and reported that high levels of leptin mid-pregnancy are indeed associated with excessive gestational weight gain [27]. Leptin is a pro-inflammatory adipokine produced mainly by adipocytes and regulated by steroids, which is involved in regulating maternal metabolism but also trophoblast invasion among many other functions [61]. Leptin levels increase during pregnancy due to placenta production of leptin, the levels usually peaking during the second-third trimester [61]. Aside from its effect on gestational weight gain, it is assumed that high levels of leptin play a role in pre-eclampsia's pathophysiology but the exact mechanisms are not fully elucidated. A case-control study conducted in Hungary associated elevated serum leptin concentrations with increased systolic blood pressure due to its effect on sympathetic activity [36]. A more recent case-control study conducted in 2020 proposed leptin as a strong predictor of pre-eclampsia [37]. However, the absence of a proper assessment of the case numbers and controls based on BMI stratification makes the interpretation of the study rather difficult. Furthermore, several reports in the literature provide contradicting results that do not support the conclusion by Bawah et al. [37]. Elucidating this point of contention and clarifying whether leptin affects the trophoblast vascular invasion directly, or indirectly through abnormalities in maternal metabolism and gestational weight gain remains a wanting topic for future research in the field. A strong linear relationship appears to exist among leptin levels, BMI, and C-reactive protein (CRP) concentrations [36]. Both BMI and CRP levels are elevated with increased leptin concentrations [36]. C-reactive protein is produced by hepatocytes following an increase in the levels of inflammatory cytokines such as interleukin-6, TNF-α, and interleukin-1β, and it is an established inflammatory marker released under various pathological conditions [62]. The inflammatory cytokines that promote CRP release are a direct result of obesity, in that the program switch that occurs under excessive adipose tissue increase results in the release of these cytokines, among others [52].

Under the conditions in which cytokines and leptin levels increase, adiponectin expression and release are reduced [52]. This applies to obesity conditions, and has major implications for the increased systolic blood pressure observed in obese individuals including obese pregnant women. Hence, adiponectin can also be considered a strong predictor of pre-eclampsia risk per se, and even more so when taken together with the changes in leptin and CRP concentrations mentioned above, as a small case-control study completed in 2020 suggests [37]. This study focused on the levels of adiponectin, leptin, resistin, visfatin, and lipids in relation to pre-eclampsia risk and incidence. The study provides further evidence that the combination of reduced adiponectin and increased leptin levels is a strong predictor for pre-eclampsia. Interestingly, decreased levels of adiponectin were the best predictors of pre-eclampsia in those cases in which confounding factors such as age, parity, BMI, and family history of hypertension were controlled. Changes in adiponectin and leptin are known to promote overweight and obesity, and exacerbate insulin resistance in those individuals, lending further support to the notion that obesity and the consequent abnormal whole-body metabolism remains the single most important predictor of the risk of pre-eclampsia. This case-control study is also one of the few studies that investigated the role of resistin and visfatin in inflammation and possibly pre-eclampsia [37]. The relation between these two cytokines and pre-eclampsia, however, is not that clear, and is consistent with previous research reporting unchanged levels of resistin and visfatin in pre-eclampsia [38]. Hence, compared to leptin and adiponectin, resistin and visfatin remain inconsistent and not fully reliable biomarkers for pre-eclampsia prediction at the present time, requiring further and more detailed investigation to attain this role. More detailed investigation is also necessary to determine how rapidly the changes in adiponectin and leptin levels promotes the onset and/or the progression of pre-eclampsia.

While leptin and adiponectin appear to be more reliable predictors of pre-eclampsia in the long-term, the well documented changes in sFlt-1: PIGF ratio are a more dependable biomarker for short-term prediction of pre-eclampsia. The anti-angiogenic sFlt-1 protein is responsible for the endothelial dysfunction observed in pre-eclampsia pathology [63]. PIGF (placenta growth factor) is significantly inhibited by sFlt-1. Hence, the higher the levels of sFlt-1, the lower the levels of PIGF will be, resulting in a high sFlt-1: PIGF ratio [64]. Multiple studies support the notion that a high sFlt-1: PIGF ratio indicates an imminent risk for pre-eclampsia [39,65], and is a very accurate, short-term predictor of pre-eclampsia, especially when the sFlt-1: PIGF ratio is significantly increased [40]. In this regard, the sFlt-1: PIGF ratio has been reported to be a far better predictor for pre-eclampsia than the individual measurements of sFlt-1 and PIGF [41].

Changes in serum lipid profile have also been considered as potential predictors of pre-eclampsia. However, the relation between lipid levels and pre-eclampsia is rather weak [37]. With the exception of a significant reduction in high density lipoprotein (HDL) levels, observed biochemical changes in lipid concentrations and pattern have been attributed to the significant oxidative stress observed in women with pre-eclampsia, which targets lipids but also proteins [37]. The modification of both lipids and proteins by oxidative stress can also explains the endothelial changes observed at the level of the glomerular filtration barrier, which give rise to proteinuria, an objective finding in support of pre-eclampsia diagnosis, and supports the idea of antioxidant therapy as a preventative measure for pre-eclampsia [42]. The effectiveness of such a therapy to reduce the severity of pre-eclampsia and its associated risks, however, remains to be demonstrated.

More recently, inhibin-A [43,44] and oxidative stress [66] have also gain relevance as predictors of pre-eclampsia onset, pre-eclampsia progression as well as fetal growth restriction. The Osredkar's group has recently reported that when assessed in combination with PIGF or with the sFlt-1/PIGF ratio, inhibin-A markedly and significantly increased the detection rate of pre-eclampsia and pre-eclampsia complicated by fetal growth restriction in most of the cases, the only limitation being cases of fetal growth restriction in deliveries earlier than 34 weeks, wherefore the diagnostic predictive value of inhibin-A was rather limited [43]. Noteworthy, the levels of inhibin-A are increased, and the levels of inhibin-B are decreased in obese women [67].

In accordance with the notion that pre-eclampsia can be caused by abnormal placentation due to the defective invasion of cytotrophoblast by the spiral arteries [53], reduced blood flow and abnormal oxygen utilization have been indicated to result in oxidative stress and dysfunction of placental endothelial nitric oxide synthase, ultimately leading to pre-eclampsia reviewed in [66]. While physiological levels of reactive oxygen species (ROS) are associated with the rapid development of the placenta, supra-physiological levels of ROS together with depletion of antioxidants and abnormalities in superoxide dismutase activity are linked to impaired trophoblast invasion, poor placentation, and pathological waves of hypoxia/reoxygenation. In turn, these conditions promote oxidative stress and production of lipoperoxides that lead to cellular dysfunction, inflammation, and apoptosis [66]. Formation of peroxynitrite (ONOO-) and inhibition of endothelial nitric oxide synthase (eNOS) enzymatic activity are among the most common abnormalities observed in the placenta following an increase in ROS formation [66]. Interestingly, adiponectin has been reported to regulate ROS formation in several tissues [68] including placenta [69].

7. Conclusions

The literature review reported here confirms the current view-point that overweight and obese women are at a greater risk for pre-eclampsia than normo-weight women. Our literature review also suggests that the metabolic changes central to obesity's pathology do pose a significant risk for pre-eclampsia and its adverse outcomes. The reviewed studies consistently support the presence of a direct relationship between BMI and risk of pre-eclampsia, gestational hypertension and gestational diabetes. The relevance of excessive gestational weight gain as a risk factor for pre-eclampsia is further corroborated by the reports indicating that the co-presence of both pre-pregnancy obesity and excessive gestational weight gain results in the highest risk for pre-eclampsia onset and progression. As obesity rates increases steadily in the US and worldwide, the prevalence of pre-eclampsia is likely to continue to increase together with the risk for adverse obstetric outcomes associated with high BMI levels and gestational weight gain. It is therefore crucial that these concerns are communicated clearly and in a timely manner to pre-pregnant and pregnant women, and that risk reduction measure are pursued in all pregnancies involving women with high BMI values.

In addition to obesity and weight gain, a few biomarkers have been identified that heighten the risk for pre-eclampsia onset when present. Among these biomarkers, the strongest short-term predictor of pre-eclampsia is the sFlt-1: PIGF ratio, as several retrospective, and prospective studies clearly indicate. Inhibin-A may further enhance the clinical utility of sFlt1: PIGF ratio in predicting fetal growth restriction in patients with pre-eclampsia [43]. Other biomarkers associated with increased risk for pre-eclampsia later in pregnancy include elevated leptin levels, decreased serum adiponectin levels and elevated C-reactive protein levels. Furthermore, it cannot be excluded that the ratio adiponectin/leptin is possibly more important rather than the individual serum levels of the two adipokines in obesity and obese women. Similarly, it is not fully elucidated how leptin affects trophoblast implantation, and to which extent this effect contributes to pre-eclampsia development. The literature review also produced a consistent body of evidence about the roles of oxidative stress, peroxynitrite production, and decreased eNOS enzymatic activity in impaired placentation and pre-pre-eclampsia development [66], further supported by the potential evidence of a regulatory role by adiponectin on ROS formation [68,69]. Predictably, supplementation with NO donors or L-arginine, and inhibition of type-5 phosphodiesterase, which normally promotes cGMP degradation and limits NO-cGMP functionality, are actively pursued as therapeutic approaches to attenuate oxidative stress and vasoconstriction, and improve uterine vascularization [66]. Furthermore, a possible interplay between leptin and adiponectin on one hand, and ROS formation and alterations in other biomarkers for pre-eclampsia on the other hand, may exist, further emphasizing the possible role of dysmetabolism in adipocytes, and the onset and progression of pre-eclampsia in overweight pregnant women.

In contrast to the robust support from the literature on the predictive relevance of adiponectin and CRP levels for pre-eclampsia, recent literature reports inconsistent results for resistin and visfatin. Hence, for these markers of interest more research is still warranted. In considering the elevated C-reactive protein levels observed in pre-eclampsia cases as a result of systemic inflammation, it would be interesting to better investigate the mechanism(s) by which C-reactive protein levels improve following exercise regimens [45]. Although the improvement observed was not statistically significant [45], it is possible that exercise in the context of a holistic approach that includes diet and weight loss could potentially reduce systemic inflammation and consequently C-reactive protein levels and ameliorate the circulating levels of leptin and adipokines, improving pregnancy health in general.

Lastly, the importance of HLA-G levels in predicting pre-eclampsia merits more in-depth investigation. The histocompatibility antigen HLA-G appears to facilitate correct invasion of the cytotrophoblast [58], and its expression is reduced in all forms of pre-eclampsia, although a recent study [46] has not identified a strong relation between expression of this antigen and the incidence of pre-eclampsia, in particular in the decidual acute atherosis pre-eclampsia subtype. The small sample size of the study (83 women only) precludes extrapolation to the broader pre-eclampsia population, and warrant further investigation of HLA-G levels in a larger study population.

In this review, we have collated recent literature which provides evidence that confirms obesity to be a major risk factor for pre-eclampsia. The literature reviewed also suggests that the abnormalities in the levels of adiponectin, leptin, and possibly other cytokines as observed under the dysmetabolic conditions associated with weight gain and frank obesity may play an important role, albeit not fully understood, in the pathophysiological processes leading to pre-eclampsia. More detailed studies conducted in larger sample sizes are certainly necessary to understand the role adipokines play in the development of pre-eclampsia in obese pregnant women or in women who experience an excessive weight gain during their pregnancies.

Funding: Departmental Funding to Andrea M. P. Romani.

Institutional Review Board Statement: Not applicable.

Informed Consent Statement: Not applicable.

Data Availability Statement: Not applicable.

Conflicts of Interest: The authors declare no conflict of interest.

Abbreviations

Terminology and Abbreviations (in alphabetical order)

BMI	Body mass index
CDC	Center for disease control
CRP	C-reactive protein
sENG	Serum Endoglin
FGR	Fetal growth restriction
sFlt-1	Circulating soluble fms-like tyrosine kinase-1
GDM	Gestational diabetes mellitus
GWG	Gestational weight gain
HDL	High Density Lipoprotein
HDP	Hypertensive disorders of pregnancy
HLA-G	Human leukocyte antigen Class I, G
HTN	Hypertension
IOM	International Organization for Migration
LDL	Low Density Lipoprotein
LGA	Large for gestational age
PAH	Pregnancy associated hypertension
PE	Pre-eclampsia
PIGF	Placental growth factor
SGA	Small for gestational age
VEGF	Vascular endothelial growth factor
WHO	World health organization
WHOMCS	WHO multi-country survey

References

1. Damsgaard, C.T.; Michaelsen, K.F.; Molbo, D.; Mortensen, E.L.; Sørensen, T.I. Trends in adult body-mass index in 200 countries from 1975 to 2014: A pooled analysis of 1698 population-based measurement studies with 19.2 million participants. *Lancet* **2016**, *387*, 1377–1396.
2. Fisher, S.C.; Kim, S.Y.; Sharma, A.J.; Rochat, R.; Morrow, B. Is obesity still increasing among pregnant women? Prepregnancy obesity trends in 20 states, 2003–2009. *Prev. Med.* **2013**, *56*, 372–378. [CrossRef] [PubMed]
3. Roth, J.; Qiang, X.; Marbán, S.L.; Redelt, H.; Lowell, B.C. The Obesity Pandemic: Where Have We Been and Where Are We Going? *Obes. Res.* **2004**, *12* (Suppl. 2), 88S–101S. [CrossRef]
4. Facts on Obesity. Available online: https://www.who.int/news-room/facts-in-pictures/detail/6-facts-on-obesity (accessed on 30 October 2020).
5. Obesity. Available online: https://www.cdc.gov/obesity/index.html (accessed on 30 October 2020).
6. Flegal, K.M.; Carroll, M.D.; Ogden, C.L.; Johnson, C.L. Prevalence and trends in obesity among US Adults, 1999–2000. *JAMA* **2002**, *288*, 1723–1727. [CrossRef]
7. U.S. Department of Health and Human Services. Overweight and Obesity: A Major Public Health Issue. Available online: https://www.ncbi.nlm.nih.gov/books/NBK44210/ (accessed on 10 April 2022).
8. Ng, M.; Fleming, T.; Robinson, M.; Thomson, B.; Graetz, N.; Margono, C.; Mullany, E.C.; Biryukov, S.; Abbafati, C.; Abera, S.F.; et al. Global, regional, and national prevalence of overweight and obesity in children and adults during 1980–2013: A systematic analysis for the Global Burden of Disease Study 2013. *Lancet* **2014**, *384*, 766–781. [CrossRef]
9. Heslehurst, N.; Simpson, H.; Ells, L.J.; Rankin, J.; Wilkinson, J.; Lang, R.; Brown, T.J.; Summerbell, C.D. The impact of maternal BMI status on pregnancy outcomes with immediate short-term obstetric resource implications: A meta-analysis. *Obes. Rev.* **2008**, *9*, 635–683. [CrossRef] [PubMed]
10. Yao, R.; Ananth, C.V.; Park, B.; Pereira, L.; Plante, L.A. Obesity and the risk of stillbirth: A population-based cohort study. *Am. J. Obstet. Gynecol.* **2014**, *210*, 457.e1–457.e9. [CrossRef] [PubMed]
11. GBD 2015 Obesity Collaborators; Afshin, A.; Forouzanfar, M.H.; Reitsma, M.B.; Sur, P.; Estep, K.; Lee, A.; Marczak, L.; Mokdad, A.H.; Moradi-Lakeh, M.; et al. Health Effects of Overweight and Obesity in 195 Countries over 25 Years. *N. Engl. J. Med.* **2017**, *377*, 13–27. [CrossRef]

12. Lisonkova, S.; Joseph, K.S. Incidence of pre-eclampsia: Risk factors and outcomes associated with early-versus late-onset diease. *Am. J. Obstet. Gynecol.* **2013**, *209*, 544.e1–544.e12. [CrossRef]
13. Rasmussen, S.; Irgens, L.M.; Espinoza, J. Maternal obesity and excess of fetal growth in pre-eclampsia. *BJOG Int. J. Obstet. Gynaecol.* **2014**, *121*, 1351–1358. [CrossRef]
14. Mbah, A.; Kornosky, J.; Kristensen, S.; August, E.; Alio, A.; Marty, P.; Belogolovkin, V.; Bruder, K.; Salihu, H. Super-obesity and risk for early and late pre-eclampsia. *BJOG Int. J. Obstet. Gynaecol.* **2010**, *117*, 997–1004. [CrossRef] [PubMed]
15. Bodnar, L.M.; Catov, J.M.; Klebanoff, M.A.; Ness, R.B.; Roberts, J.M. Prepregnancy Body Mass Index and the Occurrence of Severe Hypertensive Disorders of Pregnancy. *Epidemiology* **2007**, *18*, 234–239. [CrossRef]
16. Bodnar, L.M.; Ness, R.B.; Markovic, N.; Roberts, J.M. The risk of pre-eclampsia rises with increasing prepregnancy body mass index. *Ann. Epidemiol.* **2005**, *15*, 475–482. [CrossRef] [PubMed]
17. Abalos, E.; Cuesta, C.; Carroli, G.; Qureshi, Z.; Widmer, M.; Vogel, J.; Souza, J.; on behalf of the WHO Multicountry Survey on Maternal and Newborn Health Research Network. Pre-eclampsia, eclampsia and adverse maternal and perinatal outcomes: A secondary analysis of the World Health Organization Multicountry Survey on Maternal and Newborn Health. *BJOG Int. J. Obstet. Gynaecol.* **2014**, *121* (Suppl. 1), 14–24. [CrossRef] [PubMed]
18. Phipps, E.A.; Thadhani, R.; Benzing, T.; Karumanchi, S.A. Pre-eclampsia: Pathogenesis, novel diagnostics and therapies. *Nat. Rev. Nephrol.* **2019**, *15*, 275–289. [CrossRef] [PubMed]
19. Zhang, J.; Meikle, S.; Trumble, A. Severe Maternal Morbidity Associated with Hypertensive Disorders in Pregnancy in the United States. *Hypertens. Pregnancy* **2003**, *22*, 203–212. [CrossRef] [PubMed]
20. Funai, E.F.; Friedlander, Y.; Paltiel, O.; Tiram, E.; Xue, X.; Deutsch, L.; Harlap, S. Long-Term Mortality After Preeclampsia. *Epidemiology* **2005**, *16*, 206–215. [CrossRef]
21. Wahabi, H.A.; Fayed, A.A.; Alzeidan, R.A.; Mandil, A.A. The independent effects of maternal obesity and gestational diabetes on the pregnancy outcomes. *BMC Endocr. Disord.* **2014**, *14*, 47. [CrossRef]
22. Mrema, D.; Lie, R.T.; Østbye, T.; Mahande, M.J.; Daltveit, A.K. The association between pre pregnancy body mass index and risk of preeclampsia: A registry based study from Tanzania. *BMC Pregnancy Childbirth* **2018**, *18*, 56. [CrossRef]
23. Sohlberg, S.; Stephansson, O.; Cnattingius, S.; Wikström, A.-K. Maternal Body Mass Index, Height, and Risks of Preeclampsia. *Am. J. Hypertens.* **2012**, *25*, 120–125. [CrossRef]
24. Dantas, E.M.D.M.; Pereira, F.V.M.; Queiroz, J.W.; Dantas, D.L.D.M.; Monteiro, G.R.G.; Duggal, P.; Azevedo, M.D.F.; Jeronimo, S.M.B.; Araújo, A.C.P.F. Preeclampsia is associated with increased maternal body weight in a northeastern Brazilian population. *Bmc Pregnancy Childbirth* **2013**, *13*, 159. [CrossRef] [PubMed]
25. Shi, P.; Liu, A.; Yin, X. Association between gestational weight gain in women with gestational diabetes mellitus and adverse pregnancy outcomes: A retrospective cohort study. *BMC Pregnancy Childbirth* **2021**, *21*, 508. [CrossRef] [PubMed]
26. Shao, Y.; Qiu, J.; Huang, H.; Mao, B.; Dai, W.; He, X.; Cui, H.; Lin, X.; Lv, L.; Wang, D.; et al. Pre-pregnancy BMI, gestational weight gain and risk of preeclampsia: A birth cohort study in Lanzhou, China. *BMC Pregnancy Childbirth* **2017**, *17*, 400. [CrossRef] [PubMed]
27. Endeshaw, M.; Abebe, F.; Worku, S.; Menber, L.; Assress, M.; Assefa, M. Obesity in young age is a risk factor for preeclampsia: A facility based case-control study, northwest Ethiopia. *BMC Pregnancy Childbirth* **2016**, *16*, 237. [CrossRef]
28. Meander, L.; Lindqvist, M.; Mogren, I.; Sandlund, J.; West, C.E.; Domellöf, M. Physical activity and sedentary time during pregnancy and associations with maternal and fetal health outcomes: An epidemiological study. *BMC Pregnancy Childbirth* **2021**, *21*, 166. [CrossRef]
29. Borgen, I.; Aamodt, G.; Harsem, H.; Haugen, M.; Meltzer, H.M.; Brantsaeter, A.L. Maternal sugar consumption and risk of pre-eclampsia in nulliparous Norwegian women. *Eur. J. Clin. Nutr.* **2012**, *66*, 920–925. [CrossRef]
30. Schoenaker, D.A.J.M.; Soedamah-Muthu, S.S.; Callaway, L.K.; Mishra, G.D. Prepregnancy dietary patterns and risk of developing hypertensive disorders of pregnancy: Results from the Australian Longitudinal Study on Women's Health. *Am. J. Clin. Nutr.* **2015**, *102*, 94–101. [CrossRef]
31. Clausen, T.; Slott, M.; Solvoll, K.; Drevon, C.A.; Vollset, S.E.; Henriksen, T. High intake of energy, sucrose, and polyunsaturated fatty acids is associated with increased risk of pre-eclampsia. *Am. J. Obstet. Gynecol.* **2001**, *185*, 451–458. [CrossRef]
32. Kibret, K.T.; Chojenta, C.; Gresham, E.; Tegegne, T.K.; Loxton, D. Maternal dietary patterns and risk of adverse pregnancy (hypertensive disorders of pregnancy and gestational diabetes mellitus) and birth (preterm birth and low birth weight) outcomes: A systematic review and meta-analysis. *Public Health Nutr.* **2019**, *22*, 506–520. [CrossRef]
33. Schoenaker, D.A.J.M.; Soedamah-Muthu, S.S.; Mishra, G.D. The association between dietary factors and gestational hypertension and pre-eclampsia: A systematic review and meta-analysis of observational studies. *BMC Med.* **2014**, *12*, 157. [CrossRef]
34. Ikem, E.; Halldorsson, T.; Birgisdottir, B.E.; Rasmussen, M.A.; Olsen, S.; Maslova, E. Dietary patterns and the risk of pregnancy-associated hypertension in the Danish National Birth Cohort: A prospective longitudinal study. *BJOG Int. J. Obstet. Gynaecol.* **2019**, *126*, 663–673. [CrossRef] [PubMed]
35. Wang, D.; Darling, A.M.; McDonald, C.R.; Perumal, N.; Liu, E.; Wang, M.; Aboud, S.; Urassa, W.; Conroy, A.L.; Hayford, K.T.; et al. Plasma concentrations of leptin at mid-pregnancy are associated with gestational weight gain among pregnant women in Tanzania: A prospective cohort study. *BMC Pregnancy Childbirth* **2021**, *21*, 675. [CrossRef] [PubMed]

36. Molvarec, A.; Szarka, A.; Walentin, S.; Beko, G.; Karádi, I.; Prohászka, Z.; Rigó, J.; Rigó, J., Jr. Serum leptin levels in relation to circulating cytokines, chemokines, adhesion molecules and angiogenic factors in normal pregnancy and preeclampsia. *Reprod. Biol. Endocrinol.* **2011**, *9*, 124. [CrossRef] [PubMed]
37. Bawah, A.T.; Yeboah, F.A.; Nanga, S.; Alidu, H.; Ngala, R.A. Serum adipocytokines and adiposity as predictive indices of preeclampsia. *Clin. Hypertens.* **2020**, *26*, 19. [CrossRef] [PubMed]
38. Hu, W.; Wang, Z.; Wang, H.; Huang, H.; Dong, M. Serum visfatin levels in late pregnancy and pre-eclampsia. *Acta Obstet. Gynecol. Scand.* **2008**, *87*, 413–418. [CrossRef] [PubMed]
39. Kapustin, R.V.; Tcybuk, E.M.; Chepanov, S.V.; Alekseenkova, E.N.; Kopteeva, E.V.; Arzhanova, O.N. Evaluation of sFlt-1 and PlGF for predicting pre-eclampsia in pregnant women with diabetes mellitus. *J. Obstet. Women's Dis.* **2021**, *70*, 43–56. [CrossRef]
40. Nikuei, P.; Rajaei, M.; Roozbeh, N.; Mohseni, F.; Poordarvishi, F.; Azad, M.; Haidari, S. Diagnostic accuracy of sFlt1/PlGF ratio as a marker for preeclampsia. *BMC Pregnancy Childbirth* **2020**, *20*, 80–86. [CrossRef]
41. Andraweera, P.H.; Dekker, G.A.; Roberts, C.T. The vascular endothelial growth factor family in adverse pregnancy outcomes. *Hum. Reprod. Update* **2012**, *18*, 436–457. [CrossRef]
42. León-Reyes, G.; Maida-Claros, R.F.; Urrutia-Medina, A.X.; Jorge-Galarza, E.; Guzmán-Grenfell, A.M.; Fuentes-García, S.; Medina-Navarro, R.; Moreno-Eutimio, M.A.; Muñoz-Sánchez, J.L.; Hicks, J.J.; et al. Oxidative profiles of LDL and HDL isolated from women with preeclampsia. *Lipids Health Dis.* **2017**, *16*, 90. [CrossRef]
43. Sharabi-Nov, A.; Sršen, T.P.; Kumer, K.; Vodušek, V.F.; Fabjan, T.; Tul, N.; Meiri, H.; Nicolaides, K.H.; Osredkar, J. Maternal Serum Inhibin-A Augments the Value of Maternal Serum PlGF and of sFlt-1/PlGF Ratio in the Prediction of Preeclampsia and/or FGR Near Delivery—A Secondary Analysis. *Reprod. Med.* **2021**, *2*, 35–49. [CrossRef]
44. Kumer, K.; Sharabi-Nov, A.; Vodušek, V.F.; Sršen, T.P.; Tul, N.; Fabjan, T.; Meiri, H.; Nicolaides, K.H.; Osredkar, J. Pro- and Anti-Angiogenic Markers as Clinical Tools for Suspected Preeclampsia with and without FGR near Delivery—A Secondary Analysis. *Reprod. Med.* **2021**, *2*, 12–25. [CrossRef]
45. Hawkins, M.; Braun, B.; Marcus, B.H.; Stanek, E.; Markenson, G.; Chasan-Taber, L. The impact of an exercise intervention on C—Reactive protein during pregnancy: A randomized controlled trial. *BMC Pregnancy Childbirth* **2015**, *15*, 139. [CrossRef] [PubMed]
46. Johnsen, G.M.; Fjeldstad, H.E.S.; Drabbels, J.J.M.; Haasnoot, G.W.; Eikmans, M.; Størvold, G.L.; Alnaes-Katjavivi, P.; Jacobsen, D.P.; Scherjon, S.A.; Redman, C.W.G.; et al. A possible role for HLA-G in development of uteroplacental acute atherosis in preeclampsia. *J. Reprod. Immunol.* **2021**, *144*, 103284. [CrossRef] [PubMed]
47. Hruby, A.; Hu, F.B. The Epidemiology of Obesity: A Big Picture. *PharmacoEconomics* **2015**, *33*, 673–689. [CrossRef]
48. Alberti, K.G.M.M.; Eckel, R.H.; Grundy, S.M.; Zimmet, P.Z.; Cleeman, J.I.; Donato, K.A.; Fruchart, J.-C.; James, W.P.T.; Loria, C.M.; Smith, S.C., Jr.; et al. Harmonizing the metabolic syndrome: A joint interim statement of the International Diabetes Federation Task Force on Epidemiology and Prevention; National Heart, Lung, and Blood Institute; American Heart Association; World Heart Federation; International Atherosclerosis Society; and International Association for the Study of Obesity. *Circulation* **2009**, *120*, 1640–1645. [CrossRef]
49. Hutley, L.; Prins, J.B. Fat as an Endocrine Organ: Relationship to the Metabolic Syndrome. *Am. J. Med. Sci.* **2005**, *330*, 280–289. [CrossRef]
50. Redinger, R.N. The pathophysiology of obesity and its clinical manifestations. *Gastroenterol. Hepatol.* **2007**, *3*, 856–863.
51. Evans, R.M.; Barrish, G.D.; Wang, Y.-X. PPARs and the complex journey to obesity. *Nat. Med.* **2004**, *10*, 355–361. [CrossRef]
52. Lafontan, M. FAT CELLS: Afferent and Efferent Messages Define New Approaches to Treat Obesity. *Annu. Rev. Pharmacol. Toxicol.* **2005**, *45*, 119–146. [CrossRef]
53. Uzan, J.; Carbonnel, M.; Piconne, O.; Asmar, R.; Ayoubi, J.-M. Pre-eclampsia: Pathophysiology, diagnosis, and management. *Vasc. Health Risk Manag.* **2011**, *7*, 467–474. [CrossRef]
54. Duley, L. The Global Impact of Pre-eclampsia and Eclampsia. *Semin. Perinatol.* **2009**, *33*, 130–137. [CrossRef] [PubMed]
55. Durán-Reyes, G.; Gómez-Meléndez, M.D.R.; la Brena, G.M.-D.; Mercado-Pichardo, E.; Medina-Navarro, R.; Hicks-Gómez, J.J. Nitric oxide synthesis inhibition suppresses implantation and decreases cGMP concentration and protein peroxidation. *Life Sci.* **1999**, *65*, 2259–2268. [CrossRef]
56. Roberts, J.M. Endothelial Dysfunction in Preeclampsia. *Semin. Reprod. Med.* **1998**, *16*, 5–15. [CrossRef]
57. Genbacev, O.; DiFederico, E.; McMaster, M.; Fisher, S.J. Invasive cytotrophoblast apoptosis in pre-eclampsia. *Hum. Reprod.* **1999**, *14* (Suppl. 2), 59–66. [CrossRef] [PubMed]
58. Colbern, G.T.; Chiang, M.H.; Main, E.K. Expression of the nonclassic histocompatibility antigen HLA-G by preeclamptic placenta. *Am. J. Obstet. Gynecol.* **1994**, *170*, 1244–1250. [CrossRef]
59. Eriksson, J.G.; Forsén, T.; Tuomilehto, J.; Osmond, C.; Barker, D.J.P. Early growth and coronary heart disease in later life: Longitudinal study. *BMJ* **2001**, *322*, 949–953. [CrossRef]
60. Sibai, B.; Dekker, G.; Kupferminc, M. Pre-eclampsia. *Lancet* **2005**, *365*, 785–799. [CrossRef]
61. Henson, M.C.; Castracane, V.D. Leptin in Pregnancy: An Update1. *Biol. Reprod.* **2006**, *74*, 218–229. [CrossRef] [PubMed]
62. Vermeire, S.; Van Assche, G.; Rutgeerts, P.J. The role of C-reactive protein as an inflammatory marker in gastrointestinal diseases. *Nat. Clin. Pract. Gastroenterol. Hepatol.* **2005**, *2*, 580–586. [CrossRef]
63. Powe, C.E.; Levine, R.J.; Karumanchi, S.A. Pre-eclampsia, a disease of the maternal endothelium the role of antiangiogenic factors and implications for later cardiovascular disease. *Circulation* **2011**, *123*, 2856–2869. [CrossRef]

64. Kendall, R.L.; Thomas, K.A. Inhibition of vascular endothelial cell growth factor activity by an endogenously encoded soluble receptor. *Proc. Natl. Acad. Sci. USA* **1993**, *90*, 10705–10709. [CrossRef] [PubMed]
65. Magee, L.A.; Brown, M.A.; Hall, D.R.; Gupte, S.; Hennessy, A.; Karumanchi, S.A.; Kenny, L.L.C.; McCarthy, F.; Myers, J.; Poon, L.C.; et al. The Hypertensive Disorders of Pregnancy: The 2021 International Society for the Study of Hypertesion in Pregnancy Classification, Diagnosis & Management Recommendations for International Practice. *Pregnancy Hypertens.* **2022**, *27*, 148–169. [PubMed]
66. Guerby, P.; Tasta, O.; Swiader, A.; Pont, F.; Bujold, E.; Parant, O.; Vayssiere, C.; Salvayre, R.; Negre-Salvayre, A. Role of oxidative stress in the dysfunction of the placental endothelial nitric oxide synthase in pre-eclampsia. *Red. Biol.* **2021**, *40*, 101861. [CrossRef] [PubMed]
67. Auge, A.P.; Longui, C.A.; Prado, R.A.A.; Neves, M.R.; Hirota, A.; Busso, N.E.; Monte, O.; Aldrighi, J.M. Abnormal inhibin A and inhibin B secretion in obese women with and without insulin resistance. *Gynecol. Endocrinol.* **2005**, *21*, 7–11. [CrossRef]
68. Esmaili, S.; Hemmati, M.; Karamian, M. Physiological role of adiponectin in different tissues: A review. *Arch. Physiol. Biochem.* **2020**, *126*, 67–73. [CrossRef]
69. Adu-Gyamfi, E.A.; Fondjo, L.A.; Owiredu, W.K.B.A.; Czika, A.M.; Nelson, W.; Lamptey, J.; Want, Y.-X.; Ding, Y.-D. The role of adiponectin in placentation and pre-eclampsia. *Cell Biochem. Funct.* **2019**, *38*, 106–117. [CrossRef]

cells

MDPI

Article

G-Protein Coupled Receptor Dysregulation May Play Roles in Severe Preeclampsia—A Weighted Gene Correlation Network Analysis of Placental Gene Expression Profile

Manuel S. Vidal, Jr. [1,*], Christian Deo T. Deguit [2], Gracia Fe B. Yu [3] and Melissa D. Amosco [4]

[1] College of Medicine, University of the Philippines, Manila 1000, Philippines
[2] Institute of Human Genetics, National Institutes of Health, University of the Philippines, Manila 1000, Philippines; ctdeguit@up.edu.ph
[3] Department of Biochemistry and Molecular Biology, University of the Philippines Manila, Manila 1000, Philippines; grafebyu@gmail.com
[4] Department of Obstetrics and Gynecology, Philippine General Hospital, University of the Philippines, Manila 1000, Philippines; mdamosco@up.edu.ph
* Correspondence: msvidal1@up.edu.ph

Abstract: Preeclampsia is one of the major hypertensive diseases of pregnancy. Genetic factors contribute to abnormal placentation. The inadequate transformation of cytotrophoblasts causes failure of maternal spiral arteries' remodeling and results in narrow, atherotic-prone vessels, leading to relative placental ischemia. This study aims to explore the possibility of identifying dysregulated gene networks that may offer a potential target in the possible prevention of preeclampsia. We performed a weighted gene correlated network analysis (WGCNA) on a subset of gene expression profiles of placental tissues from severe preeclamptic pregnancies. We identified a gene module (number of genes = 402, GS = 0.35, $p = 0.02$) enriched for several G-protein-coupled receptor (GPCR)-related genes with significant protein–protein molecular interaction (number of genes = 38, FDR = 0.0007) that may play key roles in preeclampsia. Some genes are noted to play key roles in preeclampsia, including *LPAR4/5*, *CRLR*, *NPY*, *TACR1/2*, and *SFRP4/5*, whose functions generally relate to angiogenesis and vasodilation or vasoconstriction. Other upregulated genes, including olfactory and orexigenic genes, serve limited functions in the disease pathogenesis. Altogether, this study shows the utility of WGCNA in exploring possible new gene targets, and additionally reinforces the feasibility of targeting GPCRs that may offer intervention against development and disease progression among severe preeclampsia patients.

Keywords: preeclampsia; G-protein-coupled receptors; GPCR; weighted gene correlation network analysis; WGCNA

Citation: Vidal, M.S., Jr.; Deguit, C.D.T.; Yu, G.F.B.; Amosco, M.D. G-Protein Coupled Receptor Dysregulation May Play Roles in Severe Preeclampsia—A Weighted Gene Correlation Network Analysis of Placental Gene Expression Profile. *Cells* **2022**, *11*, 763. https://doi.org/10.3390/cells11050763

Academic Editors: Ana T. Palei, Mark W. Cunningham, Lorena M. Amaral and Junie P. Warrington

Received: 31 December 2021
Accepted: 18 February 2022
Published: 22 February 2022

1. Introduction

Preeclampsia is one of the major hypertensive diseases of pregnancy, affecting about 2–8% of pregnancies [1]. Hypertensive disorders are predominant pregnancy complications, accounting for about 36.7% of maternal deaths in the Philippines [2]. The disease is characterized by new-onset hypertension beyond 20 weeks of gestation, which may further evolve to involve the hematologic, pulmonary, neurologic, and hepatic systems [1]. The pathogenesis thus has been generalized into two phases, an early defect in placentation and a later maternal syndrome due to an abnormal upregulation of placental antiangiogenic factors [3].

The etiology of preeclampsia remains unclear. Several theories have been proposed, including immunologic, genetic, and environmental factors. More recently, preeclampsia has been categorized into two subtypes depending on the period of onset or recognition of the disease [4]. Early-onset preeclampsia (<34 weeks) has been characterized to progress to relative placental ischemia due to the failure of maternal spiral arteries to remodel into high-capacitance, high-flow vessels [4]. The narrower preeclamptic spiral arteries are prone to atherotic changes, with fibrinoid necrosis, foamy macrophage accumulation, and mononuclear perivascular infiltrations [5]. On the other hand, in late-onset preeclampsia ($\geq$34 weeks), there is no dramatic change in the uteroplacental vasculature and preeclampsia itself stems from maternal stress late in pregnancy, usually associated with co-existing maternal systemic inflammatory conditions [4].

The only definitive treatment for preeclampsia is the timely delivery of the fetus and the placenta. No standard therapy has yet been approved for antepartum disease [6]. Low-dose aspirin (150 mg/day) has been shown to decrease the incidence of preeclampsia in high-risk women, and is now recommended despite its unclear mechanism [1].

As previously mentioned, several genes have been implicated in the development of a diseased placenta. Therefore, we sought to determine dysregulated genes with significant protein–protein interactions in a sample subset of severe preeclampsia patients, and to determine the possible molecular targets for intervention. Sitras et al. previously showed that specific genes involved in Notch-, tumor growth factor-beta (TGFβ)-, and vascular endothelial growth factor (VEGF)-signaling pathways were differentially expressed in severe preeclampsia. However, their analysis was limited to differential gene expression; more effective distinguishers of associated molecular pathways, such as gene correlations, were not evaluated [7]. Weighted correlation network analysis (WGCNA) provides an opportunity to discover multiple networks associated with dysregulations. By determining hub genes that act as central players within the modules, biologically meaningful insights related to the clinical traits of interest can be acquired.

Accordingly, we opted to re-analyze the placental tissue gene expression data (GSE10588) from the study of Sitras et al. through WGCNA [7]. There is a dearth of studies that utilize WGCNA to discover dysregulated networks in severe preeclampsia, and so this paper aimed to uncover and describe additional molecular pathways important in the pathology of preeclampsia.

2. Materials and Methods

2.1. Population and Evaluation of Gene Expression

Gene expression profiles were downloaded, comprising of placental tissues from Caucasian patients with normal pregnancy (n = 17) and with severe preeclampsia (n = 26) from the study by Sitras et al. (2009) from the public functional genomics data repository Gene Expression Omnibus (GEO accession number: GSE10588); 30k Human Genome Survey Microarray v.2.0 (Applied Biosystems, Waltham, MA, USA) was utilized to evaluate the gene expression profiles. Although this study was published in 2009, the last update to the gene expression dataset since then was in 2016 [7]. Severe preeclampsia was defined as either (1) at least 160 mmHg systolic blood pressure and/or 110 mmHg diastolic blood pressure, measured on at least two resting occasions 6 h apart, with proteinuria $\geq$2+ on dipstick,; or (2) hemolysis, elevated liver enzymes, and low platelet syndrome after 20 weeks of gestation. Exclusion criteria included pregnant women with pre-existing chronic hypertension or gestational hypertension without proteinuria, renal disease, systemic lupus erythematosus, or diabetes [7]. All other phenotypic data of the study population are included in Table S1.

2.2. Identification of Possible Molecular Targets through WGCNA

All data preprocessing and integration procedures such as annotation of known genes, collapsing of probes with multiple gene targets, and WGCNA were performed using R statistical programming language (version 3.5.1) (Vienna, Austria) [8]. Briefly, a scale-free topology network of genes was constructed by calculating the Pearson correlation between different pairs of genes in all samples. To calculate the adjacency matrix, the soft-thresholding power β applied to the correlation data was set to 12. The adjacency matrix was transformed into a topological overlap matrix, from which the dissimilarity values were used to construct the scale-free topology network. Gene modules from the resulting network were produced by setting the minimum module size to 100 and using the dynamic hybrid tree cut algorithm by merging modules with eigengene correlation values less than 0.60. Module eigengene (ME) values for each module were then calculated, defined as the weighted average of gene expression for each module. Module membership (MM) for each gene member in a module was also determined as the absolute correlation between the expression of the gene members and the module ME. To identify the gene modules significantly associated with preeclampsia (i.e., gene modules of interest), Gene Significance (GS) values, which correlate with ME and the presence of preeclampsia, were calculated, and a p-value of <0.05 was considered statistically significant.

Gene modules of interest were then characterized via functional enrichment analysis to identify the associated molecular pathways. This analysis was performed by uploading the entire gene list of each module to the STRING 11.0 database (https://string-db.org, accessed on 10 March 2020). This online resource can perform gene-set enrichment analysis utilizing recognized classification systems, such as Gene Ontology and Reactome [9]. STRING was also used to generate protein–protein interaction (PPI) networks (required interaction score threshold = 0.9) for the gene modules of interest.

3. Results

Twenty-two discrete clusters of strongly correlated genes were identified and assigned individual color schemes. Figure 1 shows the individual modules from the R output, showing differentially expressed gene modules determined using aggregated ME values.

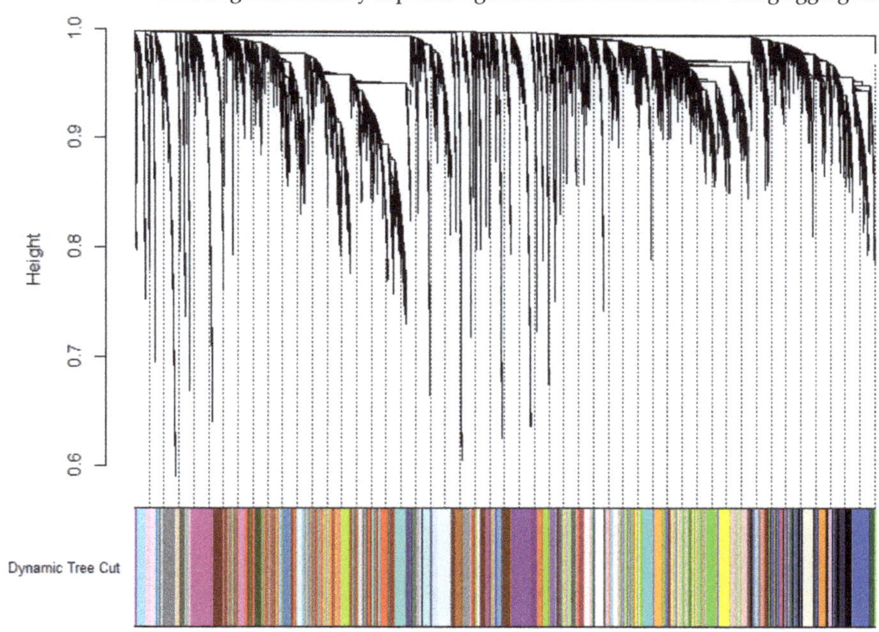

Figure 1. WGCNA results of the placental samples from severe preeclampsia patients display multiple modules of co-expressed genes, assigned by their various module colors.

Among all modules, only a handful were found to be correlated with severe preeclampsia—the dark turquoise module ($n = 402$, GS = 0.35, $p = 0.02$), dark red module ($n = 3162$, GS = 0.41, $p = 0.006$), and midnight blue module ($n = 1340$, GS = −0.69, $p = 4 \times 10^{-7}$). Figure 2 shows a summary of the modules, arranged in decreasing eigengene values. For all three modules, the dark turquoise module contained fewer genes compared to the dark red module and midnight blue module. As the latter two modules would result in a significantly large number of nodes in the resulting networks and cause further confounding, we opted to further analyze only the dark turquoise module for this study; we provide the results of the dark red and midnight blue modules in the Supplementary sections (Supplementary Tables S3 and S4, and Supplementary Figures S1 and S2). Gene enrichment analysis using the STRING database showed that most of the genes involved in the dark turquoise network belonged to the G-protein coupled receptor (GPCR) activities, as shown in Figure 3. Table 1 shows the functional enrichment analysis, while Table 2 lists genes with significant protein–protein molecular function interactions in the dark turquoise module. Individual annotations of GPCR genes are summarized in Table S1.

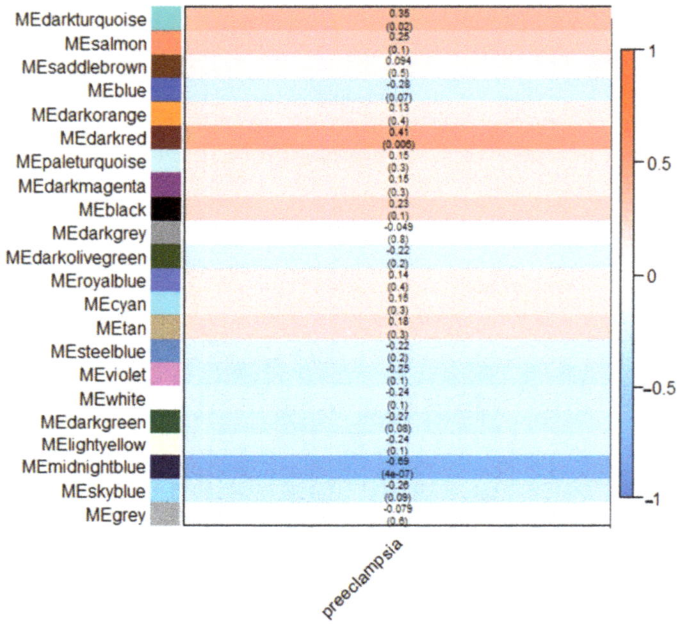

Figure 2. Module–trait relationship heatmap showing correlation of modules with severe preeclampsia. Correlation values (i.e., mean gene significance (GS)) and the corresponding p-value from correlation analysis are indicated in each cell; red shading indicates a direct correlation between the expression of the module gene members and the occurrence of severe preeclampsia, while blue shading indicates the inverse correlation.

Table 1. Selected differentially expressed gene modules, grouped according to decreasing eigengene values.

Module Color	GS Value	Number of Genes	Top Five Module Genes
Dark red	0.41 (p = 0.006)	3162	*MRPL19, FLJ12903, C2orf33, SLC25A5, ZNF627*
Dark turquoise	0.35 (p = 0.02)	402	*GPR92, GPR7, PADI3, MATN4, BAZ2B*
Midnight blue	−0.69 ($p = 4 \times 10^{-7}$)	1340	*HKE2, BANF1, CAPNS1, CREBL1, C6orf109*

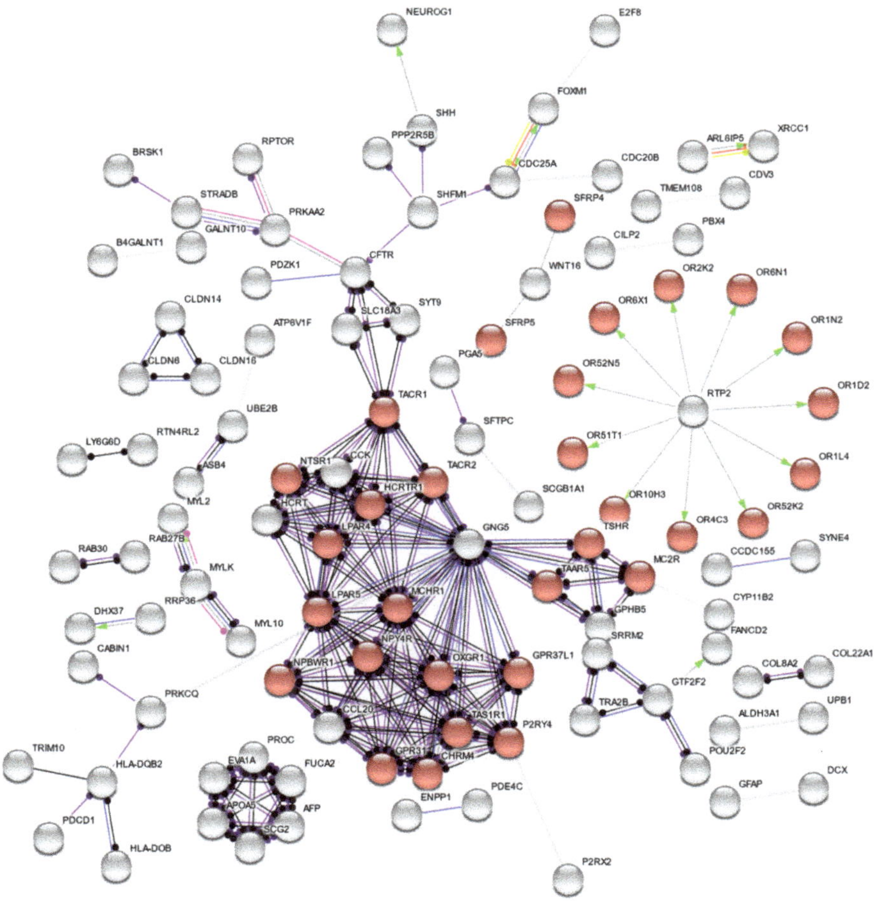

Figure 3. Protein–protein interaction network for this study, showing that the dark turquoise module is comprised of genes involved mostly in receptor activities. Highlighted in red are genes implicated in G-protein-coupled receptor activity, having the lowest false discovery rate (0.0007).

Table 2. Summary of genes with significant protein–protein molecular function interactions in the dark turquoise module, implicating genes involved mostly in receptor activities. The G-protein-coupled receptor activity network displayed the lowest false discovery rate (0.0007).

#Term ID	Description	Observed Gene Count	Background Gene Count	False Discovery Rate
GO:0004930	G-protein-coupled receptor activity	38	824	0.0007
GO:0008188	neuropeptide receptor activity	8	47	0.003
GO:0004888	transmembrane signaling receptor activity	45	1226	0.0044
GO:0008528	G-protein-coupled peptide receptor activity	12	132	0.0044
GO:0060089	molecular transducer activity	50	1483	0.0105
GO:0038023	signaling receptor activity	48	1429	0.0137

4. Discussion

The active search for a possible cure to preeclampsia is an area of great research interest. Sitras et al. utilized differential gene expression in order to determine possible gene targets for severe preeclampsia [7]; however, uncovering the molecular mechanisms underlying the disease requires a more robust bioinformatic analysis that would entail discovering significant molecular networks that would bring about the preeclamptic phenotype. In this study, we utilized WGCNA for a better noninterventional target exploration as this enables better comparisons between interacting genes instead of simply enumerating differentially expressed genes.

The majority of the upregulated genes in this study belong to the GPCR family, as seen in Figure 3 and Table 2. GPCRs have recently been implicated as potential players in the pathogenesis of preeclampsia; some have been identified to be elevated in the course of normal pregnancy and decreased in preeclamptic patients [10]. Here, we provide additional evidence that dysregulations in the GPCR profile among severe preeclampsia patients may contribute to disease progression. The GPCR activity-related genes within the dark turquoise module may also play roles in preeclampsia, as seen in Table S2.

Notable of the 47 GPCR-related genes would be the genes encoding for the lysophosphatidic acid receptors (LPAR), *LPAR4* and *LPAR5*. Lysophosphatidic acid (LPA), a product of the plasma enzyme autotaxin (ATX), exerts signaling effects via binding to its cognate receptors, LPAR1–6 [11,12]. In preeclamptic patients, interestingly, the mRNA expression levels of LPAR1–5 have been shown to be elevated in placental tissues compared to those of normal pregnant women; our findings corroborate upregulation of *LPAR4* and *LPAR5* [13]. There is a concomitant decrease in ATX production in preeclamptic patients, which has been demonstrated to occur in early-onset hypertensive diseases of pregnancy [14]. Reductions in ATX production, which translates to a reduction in LPA concentrations, in the preeclamptic placenta are involved in impaired vascular remodeling. Therefore, increasing the LPAR levels could be a compensatory action by the placenta to boost downstream Wnt activation to maintain baseline function. However, future experiments are warranted to demonstrate this hypothesis. In vivo, however, LPAR3 seems to be the only protein expressed in the placenta; however, contributions of LPAR4 and LPAR5 outside the placenta may not be dismissed [13]. LPAR4 inhibition has been demonstrated to decrease the experimental atherosclerosis elicited by adeno-associated virus expressing gain-of-function allele of the PCSK9 D377Y mutation in rats fed a fat-rich diet [15]. LPA binding to LPAR5 may result in platelet activation, and inhibition of LPAR5 activity attenuates platelet activation in vitro [16]. Therefore, the upregulations we observed in this study may be related to dysregulations outside the fetomaternal unit that may contribute to overall preeclampsia pathogenesis. Previous studies were also able to synthesize receptor-specific agonists and antagonists specific for the different LPAR subtypes [17,18]. However, these compounds are yet to be utilized for in vivo studies in diseases involving the LPARs, such as preeclampsia.

The calcitonin receptor-like receptor (CRLR) and its endogenous nonallelic vasodilatory peptide ligands, calcitonin gene-related peptide (CGRP) and adrenomedullin, mediate vasodilation in the placental endothelial and vascular smooth muscle compartments [19]. CRLR has been demonstrated to be upregulated in preeclamptic patients without changes to its native ligands in mild preeclampsia; concomitant activation of CGRP expression may increase vasodilation in the hypoxic placenta [20]. Our findings corroborate this upregulation in severe preeclampsia patients. Nitric oxide deficiency has been demonstrated to function as a switch that triggers an increase in *CRLR* expression in preeclamptic patients [21]. However, there seems to be a contradictory finding in an earlier study that found that its levels were instead downregulated in placental tissue [22]. There might be differential placental gene expression depending on the severity of preeclampsia, as seen in this study, or depending on the timing of when this gene is observed. Nonetheless, it is important to note that this gene is dysregulated in preeclampsia, and further definitive studies are needed to verify the status of its regulation in preeclampsia.

Systemic levels of neuropeptide Y (NPY), a peptide that acts as a sympathetic activator and vasoconstrictor, were found to be elevated in preeclampsia patients as compared to the normal pregnancy population [23,24]. The authors posit that decreases in the expression of NPY receptors probably leads to decreased angiogenesis and vasodilation [25]. It is plausible that NPY may function through upregulated levels of the neuropeptide Y-4 receptor (*NPY4R*) to induce vasoconstriction, although future studies need to look into the actual expression of this receptor among preeclamptic patients.

Substance P and substance K are two neurokinins that are implicated in various physiological functions, but their role in preeclampsia has not yet been established. These substances act on their specific receptors, encoded by the genes tachykinin receptor (TACR) 1 (*TACR1*) and *TACR2*, respectively. Substance P and substance K are involved in vasodilation and causes subsequent falls in blood pressure [26]. Substance P has also been shown to be upregulated following tissue injury in order to regenerate blood vessels [27]. In the context of preeclampsia, these genes encoding for tachykinin receptors may be upregulated as a compensatory mechanism in order to alleviate hypertension, although mechanistic studies are required to validate this hypothesis.

Secreted Frizzled-related proteins (SFRPs), another signaling component family, are the largest group of Wnt/β-catenin pathway inhibitors [28]. This inhibition causes eventual degradation of β-catenin, which leads to purported apoptosis of target tissues [29–31]. In particular, the SFRPs are suggested to interfere with the Wnt signal as a dominant negative mechanism [29]. Frizzled proteins may also affect angiogenesis in placental tissues, as Wnt–Fzd5 signaling has been shown to upregulate VEGF expression as well as the vascularization of primary villi [32]. *SFRP4* and *SFRP5* are observed to be upregulated in patients with severe preeclampsia [32,33]. It is suggested that a decrease in Wnt2 pathway expression in the placenta via *SFRP4* expression and inhibition may play a role in the pathophysiology of disease [31]. Interestingly, SFRP5 expression seems to be absent in the placenta in immunohistochemistry studies [33]. Therefore, it is plausible that, despite the upregulation in mRNA transcription seen in this study, *SFRP5* expression is repressed post-transcriptionally; more studies are needed to verify this observation.

Other upregulated genes that we found to have significant protein–protein interactions do not have clear physiological roles yet, or limited functions in preeclampsia and/or pregnancy at best. These genes include brain-specific angiogenesis inhibitor 3 (*BAI3*), sortilin-related VPS10 domain-containing receptor 2 (*SORCS2)*, trace amine-associated receptor 5 *(TAAR5*), and a number of G-protein-coupled receptors (GPRs) (*GPR113*), *GPR31*, *GPR37L1*, and *GPRC5D*). One gene of interest is cholinergic receptor muscarinic 4 (*CHRM4*), a choline metabolism gene that is negatively correlated with circulating soluble factor fms-like tyrosine kinase-1 (sFLT1) concentrations; it is suggested that *CHRM4* may play an as yet unknown regulatory role in sFLT1 production and preeclampsia pathogenesis [34]. Interestingly, one cluster of genes found to be significantly upregulated included a number of olfactory receptors (OR) (*OR10H3*, *OR1D2*, *OR1L4*, *OR1N2*, *OR2K2*, *OR4C3*, *OR51T1*, *OR52K2*, *OR52N5*, *OR6N1*, and *OR6X1*). No putative mechanisms have been put forth for this unusual phenomenon. It has been reported that, in the chorion portion of the fetal membranes of cases of severe preeclampsia, some olfactory receptors were upregulated compared to normal pregnancies, different from the receptors reported in this paper [35]. Although "olfactory hypersensitivity" is a known phenomenon in pregnancy or even in preeclampsia, we cannot draw any conclusions about it due to the lack of studies [36].

It would also be prudent for future studies to look into genes with significant protein–protein interactions in the dark red and midnight blue modules. As found in our Supplementary tables, these modules are enriched with functions for RNA binding and protein binding, respectively. Interestingly, immune activation in preeclampsia has been hypothesized to be, in part, caused by cell-free nucleic acids from the placenta that may bind to various Toll-like receptors (TLRs) that lead to inflammation [37]. Protein binding functions may refer to proteins such as calretinin and heparin-binding epidermal growth factor, which bind target proteins together; we hypothesize that these dysregulations have a greater effect in severe preeclampsia [38].

There are only a few studies utilizing WGCNA in the analysis of genetic predispositions toward preeclampsia. The majority of the studies in this field mainly identify gene polymorphisms in various population groups [39–42]. We reiterate that WGCNA might be a more suitable analysis compared to the analysis of differential expression of genes for a broader look into the networks responsible for the pathophysiology of the disease. Additionally, as this paper clearly focuses on the Caucasian population, the aforementioned changes may not be applicable to the general population. For instance, an analysis of multiple microarray studies in preeclampsia identified hub genes enriched in the hypoxia-inducible factor 1 (HIF-1) pathway, glutathione metabolism, and placental development [43]. Another study utilizing another group of multiple preeclampsia gene sets showed mainly immune-related genes (bone morphogenetic protein 5 (*BMP5*), cell surface glycoprotein (CD) 200R1 (*CD200R1*) and 28 (*CD28*), and TLR7) that are differentially expressed in preeclampsia [44].

Additional limitations of the study are that (1) the study population does not include a subset of patients with no proteinuria but with severe features; (2) this study generalizes severe preeclampsia gene dysregulations and does not differentiate between early- and late-onset severe preeclampsia, as well as gene predispositions that may be triggered <20 weeks of gestational age; (3) as the initial study by Sitras et al. [7] only utilized microarray studies, we are unable to do a comparative study that also uses protein levels to verify our findings at a translational level; (4) based on phenotypic data, the gestational ages between the two patient groups are significantly different (238 ± 25 weeks vs. 277 ± 9 weeks for severe preeclampsia vs. the control group), and so the initial study may have inadvertent associated changes in terms of placental gene expression that the authors cannot control for; and (5) concomitant treatment with medications such as magnesium sulfate or antihypertensive drugs may have affected the baseline gene expression and no statistical analyses were done to control for this extraneous factor. Future studies could address these five issues for a more robust look into the genetic pathophysiology of preeclampsia. Additionally, the higher incidence of preeclampsia among other ethnic groups, such as Asians and African Americans, warrants similar analyses on a larger scale in order to discover trends that may point to ethnicity-specific genes that may serve as diagnostic and prognostic markers as well as interventional points for preeclampsia.

Moreover, other publicly available gene expression datasets focusing on preeclampsia from the GEO database may be subjected to WGCNA to validate whether the modules we observed are conserved across different gene expression analysis platforms, preeclampsia clinical criteria, and tissue sources. Merging and normalization procedures can also be done on the different datasets to perform a more robust WGCNA that may point to more biologically significant pathways. Nonetheless, this study provides an example of the utilization of WGCNA to identify other genetic dysregulations that can be utilized in theoretical target modeling or experimental intervention studies.

5. Conclusions

Using WGCNA, this paper posits the role of dysregulated GPCRs in severe preeclampsia, in addition to aberrantly expressed genes in the established study by Sitras et al. [7]. Future research targeting these set of upregulated genes in severe preeclampsia must keep in mind that off-targets may result due to the ubiquitous nature of GPCRs (i.e., some genes may also be present in other organs) and that interventions must be tailored in order to specifically interfere with the relevant preeclampsia pathways. Targeting the lysophosphatidic pathway is an attractive option since LPARs act as receptors commonly activated by ATX-produced LPA, which is correspondingly expressed as an angiogenic factor for placental blood vessel formation. Similarly, targeting the CRLR and NPY pathways may also be therapeutic, since aberrant expression of both genes may be additive towards impaired LPAR expression, which leads to eventual vasoconstriction and hypoxia within endothelial cells, either in the systemic vasculature or within the placental blood supply. This study also establishes the need for a re-examination of public datasets in which WGCNA can be done in order to determine significant protein–protein interactions that may be relevant to disease progression.

Supplementary Materials: The following supporting information can be downloaded at: https://www.mdpi.com/article/10.3390/cells11050763/s1: Table S1. Phenotypic data for the patients included in the RNA microarray studies by Sitras et al. [7]. Table S2. Dark turquoise module genes related to G-protein-coupled receptor activity, and their annotated functions. Table S3. Summary of genes with significant protein–protein molecular function interactions in the dark red module. Proteins involved in RNA binding have the lowest false discovery rate (0.00098). Table S4. Summary of genes with significant protein–protein molecular function interactions in the midnight blue module. Proteins involved in protein binding have the lowest false discovery rate (0.00043). Figure S1. Protein–protein interaction network for the dark red module. The figure shows some of the connected nodes from the 1419 total nodes. Figure S2. Summary of genes with significant protein–protein molecular function interactions in the midnight blue module. The figure shows some of the connected nodes from the 1262 total nodes.

Author Contributions: Conceptualization, methodology—M.S.V.J. and C.D.T.D.; Writing—original draft—M.S.V.J.; Formal analysis—M.S.V.J. and C.D.T.D.; Writing—reviewing and editing—M.S.V.J., C.D.T.D., G.F.B.Y. and M.D.A. Supervision—M.D.A. and G.F.B.Y. All authors have read and agreed to the published version of the manuscript.

Funding: The APC was funded by the College of Medicine, University of the Philippines Manila; the MD-PhD in Molecular Medicine Committee of the College of Medicine, University of the Phil-ippines Manila; and the Department of Science and Technology—Philippine Council for Health Research and Development (DOST-PCHRD), Philippines.

Institutional Review Board Statement: Not applicable.

Informed Consent Statement: Not applicable.

Data Availability Statement: Not applicable.

Acknowledgments: M.S.V.J. would like to thank Enrico Paolo C. Banzuela, MSEd, MHPEd, FPSP for his kind assistance.

Conflicts of Interest: The authors declare no conflict of interest.

References

1. Philippine Obstetrics and Gynecologic Society. *Clinical Practice Guidelines on Hypertension in Pregnancy;* Philippine Obstetrics and Gynecologic Society: Quezon City, Philippines, 2015.
2. Department of Health—National Epidemiology Center. *The 2013 Philippine Health Statistics;* Epidemiology Bureau: Manila, Philippines, 2014. Available online: http://www.dohgovph/sites/default/files/publications/2013PHScompressed_0.pdf (accessed on 10 March 2020).
3. Romero, R.; Chaiworapongsa, T. Preeclampsia: A link between trophoblast dysregulation and an antiangiogenic state. *J. Clin. Investig.* **2013**, *123*, 2775–2777. [CrossRef] [PubMed]
4. Staff, A.C. The two-stage placental model of preeclampsia: An update. *J. Reprod. Immunol.* **2019**, *134–135*, 1–10. [CrossRef] [PubMed]
5. Herting, A. Vascular pathology in hypertensive albuminuric toxemias of pregnancy. *Clinics* **1945**, *4*, 602.
6. Rana, S.; Lemoine, E.; Granger, J.P.; Karumanchi, S.A. Preeclampsia: Pathophysiology, challenges, and perspectives. *Circ. Res.* **2019**, *124*, 1094–1112. [CrossRef]
7. Sitras, V.; Paulssen, R.; Grønaas, H.; Leirvik, J.A.; Hanssen, T.; Vårtun, Å.; Acharya, G. Differential Placental Gene Expression in Severe Preeclampsia. *Placenta* **2009**, *30*, 424–433. [CrossRef]
8. Langfelder, P.; Horvath, S. WGCNA: An R package for weighted correlation network analysis. *BMC Bioinform.* **2008**, *9*, 559. [CrossRef]
9. Szklarczyk, D.; Gable, A.L.; Lyon, D.; Junge, A.; Wyder, S.; Huerta-Cepas, J.; Simonovic, M.; Doncheva, N.T.; Morris, J.H.; Bork, P.; et al. STRING v11: Protein–protein association networks with increased coverage, supporting functional discovery in genome-wide experimental datasets. *Nucleic Acids Res.* **2019**, *47*, D607–D613. [CrossRef]
10. McGuane, J.T.; Conrad, K.P. GPCRs as potential therapeutic targets in preeclampsia. *Drug Discov. Today Dis. Model.* **2012**, *9*, e119–e127. [CrossRef]

11. Iwasawa, Y.; Fujii, T.; Nagamatsu, T.; Kawana, K.; Okudaira, S.; Miura, S.; Matsumoto, J.; Tomio, A.; Hyodo, H.; Yamashita, T.; et al. Original Article: Expression of Autotaxin, an Ectoenzyme that Produces Lysophosphatidic Acid, in Human Placenta: Autotaxin in Human Placenta. *Am. J. Reprod. Immunol.* **2009**, *62*, 90–95. [CrossRef]
12. Tokumura, A.; Kanaya, Y.; Miyake, M.; Yamano, S.; Irahara, M.; Fukuzawa, K. Increased production of bioactive lysophosphatidic acid by serum lysophospholipase D in human pregnancy. *Biol. Reprod.* **2002**, *67*, 1386–1392. [CrossRef]
13. Fujii, T.; Nagamatsu, T.; Schust, D.J.; Ichikawa, M.; Kumasawa, K.; Yabe, S.; Iriyama, T.; Hirota, Y.; Osuga, Y.; Aoki, J.; et al. Placental expression of lysophosphatidic acid receptors in normal pregnancy and preeclampsia. *Am. J. Reprod. Immunol.* **2019**, *82*, e13176. [CrossRef]
14. Ichikawa, M.; Nagamatsu, T.; Nishimori, Y.; Fujii, T.; Kumasawa, K.; Iriyama, T.; Osuga, Y.; Fujii, T. Role of placental ATX-LPA system and its pathologic relevance to Hypertensive disorders of pregnancy. *Placenta* **2019**, *87*, 64–65. [CrossRef]
15. Yang, L.; Kraemer, M.; Fang, X.F.; Angel, P.M.; Drake, R.R.; Morris, A.J.; Smyth, S.S. LPA receptor 4 deficiency attenuates experimental atherosclerosis. *J. Lipid Res.* **2019**, *60*, 972–980. [CrossRef]
16. Kozian, D.H.; Evers, A.; Florian, P.; Wonerow, P.; Joho, S.; Nazare, M. Selective non-lipid modulator of LPA5 activity in human platelets. *Bioorg. Med. Chem. Lett.* **2012**, *22*, 5239–5243. [CrossRef]
17. Tamaruya, Y.; Suzuki, M.; Kamura, G.; Kanai, M.; Hama, K.; Shimizu, K.; Aoki, J.; Arai, H.; Shibasaki, M. Identifying Specific Conformations by Using a Carbohydrate Scaffold: Discovery of Subtype-Selective LPA-Receptor Agonists and an Antagonist. *Angew. Chem. Int. Ed.* **2004**, *43*, 2834–2837. [CrossRef]
18. Kano, K.; Matsumoto, H.; Inoue, A.; Yukiura, H.; Kanai, M.; Chun, J.; Ishii, S.; Shimizu, T.; Aoki, J. Molecular mechanism of lysophosphatidic acid-induced hypertensive response. *Sci. Rep.* **2019**, *9*, 2662. [CrossRef]
19. Dong, Y.-L.; Vegiraju, S.; Chauhan, M.; Gangula, P.R.R.; Hankins, G.D.V.; Goodrum, L.; Yallampalli, C. Involvement of calcitonin gene-related peptide in control of human fetoplacental vascular tone. *Am. J. Physiol. Circ. Physiol.* **2004**, *286*, H230–H239. [CrossRef]
20. Ariza, A.C.; Bobadilla, N.; Díaz, L.; Avila, E.; Larrea, F.; Halhali, A. Placental gene expression of calcitonin gene-related peptide and nitric oxide synthases in preeclampsia: Effects of magnesium sulfate. *Magnes. Res.* **2009**, *22*, 44–49. [CrossRef]
21. Chauhan, M.; Balakrishnan, M.; Yallampalli, C. 334: Regulation of CRLR and RAMPs by nitric oxide in endothelial cells. *Am. J. Obstet. Gynecol.* **2016**, *214*, S188. [CrossRef]
22. Dong, Y.-L.; Green, K.E.; Vegiragu, S.; Hankins, G.D.V.; Martin, E.; Chauhan, M.; Thota, C.; Yallampalli, C. Evidence for Decreased Calcitonin Gene-Related Peptide (CGRP) Receptors and Compromised Responsiveness to CGRP of Fetoplacental Vessels in Preeclamptic Pregnancies. *J. Clin. Endocrinol. Metab.* **2005**, *90*, 2336–2343. [CrossRef]
23. Khatun, S.; Kanayama, N.; Belayet, H.M.; Bhuiyan, A.B.; Jahan, S.; Begum, A.; Kobayashi, T.; Terao, T. Increased concentrations of plasma neuropeptide Y in patients with eclampsia and preeclampsia. *Am. J. Obstet. Gynecol.* **2000**, *182*, 896–900. [CrossRef]
24. Paiva, S.P.C.; Veloso, C.A.; Campos, F.F.C.; Carneiro, M.M.; Tilan, J.U.; Wang, H.; Umans, J.G.; Zukowska, Z.; Kitlinska, J. Elevated levels of neuropeptide Y in preeclampsia: A pilot study implicating a role for stress in pathogenesis of the disease. *Neuropeptides* **2016**, *55*, 127–135. [CrossRef]
25. Klinjampa, R.; Sitticharoon, C.; Souvannavong-Vilivong, X.; Sripong, C.; Keadkraichaiwat, I.; Churintaraphan, M.; Chatree, S.; Lertbunnaphong, T. Placental Neuropeptide Y (NPY) and NPY receptors expressions and serum NPY levels in preeclampsia. *Exp. Biol. Med.* **2019**, *244*, 380–388. [CrossRef]
26. Hua, X.; Lundberg, J.M.; Theodorsson-Norheim, E.; Brodin, E. Comparison of cardiovascular and bronchoconstrictor effects of substance P, substance K and other tachykinins. *Naunyn-Schmiedebergs Arch. Exp. Pathol. Pharmakol.* **1984**, *328*, 196–201. [CrossRef]
27. Walsh, D.A.; McWilliams, D.F. Tachykinins and the Cardiovascular System. *Curr. Drug Targets* **2006**, *7*, 1031–1042. [CrossRef]
28. Bovolenta, P.; Esteve, P.; Ruiz, J.M.; Cisneros, E.; Lopez-Rios, J. Beyond Wnt inhibition: New functions of secreted Frizzled-related proteins in development and disease. *J. Cell Sci.* **2008**, *121*, 737–746. [CrossRef]
29. Bafico, A.; Gazit, A.; Pramila, T.; Finch, P.W.; Yaniv, A.; Aaronson, S.A. Interaction of Frizzled Related Protein (FRP) with Wnt Ligands and the Frizzled Receptor Suggests Alternative Mechanisms for FRP Inhibition of Wnt Signaling. *J. Biol. Chem.* **1999**, *274*, 16180–16187. [CrossRef]
30. Lacher, M.D.; Siegenthaler, A.; Jäger, R.; Yan, X.; Hett, S.; Xuan, L.; Saurer, S.; Lareu, R.R.; Dharmarajan, A.M.; Friis, R. Role of DDC-4/sFRP-4, a secreted Frizzled-related protein, at the onset of apoptosis in mammary involution. *Cell Death Differ.* **2003**, *10*, 528–538. [CrossRef]
31. Zhang, Z.; Zhang, L.; Zhang, L.; Jia, L.; Wang, P.; Gao, Y. Association of Wnt2 and sFRP4 Expression in the Third Trimester Placenta in Women With Severe Preeclampsia. *Reprod. Sci.* **2013**, *20*, 981–989. [CrossRef]
32. Lu, J.; Zhang, S.; Nakano, H.; Simmons, D.G.; Wang, S.; Kong, S.; Wang, Q.; Shen, L.; Tu, Z.; Wang, W.; et al. A Positive Feedback Loop Involving Gcm1 and Fzd5 Directs Chorionic Branching Morphogenesis in the Placenta. *PLoS Biol.* **2013**, *11*, e1001536. [CrossRef]
33. Zhang, Y.; Ran, Y.; Ma, Y.; Huang, H.; Chen, Y.; Qi, H. Elevated Serum SFRP5 Levels During Preeclampsia and Its Potential Association with Trophoblast Dysfunction via Wnt/β-Catenin Suppression. *Reprod. Sci.* **2022**, *29*, 163–172. [CrossRef] [PubMed]
34. Jiang, X.; Bar, H.Y.; Yan, J.; Jones, S.; Brannon, P.M.; West, A.A.; Perry, C.A.; Ganti, A.; Pressman, E.; Devapatla, S.; et al. A higher maternal choline intake among third-trimester pregnant women lowers placental and circulating concentrations of the antiangiogenic factor fms-like tyrosine kinase-1 (sFLT1). *FASEB J.* **2013**, *27*, 1245–1253. [CrossRef] [PubMed]

35. Garrido-Gomez, T.; Ona, K.; Kapidzic, M.; Gormley, M.; Simón, C.; Genbacev, O.; Fisher, S.J. Severe preeclampsia is associated with alterations in cytotrophoblasts of the smooth chorion. *Development* **2017**, *144*, 767–777. [CrossRef] [PubMed]
36. Cameron, E.L. Pregnancy and olfaction: A review. *Front. Psychol.* **2014**, *5*, 67. [CrossRef] [PubMed]
37. Konečná, B.; Lauková, L.; Vlková, B. Immune Activation by Nucleic Acids: A Role in Pregnancy Complications. *Scand. J. Immunol.* **2018**, *87*, e12651. [CrossRef] [PubMed]
38. Shu, C.; Liu, Z.; Cui, L.; Wei, C.; Wang, S.; Tang, J.J.; Cui, M.; Lian, G.; Li, W.; Liu, X.; et al. Protein Profiling of Preeclampsia Placental Tissues. *PLoS ONE* **2014**, *9*, e112890. [CrossRef] [PubMed]
39. Liu, J.; Song, G.; Zhao, G.; Meng, T. Gene polymorphism associated with TGF-β1 and susceptibility to preeclampsia: A meta-analysis and trial sequential analysis. *J. Obstet. Gynaecol. Res.* **2021**, *47*, 2031–2041. [CrossRef]
40. Liu, Y.; Li, P. Meta-Analysis of Angiotensin-Converting Enzyme Insersion/Delection Polymorphism and Pre-Eclampsia Susceptibility. *J. Obstet. Gynaecol. Res.* **2020**, *46*, 1744–1757. [CrossRef]
41. Wang, X.; Sun, T.; Chen, G.; Gao, H. Association between Vascular Endothelial Growth Factor Gene Polymorphisms and Pre-Eclampsia Susceptibility: An Updated Meta-Analysis. *Immunol. Investig.* **2020**, *49*, 120–133. [CrossRef]
42. Miller, A.K.; Azhibekov, T.; O'Toole, J.F.; Sedor, J.R.; Williams, S.M.; Redline, R.W.; Bruggeman, L.A. Association of preeclampsia with infant APOL1 genotype in African Americans. *BMC Med. Genet.* **2020**, *21*, 110. [CrossRef]
43. Kang, Q.; Li, W.; Xiao, J.; Yu, N.; Fan, L.; Sha, M.; Ma, S.; Wu, J.; Chen, S. Integrated Analysis of Multiple Microarray Studies to Identify Novel Gene Signatures in Preeclampsia. *Placenta* **2021**, *105*, 104–118. [CrossRef]
44. Meng, Y.; Li, C.; Liu, C.-X. Immune Cell Infiltration Landscape and Immune Marker Molecular Typing in Preeclampsia. *Bioengineered* **2021**, *12*, 540–554. [CrossRef]

cells

Article

Progesterone Induced Blocking Factor Reduces Hypertension and Placental Mitochondrial Dysfunction in Response to sFlt-1 during Pregnancy

Evangeline Deer [1], Jalisa Jones [1], Denise C. Cornelius [2], Kyleigh Comley [1], Owen Herrock [1], Nathan Campbell [1], Sarah Fitzgerald [1], Tarek Ibrahim [1], Babbette LaMarca [1,3] and Lorena M. Amaral [1,*]

1 Department of Pharmacology and Toxicology, University of Mississippi Medical Center, Jackson, MS 39216, USA; edeer@umc.edu (E.D.); jonespjalisa@gmail.com (J.J.); kmcomley@go.olemiss.edu (K.C.); oherrock@umc.edu (O.H.); ncampbell@umc.edu (N.C.); sjfitzgerald@umc.edu (S.F.); ibrahimt@uthscsa.edu (T.I.); bblamarca@umc.edu (B.L.)
2 Department of Emergency Medicine, University of Mississippi Medical Center, Jackson, MS 39126, USA; dcornelius@umc.edu
3 Department of Obstetrics and Gynecology, University of Mississippi Medical Center, Jackson, MS 39216, USA
* Correspondence: lmamaral@umc.edu

Abstract: Preeclampsia (PE) is characterized by new onset hypertension in association with placental ischemia, reduced fetal weight, elevated soluble fms-like tyrosine kinase-1 (sFlt-1), and placental mitochondrial (mt) dysfunction and oxidative stress (ROS). Progesterone induced blocking factor (PIBF) is a product of progesterone signaling that blocks inflammatory processes and we have previously shown PIBF to lower mean arterial blood pressure (MAP) and sFlt-1 in a rat model of PE. Infusion of sFlt-1 causes hypertension and many characteristics of PE in pregnant rodents, however, its role in causing mt dysfunction is unknown. Therefore, we hypothesize that PIBF will improve mt function and MAP in response to elevated sFlt-1 during pregnancy. We tested our hypothesis by infusing sFlt-1 via miniosmotic pumps in normal pregnant (NP) Sprague-Dawley rats ($3.7\ \mu g \cdot kg^{-1} \cdot day^{-1}$) on gestation days (GD) 13–19 in the presence or absence of PIBF (2.0 μg/mL) injected intraperitoneally on GD 15 and examined mean arterial blood pressure (MAP) and placental mt ROS on GD 19. sFlt-1 increased MAP to 112 + 2 (n = 11) compared to NP rats (98 + 2 mmHg, n = 15, $p < 0.05$), which was lowered in the presence of sFlt-1 (100 + 1 mmHg, n = 5, $p < 0.05$). Placental mtATP was reduced in sFlt-1 infused rats versus NP controls, but was improved with PIBF. Placental mtROS was elevated with sFlt-1 compared to NP controls, but was reduced with PIBF. Sera from NP + sFlt-1 increased endothelial cell mtROS, which was attenuated with PIBF. These data demonstrate sFlt-1 induced HTN during pregnancy reduces placental mt function. Importantly, PIBF improved placental mt function and HTN, indicating the efficacy of improved progesterone signaling as potential therapeutics for PE.

Keywords: hypertension; preeclampsia; sFlt-1; oxidative stress; placental ischemia

Citation: Deer, E.; Jones, J.; Cornelius, D.C.; Comley, K.; Herrock, O.; Campbell, N.; Fitzgerald, S.; Ibrahim, T.; LaMarca, B.; Amaral, L.M. Progesterone Induced Blocking Factor Reduces Hypertension and Placental Mitochondrial Dysfunction in Response to sFlt-1 during Pregnancy. *Cells* **2021**, *10*, 2817. https://doi.org/10.3390/cells10112817

Academic Editor: Agniezka Rak

Received: 7 August 2021
Accepted: 10 October 2021
Published: 20 October 2021

1. Introduction

Preeclampsia (PE) is defined as new onset hypertension during pregnancy after the 20th week of gestation. It is associated with or without chronic immune activation, proteinuria, fetal growth restriction, and maternal endothelial dysfunction [1]. PE is a major cause of maternal, fetal, and neonatal morbidity and mortality affecting 5–7% of pregnancies worldwide [2]. Unfortunately, pregnant women with PE are at a higher risk of developing severe complications such as placental abruption, cerebrovascular dysfunction, multiorgan dysfunction, and disseminated intravascular coagulation. The fetus is at risk for intrauterine growth restriction, prematurity, and intrauterine death [3]. Despite being a leading cause of maternal death and maternal and perinatal morbidity, the mechanisms responsible for the pathogenesis of PE are unclear. Therefore, additional investigation into the pathophysiological mechanisms resulting in the development of PE is needed in order to elucidate potential therapies.

In PE, hypertension develops during the late second or third trimesters of pregnancy and remits after delivery in most cases, thereby implicating the placenta as the primary culprit in the disease. Studies have indicated that inflammatory cytokines increase and that there is an imbalance of regulatory and effector T cells and reactive oxygen species. Studies have implicated an imbalance in angiogenic factors, such as increased soluble fms-like tyrosine kinase 1 (sFlt-1) and decreased placental and vascular endothelial growth factor. These factors have been shown to mediate endothelial dysfunction causing the release of endothelin-1 (ET-1) and a decrease of vasodilators, such as nitric oxide, which contributes to the development of hypertension [4–12]. Increased sFlt-1 is known to decrease renal function and cause hypertension in non-pregnant women and during pregnancy [13]. Additionally, circulating and placental levels of sFlt-1 mRNA have been notably higher in women who have PE compared to normal pregnant women [14]. Tam Tam et al. demonstrated that increases in blood pressure in response to chronic sFlt-1 in pregnant rats is associated with increases in ROS in the placenta, kidney, and aorta, which was attenuated with Tempol, demonstrating that ROS is an important mediator of hypertension in response to sFlt-1 during pregnancy [15]. Although many studies demonstrate the importance of sFlt-1 to cause detrimental effects to the fetus and maternal hypertension during pregnancy, we do not know if mt dysfunction is an addition mechanism compromised by elevated sFlt-1 during PE.

Oxidative stress is an imbalance between reactive oxygen species (ROS) and antioxidant defense in the cell and occurs in normal pregnancies, however, overproduction of ROS leads to a decline of antioxidants, resulting in adverse pregnancy outcomes [16–18]. ROS have a major role in cell physiology as second messengers in signaling pathways, but also have a central role in the initiation of placental and endothelial dysfunction [16]. A central function of mitochondria (mt) is to produce energy in the form of adenosine triphosphate (ATP) making mt integral to cell survival. Changes in mitochondrial metabolic and bioenergetics, such as elevations in mt ROS and a decrease in respiration, can cause mitochondrial structural changes or death [19]. This decrease in ATP and electron transport activity (ETC) reduces important cellular functions, leading to tissue damage and a change in organ homeostasis [20]. Our previous studies have demonstrated a role for mitochondrial dysfunction and oxidative stress in the kidney and placenta of the pregnant RUPP rat model of PE or in response to RUPP CD4+ T cells or TNF-α during pregnancy, indicating an important role for inflammation to mediate multi-organ mt ROS and dysfunction in association with hypertension in pregnant rats [21–24].

A substantial amount of evidence indicates that progesterone may balance the inflammatory environment in the early trimesters of pregnancy [25]. In the presence of progesterone, lymphocytes release a protein known as progesterone induced blocking factor (PIBF), which mediates anti-inflammatory effects [26–28]. Previously, we have shown that circulating progesterone is decreased in PE patients and that progesterone decreased ET-1 secretion in vascular endothelial cells treated with serum from preeclamptic patients [29]. We have shown that PIBF improves inflammatory cytokines and decreases NK cell activation and CD4+ T cells in association with lowered ET-1 and sFLT-1 and, thus, hypertension in RUPP rats [30]. Importantly, PIBF normalized signs of PE, such as sFlt-1, which is known to play a role in hypertension and reducing fetal weight in PE [14,31]. However, it is unknown if PIBF can affect the role of mt mediated ROS in response to sFlt-1 induced hypertension during pregnancy.

2. Materials and Methods

2.1. Animals

Pregnant female Sprague Dawley (SD) rats weighing approximately 200–250 g were purchased from Envigo (Indianapolis, IN, USA) for use in the study. Rats were housed in a temperature-controlled room (23 °C) with a 12:12 hour light/dark cycle with free access to standard rat chow and water. The experiments were in compliance with the guidelines of the University of Mississippi Medical Center, and the animals were handled based on the principles in the National Institutes of Health Guide for the Care of Animals and the Institutional Animal Care and Use Committee (IACUC).

2.2. *Infusion of sFlt-1/PIBF into Pregnant Sprague Dawley Rats*

Rats were divided into two control groups, consisting of normal pregnant (NP, n = 10) and sFlt-1 induced hypertensive rats (NP+ sFlt-1, n = 9). On gestational day (GD) 13, sFlt-1 was infused into NP Sprague Dawley rats (3.7 $\mu g \cdot kg^{-1} \cdot day^{-1}$) for 6 days via mini-osmotic pumps (model 2001, Alzet Scientific Corporation, Cupertino, CA, USA) as previously described (11). PIBF (2.0 μg/mL) (NP + PIBF) (n = 9) was injected intraperitoneally on GD 15 to sFlt-1 induced hypertensive pregnant rats. On day 18, indwelling carotid catheters were inserted in all groups for blood pressure measurements. Using isoflurane anesthesia (Webster, Sterling, MA, USA), catheters inserted were V3 tubing (Scientific Commodities, Inc., Lake Havasu City, AZ, USA), and were tunneled to the back of the neck and exteriorized. Carprofen (5 mg/kg) was administered via subcutaneous injection immediately following surgical procedure. On GD 19, blood pressure was measured with a pressure transducer (Cobe II tranducer CDX Sema, Aurora, CO, USA) and recorded continuously for one-hour after a 30 min stabilization period as previously described (11). Subsequently, placental and fetal weights were measured and blood and placentas were collected for mitochondrial function analysis.

2.3. *Isolation of Mitochondria*

Placental mitochondria were isolated using the differential centrifugation method [21,32]. Tissues were rinsed and homogenized using a Dounce homogenizer. The homogenate was centrifuged at 4000 rpm for 3 min at 4 °C. The supernatant was centrifuged at 10,000 rpm for 10 min at 4 °C, and the pellet was collected, suspended in 1 mL of Mito I buffer (250 mM sucrose, 10 mM HEPES, 1 mM EGTA 0.1% BSA, pH 7.2), and centrifuged at 10,000 rpm for 10 min at 4 °C. Afterwards, the pellet was suspended in 1 mL of Mito II (250 mM sucrose, 10 mM HEPES, 0.1% BSA, pH 7.2), and centrifuged at 10,000 rpm for 10 min at 4 °C. The final pellet was collected and suspended in 200 μL mL of Mito II buffer and used for respiration and ROS experiments.

2.4. *Mitochondrial Respiration*

Respiration in isolated placental mitochondrial was measured using the Oxgraph 2K (Oroboros Instruments, Innsbruck, Austria). The basal, state 2, state 3, state 4, and uncoupled respiration rates were measured using glutamate/malate, ADP, oligomycin, and FCCP (carbonyl cyanide-4-[trifluoromethoxy]phenylhydrazone), respectively [21]. Non-mitochondrial respiration was measured with the use of rotenone and antimycin A. The collected data were analyzed and expressed as pmol of oxygen consumed per second per milligram of mitochondrial protein.

2.5. *Mitochondrial ROS*

Mitochondrial hydrogen oxygen peroxide (H_2O_2) production in placental mitochondria were determined using the amplex red assay [21,33]. Mitochondria (0.4 mg/mL) were incubated in a 96 well plate containing respiration buffer, superoxide dismutase (40 U/mL), horseradish peroxidase (4 U/mL), and succinate (10 mM). Amplex red (10 μM) was added to the wells last to initiate the reaction. The real-time production of H_2O_2 was measured using a plate reader at 555/581 nm excitation/emission for 30 min at 25 °C. Sample controls (blanks without amplex red or mitochondrial protein) were included in the assay.

2.6. *Endothelial Mitochondrial ROS*

MitoSOX red, a fluorogenic dye that targets the mitochondria in live cells, was used to measure mitochondrial specific reactive oxygen species. To determine vascular endothelial cell mt function, human umbilical venous endothelial cells (HUVECs) (ATCC) were grown in HUVEC media (Medium 199-DMEM (50:50), 10% FBS, and 1% antimycotic/antibiotic) to a 70% confluence on gelatin coated T25 culture flasks in a humidified atmosphere of 5% CO_2 at 37 °C. At passage 4, the HUVEC cells were cultured in 6-well plates and incubated overnight. Cells were serum starved for 4h and incubated with 10% experimental NP serum (n = 6), NP + sFlt-1 serum (n = 5), or NP + sFlt-1+ PIBF serum (n = 5) overnight. Our lab has established that there is a significant increase in HUVEC mtROS after incubation with 10% RUPP rat serum [21]. After overnight incubation, experimental serums were washed off and cells were incubated with MitoSOX red (5 µM) for 30 min at 37 °C. Antimycin A (100 µM) was used as a positive control for the experiment. After washing the cells with DPBS twice, serum free medium was added and the cells were incubated for 4 hours. In the final stage, cells were collected and analyzed using flow cytometry in the FL2 channel (Miltenyi MACSQuant Analyzer 10, San Diego, CA, USA).

2.7. Statistical Analysis

Statistical analyses were performed using GraphPad Prism 7.02 software (GraphPad Software, San Diego, CA, USA). A one-way ANOVA with Bonferroni multiple comparisons test as post-hoc analysis were conducted for normally distributed variables. The Mann-Whitney U test or Kruskal–Wallis one-way ANOVA with Dunn's multiple comparison post-hoc test were used for comparison of non-normally distributed variables. Results were reported as means ± SEM and were considered as statistically significant when $p < 0.05$.

3. Results

Mean arterial pressure (MAP) was increased in sFlt-1 infused rats to 112 ± 2 (n = 9) compared to control NP rats 98 ± 2 mmHg (n = 10, $p < 0.05$), Figure 1. Administration of PIBF reduced MAP to 100 ± 1 in the presence of sFlt1 (n = 9, $p < 0.05$). There were no significant changed in the body weights, placental weights, and fetal weights (Table 1).

Although not significant, placental mitochondrial state 3 respiration was lower in NP + sFlt-1 infused rats (191 ± 54 pmol of O2/s/mg, n = 5) compared to NP controls (263 ± 18 pmol of of O2/s/mg, n = 4) and was normalized in NP + sFlt-1+PIBF (309 ± 153 pmol of O2/s/mg, n = 5), (Figure 2A). Uncoupled respiration also decreased in NP + sFlt-1 infused rats (96 ± 34 pmol of O2/s/mg, n = 5) compared to NP controls (159 ± 16 pmol of O2/s/mg, n = 4) and was normalized with PIBF (199 ± 93 pmol of O2/s/mg, n = 5) (Figure 2B).

Moreover, placental mitochondrial ROS in NP + sFlt-1 was 429 ± 32% fold (n = 5, $p < 0.05$) compared to NP controls (100 ± 6% fold, n = 5), but was normalized in NP + sFlt-1 + PIBF (234 ± 15 % fold, n = 5, $p < 0.05$) (Figure 3). Additionally, HUVECs incubated with 10% sera from NP + sFlt-1 rats exhibited increased maximal respiration (2.88 ± 0.18% gated, n = 5) compared to HUVECS treated with NP sera (1.26 ± 0.59% gated, n = 6), but HUVECS treated with sera from NP + sFlt-1 with PIBF was normalized (0.68 ± 0.34% gated, n = 5, $p < 0.05$) as shown in Figure 4.

Figures, Tables, and Schemes

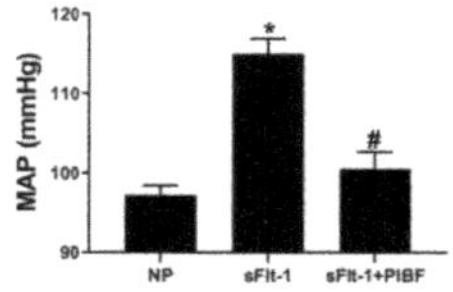

Figure 1. Infusion of sFlt-1 into NP rats increased MAP (n = 9) when compared with NP rats (n = 10, $p < 0.05$). Treatment with PIBF in the presence of sFlt-1 reduced MAP (n = 9, $p < 0.05$). Statistical differences were established using a one-way ANOVA. Results were reported as means ± SEM and considered statistically significant when $p < 0.05$ with using one-ANOVA. * $p < 0.05$ vs. NP control; # $p < 0.05$ vs. sFlt-1.

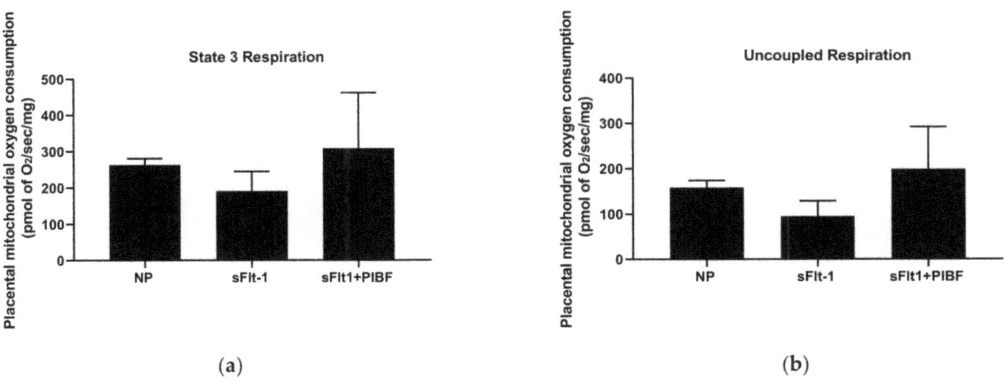

Figure 2. (**a**) State 3 placental mitochondrial respiration was reduced in placentas of sFlt-1 (n = 5) infused rats when compared to NP controls (n = 4), but was improved with PIBF (n = 5). (**b**) Uncoupled placental mitochondrial respiration was reduced in placentas of sFlt-1 infused rats (n = 5) when compared to NP controls (n = 4), but was improved with PIBF (n = 5). Statistical differences were established using a Kruskal–Wallis one-way ANOVA with Dunn's multiple comparison post-hoc test. Results were reported as means ± SEM and considered statistically significant when $p < 0.05$.

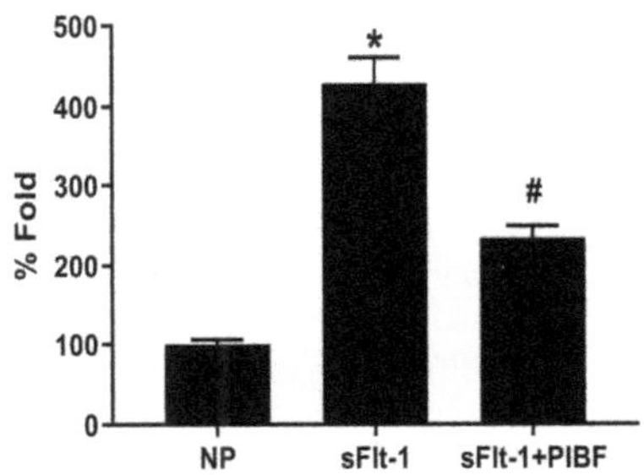

Figure 3. Mitochondrial ROS was significantly elevated in NP + sFlt-1 (n = 5, $p < 0.05$) but was lowered in both NP + sFlt-1+ PIBF (n = 5, $p < 0.05$) and NP controls (n = 5). Statistical differences were established using a Kruskal–Wallis one-way ANOVA with Dunn's multiple comparison post-hoc test. Results were reported as means ± SEM and considered statistically significant when $p < 0.05$. * $p < 0.05$ vs. NP control; # $p < 0.05$ vs. sFlt-1 Kruskal–Wallis.

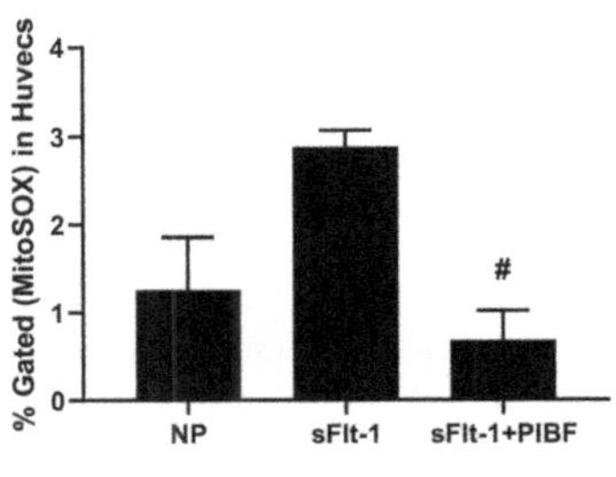

Figure 4. Sera from NP + sFlt-1 supplemented with PIBF attenuated endothelial cell mitochondrial ROS (n = 5, $p < 0.05$) compared to sera from NP + sFlt-1 (n = 5). Also, endothelial mt function was significantly reduced in NP + sFLT-1 compared to NP controls (n = 6). Statistical differences were established using a Kruskal–Wallis one-way ANOVA with Dunn's multiple comparison post-hoc test. Results were reported as means ± SEM and considered statistically significant when $p < 0.05$. # $p < 0.05$ vs. sFlt-1 Kruskal–Wallis.

Table 1. Body, placental, and fetal weights.

	Body Weights (g)	Placental Weights (g)	Fetal Weights (g)
NP	317 ± 9.32	0.57 ± 0.02	2.27 ± 0.06
sFlt-1	322 ± 8.07	0.56 ± 0.01	2.31 ± 0.07
sFlt-1+PIBF	303 ± 6	0.54 ± 0.02	2.19 ± 0.06

4. Discussion

Because current treatments for PE are not continuously effective to improve maternal and fetal outcomes, there is an emphasis on drug discovery for the treatment of PE. The current therapeutics used for the development of severe preterm PE during pregnancy, prior to 34 weeks gestation, include intravenously-infused magnesium sulfate to slow disease progression and prevent maternal seizure, potent glucocorticoids to enhance fetal lung maturation, and anti-hypertensives to prevent stroke. Notably, we have utilized various anti-inflammatory agents, such as the T cell suppressor abatecept (Orencia) [34], tumor necrosis factor-α inhibitor etanercept [35], B lymphocyte-depleting agent rituximab [36], and IL-17RC [37] to decrease hypertension and circulating factors in RUPP rat model of PE that are also elevated during preeclampsia. Furthermore, all of these anti-inflammatory agents were able to effectively decrease blood pressure and inflammation in our RUPP model of PE, but have not been considered obstetrically safe for use in PE women. Considering our previously published data that progesterone is lower in PE women at out hospital compared to those with uneventful pregnancies, we believe that the utilization of progesterone in preeclamptic patients, or those who have developed clinical manifestations of preterm PE, is an area that requires additional focus [29]. We have shown that progesterone in the form of 17-OHPC administered after placental ischemia improved ET-1, uterine artery resistance, intrauterine growth restriction (IUGR), and blood pressure in rat models of PE [38,39], thereby further establishing proof of concept supporting that progesterone could be a beneficial treatment for PE. However, the mechanisms of action whereby progesterone supplementation could be beneficial for PE women was lacking. Ours and other labs have shown that PIBF, a product of progesterone signaling, lowers inflammation and vasoactive factors such as sFlt-1 and, thus, hypertension in the RUPP rat model of PE.

sFlt-1 is notoriously associated with PE, reduced renal function, and IUGR during pregnancy. We and others have shown that sFlt-1 induced hypertension is associated with reactive oxygen species (ROS) in pregnant rodents. Similar to sFlt-1, oxidative stress has an important role in the pathophysiology of PE. Therefore, in this study we tested the effect of sFlt-1 to cause mitochondrial function and the role of PIBF to normalize it and hypertension in a rat model of sFlt-1 induced hypertension during pregnancy. We demonstrated that PIBF lowers blood pressure in response to sFlt-1 and normalized mitochondrial function in the placenta of sFlt-1 induced hypertensive rats.

Mitochondrial oxidative stress has become an area of emphasis for many labs investigating the pathology of PE. We have shown a reduction of vascular endothelial mt respiration and increased mtROS to be caused by various inflammatory mediators, such as TNF-α, agonistic autoantibodies to the angiotensin II type I receptor, and CD4+T cells in pregnant rats. However, we do not know the effect of a broad anti-inflammatory agent on mt function and hypertension rat models of PE. Deleterious effects of free radicals include oxidative damage of biomolecules, initiation of lipid peroxidation, and cellular dysfunction, and are suggested to initiate maternal leukocyte activation and endothelial dysfunction in PE [40]. Typically, hypertension results from increases in the production of ROS in the vascular wall, thereby causing increases in oxidative stress and mediation of vascular diseases [41]. A major source of ROS production occurs in mitochondria and is produced by the complexes of the electron transport chain. Additionally, ROS is able to interact and inactivate nitric oxide to cause the endothelial dysfunction that occurs during PE, and, therefore, identifying agents that can protect mitochondria is necessary to prevent cellular damage. In this study, we demonstrate that PIBF released after progesterone signaling through its receptor can protect placental mitochondrial function and thereby improve overall organ function, thus resulting in lower blood pressure in the presence of sFlt-1.

McCarthy et al. showed that vascular mtROS and decreased respiration in HUVECs is caused by the release of soluble factors released in the circulation of PE women [42]. Sanchez-Aranguren et al. demonstrated an alteration in the mitochondrial bioenergetics that were induced from increases in sFlt-1 levels in HUVECS treated with plasma from PE women compared to normal pregnant women. [43]. Zhai et al showed that the proliferation of HUVECs was damaged and apoptosis was up-regulated when stimulated by sFlt-1, which further demonstrates that sFlt-1 contributes to endothelial dysfunction [44].

Endothelial dysfunction is a customary phenotype of PE characterized by vasoconstriction, reduced blood flow to organs, and the release of placental factors in response to placental ischemia [1,45–47]. Studies have previously indicated the importance of the maternal vascular endothelium in regulation vascular inflammation, and oxidative stress [48]. We have demonstrated that progesterone attenuated ET-1 secretion in HUVECS exposed to RUPP rat sera [29]. In this current study, we exposed HUVECs to sera from sFlt-1 induced hypertensive rats before and after supplementation with PIBF, and found a significant decrease in endothelial cell mtROS compared to those treated with NP+sFlt-1 serum. Therefore, these results coupled with our previous results indicate that progesterone could provide protection from soluble factors released in response to placental ischemia [29].

Studies have shown that placental overexpression of sFlt-1 is induced by hypoxia, suggesting that placental ischemia would promote sflt-1 expression and result in an imbalance between pro- and antiangiogenic factors in preeclampsia [49]. Furthermore, increased levels of sFlt-1 correlate with decreased NO formation in women with preeclampsia, which may additively work together to induce endothelial dysfunction. This data further implicates sFlt-1 as having detrimental effects in PE women and playing a key role in metabolic modulation and reprogramming in placenta and endothelium during pregnancy [43]. Importantly, we demonstrate PIBF to play a protective role against not only hypertension for placental and endothelial function during pregnancy.

Collectively, these findings further demonstrate that sFlt-1 causes hypertension and increases in mitochondrial dysfunction, but that these changes can be mitigated with progesterone. Therefore, the data suggests progesterone supplementation as a strategy to assist in the management of PE by correcting imbalances of hormones during pregnancy.

5. Conclusions

In conclusion, these results ultimately indicate that normal mitochondrial activity is important in normal pregnancies. Our data demonstrates that mt dysfunction could be normalized with certain therapeutic treatments, such as progesterone or progesterone products during pregnancy. Although the use of pharmacological inhibitors utilizing inflammatory cytokines or lymphocyte therapies are not well understood, our results reveal that administration of PIBF reduces hypertension, oxidative stress, and endothelial dysfunction during pregnancy and thereby indicates that there are exciting possibilities for obstetric use of progesterone in the future for preeclamptic pregnancies. Although our data demonstrates that PIBF can improve hypertension and mt function during pregnancy, the exact mechanism by which this occurs needs to be further examined.

Author Contributions: Conceptualization, E.D., B.L., and L.M.A.; data curation, E.D., J.J., D.C.C., K.C., O.H., N.C., T.I., B.L., and L.M.A.; funding acquisition, E.D., B.L., and L.M.A.; methodology, E.D., J.J., D.C.C., K.C., O.H., N.C., T.I., B.L., and L.M.A.; project administration, E.D., B.L., and L.M.A.; resources, B.L., and L.M.A.; supervision, E.D., B.L., and L.M.A.; validation, E.D., B.L., and L.M.A.; visualization, E.D., B.L., and L.M.A.; writing—original draft, E.D., B.L., and L.M.A.; writing—review and editing, E.D., B.L., and L.M.A., investigation, S.F. All authors have read and agreed to the published version of the manuscript.

Funding: This study was supported by NIH grants RO1HD067541-10 (B.L.), P20GM121334 (L.M.A., B.L.) and 1U54GM1155428, American Heart Association (AHA) early career award 19CDA34670055 (L.M.A.), and the T32 Trainee Grant T32-HL105324 (E.D.).

Institutional Review Board Statement: The study was conducted according to the guidelines of the University of Mississippi Medical Center, and the animals were handled according to the guiding principles published in the National Institutes of Health Guide for the Care of Animals and the Institutional Animal Care and Use Committee.

Informed Consent Statement: Not applicable.

Data Availability Statement: All data relevant to the study are included in the article.

Conflicts of Interest: The authors declare no conflict of interest.

References

1. Cornelius, D.C.; Cottrell, J.; Amaral, L.M.; LaMarca, B. Inflammatory mediators: A causal link to hypertension during preeclampsia. *Br. J. Pharmacol.* **2019**, *176*, 1914–1921. [CrossRef]
2. Herse, F.; LaMarca, B. Angiotensin II Type 1 Receptor Autoantibody (AT1-AA)-Mediated Pregnancy Hypertension. *Am. J. Reprod. Immunol.* **2012**, *69*, 413–418. [CrossRef] [PubMed]
3. Cífková, R. Hypertension in pregnancy. In *Manual of Hypertension of the European Society of Hypertension*; CRC Press: Boca Raton, FL, USA, 2021; ISBN 9780367778613.
4. Roberts, J.M.; Pearson, G.; Cutler, J.; Lindheimer, M. Summary of the NHLBI working group on research on hypertension during pregnancy. *Hypertension* **2003**, *41*, 437–445. [CrossRef]
5. Sibai, B.; Dekker, G.; Kupferminc, M. Pre-eclampsia. *Lancet* **2005**, *365*, 785–799. [CrossRef]
6. Matsubara, K.; Higaki, T.; Matsubara, Y.; Nawa, A. Nitric oxide and reactive oxygen species in the pathogenesis of preeclampsia. *Int. J. Mol. Sci.* **2015**, *16*, 4600–4614. [CrossRef] [PubMed]
7. Redman, C.W.; Sargent, I.L. Latest advances in understanding preeclampsia. *Science* **2005**, *308*, 1592–1594. [CrossRef]
8. Moncada, S.; Higgs, A. The L-arginine-nitric oxide pathway. *N. Engl. J. Med.* **1993**, *329*, 2002–2012. [PubMed]
9. Sandrim, V.C.; Palei, A.C.T.; Luizon, M.; Izidoro-Toledo, T.; Cavalli, R.; Tanus-Santos, J.E. eNOS haplotypes affect the responsiveness to antihypertensive therapy in preeclampsia but not in gestational hypertension. *Pharm. J.* **2009**, *10*, 40–45. [CrossRef]
10. Brewer, J.; Liu, R.; Lu, Y.; Scott, J.; Wallace, K.; Wallukat, G.; Moseley, J.; Herse, F.; Dechend, R.; Martin, J.N.; et al. Endothelin-1, oxidative stress, and endogenous angiotensin II: Mechanisms of angiotensin II type I receptor autoantibody–enhanced renal and blood pressure response during pregnancy. *Hypertension* **2013**, *62*, 886–892. [CrossRef]
11. Murphy, S.R.; LaMarca, B.B.D.; Parrish, M.; Cockrell, K.; Granger, J.P. Control of soluble fms-like tyrosine-1 (sFlt-1) production response to placental ischemia/hypoxia: Role of tumor necrosis factor-α. *Am. J. Physiol. Regul. Integr. Comp. Physiol.* **2013**, *304*, R130–R135. [CrossRef]
12. Wang, A.; Rana, S.; Karumanchi, S.A. Preeclampsia: The Role of Angiogenic Factors in Its Pathogenesis. *Physiology* **2009**, *24*, 147–158. [CrossRef] [PubMed]
13. Amaral, L.M.; Cunningham, M.W., Jr.; Cornelius, D.C.; LaMarca, B. Preeclampsia: Long-term consequences for vascular health. *Vasc. Health Risk Manag.* **2015**, *11*, 403.
14. Maynard, S.E.; Min, J.-Y.; Merchan, J.; Lim, K.-H.; Li, J.; Mondal, S.; Libermann, T.A.; Morgan, J.P.; Sellke, F.W.; Stillman, I.E.; et al. Excess placental soluble fms-like tyrosine kinase 1 (sFlt1) may contribute to endothelial dysfunction, hypertension, and proteinuria in preeclampsia. *J. Clin. Investig.* **2003**, *111*, 649–658. [CrossRef] [PubMed]
15. Tam, K.B.T.; Lamarca, B.; Arany, M.; Cockrell, K.; Fournier, L.; Murphy, S.; Martin, J.N.; Granger, J.P. Role of reactive oxygen species during hypertension in response to chronic antiangiogenic factor (sFlt-1) excess in pregnant rats. *Am. J. Hypertens.* **2011**, *24*, 110–113. [CrossRef]
16. Aouache, R.; Biquard, L.; Vaiman, D.; Miralles, F. Oxidative stress in preeclampsia and placental diseases. *Int. J. Mol. Sci.* **2018**, *19*, 1496. [CrossRef] [PubMed]
17. Birben, E.; Sahiner, U.M.; Sackesen, C.; Erzurum, S.; Kalayci, O. Oxidative stress and antioxidant defense. *World Allergy Organ. J.* **2012**, *5*, 9–19. [CrossRef]
18. Sultana, Z.; Maiti, K.; Aitken, J.; Morris, J.; Dedman, L.; Smith, R. Oxidative stress, placental ageing-related pathologies and adverse pregnancy outcomes. *Am. J. Reprod. Immunol.* **2017**, *77*, e12653. [CrossRef]
19. Eirin, A.; Lerman, A.; Lerman, L.O. Mitochondrial injury and dysfunction in hypertension-induced cardiac damage. *Eur. Heart J.* **2014**, *35*, 3258–3266. [CrossRef]
20. Redza-Dutordoir, M.; Averill-Bates, D.A. Activation of apoptosis signalling pathways by reactive oxygen species. *Biochim. Biophys. Acta (BBA) Mol. Cell Res.* **2016**, *1863*, 2977–2992. [CrossRef]
21. Vaka, V.R.; McMaster, K.M.; Cunningham, M.W., Jr.; Ibrahim, T.; Hazlewood, R.; Usry, N.; Cornelius, D.C.; Amaral, L.M.; Lamarca, B. Role of mitochondrial dysfunction and reactive oxygen species in mediating hypertension in the reduced uterine perfusion pressure rat model of preeclampsia. *Hypertension* **2018**, *72*, 703–711. [CrossRef]

22. Cunningham, M.W.; Jayaram, A.; Deer, E.; Amaral, L.M.; Vaka, V.R.; Ibrahim, T.; Cornelius, D.C.; Lamarca, B. Tumor necro-sis factor alpha (TNF-α) blockade improves natural killer cell (NK) activation, hypertension, and mitochondrial oxidative stress in a preclinical rat model of preeclampsia. *Hypertens. Pregnancy* **2020**, *39*, 399–404. [CrossRef] [PubMed]
23. Jayaram, A.; Deer, E.; Amaral, L.M.; Campbell, N.; Vaka, V.R.; Cunningham, M.; Ibrahim, T.; Cornelius, D.C.; LaMarca, B.B. The role of tumor necrosis factor in triggering activation of natural killer cell, multi-organ mitochondrial dysfunction and hypertension during pregnancy. *Pregnancy Hypertens.* **2021**, *24*, 65–72. [CrossRef] [PubMed]
24. Deer, E.; Reeve, K.E.; Amaral, L.M.; Vaka, V.R.; Franks, M.; Campbell, N.; Fitzgerald, S.; Herrock, O.T.; Ibrahim, T.; Cornelius, D.C.; et al. CD4+ T cells cause renal and placental mitochondrial oxidative stress as mechanisms of hypertension in response to placental ischemia. *Am. J. Physiol. Physiol.* **2021**, *320*, F47–F54. [CrossRef]
25. Raghupathy, R.; Kalinka, J. Cytokine imbalance in pregnancy complications and its modulation. *Front. Biosci.* **2008**, *13*, 985–994. [CrossRef] [PubMed]
26. Szekeres-Bartho, J.; Autran, B.; Debre, P.; Andreu, G.; Denver, L.; Chaouat, G. Immunoregulatory effects of a suppressor factor from healthy pregnant women's lymphocytes after progesterone induction. *Cell. Immunol.* **1989**, *122*, 281–294. [CrossRef]
27. Szekeres-Bartho, J.; Faust, Z.; Varga, P.; Szereday, L.; Kelemen, K. The immunological pregnancy protective effect of progesterone is manifested via controlling cytokine production. *Am. J. Reprod. Immunol.* **1996**, *35*, 348–351. [CrossRef]
28. Szekeres-Bartho, J.; Barakonyi, A.; Par, G.; Polgar, B.; Palkovics, T.; Szereday, L. Progesterone as an immunomodulatory molecule. *Int. Immunopharmacol.* **2001**, *1*, 1037–1048. [CrossRef]
29. Kiprono, L.V.; Wallace, K.; Moseley, J.; Martin, J., Jr.; LaMarca, B. Progesterone blunts vascular endothelial cell secretion of endothelin-1 in response to placental ischemia. *Am. J. Obstet. Gynecol.* **2013**, *209*, 44.e1–44.e6. [CrossRef]
30. Cottrell, J.N.; Witcher, A.C.; Comley, K.M.; Cunningham, M.W., Jr.; Ibrahim, T.; Cornelius, D.C.; LaMarca, B.D.; Amaral, L.M. Progesterone-induced blocking factor improves blood pressure, inflammation, and pup weight in response to reduced uterine perfusion pressure (RUPP). *Am. J. Physiol. Integr. Comp. Physiol.* **2021**, *320*, R719–R727. [CrossRef]
31. Baltajian, K.; Hecht, J.L.; Wenger, J.B.; Salahuddin, S.; Verlohren, S.; Perschel, F.H.; Zsengeller, Z.K.; Thadhani, R.; Karumanchi, S.A.; Rana, S. Placental lesions of vascular insufficiency are associated with anti-angiogenic state in women with preeclampsia. *Hypertens. Pregnancy* **2014**, *33*, 427–439. [CrossRef]
32. Rebelato, H.J.; Esquisatto, M.A.M.; Moraes, C.; Amaral, M.E.C.; Catisti, R. Gestational protein restriction induces alterations in placental morphology and mitochondrial function in rats during late pregnancy. *J. Mol. Histol.* **2013**, *44*, 629–637. [CrossRef] [PubMed]
33. LaMarca, B.B.D.; Cockrell, K.; Sullivan, E.; Bennett, W.; Granger, J.P. Role of Endothelin in Mediating Tumor Necrosis Factor-Induced Hypertension in Pregnant Rats. *Hypertension* **2005**, *46*, 82–86. [CrossRef]
34. Novotny, S.; Wallace, K.; Herse, F.; Moseley, J.; Darby, M.; Health, J.; Gill, J.; Wallukat, G.; Martin, J.N.; Dechend, R.; et al. CD4+ T cells play a critical role in mediating hypertension in response to placental ischemia. *J. Hypertens. Open Access* **2013**, *2*, 14873.
35. LaMarca, B.; Speed, J.; Fournier, L.; Babcock, S.A.; Berry, H.; Cockrell, K.; Granger, J.P. Hypertension in response to chronic reductions in uterine perfusion in pregnant rats: Effect of tumor necrosis factor-α blockade. *Hypertension* **2008**, *52*, 1161–1167. [CrossRef] [PubMed]
36. LaMarca, B.; Wallace, K.; Herse, F.; Wallukat, G.; Martin, J.N., Jr.; Weimer, A.; Dechend, R. Hypertension in response to placental ischemia during pregnancy: Role of B lymphocytes. *Hypertension* **2011**, *57*, 865–871. [CrossRef]
37. Cornelius, D.C.; Hogg, J.P.; Scott, J.; Wallace, K.; Herse, F.; Moseley, J.; Wallukat, G.; Dechend, R.; Lamarca, B. Administration of interleukin-17 soluble receptor C suppresses TH17 cells, oxidative stress, and hypertension in response to placental ischemia during pregnancy. *Hypertension* **2013**, *62*, 1068–1073. [CrossRef] [PubMed]
38. Amaral, L.M.; Cornelius, D.C.; Harmon, A.; Moseley, J.; Martin, J.N., Jr.; LaMarca, B. 17-hydroxyprogesterone caproate significantly improves clinical characteristics of preeclampsia in the reduced uterine perfusion pressure rat model. *Hypertension* **2015**, *65*, 225–231. [CrossRef]
39. Veillon, E.W., Jr.; Keiser, S.D.; Parrish, M.R.; Bennett, W.; Cokrell, K.; Ray, L.F.; Granger, J.P.; Martin, J.N.; Lamarca, B. 17-Hydroxyprogesterone blunts the hypertensive response associated with reductions in uterine perfusion pressure in pregnant rats. *Am. J. Obstet. Gynecol.* **2009**, *201*, 324.e1–324.e6. [CrossRef]
40. Raijmakers, M.T.; Dechend, R.; Poston, L. Oxidative stress and preeclampsia: Rationale for antioxidant clinical trials. *Hypertension* **2004**, *44*, 374–380. [CrossRef]
41. Rodrigo, R.; González, J.; Paoletto, F. The role of oxidative stress in the pathophysiology of hypertension. *Hypertens. Res.* **2011**, *34*, 431–440. [CrossRef]
42. McCarthy, C.; Kenny, L.C. Therapeutically targeting mitochondrial redox signalling alleviates endothelial dysfunction in preeclampsia. *Sci. Rep.* **2016**, *6*, 32683. [CrossRef]
43. Sánchez-Aranguren, L.C.; Espinosa-González, C.T.; González-Ortiz, L.M.; Sanabria-Barrera, S.M.; Riaño-Medina, C.E.; Nuñez, A.F.; Ahmed, A.; Vasquez-Vivar, J.; López, M. Soluble Fms-like tyrosine kinase-1 alters cellular metabolism and mitochondrial bioenergetics in preeclampsia. *Front. Physiol.* **2018**, *9*, 83. [CrossRef]
44. Zhai, Y.; Liu, Y.; Qi, Y.; Long, X.; Gao, J.; Yao, X.; Chen, Y.; Wang, X.; Lu, S.; Zhao, Z. The soluble VEGF receptor sFlt-1 contributes to endothelial dysfunction in IgA nephropathy. *PLoS ONE* **2020**, *15*, e0234492. [CrossRef]
45. Amaral, L.M.; Wallace, K.; Owens, M.; Lamarca, B. Pathophysiology and Current Clinical Management of Preeclampsia. *Curr. Hypertens. Rep.* **2017**, *19*, 61. [CrossRef]

46. Granger, J.P.; Alexander, B.T.; Llinás, M.T.; Bennett, W.A.; Khalil, R.A. Pathophysiology of Hypertension During Preeclampsia Linking Placental Ischemia With Endothelial Dysfunction. *Hypertension* **2001**, *38*, 718–722. [CrossRef] [PubMed]
47. Granger, J.P. Inflammatory cytokines, vascular function, and hypertension. *Am. J. Physiol.-Regul. Integr. Comp. Physiol.* **2004**, *286*, R989–R990. [CrossRef] [PubMed]
48. Lamarca, B. Endothelial dysfunction. An important mediator in the pathophysiology of hypertension during pre-eclampsia. *Minerva Ginecol.* **2012**, *64*, 309–320. [PubMed]
49. Brennan, L.J.; Morton, J.S.; Davidge, S.T. Vascular Dysfunction in Preeclampsia. *Microcirculation* **2014**, *21*, 4–14. [CrossRef]

cells

MDPI

Review

Pathological AT1R-B2R Protein Aggregation and Preeclampsia

Ursula Quitterer [1,2,*] and Said AbdAlla [1]

[1] Molecular Pharmacology, Department of Chemistry and Applied Biosciences, ETH Zurich, Winterthurerstrasse 190, CH-8057 Zürich, Switzerland; said.abdalla@pharma.ethz.ch
[2] Department of Medicine, Institute of Pharmacology and Toxicology, University of Zurich, Winterthurerstrasse 190, CH-8057 Zürich, Switzerland
* Correspondence: ursula.quitterer@pharma.ethz.ch; Tel.: +41-44-6329801

Abstract: Preeclampsia is one of the most frequent and severe complications of pregnancy. Symptoms of preeclampsia usually occur after 20 weeks of pregnancy and include hypertension and kidney dysfunction with proteinuria. Up to now, delivery of the infant has been the most effective and life-saving treatment to alleviate symptoms of preeclampsia because a causative treatment does not exist, which could prolong a pregnancy complicated with preeclampsia. Preeclampsia is a complex medical condition, which is attributed to a variety of different risk factors and causes. Risk factors account for insufficient placentation and impaired vasculogenesis and finally culminate in this life-threatening condition of pregnancy. Despite progress, many pathomechanisms and causes of preeclampsia are still incompletely understood. In recent years, it was found that excessive protein complex formation between G-protein-coupled receptors is a common sign of preeclampsia. Specifically, the aberrant heteromerization of two vasoactive G-protein-coupled receptors (GPCRs), the angiotensin II AT1 receptor and the bradykinin B2 receptor, is a causative factor of preeclampsia symptoms. Based on this knowledge, inhibition of abnormal GPCR protein complex formation is an experimental treatment approach of preeclampsia. This review summarizes the impact of pathological GPCR protein aggregation on symptoms of preeclampsia and delineates potential new therapeutic targets.

Keywords: preeclampsia; angiotensin II; AGTR1 (angiotensin II receptor type 1); bradykinin; BDKRB2 (bradykinin receptor B2); AT1R-B2R heteromer (protein complex formed of AT1R-B2R); G-protein-coupled receptor; protein aggregation; ARRB (beta-arrestin)

Citation: Quitterer, U.; AbdAlla, S. Pathological AT1R-B2R Protein Aggregation and Preeclampsia. *Cells* **2021**, *10*, 2609. https://doi.org/10.3390/cells10102609

Academic Editors: Junie P. Warrington, Ana T. Palei, Mark W. Cunningham and Lorena M. Amaral

Received: 5 September 2021
Accepted: 27 September 2021
Published: 1 October 2021

1. Introduction

Preeclampsia is a life-threatening complication of pregnancy, which affects about 2–8% of pregnancies world-wide [1,2]. Risk factors of preeclampsia cause insufficient placentation with defective blood vessels and impaired oxygen delivery to the unborn. As a consequence of defective placentation and impaired vasculogenesis, maternal symptoms of preeclampsia evolve, usually after 20 weeks of pregnancy. Monitoring of blood pressure during pregnancy is an important measure of diagnosis because high blood pressure which exceeds 140/90 mm Hg is a common symptom of preeclampsia [3]. Currently, the only treatment of preeclampsia is (premature) delivery of the infant [4].

However, preterm birth is associated with a high risk to the infant and can cause severe complications, which include respiratory dysfunction due to immature lung development and other health problems [5]. Search for pathomechanisms is necessary to develop a mechanistic basis of preeclampsia treatment, which could minimize life-threatening risks and prolong safe pregnancy. While acute preeclampsia symptoms subside after delivery, a pregnancy complicated with preeclampsia is associated with an increased risk of cardiovascular disease, hypertension and renal dysfunction later in life, for the mother and the infant [6,7]. Therefore, treatment approaches are needed, which leverage not only acute preeclampsia symptoms but also minimize late preeclampsia complications of the cardiovascular and renal systems.

Research efforts worldwide identified several factors associated with preeclampsia. An increase in circulating levels of sFlt1 (soluble FMS-like tyrosine kinase 1) and down-regulation of PIGF (placental growth factor) levels are diagnostic markers, which assist in the early diagnosis of preeclampsia [8,9]. The involvement of sFlt1 in symptoms of preeclampsia is documented [10] whereas deficiency of PIGF is redundant and does not cause preeclampsia symptoms [11]. Administration of sFlt1 to pregnant rats causes major symptoms of preeclampsia such as endothelial dysfunction, hypertension and proteinuria [10]. However, the primary trigger of abnormal placentation and sFlt1 up-regulation in preeclampsia still remains to be found.

The search for pathomechanisms of preeclampsia elucidated that pathological protein aggregation of vasoactive G-protein-coupled receptors (GPCRs) characterizes pregnancies complicated with preeclampsia [12–14]. Aberrant GPCR protein aggregation affects major target organs of preeclampsia, i.e., the placenta, the vasculature, the kidney and circulating blood cells [12–14]. Specifically, a protein complex formed by GPCR heteromerization was documented to be a sufficient cause of preeclampsia [14]. This protein complex consists of two G-protein-coupled receptors, which are known to regulate vascular tone, i.e., the angiotensin II AT1 receptor, AGTR1, and the bradykinin B2 receptor, BDKRB2 [15,16]. Targeting this protein complex formation has been shown to be effective in treating symptoms of experimental preeclampsia [14]. Notably, the protein beta-arrestin1 could support the down-regulation of pathological GPCR aggregates by a beta-arrestin-mediated mechanism [14].

Based on these findings, ongoing research efforts aim to develop a treatment of acute preeclampsia symptoms and/or late cardiovascular preeclampsia complications by inhibition of disease-causing GPCR protein aggregation. This review gives an overview of approaches to target symptoms of preeclampsia by interference with pathological GPCR protein aggregation.

2. Preeclampsia: Risk Factors, Symptoms and Diagnosis

2.1. Risk Factors of Preeclampsia

There are several established risk factors, which are associated with a significantly increased risk of preeclampsia [6,17–22]. Frequent risk factors of preeclampsia, which predispose to a high preeclampsia risk, include major cardiovascular and metabolic diseases such as chronic hypertension, chronic kidney disease, hypertensive disease during a previous pregnancy, type 1 and type 2 diabetes, and obesity with a maternal body mass index of >30 kg/m^2 [17–19]. A high preeclampsia risk is also associated with autoimmune diseases, which cause damage to blood vessels, such as systemic lupus erythematosus, or hypercoagulation such as the antiphospholipid syndrome [17–19]. Other risk factors that predispose to moderate risk are advanced age (older than 40 years), multiple pregnancies with twins, triplets or other multiples, the first pregnancy, a long interval between two pregnancies of more than ten years, a new paternity with a long interval between two pregnancies, and a family history of preeclampsia [17–19]. In addition, ethnicity can affect the risk of preeclampsia because black women have a higher risk compared to other races [20]. Finally, pregnancies that arise from in vitro fertilization with oocyte donation are also associated with an increased risk of preeclampsia [21,22]. Taken together, the preeclampsia risk is increased in women with major cardiovascular and metabolic diseases and/or autoimmune diseases, which predispose them to hypercoagulation and damage to blood vessels by autoimmune reaction.

2.2. Major Symptoms of Preeclampsia

Preeclampsia is a hypertensive disorder of pregnancy. Consequently, one of the most important symptoms of preeclampsia is a rise in blood pressure above 140/90 mm Hg, usually after 20 weeks of pregnancy, in women with previously normal blood pressure [18,23]. In addition, hypertensive disorders such as chronic hypertension that were present before pregnancy or before the 20th week of pregnancy often advance to preeclampsia. This increase in blood pressure can be associated with severe headache, and visual disturbances such as a change in vision, blurred vision or loss of vision [23]. Additional warning signs are epigastric pain, nausea and vomiting [23]. These symptoms are directly linked to high blood pressure and pathophysiological mechanisms of preeclampsia. Headache can be attributed to vasospasm of cerebral arteries and/or cerebral edema [24,25]. Visual disturbances are a consequence of impaired blood flow and vasospasm of retinal arteries causing ischemic injury [26]. Epigastric pain, nausea and vomiting are a consequence of hepatic dysfunction as a result of obstructed blood flow to hepatic sinusoids due to vascular constriction and fibrin-like vascular deposits [27,28].

However, hypertension is not the sole symptom of preeclampsia. Concomitantly with hypertension, preeclampsia causes damage to the microvasculature of other organs, such as the kidney [29–31]. Vice versa, kidney disease is causally linked to preeclampsia because preexisting kidney disease is a strong risk factor of preeclampsia [30,31]. Preeclampsia-associated renal dysfunction accounts for proteinuria with an increased albumin-creatinine ratio and/or other markers of acute kidney injury [30–32]. Together with hypertension, proteinuria is one of the most frequent symptoms of preeclampsia [33]. Notably, the severity of proteinuria is an indicator of adverse outcome of pregnancies complicated with preeclampsia [33]. As a consequence of kidney dysfunction with decreased urine output, edema with shortness of breath can develop [32].

Additional symptoms of severe preeclampsia are liver dysfunction together with abdominal pain, nausea and vomiting [34]. Another frequent symptom of preeclampsia relates to dysfunctional coagulation, which can be a consequence of thrombotic microangiopathy [35,36], Ensuing thrombocytopenia and hemolysis are characteristic features, which are frequently associated with severe preeclampsia [35,36]. In this context, a serious condition, which coexists with preeclampsia in 70–80% of cases, is the HELLP syndrome. HELPP encompasses hemolysis, elevated liver enzymes and low platelet count [37]. Finally, preeclampsia also affects the unborn. As a consequence of insufficient oxygen delivery, fetal growth restriction is frequently associated with preeclampsia [38,39].

Taken together, preeclampsia is characterized by a rise in blood pressure. Associated symptoms are causally linked to the blood pressure rise and the underlying pathophysiology, which triggers a dysfunction of microvascular beads, specifically the kidney, the brain and the liver [29].

2.3. Diagnosis of Preeclampsia

To facilitate early diagnosis of preeclampsia, blood pressure monitoring is a well-established measure of prenatal care [3,40,41]. A rise in blood pressure above 140/90 mm Hg usually after 20 weeks of pregnancy in women with previously normal blood pressure is an important diagnostic criterion of preeclampsia. The rise in blood pressure needs to be confirmed twice, at least 4 h apart [3,40,41].

Hypertension is not the sole diagnostic criterion of preeclampsia [42]. There are other hypertensive disorders of pregnancy that need to be differentiated from preeclampsia, e.g., gestational hypertension or chronic hypertension. Gestational hypertension is hypertension, which occurs for the first time during the second half of pregnancy without proteinuria and organ damage. Preeclampsia is also differentiated from chronic hypertension, which is elevated blood pressure greater than 140/90 mm Hg that was present before pregnancy or before the 20th week of pregnancy. Frequently, these hypertensive disorders of pregnancy advance to preeclampsia [18,42].

In preeclampsia, hypertension is associated with other signs of end organ damage. Thus, the diagnosis of preeclampsia with new-onset hypertension is usually accompanied by one or more additional symptoms of organ dysfunction [18,41]. Kidney dysfunction with proteinuria is such a characteristic but not mandatory feature of preeclampsia [18,41]. Therefore, proteinuria with ≥300 mg in a 24-h urine or a spot urine protein-to-creatinine ratio of ≥30 mg/mmol is often used as a major accompanying diagnostic criterion of preeclampsia [41]. Other diagnostic features of organ dysfunction are acute kidney injury with creatinine of ≥90 micromol/L, or symptoms of liver dysfunction with elevated transaminases with or without abdominal pain. Neurological and hematological complications can also reflect organ dysfunction in preeclampsia. The impaired utero-placental blood flow is diagnosed by uterine artery Doppler waveform analysis [43]. Utero-placental dysfunction is a frequent cause of fetal growth restriction or stillbirth [41,43].

Apart from symptom-based diagnosis, the measurement of circulating angiogenic biomarkers of preeclampsia further facilitates the diagnosis. Of specific value is the determination of circulating levels of sFlt1 and PIGF [8,9,44,45]. In preeclampsia, circulating levels of sFlt1 tend to be increased while circulating PIGF levels are reduced [8,9]. The measurement of PIGF can be used for risk stratification and substantially reduces the time to diagnosis of preeclampsia [46–49]. In some countries, the ratio of sFlt1/PIGF is recommended as diagnostic criterion to differentiate preeclampsia cases from non-preeclamptic pregnancies [50]. Measurement of the sFlt1/PIGF ratio significantly improves clinical precision but the test has a lower power to differentiate between severe or early onset forms of the condition [44,51]. Likewise, low PIGF should not be considered as a sufficient criterion to perform immediate delivery [52]. Taken together, angiogenic biomarker testing combined with risk stratification substantially reduces the time to diagnosis so that suspected preeclampsia can be considered much earlier and even before proteinuria develops [49].

2.4. Prevention Therapy of Preeclampsia with Aspirin

Up to now, there has been no treatment for preeclampsia. The only measure to relieve preeclampsia symptoms is the premature delivery of the infant. Therefore, many guidelines world-wide focus on prevention to delay early onset preeclampsia in high-risk pregnancies. In this context, low-dose aspirin is the most widely recommended prevention therapy of preeclampsia [53–56]. The recommended therapy for pregnancies with high preeclampsia risk consists of the daily intake of aspirin at a dose of ≥100 mg/d starting from week 11 to 14 of gestation [41,53–56]. In high-risk pregnancies, aspirin therapy can significantly reduce preterm preeclampsia whereas term preeclampsia is not affected [56]. Despite these recommendations, there is still uncertainty regarding the recommended aspirin dose and the identification of those pregnancies who will profit most from the prevention therapy [53].

3. Pathomechanisms of Preeclampsia Leading to Hypertension

3.1. The Interrelationship between Impaired Arterial Function, Dysfunctional Placentation, and Hypertension

To identify potential targets for preeclampsia treatment, research activities focus on pathomechanisms, which account for major preeclampsia symptoms. Hypertension after week 20 of pregnancy is one of the major symptoms of preeclampsia. The current concept considers defective placentation with ensuing insufficient uteroplacental perfusion as an initiating event of preeclampsia and preeclampsia hypertension [4,57–59]. Dysfunctional placentation is thought to develop as a result of insufficient cytotrophoblast invasion of spiral arterioles [59]. The reduced uteroplacental perfusion accounts for placental ischemia [57,58]. Insufficient oxygen delivery to the placenta with hypoxia and cellular ischemia leads to a widespread dysfunction of the maternal vascular endothelium with increased formation of vasoconstrictors, endothelin and thromboxane, and hypersensitivity to angiotensin II AT1 receptor stimulation [57–59]. Endothelial dysfunction is accompanied by a decreased formation of vasodilators, NO and prostacyclin [57–59]. Systemic endothelial dysfunction and hypoxia also trigger a disbalance of pro-angiogenic and anti-angiogenic factors with predominance of the anti-angiogenic factor sFlt1 [10,60]. The sFlt1 is the soluble receptor for the vascular endothelial growth factor (soluble VEGFR-1), which antagonizes pro-angiogenic effects of VEGF [10]. The net result is exaggerated vasoconstriction and hypertension. Due to insufficient placentation, the high blood pressure becomes detrimental not only to the mother but also does not improve the oxygen delivery to the developing fetus.

The unresolved question remains: What triggers placental dysfunction that ultimately causes preeclampsia hypertension? Emerging evidence suggests that impaired arterial function could in fact precede placental dysfunction [61]. Consequently, placental impairment would be secondary to cardiovascular dysfunction during the pathogenesis of preeclampsia [61,62]. In agreement with this concept, major risk factors of preeclampsia such as cardiovascular and metabolic diseases, cause damage to the vascular bed. Thus, it is the interplay between primary cardiovascular dysfunction and ensuing placental dysfunction, which finally culminates in insufficient uteroplacental perfusion and placental ischemia [62]. The consequence is a widespread damage of the maternal endothelium and endothelial dysfunction. Endothelial dysfunction is a well-known cause of hypertension and end-organ damage, i.e., the typical symptoms of preeclampsia.

3.2. The Angiotensin II AT1 Receptor Hypersensitivity of Preeclampsia

Pathologic alterations of the renin angiotensin aldosterone system (RAAS) are important contributors to preeclampsia. Under physiological conditions, the renin angiotensin aldosterone system exerts an indispensable role in the regulation of vascular tone, blood pressure, kidney function, and salt and water homeostasis [63–65]. By cleavage of angiotensinogen, the protease renin liberates angiotensin I, which is further processed by the angiotensin-converting enzyme (ACE) into the octapeptide angiotensin II. Angiotensin II is a major vasoconstrictor by activation of the AT1 receptor, AGTR1 [15,63,64]. In addition, the AT1 receptor regulates salt and water homeostasis, either directly or indirectly by stimulation of the release of the mineral corticoid aldosterone [63,64]. Apart from essential physiological functions, overactivation of RAAS has fundamental pathological consequences and contributes to major cardiovascular and metabolic diseases such as hypertension, atherosclerosis, kidney dysfunction, heart failure, diabetes mellitus, and neurodegeneration [64–66]. In agreement with these pathological roles, inhibition of AT1 receptor activation by an ACE inhibitor or an AT1 receptor antagonist are well-documented, evidence-based treatment approaches of cardiovascular diseases with or without concomitant metabolic diseases [63–65].

Because cardiovascular and metabolic diseases are also major risk factors of preeclampsia, it is not surprising that dysfunctional RAAS and exaggerated AT1 receptor activation exert major pathological roles in preeclampsia [13,66]. Notably, an increased sensitivity of angiotensin II AT1 receptor-stimulated responses is a long-known and well-established feature of pregnancies complicated with preeclampsia [12–14,67–70]. The angiotensin II AT1 receptor hypersensitivity is documented on vascular specimens and circulating blood platelets of pregnant women with preeclampsia [12–14,67–70]. Vascular angiotensin II AT1 receptor hyperactivation not only contributes to hypertension but also to renal dysfunction with proteinuria and placental dysfunction due to impaired vasculogenesis [12,14,67–70].

While circulating levels of angiotensin II are low and apparently "exhausted" in preeclampsia [71], pregnant women with preeclampsia often develop agonistic autoantibodies against the AT1 receptor [72]. Through the activation of hypersensitive AT1 receptors, these agonistic autoantibodies contribute to preeclampsia hypertension and preeclampsia symptoms [14,72–74]. Causality between agonistic autoantibodies to the AT1 receptor and preeclampsia symptoms was confirmed by several experimental models [73,74]. During the pathogenesis of preeclampsia, agonistic AT1 receptor autoantibodies are triggered by hypersensitive AT1 receptors and/or high blood pressure [14,74,75]. Taken together, hypersensitive and hyperactive AT1 receptors account for exaggerated vasoconstriction and high blood pressure. In addition, agonistic AT1 receptor autoantibodies act as potent and sustained activators of hypersensitive AT1 receptors during the pathogenesis of preeclampsia [14,74].

4. Pathologic GPCR Protein Aggregation and Preeclampsia

4.1. Pathologic Protein Complex Formation between the AT1 Receptor and the B2 Receptor Causes Angiotensin II AT1 Receptor Hypersensitivity of Preeclampsia

In view of the causative role of angiotensin II AT1 receptor hypersensitivity in symptoms of preeclampsia, the question arises what causes the angiotensin II hypersensitivity of preeclampsia.

Several studies showed that angiotensin II AT1 receptor hypersensitivity is caused by aberrant protein complex formation between the AT1 receptor with the bradykinin B2 receptor [12–14].

The bradykinin B2 receptor is a ubiquitously expressed G-protein-coupled receptor, whose in vivo functions are dispensable [76–79]. Physiologic effects of the bradykinin B2 receptor include the release of nitric oxide and prostaglandins from vascular endothelial cells and vasodilation [76–79]. These features of the bradykinin B2 receptor contribute to blood pressure lowering and cardioprotection, preferably under conditions with increased bradykinin, e.g., when the ACE-mediated degradation of bradykinin is suppressed by an ACE inhibitor [79]. However, under pathological conditions with endothelial dysfunction and endothelial nitric oxide synthase (eNOS) uncoupling, e.g., preeclampsia and atherosclerosis, the endothelial bradykinin B2 receptor stimulates the generation of reactive oxygen species and thereby loses its cardioprotective functions [80–82]. As a consequence of endothelial dysfunction, bradykinin B2 receptor-mediated relaxation of myometrial vessels is impaired in preeclampsia [83,84]. Concomitant with endothelial bradykinin B2 receptor dysfunction, systemic levels of the bradykinin-generating enzyme kallikrein as measured by urinary kallikrein are reduced in preeclampsia [85–87]. Furthermore, dysregulated bradykinin B2 receptor and endothelial nitric oxide synthase systems were also documented at the feto-maternal interface of clinical preeclampsia specimens, which showed increased bradykinin B2 receptor levels in extravillous trophoblasts [88]. Local increases in bradykinin B2 receptor protein levels in preeclampsia could be caused by decreased bradykinin B2 receptor stimulation with ensuing depressed B2 receptor down-regulation as a consequence of low systemic kallikrein and bradykinin levels in preeclampsia [85–87]. Taken together, endothelial bradykinin B2 receptor-stimulated vasodilation is impaired, and kallikrein and bradykinin levels are reduced in preeclampsia hypertension.

In addition, under pathological conditions of preeclampsia, there are increased levels of a heteromeric protein complex between the angiotensin II AT1 receptor and the bradykinin B2 receptor on vascular smooth muscle cells and platelets [12–14]. This protein complex formation leads to enhanced G-protein coupling and activation of the AT1 receptor [12–14,89]. Notably, the heteromeric AT1R-B2R protein complex between the AT1 receptor and the B2 receptor forms a platform, which facilitates enhanced G-protein activation (Figure 1).

The hypersensitive AT1R-B2R protein complex is distinguished from AT1 receptor monomers by enhanced G-protein activation and a strongly increased mechanosensitivity [14,91]. Consequently, AT1R-B2R can be activated by mechanical forces independently of the agonist angiotensin II. Enhanced G-protein-mediated signal generation by AT1R-B2R accounts for exaggerated calcium signaling (Figure 1). The increased calcium response leads to an increased Na^+-H^+-exchanger activation with ensuing increased extracellular acidification [14], which contributes to smooth muscle cell contraction and blood pressure rise [92]. Exaggerated signaling triggered by hypersensitive AT1R-B2R complexes also promotes the increased generation of reactive oxygen species and oxidative stress [12]. Moreover, the enhanced AT1R-B2R-stimulated signaling cascade could be directly involved in beta-arrestin1 (ARRB1) dysfunction of preeclampsia [14]. Dysfunctional ARRB1 not only leads to sustained signaling but also impairs AT1R-B2R down-regulation. Impaired receptor protein down-regulation confers to AT1R-B2R a major feature of pathological protein aggregates, which is dysfunctional protein degradation and impaired protein clearance (Figure 1).

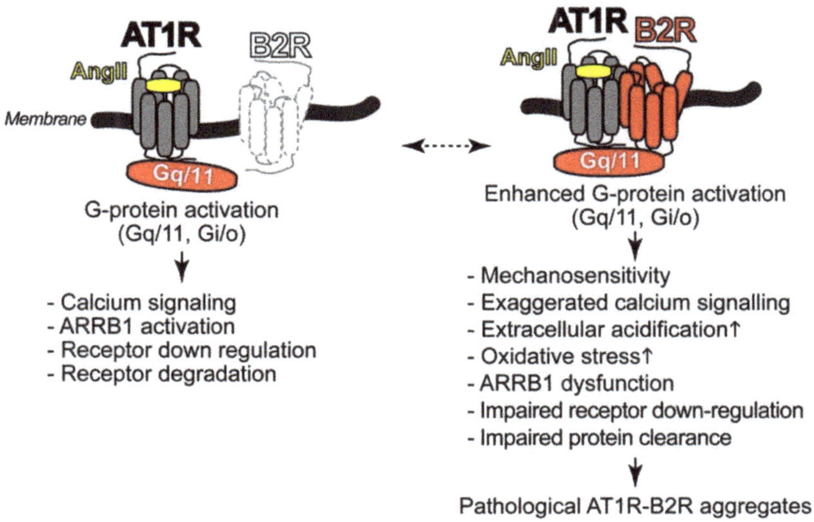

Figure 1. The AT1R-B2R protein complex forms a platform for enhanced G-protein activation. The scheme illustrates how protein complex formation between AT1R (angiotensin II receptor type 1) and B2R (bradykinin B2 receptor) facilitates the interaction with heterotrimeric G-proteins such as Gq/11 (heterotrimeric GTP-binding protein of Gq/11 family) and Gi/o (heterotrimeric GTP-binding protein of Gi/o family). As a consequence of stabilized G-protein interaction, stimulation of the AT1R-B2R protein complex with angiotensin II (AngII) or mechanical stimulation leads to enhanced G-protein activation and signaling, and beta-arrestin1 (ARRB1) dysfunction (**right side**) compared to the monomeric AT1 receptor (**left side**). This Figure was adapted from [90].

Several lines of evidence support that the B2R-mediated enhancement of angiotensin II-stimulated G-protein activation is mediated by a direct AT1R-B2R protein interaction. First, agonist stimulation of B2R is not essential for the B2R-mediated enhancement of AT1R-stimulated signaling because a mutated B2R (B2R-F297A) with defective agonist binding site also enhances AT1R-stimulated signaling [89]. Second, a mutated B2R (B2R-Y157A) with impaired G-protein activation due to the mutation of Y157A in the conserved DRY motif of B2R is incapable to enhance AT1R-stimulated signaling [89]. Third, shielding of the DRY motif of B2R by site-directed antibodies against the connecting loop between membrane domains III-IV of B2R prevented the angiotensin II-stimulated Gαq/11 (Gq/11 protein alpha subunit) activation and redistribution on maternal vessels of pregnancies complicated with preeclampsia [12]. These preeclamptic vessels were characterized by increased AT1R-B2R protein complexes. Thus, the B2R protein is a sufficient cause of enhanced angiotensin II AT1R-stimulated G-protein activation and signaling.

Some of the above-mentioned features of AT1R-B2R heteromers are reminiscent of obligatory GPCR heteromers such as metabotropic glutamate receptor heteromers or GABAB receptor heteromers [93]. However, there is a major difference between AT1R-B2R and these obligatory GPCR heteromers: In the AT1R-B2R complex, each receptor is fully functional, and acts additively or even synergistically with the other binding partner. Therefore, exaggerated signal generation by AT1R-B2R occurs under pathological conditions with increased AT1R-B2R aggregates such as preeclampsia hypertension. The AT1R can also heteromerize with other GPCRs, such as the angiotensin II receptor type 2, AT2R (AGTR2), or the MAS receptor (MAS1). In contrast to the AT1R signal-sensitizing effect mediated by B2R, the AT2R exerts an antagonistic function and inhibits AT1R-stimulated signaling [94]. Similarly, the interaction of MAS1 with AT1R causes inhibition of AT1R-stimulated functions [95]. On the other hand, the B2R not only enhances signaling stimulated by AT1R via AT1R-B2R heteromerization but also AT2R-mediated effects via

AT2R-B2R heteromerization [96]. Depending on the receptor "armamentarium" of a given cell, GPCR heteromerization can thus enhance or depress AT1R-stimulated functions. In preeclampsia hypertension, GPCR heteromerization enhances AT1R-mediated effects.

As a consequence of enhanced signaling, the AT1R-B2R heteromer accounts for increased angiotensin II-stimulated vasoconstriction [12–14,89]. Exaggerated vasoconstriction by increased maternal AT1R-B2R heteromers leads to a rise in blood pressure and hypertension at the end of pregnancy in AT1R-B2R-transgenic mice with smooth muscle-specific expression of AT1R-B2R [12–14]. Vice versa, mice deficient in B2R (Bdkrb2$^{-/-}$) show depressed angiotensin II AT1R-stimulated vasoconstriction [14,97]. These observations with Bdkrb2$^{-/-}$ mice indicate that under physiological conditions, the B2R may also contribute to angiotensin II AT1R-stimulated vasoconstriction by direct AT1R-B2R interaction. However, the physiological AT1R-B2R heteromerization with endogenously expressed receptor protein levels is different from pathological AT1R-B2R protein complex formation of preeclampsia because AT1R-B2R aggregation in preeclampsia is sustained and practically irreversible. Notably, in preeclampsia, down-regulation of AT1R-B2R protein complexes is impaired, in part due to beta-arrestin1 (ARRB1) dysfunction [12–14]. Consequently, the AT1R-B2R protein complex of preeclampsia fulfills criteria of pathological protein aggregates, which accumulate because of dysfunctional protein clearance and/or impaired protein degradation (Figure 1).

Hyperactive AT1R-B2R heteromers trigger enhanced down-stream effects such as the increased generation of vasoconstrictors and endothelin-1 [14,98], which are elevated in preeclampsia [99], and the anti-angiogenic sFlt1, which is a major feature of preeclampsia [8,14]. On placental vessels, the hyperactive AT1-B2 receptor complex contributes to impaired vasculogenesis and placental dysfunction of preeclampsia [12,14]. In the kidney, exaggerated AT1 receptor signaling accounts for kidney dysfunction and proteinuria [14,98]. Preeclampsia symptoms triggered by smooth muscle-specific AT1R-B2R expression also induce hemolysis and low platelet count [14]. Taken together, increased levels of hyperactive heteromeric AT1R-B2R aggregates on target organs of preeclampsia account for angiotensin II hypersensitivity of pregnancies complicated with preeclampsia. Angiotensin II hypersensitivity is a well-established feature of preeclampsia, which has been known for almost fifty years [67].

4.2. AT1R-B2R Heteromeric Protein Complexes Trigger Major Symptoms of Preeclampsia

Levels of hyperactive AT1R-B2R protein complexes are high in the maternal vasculature of women with pregnancies complicated by preeclampsia [12]. As detailed above, those hyperactive AT1R-B2R protein aggregates on blood vessels account for enhanced angiotensin II-stimulated G-protein-mediated signaling [12–14]. Enhanced vascular G-protein-stimulated signaling stimulated by angiotensin II is responsible for vasoconstriction, increased blood pressure and vascular remodeling in animal models and humans [100–102]. Increased vascular AT1R-B2R complex levels in transgenic mice with smooth muscle-specific AT1R-B2R expression accounted for a strong blood pressure rise and renal dysfunction with proteinuria at the end of pregnancy [14]. These symptoms are the main features of pregnancies complicated with preeclampsia (Figure 2).

In conjunction with major preeclampsia symptoms (Figure 2), AT1R-B2R protein aggregates triggered a rise in circulating levels of sFlt1 [14]. An increase in sFlt1 is a well-established marker of human preeclampsia [8–10,50]. The sFlt1 could be directly induced by AT1R-B2R-stimulated calcium signaling and/or oxidative stress [103,104]. Oxidative stress at the maternal-fetal interface is a typical symptom of preeclampsia [80,105]. Hyperactivation of AT1R and/or AT1R-B2R triggers enhanced calcium signaling, oxidative stress and sFlt1 [14,103,104]. Concomitantly with sFlt1, AT1R-B2R also increased other vasoactive peptides, such as endothelin-1 [14,98]. Increased circulating endothelin-1 levels contribute to preeclampsia symptoms and hypertension [99].

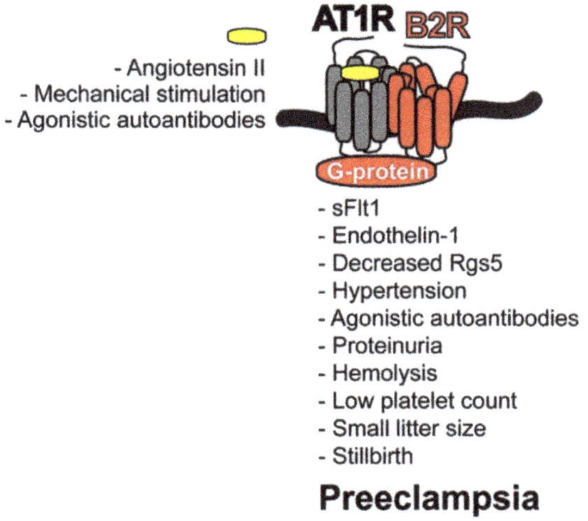

Figure 2. Overview of preeclampsia symptoms triggered by AT1R-B2R. AT1R-B2R triggers major preeclampsia symptoms. Causality between increased AT1R-B2R protein complex formation and preeclampsia symptoms was documented in transgenic mice with smooth muscle-specific AT1R-B2R expression. AT1R-B2R could also cause preeclampsia symptoms in humans because preeclamptic pregnancies are characterized by increased vascular contents of AT1R-B2R aggregation. During the pathogenesis of preeclampsia, AT1R-B2R is activated by angiotensin II, mechanical stimulation and agonistic autoantibodies against AT1R.

AT1R-B2R also led to reduced vascular Rgs5 (Regulator of G-protein signaling 5) levels [14]. Of note, decreased vascular and placental RGS5 levels are typical features of human pregnancies complicated with preeclampsia [14,106]. The decreased RGS5 level could further enhance the angiotensin II hypersensitivity and preeclampsia hypertension [106].

Together with hypertension, AT1R-B2R triggered AT1R-reactive, agonistic autoantibodies (Figure 2). In pregnant women with preeclampsia, the increased pressure and preeclampsia symptoms are also accompanied by the induction of agonistic autoantibodies against the AT1 receptor [72–75]. Agonistic AT1 receptor autoantibodies, which are triggered by increased vascular AT1R-B2R heteromers, are most likely induced as a consequence of the preeclampsia-related rise in blood pressure [14].

In addition to hypertension and kidney dysfunction with proteinuria, typical preeclampsia symptoms and end-organ damage were further documented in pregnant AT1R-B2R-transgenic mice by the presence of hemolysis and low platelet count (Figure 2).

In concert with increased levels of circulating sFlt1, enhanced calcium signaling by vascular and placental AT1R-B2R could impair vasculogenesis and cause defective placentation [14]. Likewise, increased vascular AT1R-B2R led to additional preeclampsia symptoms such as intrauterine growth retardation with reduced embryo weights and small litter size [14]. Severe preeclampsia symptoms caused by smooth muscle-specific AT1R-B2R expression also led to stillbirth, as documented by a strongly increased number of dead embryos [14].

Taken together, increased levels of pathological AT1R-B2R protein complexes are a sufficient cause of major preeclampsia symptoms with concomitant rise in antiangiogenic and vasoactive peptides, hypertension, renal dysfunction and end-organ damage in vivo, in transgenic mice with smooth muscle-specific AT1R-B2R expression (Figure 2). Because hyperactive AT1R-B2R heteromers are also increased on maternal and placental vessels of pregnant women with preeclampsia [12–14], the AT1R-B2R protein complex could also account for preeclampsia symptoms in human pregnancies complicated with preeclampsia.

4.3. Hypersensitive AT1R-B2R Heteromers Are Activated by Mechanical Forces, Which Increase during Pregnancy

As a GPCR heteromer, the heteromeric protein complex between the AT1 receptor and the B2 receptor (AT1R-B2R) is activated by the agonist angiotensin II [12–14,89,98]. However, circulating angiotensin II levels are depressed in preeclampsia [71]. In addition, plasma levels of the angiotensin II-generating enzyme, ACE, are decreased in pregnancy [107]. Therefore, the question arises, how pathological AT1R-B2R heteromers are activated in pregnancy, when preeclampsia symptoms evolve (Figure 3).

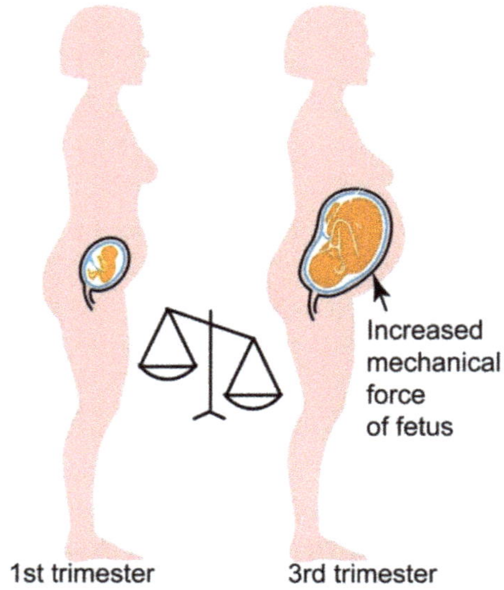

Figure 3. Mechanical forces are high in late-stage pregnancy. Due to the increasing fetal weight, mechanical forces are continuously rising in pregnancy. Consequently, mechanical forces are low in the 1st trimester (**left**), whereas in late-stage pregnancy (3rd trimester), the mechanical force induced by the increased fetal weight is high (**right**).

In search for an activator of the AT1R-B2R heteromeric protein complex at the onset of preeclampsia, we found that the ATR-B2R heteromer is also activated without agonist, solely by mechanical forces [14]. Because mechanical activation of AT1R-B2R is agonist-independent, hyperactivation of AT1R-B2R can occur even with low levels of circulating angiotensin II when mechanical forces are high.

During late-stage pregnancy, without and with preeclampsia, mechanical forces are constantly rising due to the increasing weight of the growing fetus [108,109]. Consequently, the AT1R-B2R protein complex is substantially activated by mechanical forces at late-stage pregnancy when mechanical forces are high due to the increased fetal weight (Figure 3).

The third trimester is also the time of onset of preeclampsia symptoms, notably the increased blood pressure. In agreement with involvement of AT1R-B2R in preeclampsia hypertension, a transgenic model with vascular AT1R-B2R expression developed preeclampsia symptoms with elevated blood pressure at late-stage pregnancy [14]. Moreover, twin pregnancies or multifetal pregnancies have an increased preeclampsia risk [109].

By enhancement of AT1R-B2R-mediated signaling, mechano-stimulation of AT1R-B2R by the growing fetus could also actively promote an increase in the preeclampsia-related anti-angiogenic sFlt1 [103,104] and vasoactive endothelin-1 [98,99].

With the increased blood pressure, autoantibodies against the AT1R are triggered, which further stimulate the AT1R-B2R protein complex [14,72–75]. Thus, the interplay between increasing mechanical forces, agonistic autoantibodies and circulating angiotensin

II levels culminate in AT1R-B2R-mediated symptoms of preeclampsia and preeclampsia hypertension at end-stage pregnancy (Figure 3).

4.4. Inducers of AT1R-B2R Heteromerization in Preeclampsia

What triggers AT1R-B2R heteromerization? Anton et al. showed that inflammatory stimuli induced by endotoxin application triggered the formation of the AT1R-B2R protein complex [110]. In agreement with a causal involvement of AT1R-B2R in hypertension, the inflammation-induced AT1R-B2R also led to high blood pressure in the experimental rat model [110]. Inflammation could also trigger the AT1R-B2R heteromer in preeclampsia (Figure 4) because inflammation plays a fundamental role in preeclampsia, and inflammatory markers are high in pregnancies complicated with preeclampsia [111,112].

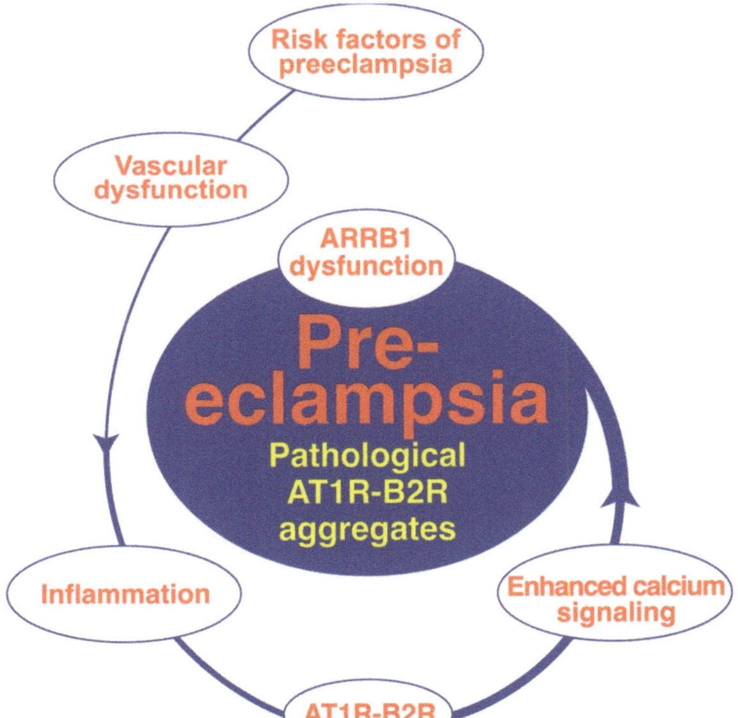

Figure 4. Accumulation of pathological AT1R-B2R aggregates during the pathogenesis of preeclampsia. Pregnancy-induced inflammation triggers AT1R-B2R protein complexes in individuals with major risk factors of preeclampsia and vascular dysfunction. AT1R-B2R stimulates enhanced calcium signaling, aggravates vascular dysfunction and promotes accumulation of pathological AT1R-B2R protein aggregates, in part by ARRB1 dysfunction. At later stages of pregnancy, when mechanical forces are high, pathological AT1R-B2R protein complexes are activated by mechanical forces without angiotensin II, trigger agonistic AT1R autoantibodies and symptoms of preeclampsia.

Concomitant with inflammation-dependent induction of AT1R-B2R, hypersensitive AT1R-B2R aggregates are stabilized during the pathogenesis of preeclampsia because preeclampsia and AT1R-B2R promote the dysfunction of beta-arrestin1 (ARRB1). ARRB1 is known for its involvement in desensitization and down-regulation of AT1R and AT1R-B2R [113,114]. AT1R-B2R could directly stimulate ARRB1 dysfunction by ERK1/2-mediated phosphorylation on serine 412 [14,115]. ERK1/2-stimulated signaling and levels of dysfunctional ARRB1 are increased on vascular cells of the experimental AT1R-B2R-induced preeclampsia model, and on placentas of pregnancies complicated with preeclampsia [14,102]. Consequently, during the pathogenesis of preeclampsia, inflammation-induced AT1R-B2R protein complexes stimulate ARRB1 dysfunction, which in turn promotes pathological AT1R-B2R protein aggregation by protein stabilization (Figure 4).

In addition, the AT1R-B2R-mediated signal enhancement further increases levels of the AT1R-B2R protein complex because the bradykinin B2 receptor gene is a direct target of angiotensin II-stimulated AT1R-mediated signaling [116]. With local B2 receptor induction in cells with AT1 receptor expression and signaling, AT1R-B2R protein complex levels also increase [12,89]. Thus, in frame of the pathogenesis of preeclampsia, AT1R-B2R protein complex formation is initially triggered by inflammation. The thereby enforced AT1R-B2R-mediated signaling triggers more AT1R-B2R by B2R induction and AT1R-B2R protein stabilization as a consequence of AT1R-B2R-mediated ARRB1 impairment. Ensuing accumulation of pathological AT1R-B2R aggregates triggers vascular and placental dysfunction and finally symptoms of preeclampsia (Figure 4).

How is AT1R-B2R protein complex formation related to established risk factors of preeclampsia? The transgenic mouse model of AT1R-B2R-induced preeclampsia develops vascular dysfunction [14]. In this respect, AT1R-B2R-transgenic mice mimic major established risk factors of preeclampsia, which also trigger vascular dysfunction, e.g., hypertension, diabetes, obesity, atherosclerosis, and autoimmune diseases, which cause damage to blood vessels, such as systemic lupus erythematosus, or hyper-coagulation such as the antiphospholipid syndrome [17–19]. Vascular dysfunction could impair spiral artery remodeling in early pregnancy [14,117]. Pregnancy-induced inflammation could trigger AT1R-B2R [110], and thereby further aggravate vascular dysfunction. Increased AT1R-B2R stimulates enhances calcium signaling, which promotes the transformation of AT1R-B2R complexes into pathological AT1R-B2R protein aggregates, in part by ARRB1 dysfunction. These pathological AT1R-B2R protein complexes are activated independently of angiotensin II by mechanical forces, when mechanical forces are high at later stages of pregnancy. Thereby, pathological AT1R-B2R complexes promote the preeclampsia-related rise in blood pressure, trigger agonistic AT1R autoantibodies and symptoms of preeclampsia (Figure 4).

5. Treatment Approaches of Preeclampsia

5.1. *Aspirin for Prevention of Preeclampsia*

There are very few treatment options for preeclampsia symptoms with documented efficacy. The most effective treatment for acute preeclampsia symptoms is still premature delivery of the baby. Nevertheless, numerous efforts and clinical studies were performed worldwide, which aim to reduce the risk of preeclampsia by pharmacological approaches. Currently, one of the best documented options is the use of low-dose aspirin for the extension of pregnancies at high risk of preeclampsia [53–56]. The introduction of aspirin was however highly controversial. Early data from 1985 showed the first evidence that low-dose aspirin could retard symptoms of preeclampsia [118]. It took more than 30 years until large placebo-controlled trials could confirm the therapeutic benefit of aspirin for prevention of preeclampsia in high-risk pregnancies with a history of preeclampsia [119]. According to currently available clinical studies, aspirin at a low dose of 80 to 150 mg/d started at 12 weeks of gestation could reduce the risk of preeclampsia in frame of secondary prevention in pregnant women with a history of preeclampsia [53–56]. Treatment side effects such as abruptio placentae could be a problem in nulliparous women [53–56]. Ongoing efforts aim to define the optimum aspirin dose, which could be higher than the currently recommended dose [120].

The mechanism of action of aspirin includes the inhibition of thromboxane A2 and platelet aggregation [119]. When given at a low dose, aspirin selectively inhibits platelet-derived thromboxane whereas endothelium-derived prostacyclin is maintained [119]. Notably, preeclampsia is characterized by increased thromboxane A2 and depressed prostacyclin levels [119]. This imbalance of thromboxane A2 versus prostacyclin is present from 13 weeks of gestation on [119]. Therefore, aspirin shows the highest benefit when it starts at about 12–13 weeks of gestation [53–56,119]. It seems to be of little value when started later, at 16 weeks of gestation [53–56,119]. Taken together, aspirin is currently one of the best documented pharmacological treatment options for risk reduction in preeclampsia and extension of pregnancy duration in pregnancies at high risk of preeclampsia.

Increased AT1R-B2R protein complex formation also exerts a major role in dysfunctional platelet aggregation of preeclamptic women. Platelets isolated from women with pregnancies complicated by preeclampsia display angiotensin II AT1R hypersensitivity [68], and increased AT1R-B2R heteromerization accounts for angiotensin II AT1R hypersensitivity of platelets [12]. Causality between AT1R-B2R and platelet dysfunction is further documented by the fact that the AT1R-B2R-induced preeclampsia model develops not only preeclampsia symptoms but also low platelet count and hemolysis [14]. Furthermore, the platelet aggregation-enhancing function of angiotensin II AT1R-stimulated signaling is well-documented in vitro and in vivo [121–123]. Therefore, preeclampsia prevention with aspirin is also expected to counteract the AT1R-B2R-enhanced platelet dysfunction of preeclampsia.

5.2. Targeting of Preeclampsia Symptoms and AT1R-B2R by Beta-Arrestin-Biased Agonism at the AT1 Receptor

Based on the causal role of AT1R-B2R in the pathogenesis of preeclampsia, the AT1R-B2R protein complex is an emerging therapeutic target of preeclampsia (Figure 5). Several targeting strategies are conceivable. The most straightforward approach would be the direct inhibition of the angiotensin II-stimulated AT1R-B2R by an AT1R antagonist. However, direct inhibition of the angiotensin II system is not feasible in preeclampsia because the RAAS exerts an indispensable role in fetal development, notably in the second and third trimester [124,125]. Therefore, direct inhibition of angiotensin II by an ACE inhibitor or AT1R antagonist is contraindicated in pregnancy because of fetal side effects such as oligohydramnios [124,125].

A different approach of AT1R-B2R targeting is based on the role of beta-arrestin, ARRB (1), in GPCR and AT1R-B2R down-regulation [14,113]. The concept relies on beta-arrestin-biased agonism [126–130]. Among other effects, beta-arrestin-biased agonists stimulate beta-arrestin-mediated GPCR internalization and down-regulation [126–130]. By stimulating beta-arrestin-mediated GPCR down-regulation, beta-arrestin-biased agonists have a bias for beta-arrestin but do not activate G-protein-stimulated signaling (Figure 5). With these characteristics, a beta-arrestin-biased agonist at the AT1 receptor stimulates AT1R-B2R downregulation and prevents hyperactive angiotensin II AT1R-B2R-stimulated G-protein activation and signaling [14,113]. By AT1R-B2R down-regulation, the beta-arrestin-biased agonist, [Sar^1,Ile^4,Ile^8]-AngII, also blocks stimulation of AT1R-B2R by mechanical forces [14]. For comparison, the non-biased AT1 receptor antagonist, losartan, not only blocks AT1R-B2R-stimulated receptor signaling but also prevents beta-arrestin-dependent receptor down-regulation [14]. Moreover, and in contrast to the biased agonist, the unbiased antagonist losartan does not block signaling of AT1R-B2R aggregates stimulated by mechanical forces [14].

In contrast to unbiased AT1R antagonists, beta-arrestin-biased agonists/antagonists at the AT1 receptor not only promote AT1R-B2R co-internalization and down-regulation in vitro, in cultured cells, but also in vivo [14,113]. The beta-arrestin-biased AT1R agonist, TRV027, prevented preeclampsia-induced hypertension and AT1R-B2R hyperactivation-induced symptoms of preeclampsia in an experimental AT1R-B2R-induced preeclampsia model [14]. The beta-arrestin-biased AT1R agonist, TRV027, also lowered blood pressure in several other experimental models of hypertension [131,132]. Thus, blood pressure-lowering and down-regulation of AT1R-B2R protein aggregates in vivo is feasible by a beta-arrestin-biased AT1R agonist [14,131,132]. In addition, targeting of AT1R-B2R by beta-arrestin-biased agonism not only lowers blood pressure but also prevents symptoms of preeclampsia [14].

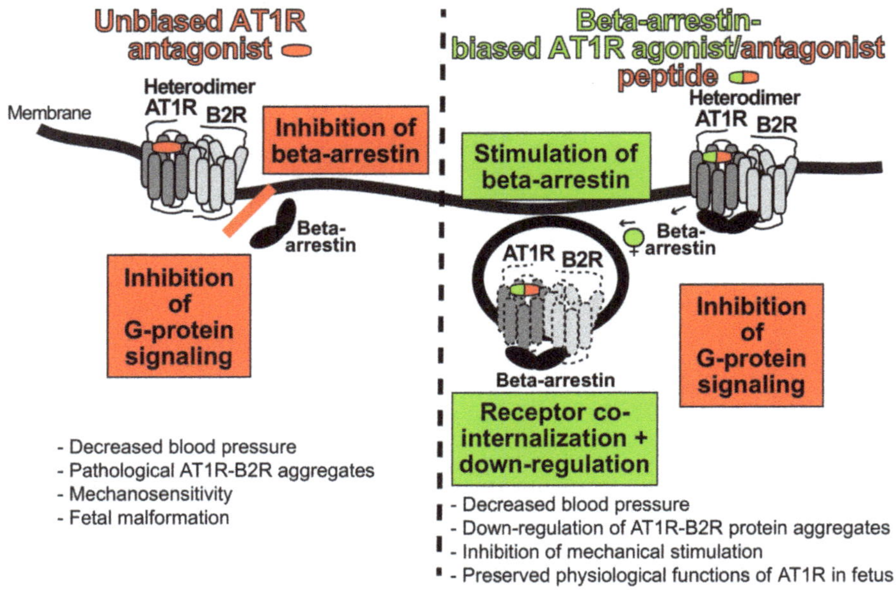

Figure 5. Major differences between unbiased AT1R antagonists and beta-arrestin-biased AT1R agonists/antagonists. An unbiased AT1R antagonist such as losartan inhibits (red color) both, G-protein signaling and beta-arrestin recruitment to the AT1R-B2R complex (**left side**). A beta-arrestin-biased AT1R agonist/antagonist peptide inhibits G-protein signaling (red color) but stimulates beta-arrestin recruitment (green color) to the AT1R-B2R complex. Thereby, the beta-arrestin-biased AT1R agonist/antagonist stimulates AT1R-B2R co-internalization (indicated by arrows) and down-regulation (**right side**). This Figure was adapted from [90].

Currently available beta-arrestin-biased AT1R agonists are peptides with a short half-life and a low oral bioavailability. The first beta-arrestin-biased AT1R agonist tested for clinical use, TRV027, has a favorable side-effect profile in humans, in a phase-II clinical study [133]. These features could be relevant for the development of biased AT1 receptor agonists for use in preeclampsia because due to the short half-life and low bioavailability, the indispensable fetal AT1 receptor is not fully blocked, and physiological functions of AT1R in the fetus will be preserved.

5.3. Inhibition of Exaggerated AT1R-B2R-Stimulated Calcium Signaling of Preeclampsia

Hyperactive AT1R-B2R aggregates stimulate excessive calcium signaling, which contributes to vasoconstriction, aberrant vascular remodeling and dysfunctional placentation during the pathogenesis of preeclampsia [12–14]. Inhibition of cellular calcium overload-induced vasoconstriction by an L-type calcium channel antagonist is frequently used to treat preeclampsia-induced hypertension [134]. The comparison of different L-type calcium channel antagonists showed that the long-acting amlodipine not only lowers blood pressure but also decreases AT1R-B2R protein complexes by restoration of beta-arrestin1 function in experimental and human preeclampsia cases [14]. However, due to their powerful antihypertensive activity, L-type calcium channel antagonists are only applicable for preeclampsia cases with high blood pressure increases [134].

An alternative approach could target AT1R (AT1R-B2R), and the L-type calcium channel simultaneously [135]. The experimental strategy uses a bivalent vaccine against the AT1 receptor and the L-type calcium channel, and thereby simultaneously inhibits the AT1 receptor and the Cav 1.2 channels as the major L-type calcium channel in humans [135]. The bivalent vaccine effectively lowered blood pressure and prevented end-organ damage in experimental models of hypertension in mice and rats, without major side effects [135].

A similar approach is conceivable to target the hyperactive AT1R-B2R of individuals at high risk of developing preeclampsia hypertension.

Together, these findings indicate that exaggerated AT1R-B2R protein complex formation of preeclampsia could be targeted by different approaches and chemical entities, e.g., by beta-arrestin-biased AT1R agonists, by inhibition of exaggerated calcium signaling or both. Research efforts are ongoing, which aim to develop specific targeting approaches of pathological AT1R-B2R aggregates and novel AT1R-inhibitory compounds [136]. These strategies could be further exploited to develop a therapy for preeclampsia symptoms, which is safe for the mother and infant.

6. Long-Term Complications of Preeclampsia

6.1. Preeclampsia Increases the Risk of Cardiovascular and Renal Dysfunction Later in Life

Despite enormous efforts worldwide, delivery is still the most effective therapy for preeclampsia symptoms. (Preterm) delivery usually treats the acute maternal symptoms of preeclampsia. However, pregnancies complicated with preeclampsia are associated with an increased risk of cardiovascular disease and kidney dysfunction later in life [7,137,138]. This association was already recognized early in the 19th century and in the 1960s and 1970s of the 20th century [137]. Meanwhile, a panoply of different studies worldwide document that a woman with a pregnancy complicated with preeclampsia is at increased risk to develop chronic hypertension and renal dysfunction during lifetime [137,138]. With a history of preeclampsia, the risk of cardiovascular disease is increased 2-fold, and the risk of end-stage renal disease is increased up to 5–10-fold [138]. The increased risk of cardiovascular and/or renal disease after preeclampsia is not fully understood. On one hand, pregnancy complications such as preeclampsia, could unmask a preexisting elevated risk of cardiovascular damage in these women [139]. On the other hand, the preeclampsia-induced damage could initiate pathophysiological changes that trigger cardiovascular disease later in life [139].

Preeclampsia does not only affect the cardiovascular health of the mother but also of the infant. Children born after a pregnancy complicated with preeclampsia usually have a low birth-weight. Preterm birth with low birth weight is known to be associated with a high risk of cardiovascular disease later in life, and an increased cardiovascular mortality [6,140]. Apart from prematurity, increasing evidence suggests that preeclampsia has an extra impact on the cardiovascular health of the offspring [140,141]. The increased risk of hypertension and stroke later in life indicates that preeclampsia has a yet unexplored impact on cardiovascular health of the infant, which cannot solely be explained by premature birth alone. Most likely, preeclampsia is associated with substantial cardiac and vascular alterations in the mother and the offspring, all of which are poorly understood.

6.2. AT1R-B2R Aggregation Increases the Risk of Renal Dysfunction as a Long-Term Complication of Preeclampsia

Although delivery of the infant relieves acute symptoms of preeclampsia, women with a pregnancy complicated by preeclampsia have an increased risk of developing cardiovascular and renal disease later in life.

Notably, a major long-term complication of preeclampsia is an increased risk of renal failure [138,142–146]. Several recent meta-analyses provide strong evidence that women with a pregnancy complicated with preeclampsia have a strong, up to 5–12-fold increased risk to develop end-stage kidney disease [139,142–146]. Until now, data came from meta-analyses. Those study results suggest that a subset of women with a history of preeclampsia have a highly increased risk of kidney disease [144]. The risk of kidney disease could be associated with the severity and/or subtype of preeclampsia because early onset preeclampsia seems to be associated with a higher risk of renal disease within 5 years after preeclampsia than that of late-onset preeclampsia [144]. The increased risk of renal failure after preeclampsia is at least partially associated with chronic hypertension because preeclampsia superimposed by chronic hypertension confers a substantially increased risk of end-stage renal disease compared to preeclampsia alone [145]. In addition, the risk of end-stage kidney disease is substantially enhanced after more than one pregnancy complicated with preeclampsia [146].

The experimental model of AT1R-B2R-induced preeclampsia also shows renal complications in the long term. In situ examination and histological analyses found that 30% of female mice with increased vascular AT1R-B2R heteromerization and several pregnancies complicated with preeclampsia had developed kidney atrophy (Figure 6). These findings complement data from human studies, which show that the risk of end-stage kidney disease rises substantially after several pregnancies complicated with preeclampsia [146]. Therefore, therapeutic approaches to inhibit and down-regulate AT1R-B2R aggregates aim not only to alleviate acute preeclampsia symptoms but also counteract and prevent long-term complications of preeclampsia, and other disorders associated with increased pathological AT1R-B2R aggregation.

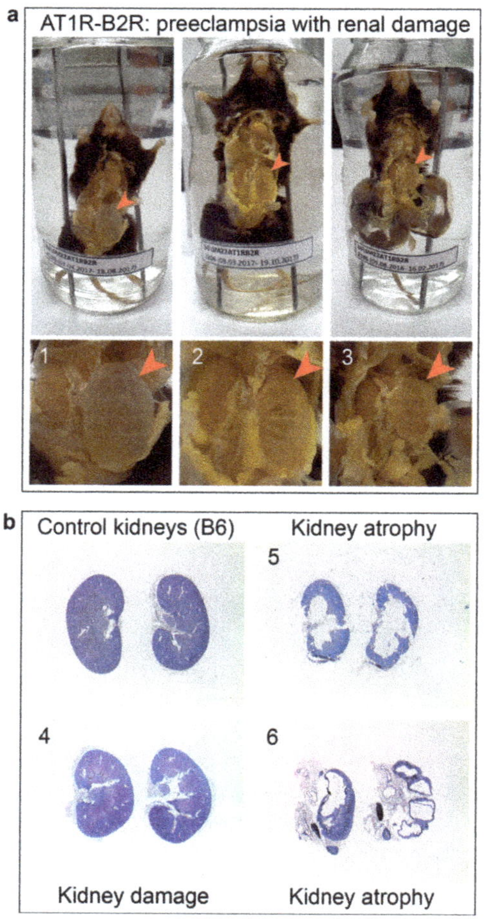

Figure 6. Kidney damage after multiple preeclamptic pregnancies in AT1R-B2R-transgenic mice. (**a**) In situ examination shows preeclampsia symptoms with uterine hemorrhage and abortion (upper panels), and kidney damage (lower panels) of female AT1R-B2R-transgenic mice (no. 1–3; age: 4–7 months); (**b**) Kidney damage and renal atrophy was detected by histological analysis of hematoxylin-eosin-stained kidney specimens of three female AT1R-B2R-transgenic mice after multiple preeclamptic pregnancies (no. 4–6; age 4–7 months). The left upper panel shows control kidney specimens of a non-transgenic, age-matched, female B6 mouse with uncomplicated pregnancies.

6.3. Outlook

Increased pathological AT1R-B2R aggregation is a major contributor to angiotensin II hypersensitivity of preeclampsia and a sufficient cause of major preeclampsia symptoms in vivo. Because biopsy specimens of women with pregnancies complicated by preeclampsia are also characterized by an increased AT1R-B2R protein complex formation [12,14], a causative role of AT1R-B2R in human preeclampsia is strongly suggested. Based on the causal involvement in major preeclampsia symptoms, pathological AT1R-B2R protein complexes could be exploited as a pharmacological target to develop a treatment strategy. Such a treatment could delay the onset of acute preeclampsia symptoms in individuals with increased AT1R-B2R aggregation. In addition, inhibition of pathological AT1R-B2R aggregates by pharmacological approaches could also constitute a potential strategy to prevent long-term complications of preeclampsia in the mother and offspring.

Author Contributions: Conceptualization, U.Q. and S.A.; investigation, U.Q. and S.A.; resources, U.Q. and S.A.; data curation, U.Q.; writing—original draft preparation, U.Q. and S.A.; visualization, U.Q. and S.A.; project administration, U.Q.; funding acquisition, U.Q. All authors have read and agreed to the published version of the manuscript.

Funding: This research was funded by ETH, grant number ETH-18 14-2.

Institutional Review Board Statement: Generation and phenotyping of transgenic mice was conducted according to NIH and Swiss guidelines, and approved by the Cantonal Veterinary Office (145-G, date of approval 14 February 2013 to S.A.).

Informed Consent Statement: Not applicable.

Data Availability Statement: All data presented are contained within the article.

Conflicts of Interest: The authors declare no conflict of interest.

References

1. Abalos, E.; Cuesta, C.; Grosso, A.L.; Chou, D.; Say, L. Global and regional estimates of preeclampsia and eclampsia: A systematic review. *Eur. J. Obstet. Gynecol. Reprod. Biol.* **2013**, *170*, 1–7. [CrossRef]
2. American College of Obstetricians and Gynecologists. Gestational hypertension and preeclampsia: ACOG practice bulletin, number 222. *Obstet. Gynecol.* **2020**, *135*, e237–e260. [CrossRef]
3. Bibbins-Domingo, K.; Grossman, D.C.; Curry, S.J.; Barry, M.J.; Davidson, K.W.; Doubeni, C.A.; Epling, J.W.; Kemper, A.R.; Krist, A.H.; Kurth, A.E.; et al. Screening for preeclampsia: US Preventive Services Task Force recommendation statement. *JAMA* **2017**, *317*, 1661–1667. [CrossRef] [PubMed]
4. Chappell, L.C.; Cluver, C.A.; Kingdom, J.; Tong, S. Pre-eclampsia. *Lancet* **2021**, *398*, 341–354. [CrossRef]
5. Spong, C.Y.; Mercer, B.M.; D'Alton, M.; Kilpatrick, S.; Blackwell, S.; Saade, G. Timing of indicated late-preterm and early-term birth. *Obstet. Gynecol.* **2011**, *118*, 323–333. [CrossRef] [PubMed]
6. Fox, R.; Kitt, J.; Leeson, P.; Aye, C.Y.L.; Lewandowski, A.J. Preeclampsia: Risk factors, diagnosis, management, and the cardiovascular impact on the offspring. *J. Clin. Med.* **2019**, *8*, 1625. [CrossRef] [PubMed]
7. Amaral, L.M.; Cunningham, M.W.; Cornelius, D.C.; LaMarca, B. Preeclampsia: Long-term consequences for vascular health. *Vasc. Health Risk Manag.* **2015**, *11*, 403–415. [CrossRef]
8. Levine, R.J.; Maynard, S.E.; Qian, C.; Lim, K.H.; England, L.J.; Yu, K.F.; Schisterman, E.F.; Thadhani, R.; Sachs, B.P.; Epstein, F.H.; et al. Circulating angiogenic factors and the risk of preeclampsia. *N. Engl. J. Med.* **2004**, *350*, 672–683. [CrossRef]
9. Liu, Y.; Zhao, Y.; Yu, A.; Zhao, B.; Gao, Y.; Niu, H. Diagnostic accuracy of the soluble Fms-like tyrosine kinase-1/placental growth factor ratio for preeclampsia: A meta-analysis based on 20 studies. *Arch. Gynecol. Obstet.* **2015**, *292*, 507–518. [CrossRef]
10. Maynard, S.E.; Min, J.Y.; Merchan, J.; Lim, K.H.; Li, J.; Mondal, S.; Libermann, T.A.; Morgan, J.P.; Sellke, F.W.; Stillman, I.E.; et al. Excess placental soluble fms-like tyrosine kinase1 (sFlt1) may contribute to endothelial dysfunction, hypertension, and proteinuria in preeclampsia. *J. Clin. Investig.* **2003**, *111*, 649–658. [CrossRef]
11. Carmeliet, P.; Moons, L.; Luttun, A.; Vincenti, V.; Compernolle, V.; De Mol, M.; Wu, Y.; Bono, F.; Devy, L.; Beck, H.; et al. Synergism between vascular endothelial growth factor and placental growth factor contributes to angiogenesis and plasma extravasation in pathological conditions. *Nat. Med.* **2001**, *7*, 575–583. [CrossRef]
12. AbdAlla, S.; Lother, H.; el Massiery, A.; Quitterer, U. Increased AT(1) receptor heterodimers in preeclampsia mediate enhanced angiotensin II responsiveness. *Nat. Med.* **2001**, *7*, 1003–1009. [CrossRef]
13. Quitterer, U.; Lother, H.; Abdalla, S. AT1 receptor heterodimers and angiotensin II responsiveness in preeclampsia. *Semin. Nephrol.* **2004**, *24*, 115–119. [CrossRef]

14. Quitterer, U.; Fu, X.; Pohl, A.; Bayoumy, K.M.; Langer, A.; AbdAlla, S. Beta-arrestin1 prevents preeclampsia by downregulation of mechanosensitive AT1-B2 receptor heteromers. *Cell* **2019**, *176*, 318–333. [CrossRef]
15. Ito, M.; Oliverio, M.I.; Mannon, P.J.; Best, C.F.; Maeda, N.; Smithies, O.; Coffman, T.M. Regulation of blood pressure by the type 1A angiotensin II receptor gene. *Proc. Natl. Acad. Sci. USA* **1995**, *92*, 3521–3525. [CrossRef]
16. Alfie, M.E.; Sigmon, D.H.; Pomposiello, S.I.; Carretero, O.A. Effect of high salt intake in mutant mice lacking bradykinin-B2 receptors. *Hypertension* **1997**, *29*, 483–487. [CrossRef] [PubMed]
17. Hypertension in Pregnancy: Diagnosis and Management. Available online: https://www.nice.org.uk/guidance/ng133 (accessed on 29 September 2021).
18. Brown, M.A.; Magee, L.A.; Kenny, L.C.; Karumanchi, S.A.; McCarthy, F.P.; Saito, S.; Hall, D.R.; Warren, C.E.; Adoyi, G.; Ishaku, S. International Society for the Study of Hypertension in Pregnancy (ISSHP). Hypertensive disorders of pregnancy: ISSHP classification, diagnosis, and management recommendations for international practice. *Hypertension* **2018**, *72*, 24–43. [CrossRef] [PubMed]
19. Bartsch, E.; Medcalf, K.E.; Park, A.L.; Ray, J.G.; High Risk of Pre-eclampsia Identification Group. Clinical risk factors for pre-eclampsia determined in early pregnancy: Systematic review and meta-analysis of large cohort studies. *BMJ* **2016**, *353*, i1753. [CrossRef]
20. Ghosh, G.; Grewal, J.; Männistö, T.; Mendola, P.; Chen, Z.; Xie, Y.; Laughon, S.K. Racial/ethnic differences in pregnancy-related hypertensive disease in nulliparous women. *Ethn. Dis.* **2014**, *24*, 283–289. [PubMed]
21. Levron, Y.; Dviri, M.; Segol, I.; Yerushalmi, G.M.; Hourvitz, A.; Orvieto, R.; Mazaki-Tovi, S.; Yinon, Y. The 'immunologic theory' of preeclampsia revisited: A lesson from donor oocyte gestations. *Am. J. Obstet. Gynecol.* **2014**, *211*, e1–e5. [CrossRef]
22. Blazquez, A.; Garcia, D.; Rodriguez, A.; Vassena, R.; Figueras, F.; Vernaeve, V. Is oocyte donation a risk factor for preeclampsia? A systematic review and meta-analysis. *J. Assist. Reprod. Genet.* **2016**, *33*, 855–863. [CrossRef] [PubMed]
23. Thangaratinam, S.; Gallos, I.D.; Meah, N.; Usman, S.; Ismail, K.M.; Khan, K.S.; TIPPS (Tests in Prediction of Pre-eclampsia's Severity) Review Group. How accurate are maternal symptoms in predicting impending complications in women with preeclampsia? A systematic review and meta-analysis. *Acta Obstet Gynecol. Scand.* **2011**, *90*, 564–573. [CrossRef] [PubMed]
24. Matsuda, H.; Sakaguchi, K.; Shibasaki, T.; Takahashi, H.; Kawakami, Y.; Furuya, K.; Kikuchi, Y. Cerebral edema on MRI in severe preeclamptic women developing eclampsia. *J. Perinat. Med.* **2005**, *33*, 199–205. [CrossRef]
25. Judy, A.E.; McCain, C.L.; Lawton, E.S.; Morton, C.H.; Main, E.K.; Druzin, M.L. Systolic hypertension, preeclampsia-related mortality, and stroke in California. *Obstet. Gynecol.* **2019**, *133*, 1151–1159. [CrossRef]
26. Cunningham, F.G.; Fernandez, C.O.; Hernandez, C. Blindness associated with preeclampsia and eclampsia. *Am. J. Obstet. Gynecol.* **1995**, *172*, 1291–1298. [CrossRef]
27. Arias, F.; Mancilla-Jimenez, R. Hepatic fibrinogen deposits in preeclampsia. Immunofluorescent evidence. *N. Engl. J. Med.* **1976**, *295*, 578–582. [CrossRef]
28. Schneider, H. Leberpathologie im Rahmen des HELLP-Syndroms. *Arch. Gynecol. Obstet.* **1994**, *255* (Suppl. 2), S245–S254. [CrossRef] [PubMed]
29. Stanhewicz, A.E.; Nuckols, V.R.; Pierce, G.L. Maternal microvascular dysfunction during preeclamptic pregnancy. *Clin. Sci.* **2021**, *135*, 1083–1101. [CrossRef] [PubMed]
30. Kattah, A. Preeclampsia and kidney disease: Deciphering cause and effect. *Curr. Hypertens. Res.* **2020**, *22*, 91. [CrossRef] [PubMed]
31. Prakash, J.; Ganiger, V.C. Acute kidney injury in pregnancy-specific disorders. *Indian J. Nephrol.* **2017**, *27*, 258–270. [CrossRef] [PubMed]
32. Thornton, C.E.; Makris, A.; Ogle, R.F.; Tooher, J.M.; Hennessy, A. Role of proteinuria in defining pre-eclampsia: Clinical outcomes for women and babies. *Clin. Exp. Pharmacol. Physiol.* **2010**, *37*, 466–470. [CrossRef]
33. Lei, T.; Qiu, T.; Liao, W.; Li, K.; Lai, X.; Huang, H.; Yuan, R.; Chen, L. Proteinuria may be an indicator of adverse pregnancy outcomes in patients with preeclampsia: A retrospective study. *Reprod. Biol. Endocrinol.* **2021**, *19*, 71. [CrossRef]
34. Sasamori, Y.; Tanaka, A.; Ayabe, T. Liver disease in pregnancy. *Hepatol. Res.* **2020**, *50*, 1015–1023. [CrossRef] [PubMed]
35. Han, C.; Huang, P.; Lyu, M.; Dong, J. Oxidative stress and preeclampsia-associated prothrombotic state. *Antioxidants.* **2020**, *9*, 1139. [CrossRef] [PubMed]
36. Fakhouri, F.; Scully, M.; Provot, F.; Blasco, M.; Coppo, P.; Noris, M.; Paizis, K.; Kavanagh, D.; Pene, F.; Quezada, S.; et al. Management of thrombotic microangiopathy in pregnancy and postpartum: Report from an international working group. *Blood* **2020**, *136*, 2103–2117. [CrossRef] [PubMed]
37. Abildgaard, U.; Heimdal, K. Pathogenesis of the syndrome of hemolysis, elevated liver enzymes, and low platelet count (HELLP): A review. *Eur. J. Obstet. Gynecol. Reprod. Biol.* **2013**, *166*, 117–123. [CrossRef] [PubMed]
38. Chew, L.C.; Verma, R.P. *Fetal Growth Restriction*; StatPearls Publishing: Treasure Island, FL, USA, 14 August 2021.
39. Mecacci, F.; Avagliano, L.; Lisi, F.; Clemenza, S.; Serena, C.; Vannuccini, S.; Rambaldi, M.P.; Simeone, S.; Ottanelli, S.; Petraglia, F. Fetal growth restriction: Does an integrated maternal hemodynamic-placental model fit better? *Reprod. Sci.* **2021**, *28*, 2422–2435. [CrossRef]
40. Scott, G.; Gillon, T.E.; Pels, A.; von Dadelszen, P.; Magee, L.A. Guidelines-similarities and dissimilarities: A systematic review of international clinical practice guidelines for pregnancy hypertension. *Am. J. Obstet. Gynecol.* **2020**, S0002-9378(20)30846-2. [CrossRef]

41. Poon, L.C.; Shennan, A.; Hyett, J.A.; Kapur, A.; Hadar, E.; Divakar, H.; McAuliffe, F.; da Silva Costa, F.; von Dadelszen, P.; McIntyre, H.D.; et al. The International Federation of Gynecology and Obstetrics (FIGO) initiative on pre-eclampsia: A pragmatic guide for first-trimester screening and prevention. *Int. J. Gynaecol. Obstet.* **2019**, *145*, 1–33. [CrossRef]
42. Wilkerson, R.G.; Ogunbodede, A.C. Hypertensive disorders of pregnancy. *Emerg. Med. Clin. N. Am.* **2019**, *37*, 301–316. [CrossRef] [PubMed]
43. Papageorghiou, A.T.; Yu, C.K.; Bindra, R.; Pandis, G.; Nicolaides, K.H.; Fetal Medicine Foundation Second Trimester Screening Group. Multicenter screening for pre-eclampsia and fetal growth restriction by transvaginal uterine artery Doppler at 23 weeks of gestation. *Ultrasound Obstet. Gynecol.* **2001**, *18*, 441–449. [CrossRef] [PubMed]
44. Nikuei, P.; Rajaei, M.; Roozbeh, N.; Mohseni, F.; Poordarvishi, F.; Azad, M.; Haidari, S. Diagnostic accuracy of sFlt1/PlGF ratio as a marker for preeclampsia. *BMC Pregnancy Childbirth* **2020**, *20*, 80. [CrossRef] [PubMed]
45. Hurrell, A.; Beardmore-Gray, A.; Duhig, K.; Webster, L.; Chappell, L.C.; Shennan, A.H. Placental growth factor in suspected preterm pre-eclampsia: A review of the evidence and practicalities of implementation. *BJOG* **2020**, *127*, 1590–1597. [CrossRef] [PubMed]
46. Duhig, K.E.; Myers, J.; Seed, P.T.; Sparkes, J.; Lowe, J.; Hunter, R.M.; Shennan, A.H.; Chappell, L.C.; PARROT Trial Group. Placental growth factor testing to assess women with suspected pre-eclampsia: A multicentre, pragmatic, stepped-wedge cluster-randomized controlled trial. *Lancet* **2019**, *393*, 1807–1818. [CrossRef]
47. Parchem, J.G.; Brock, C.O.; Chen, H.Y.; Kalluri, R.; Barton, J.R.; Sibai, B.M. Preeclampsia Triage by Rapid Assay Trial (PETRA) Investigators. Placental Growth Factor and the risk of adverse neonatal and maternal outcomes. *Obstet. Gynecol.* **2020**, *135*, 665–673. [CrossRef]
48. Barton, J.R.; Woelkers, D.A.; Newman, R.B.; Combs, C.A.; How, H.Y.; Boggess, K.A.; Martin, J.N.; Kupfer, K.; Sibai, B.M.; for the PETRA (Preeclampsia Triage by Rapid Assay) Trial. Placental growth factor predicts time to delivery in women with signs or symptoms of early preterm preeclampsia: A prospective multicenter study. *Am. J. Obstet. Gynecol.* **2020**, *222*, e1–e259. [CrossRef]
49. Hurrell, A.; Duhig, K.; Vandermolen, B.; Shennan, A.H. Recent advances in the diagnosis and management of pre-eclampsia. *Fac. Rev.* **2020**, *9*, 10. [CrossRef]
50. PlGF-Based Testing to Help Diagnose Suspected Pre-Eclampsia (Triage PlGF Test, Elecsys Immunoassay sFlt-1/PlGF Ratio, DELFIA Xpress PlGF 1-2-3 Test, and BRAHMS sFlt-1 Kryptor/BRAHMS PlGF Plus Kryptor PE Ratio). Available online: https://www.nice.org.uk/guidance/dg23 (accessed on 29 September 2021).
51. Cerdeira, A.S.; O'Sullivan, J.; Ohuma, E.O.; Harrington, D.; Szafranski, P.; Black, R.; Mackillop, L.; Impey, L.; Greenwood, C.; James, T.; et al. Randomized interventional study on prediction of preeclampsia/eclampsia in women with suspected preeclampsia: INSPIRE. *Hypertension* **2019**, *74*, 983–990. [CrossRef]
52. Ormesher, L.; Johnstone, E.D.; Shawkat, E.; Dempsey, A.; Chmiel, C.; Ingram, E.; Higgins, L.E.; Myers, J.E. A clinical evaluation of placental growth factor in routine practice in high-risk women presenting with suspected pre-eclampsia and/or fetal growth restriction. *Pregnancy Hypertens.* **2018**, *14*, 234–239. [CrossRef]
53. Mather, A.R.; Dom, A.M.; Thorburg, L.L. Low-dose aspirin in pregnancy: Who? when? how much? and why? *Curr. Opin. Obstet. Gynecol.* **2021**, *33*, 65–71. [CrossRef]
54. Wright, D.; Nicolaides, K.H. Aspirin delays the development of preeclampsia. *Am. J. Obstet. Gynecol.* **2019**, *220*, e1–e580. [CrossRef] [PubMed]
55. ACOG Committee Opinion No. 743. Low-dose aspirin use during pregnancy. *Obstet. Gynecol.* **2018**, *132*, e44–e52. [CrossRef]
56. Walsh, S.W.; Strauss, J.F. The road to low-dose aspirin therapy for the prevention of preeclampsia began with the placenta. *Int. J. Mol. Sci.* **2021**, *22*, 6985. [CrossRef] [PubMed]
57. Granger, J.P.; Alexander, B.T.; Llinas, M.T.; Bennett, W.A.; Khalil, R.A. Pathophysiology of hypertension during preeclampsia linking placental ischemia with endothelial dysfunction. *Hypertension* **2001**, *38*, 718–722. [CrossRef] [PubMed]
58. Qu, H.; Khalil, R.A. Vascular mechanisms and molecular targets in hypertensive pregnancy and preeclampsia. *Am. J. Physiol. Heart Circ. Physiol.* **2020**, *319*, H661–H681. [CrossRef] [PubMed]
59. Ives, C.W.; Sinkey, R.; Rajapreyar, I.; Tita, A.T.N.; Oparil, S. Preeclampsia-pathophysiology and clinical presentations: JACC State-of-the-art review. *J. Am. Coll. Cardiol.* **2020**, *76*, 1690–1702. [CrossRef]
60. Hong, K.; Park, H.J.; Cha, D. Clinical implications of placenta-derived angiogenic/anti-angiogenic biomarkers in preeclampsia. *Biomark. Med.* **2021**, *15*, 523–536. [CrossRef] [PubMed]
61. Pereira, M.M.; Torrado, J.; Sosa, C.; Zocalo, Y.; Bia, D. Role of arterial impairment in preeclampsia: Should the paradigm shift? *Am. J. Physiol. Heart Circ. Physiol.* **2021**, *320*, H2011–H2030. [CrossRef]
62. Melchiorre, K.; Giorgione, V.; Thilaganathan, B. The placenta and preeclampsia: Villain or victim? *Am. J. Obstet. Gynecol* **2021**, S0002-9378(20)31198-4. [CrossRef]
63. Mirabito Colafella, K.M.; Bovée, D.M.; Danser, A.H.J. The renin-angiotensin-aldosterone system and its therapeutic targets. *Exp Eye Res.* **2019**, *186*, 107680. [CrossRef]
64. Atlas, S.A. The renin-angiotensin aldosterone system: Pathophysiological role and pharmacologic inhibition. *J. Manag. Care Pharm.* **2007**, *13*, 9–20. [CrossRef] [PubMed]
65. Quitterer, U.; AbdAlla, S. Improvements of symptoms of Alzheimer's disease by inhibition of the angiotensin system. *Pharmacol. Res.* **2020**, *154*, 104230. [CrossRef] [PubMed]
66. Shah, D.M. The role of RAS in the pathogenesis of preeclampsia. *Curr. Hypertens. Rep.* **2006**, *8*, 144–152. [CrossRef] [PubMed]

67. Gant, N.F.; Daley, G.L.; Chand, S.; Whalley, P.J.; MacDonald, P.C. A study of angiotensin II pressor response throughout primigravid pregnancy. *J. Clin. Investig.* **1973**, *52*, 2682–2689. [CrossRef] [PubMed]
68. Haller, H.; Oeney, T.; Hauck, U.; Distler, A.; Philipp, T. Increased intracellular free calcium and sensitivity to angiotensin II in platelets of preeclamptic women. *Am. J. Hypertens.* **1989**, *2*, 238–243. [CrossRef] [PubMed]
69. Burke, S.D.; Zsengellér, Z.K.; Khankin, E.V.; Lo, A.S.; Rajakumar, A.; DuPont, J.J.; McCurley, A.; Moss, M.E.; Zhang, D.; Clark, C.D.; et al. Soluble fms-like tyrosine kinase 1 promotes angiotensin II sensitivity in preeclampsia. *J. Clin. Investig.* **2016**, *126*, 2561–2574. [CrossRef]
70. Stanhewicz, A.E.; Jandu, S.; Santhanam, L.; Alexander, L.M. Increased angiotensin II sensitivity contributes to microvascular dysfunction in women who have had preeclampsia. *Hypertension* **2017**, *70*, 382–389. [CrossRef] [PubMed]
71. Irani, R.A.; Xia, Y. Renin angiotensin signaling in normal pregnancy and preeclampsia. *Semin. Nephrol.* **2011**, *31*, 47–58. [CrossRef] [PubMed]
72. Wallukat, G.; Homuth, V.; Fischer, T.; Lindschau, C.; Horstkamp, B.; Jüpner, A.; Baur, E.; Nissen, E.; Vetter, K.; Neichel, D.; et al. Patients with preeclampsia develop agonistic autoantibodies against the angiotensin AT1 receptor. *J. Clin. Investig.* **1999**, *103*, 945–952. [CrossRef]
73. Dechend, R.; Gratze, P.; Wallukat, G.; Shagdarsuren, E.; Plehm, R.; Bräsen, J.H.; Fiebeler, A.; Schneider, W.; Caluwaerts, S.; Vercruysse, L.; et al. Agonistic autoantibodies to the AT1 receptor in a transgenic rat model of preeclampsia. *Hypertension* **2005**, *45*, 742–746. [CrossRef]
74. Xia, Y.; Kellems, R.E. Angiotensin receptor agonistic autoantibodies and hypertension: Preeclampsia and beyond. *Circ. Res.* **2013**, *113*, 78–87. [CrossRef]
75. Chen, R.F.; Wang, J.; Jiao, X.Y.; Liu, H.R.; Zhao, R.R.; Zhi, J.M. Detection of serum autoantibodies against AT1A-receptor during the development of the four types of hypertensive rat models. *Sheng Li Xue Bao* **2006**, *58*, 90–94.
76. McEachern, A.E.; Shelton, E.R.; Bhakta, S.; Obernolte, R.; Bach, C.; Zuppan, P.; Fujisaki, J.; Aldrich, R.W.; Jarnagin, K. Expression cloning of a rat B2 bradykinin receptor. *Proc. Natl. Acad. Sci. USA* **1991**, *88*, 7724–7728. [CrossRef] [PubMed]
77. Borkowski, J.A.; Ransom, R.W.; Seabrook, G.R.; Trumbauer, H.; Chen, H.; Hill, R.G.; Strader, C.D. Targeted disruption of a B2 bradykinin receptor gene in mice eliminates bradykinin action in smooth muscle and neurons. *J. Biol. Chem.* **1995**, *270*, 13706–13710. [CrossRef]
78. Girolami, J.P.; Bouby, N.; Richer-Giudicelli, C.; Alhenc-Gelas, F. Kinins and kinin receptors in cardiovascular and renal diseases. *Pharmaceuticals* **2021**, *14*, 240. [CrossRef]
79. Gainer, J.V.; Morrow, J.D.; Loveland, A.; King, D.J.; Brown, N.J. Effect of bradykinin-receptor blockade on the response to angiotensin-converting-enzyme inhibitor in normotensive and hypertensive subjects. *N. Engl. J. Med.* **1998**, *339*, 1285–1292. [CrossRef] [PubMed]
80. Guerby, P.; Tasta, O.; Swiader, A.; Pont, F.; Bujold, E.; Parant, O.; Vayssiere, C.; Salvayre, R.; Negre-Salvayre, A. Role of oxidative stress in the dysfunction of the placental endothelial nitric oxide synthase in preeclampsia. *Redox. Biol.* **2021**, *40*, 101861. [CrossRef]
81. Kawashima, S.; Yokoyama, M. Dysfunction of endothelial nitric oxide synthase and atherosclerosis. *Arterioscler. Thromb. Vasc. Biol.* **2004**, *24*, 998–1005. [CrossRef]
82. Perhal, A.; Wolf, S.; Jamous, Y.F.; Langer, A.; Abd Alla, J.; Quitterer, U. Increased reactive oxygen species generation contributes to the atherogenic activity of the B2 bradykinin receptor. *Front. Med.* **2019**, *6*, 32. [CrossRef]
83. Ashworth, J.R.; Warren, A.Y.; Baker, P.N.; Johnson, I.R. Loss of endothelium-dependent relaxation in myometrial resistance arteries in pre-eclampsia. *Br. J. Obstet. Gynaecol.* **1997**, *104*, 1152–1158. [CrossRef]
84. Svedas, E.; Nisell, H.; Vanwijk, M.J.; Nikas, Y.; Kublickiene, K.R. Endothelial dysfunction in uterine circulation in preeclampsia: Can estrogens improve it. *Am. J. Obstet. Gynecol.* **2002**, *187*, 1608–1616. [CrossRef] [PubMed]
85. Sipilä, R.; Jalkanen, M.; Huotari, K. Urinary kallikrein excretion in normal and hypertensive pregnancy at term. *Ann. Clin. Res.* **1986**, *18*, 208–210. [PubMed]
86. Millar, J.G.; Campbell, S.K.; Albano, J.D.; Higgins, B.R.; Clark, A.D. Early prediction of pre-eclampsia by measurement of kallikrein and creatinine on a random urine sample. *Br. J. Obstet. Gynaecol.* **1996**, *103*, 421–426. [CrossRef] [PubMed]
87. Kyle, P.M.; Campbell, S.; Buckley, D.; Kissane, J.; de Swiet, M.; Albano, J.; Millar, J.G.; Redman, C.W. A comparison of the inactive urinary kallikrein:creatinine ratio and the angiotensin sensitivity test for the prediction of pre-eclampsia. *Br. J. Obstet. Gynaecol.* **1996**, *103*, 981–987. [CrossRef]
88. Corthorn, J.; Germain, A.A.; Chacon, C.; Rey, S.; Soto, G.X.; Figueroa, C.D.; Müller-Esterl, W.; Duarte, I.; Valdes, G. Expression of kallikrein, bradykinin b2 receptor, and endothelial nitric oxide synthase in placenta in normal gestation, preeclampsia, and placenta accreta. *Endocrine* **2006**, *29*, 491–499. [CrossRef]
89. AbdAlla, S.; Lother, H.; Quitterer, U. AT1-receptor heterodimers show enhanced G-protein activation and altered receptor sequestration. *Nature* **2000**, *407*, 94–98. [CrossRef]
90. Quitterer, U.; AbdAlla, S. Discovery of pathologic GPCR aggregation. *Front. Med.* **2019**, *6*, 9. [CrossRef]
91. Zou, Y.; Akazawa, H.; Qin, Y.; Sano, M.; Takano, H.; Minamino, T.; Makita, N.; Iwanaga, K.; Zhu, W.; Kudoh, S.; et al. Mechanical stress activates angiotensin II type 1 receptor without the involvement of angiotensin II. *Nat. Cell Biol.* **2004**, *6*, 499–506. [CrossRef]

92. Berk, B.C.; Aronow, M.S.; Brock, T.A.; Cragoe, E.; Gimbrone, M.A.; Alexander, R.W. Angiotensin II-stimulated Na^+/H^+ exchange in cultured vascular smooth muscle cells. Evidence for protein kinase C-dependent and independent pathways. *J. Biol. Chem.* **1987**, *262*, 5057–5064. [CrossRef]
93. Bulenger, S.; Marullo, S.; Bouvier, M. Emerging role of homo- and heterodimerization in G-protein-coupled receptor biosynthesis and maturation. *Trends Pharmacol. Sci.* **2005**, *26*, 131–137. [CrossRef]
94. AbdAlla, S.; Lother, H.; Abdel-tawab, A.M.; Quitterer, U. The angiotensin II AT2 receptor is an AT1 receptor antagonist. *J. Biol. Chem.* **2001**, *276*, 39721–39726. [CrossRef] [PubMed]
95. Kostenis, E.; Milligan, G.; Christopoulos, A.; Sanchez-Ferrer, C.F.; Heringer-Walther, S.; Sexton, P.M.; Gembardt, F.; Kellett, E.; Martini, L.; Vanderheyden, P.; et al. G-protein-coupled receptor Mas is a physiological antagonist of the angiotensin II type 1 receptor. *Circulation* **2005**, *111*, 1806–1813. [CrossRef] [PubMed]
96. Abadir, P.M.; Periasamy, A.; Carey, R.M.; Siragy, H.M. Angiotensin II type 2 receptor-bradykinin B2 receptor functional heterodimerization. *Hypertension* **2006**, *48*, 316–322. [CrossRef] [PubMed]
97. Alfie, M.E.; Yang, X.P.; Hess, F.; Carretero, O.A. Salt-sensitive hypertension in bradykinin B2 receptor knockout mice. *Biochem. Biophys. Res. Commun.* **1996**, *224*, 625–630. [CrossRef]
98. AbdAlla, S.; Abdel-Baset, A.; Lother, H.; el Massiery, A.; Quitterer, U. Mesangial AT1/B2 receptor heterodimers contribute to angiotensin II hyperresponsiveness in experimental hypertension. *J. Mol. Neurosci.* **2005**, *26*, 185–192. [CrossRef]
99. Saleh, L.; Verdonk, K.; Visser, W.; van den Meiracker, A.H.; Danser, A.H. The emerging role of endothelin-1 in the pathogenesis of pre-eclampsia. *Ther. Adv. Cardiovasc. Dis.* **2016**, *10*, 282–293. [CrossRef]
100. Somlyo, A.P.; Somlyo, A.V. Ca^{2+} sensitivity of smooth muscle and nonmuscle myosin II: Modulated by G proteins, kinases, and myosin phosphatase. *Physiol. Rev.* **2003**, *83*, 1325–1358. [CrossRef]
101. Wirth, A.; Benyo, Z.; Lukasova, M.; Leutgeb, B.; Wettschureck, N.; Gorbey, S.; Orsy, P.; Horvath, B.; Maser-Gluth, C.; Greiner, E.; et al. G12-G13-LARG-mediated signaling in vascular smooth muscle is required for salt-induced hypertension. *Nat. Med.* **2008**, *14*, 64–68. [CrossRef]
102. Zhao, M.; Li, L.; Yang, X.; Cui, J.; Li, H. FN1, FOS, and ITGA5 induce preeclampsia: Abnormal expression and methylation. *Hypertens. Pregnancy* **2017**, *36*, 302–309. [CrossRef]
103. Murphy, S.R.; Cockrell, K. Regulation of soluble fms-like tyrosine kinase-1 production in response to placental ischemia/hypoxia: Role of angiotensin II. *Physiol. Rep.* **2015**, *3*, e12310. [CrossRef]
104. Zhou, C.C.; Ahmad, S.; Mi, T.; Xia, L.; Abbasi, S.; Hewett, P.W.; Sun, C.; Ahmed, A.; Kellems, R.E.; Xia, Y. Angiotensin II induces soluble fms-Like tyrosine kinase-1 release via calcineurin signaling pathway in pregnancy. *Circ. Res.* **2007**, *100*, 88–95. [CrossRef]
105. Matsubara, K.; Higaki, T.; Matsubara, Y.; Nawa, A. Nitric oxide and reactive oxygen species in the pathogenesis of preeclampsia. *Int. J. Mol. Sci.* **2015**, *16*, 4600–4614. [CrossRef] [PubMed]
106. Holobotovskyy, V.; Chong, Y.S.; Burchell, J.; He, B.; Phillips, M.; Leader, L.; Murphy, T.V.; Sandow, S.L.; McKitrick, D.J.; Charles, A.K.; et al. Regulator of G protein signaling 5 is a determinant of gestational hypertension and preeclampsia. *Sci. Transl. Med.* **2015**, *7*, 290ra88. [CrossRef] [PubMed]
107. Oats, J.N.; Broughton Pipkin, F.; Symonds, E.M.; Craven, D.J. A prospective study of plasma angiotensin-converting enzyme in normotensive primigravidae and their infants. *Br. J. Obstet. Gynaecol.* **1981**, *88*, 1204–1210. [CrossRef] [PubMed]
108. Burton, G.J.; Woods, A.W.; Jauniaux, E.; Kingdom, J.C. Rheological and physiological consequences of conversion of the maternal spiral arteries for uteroplacental blood flow during human pregnancy. *Placenta* **2009**, *30*, 473–482. [CrossRef] [PubMed]
109. Bdolah, Y.; Lam, C.; Rajakumar, A.; Shivalingappa, V.; Mutter, W.; Sachs, B.P.; Lim, K.H.; Bdolah-Abram, T.; Epstein, F.H.; Karumanchi, S.A. Twin pregnancy and the risk of preeclampsia: Bigger placenta or relative ischemia. *Am. J. Obstet. Gynecol.* **2008**, *198*, 428.e1–428.e6. [CrossRef]
110. Anton, E.L.; Fernandes, D.; Assreuy, J.; da Silva-Santos, J.E. Bradykinin increases BP in endotoxemic rat: Functional and biochemical evidence of angiotensin II AT1/bradykinin B2 receptor heterodimerization. *Br. J. Pharmacol.* **2019**, *176*, 2608–2626. [CrossRef]
111. Borzychowski, A.M.; Sargent, I.L.; Redman, C.W. Inflammation and pre-eclampsia. *Semin. Fetal Neonatal Med.* **2006**, *11*, 309–316. [CrossRef]
112. Cornelius, D.C.; Cottrell, J.; Amaral, L.M.; LaMarca, B. Inflammatory mediators: A causal link to hypertension during preeclampsia. *Br. J. Pharmacol.* **2019**, *176*, 1914–1921. [CrossRef]
113. Wilson, P.C.; Lee, M.H.; Appleton, K.M.; El-Shewy, H.M.; Morinelli, T.A.; Peterson, Y.K.; Luttrell, L.M.; Jaffa, A.A. The arrestin-selective angiotensin AT1 receptor agonist [Sar1,Ile4,Ile8]-AngII negatively regulates bradykinin B2 receptor signaling via AT1-B2 receptor heterodimers. *J. Biol. Chem.* **2013**, *288*, 18872–18884. [CrossRef]
114. Lee, M.H.; El-Shewy, H.M.; Luttrell, D.K.; Luttrell, L.M. Role of beta-arrestin-mediated desensitization and signaling in the control of angiotensin AT1a receptor-stimulated transcription. *J. Biol. Chem.* **2008**, *283*, 2088–2097. [CrossRef]
115. Lin, F.T.; Krueger, K.M.; Kendall, H.E.; Daaka, Y.; Fredericks, Z.L.; Pitcher, J.A.; Lefkowitz, R.J. Clathrin-mediated endocytosis of the beta-adrenergic receptor is regulated by phosphorylation/dephosphorylation of beta-arrestin1. *J. Biol. Chem.* **1997**, *272*, 31051–31057. [CrossRef]
116. Shen, B.; Harrison-Bernard, L.M.; Fuller, A.J.; Vanderpool, V.; Saifudeen, Z.; El-Dahr, S.S. The Bradykinin B2 receptor gene is a target of angiotensin II type 1 receptor signaling. *J. Am. Soc. Nephrol.* **2007**, *18*, 1140–1149. [CrossRef]

117. Magnussen, E.B.; Vatten, L.J.; Lund-Nilsen, T.I.; Salvesen, K.A.; Davey Smith, G.; Romundstad, P.R. Prepregnancy cardiovascular risk factors as predictors of pre-eclampsia: Population based cohort study. *BMJ* **2007**, *335*, 978. [CrossRef] [PubMed]
118. Beaufils, M.; Uzan, S.; Donsimoni, R.; Colau, J.C. Prevention of pre-eclampsia by early antiplatelet therapy. *Lancet* **1985**, *1*, 840–842. [CrossRef]
119. Atallah, A.; Lecarpentier, E.; Goffinet, F.; Doret-Dion, M.; Gaucherand, P.; Tsatsaris, V. Aspirin for prevention of preeclampsia. *Drugs* **2017**, *77*, 1819–1831. [CrossRef] [PubMed]
120. Van Doorn, R.; Mukhtarova, N.; Flyke, I.P.; Lasarev, M.; Kim, K.; Hennekens, C.H.; Hoppe, K.K. Dose of aspirin to prevent preterm preeclampsia in women with moderate or high-risk factors: A systematic review and meta-analysis. *PLoS ONE* **2021**, *16*, e0247782. [CrossRef]
121. Ding, Y.A.; MacIntyre, D.E.; Kenyon, C.J.; Semple, P.F. Potentiation of adrenaline-induced platelet aggregation by angiotensin II. *Thromb. Haemost.* **1985**, *54*, 717–720. [CrossRef]
122. Swartz, S.L.; Moore, T.J. Effect of angiotensin II on collagen-induced platelet activation in normotensive subjects. *Thromb. Haemost.* **1990**, *63*, 87–90. [CrossRef]
123. Kalinowski, L.; Matys, T.; Chabielska, E.; Buczko, W.; Malinski, T. Angiotensin II AT1 receptor antagonists inhibit platelet adhesion and aggregation by nitric oxide release. *Hypertension* **2002**, *40*, 521–527. [CrossRef]
124. Quan, A. Fetopathy associated with exposure to angiotensin converting enzyme inhibitors and angiotensin receptor antagonists. *Early Hum. Dev.* **2006**, *82*, 23–28. [CrossRef] [PubMed]
125. Shimada, C.; Akaishi, R.; Cho, K.; Morikawa, M.; Kaneshi, Y.; Yamda, T.; Minakami, H. Outomes of 83 fetuses exposed to angiotensin receptor blockers during the second or third trimesters: A literature review. *Hypertens. Res.* **2015**, *38*, 308–313. [CrossRef]
126. Thomas, W.G.; Qian, H.; Chang, C.S.; Karnik, S. Agonist-induced phosphorylation of the angiotensin II (AT(1A)) receptor requires generation of a conformation that is distinct from the inositol phosphate-signaling state. *J. Biol. Chem.* **2000**, *275*, 2893–2900. [CrossRef]
127. Reiter, E.; Ahn, S.; Shukla, A.K.; Lefkowitz, R.J. Molecular mechanism of beta-arrestin-biased agonism at seven-transmembrane receptors. *Annu. Rev. Pharmacol. Toxicol.* **2012**, *522*, 179–197. [CrossRef]
128. Luttrell, L.M. Minireview: More than just a hammer: Ligand "bias" and pharmaceutical discovery. *Mol. Endocrinol.* **2014**, *28*, 281–294. [CrossRef]
129. Ikeda, Y.; Kumagai, H.; Motozawa, Y.; Suzuki, J.; Komuro, I. Biased agonism of the angiotensin II type I receptor. *Int. Heart J.* **2015**, *56*, 485–488. [CrossRef]
130. Takezako, T.; Unal, H.; Karnik, S.S.; Node, K. Current topics in angiotensin II type 1 receptor research: Focus on inverse agonism, receptor dimerization and biased agonism. *Pharmacol. Res.* **2017**, *123*, 40–50. [CrossRef]
131. Carvalho-Galvao, A.; Ogunlade, B.; Xu, J.; Silva-Alves, C.R.A.; Mendes-Junior, L.G.; Guimaraes, D.D.; Cruz, J.C.; Queiroz, T.M.; Balarini, C.M.; Braga, V.A.; et al. Central administration of TRV027 improves baroreflex sensitivity and vascular reactivity in spontaneously hypertensive rats. *Clin. Sci.* **2018**, *132*, 1513–1527. [CrossRef]
132. Zanaty, M.; Seara, F.A.C.; Nakagawa, P.; Deng, G.; Mathieu, N.M.; Balapattabi, K.; Karnik, S.S.; Grobe, J.L.; Sigmund, C.D. Beta-arrestin-biased agonist targeting the brain AT1R (angiotensin II type 1 receptor) increases aversion to saline and lowers blood pressure in deoxycorticosterone acetate-salt hypertension. *Hypertension* **2021**, *77*, 420–431. [CrossRef] [PubMed]
133. Cotter, G.; Davison, B.A.; Butler, J.; Collins, S.P.; Ezekowitz, J.A.; Felker, G.M.; Filippatos, G.; Levy, P.D.; Metra, M.; Ponikowski, P.; et al. Relationship between baseline systolic blood pressure and long-term outcomes in acute heart failure patients treated with TRV027: An exploratory subgroup analysis of BLAST-AHF. *Clin. Res. Cardiol.* **2018**, *107*, 170–181. [CrossRef] [PubMed]
134. Jiang, N.; Liu, Q.; Liu, L.; Yang, W.W.; Zeng, Y. The effect of calcium channel blockers on prevention of preeclampsia in pregnant women with chronic hypertension. *Clin, Exp. Obstet. Gynecol.* **2015**, *42*, 79–81.
135. Wu, H.; Wang, Y.; Wang, G.; Qiu, Z.; Hu, X.; Zhang, H.; Yan, X.; Ke, F.; Zou, A.; Wang, W.; et al. A bivalent antihypertensive vaccine targeting L-type calcium channels and angiotensin AT1 receptors. *Br. J. Pharmacol.* **2020**, *177*, 402–419. [CrossRef] [PubMed]
136. Singh, K.D.; Jara, Z.P.; Harford, T.; Saha, P.P.; Pardhi, T.R.; Desnoyer, R.; Karnik, S.S. Novel allosteric ligands of the angiotensin receptor AT1R as autoantibody blockers. *Proc. Natl. Acad. Sci. USA* **2021**, *118*, e2019126118. [CrossRef]
137. Craici, I.; Wagner, S.; Garovic, V.D. Preeclampsia and future cardiovascular risk: Formal risk factor or failed stress test? *Ther. Adv. Cardiovasc. Dis.* **2008**, *2*, 249–259. [CrossRef] [PubMed]
138. Paauw, N.D.; Luijken, K.; Franx, A.; Verhaar, M.C.; Lely, A.T. Long-term renal and cardiovascular risk after preeclampsia: Towards screening and prevention. *Clin. Sci.* **2016**, *130*, 239–246. [CrossRef]
139. Hauspurg, A.; Ying, W.; Hubel, C.A.; Michos, E.D.; Ouyang, P. Adverse pregnancy outcomes and future maternal cardiovascular disease. *Clin. Cardiol.* **2018**, *41*, 239–246. [CrossRef]
140. Frost, A.L.; Suriano, K.; Aye, C.Y.L.; Leeson, P.; Lewandowski, A.J. The immediate and long-term impact of preeclampsia on offspring vascular and cardiac physiology in the preterm infant. *Front. Pediatr.* **2021**, *9*, 625726. [CrossRef]
141. Karatza, A.A.; Dimitriou, G. Preeclampsia emerging as a novel risk factor for cardiovascular disease in the offspring. *Curr. Pediatr. Rev.* **2020**, *16*, 194–199. [CrossRef]

142. Barrett, P.M.; McCarthy, F.P.; Kublickiene, K.; Cormican, S.; Judge, C.; Evans, M.; Kublickas, M.; Perry, I.J.; Stenvinkel, P.; Kashan, A.S. Adverse pregnancy outcomes and long-term maternal kidney disease: A systematic review and meta-analysis. *JAMA Netw. Open* **2020**, *3*, e1920964. [CrossRef] [PubMed]
143. Khashan, A.S.; Evans, M.; Kublickas, M.; McCarthy, F.P.; Kenny, L.C.; Stenvinkel, P.; Fitzgerald, T.; Kublickiene, K. Preeclampsia and risk of end stage kidney disease: A Swedish nationwide cohort study. *PLoS Med.* **2019**, *16*, e1002875. [CrossRef]
144. Kristensen, J.H.; Basit, S.; Wohlfahrt, J.; Damholt, M.B.; Boyd, H.A. Pre-eclampsia and risk of later kidney disease: Nationwide cohort study. *BMJ* **2019**, *365*, l1516. [CrossRef] [PubMed]
145. Wu, C.C.; Chen, S.H.; Ho, C.H.; Liang, F.W.; Chu, C.C.; Wang, H.Y.; Lu, Y.H. End-stage renal disease after hypertensive disorders in pregnancy. *Am. J. Obstet. Gynecol.* **2014**, *210*, e1–e8. [CrossRef] [PubMed]
146. Vikse, B.E.; Irgens, L.M.; Leivestad, T.; Skjaerven, R.; Iversen, B.M. Preeclampsia and the risk of end-stage renal disease. *N. Engl. J. Med.* **2008**, *359*, 800–809. [CrossRef] [PubMed]

cells

MDPI

Communication

Nebivolol Increases Nitric Oxide Synthase via β_3 Adrenergic Receptor in Endothelial Cells Following Exposure to Plasma from Preeclamptic Patients

Thaina Omia Bueno-Pereira , Priscila Rezeck Nunes , Mariana Bertozzi Matheus , Ananda Lini Vieira da Rocha and Valeria Cristina Sandrim *

Department of Biophysics and Pharmacology, Institute of Biosciences of Botucatu, Universidade Estadual Paulista (UNESP), Distrito Rubiao Junior, Botucatu 18680-000, Sao Paulo, Brazil; thaina.omia@unesp.br (T.O.B.-P.); priscila.nunes@unesp.br (P.R.N.); mariana.bertozzi@unesp.br (M.B.M.); ananda.rocha@unesp.br (A.L.V.d.R.)

* Correspondence: valeria.sandrim@unesp.br

Abstract: Background: Low bioavailability of nitric oxide (NO) is related to the pathophysiology of preeclampsia (PE). In the present study, we investigated the effect of nebivolol (NEB), a β_3-receptor agonist with vasodilator properties, on the NO synthesis in endothelial cells incubated with plasma from preeclamptic patients. Methods and results: Human umbilical vein endothelial cells (HUVECs) were incubated with plasma from healthy pregnant (HP) and PE women; NO quantification was assessed by a fluorescence compound. We found that endothelial cells incubated with plasma from women with PE show lower NO levels compared with the HP group ($p < 0.0001$). However, NEB treatment increases NO levels, partially, mediated by β_3 adrenergic receptors ($p < 0.0001$) and through eNOS activation ($p < 0.0001$). Conclusions: Our results suggest that NEB acts in NO synthesis through eNOS activation and β_3 adrenergic receptors in the endothelium. However, further studies will be needed to understand this molecule.

Keywords: nebivolol; nitric oxide; adrenoceptor; preeclampsia; endothelial dysfunction

Citation: Bueno-Pereira, T.O.; Nunes, P.R.; Matheus, M.B.; Vieira da Rocha, A.L.; Sandrim, V.C. Nebivolol Increases Nitric Oxide Synthase via β_3 Adrenergic Receptor in Endothelial Cells Following Exposure to Plasma from Preeclamptic Patients. *Cells* **2022**, *11*, 883. https://doi.org/10.3390/cells11050883

Academic Editors: Junie P Warrington, Ana T. Palei, Mark W. Cunningham and Lorena M. Amaral

Received: 12 January 2022
Accepted: 2 February 2022
Published: 4 March 2022

1. Introduction

Nebivolol (NEB) is a third-generation beta-blocker combining highly selectivity for the β_1 adrenoceptors and the ability to release nitric oxide (NO) through the endothelium [1]. It is a racemic mixture of two enantiomers, D- and L-nebivolol, where L-nebivolol behaves as a β_3 adrenergic receptor agonist, increasing Ca^{+2} efflux into endothelial cells by activating eNOS and consequently increasing NO bioavailability [2]. Preeclampsia (PE) is associated with a decrease in the bioavailability of NO [3]; the incubation of plasma or serum from women with PE on endothelial cells promotes changes in endothelial cell functions [4].

Therefore, in the present study, we evaluated the effect of NEB addition on NO levels in endothelial cells incubated with plasma from healthy pregnant (HP) women and with PE in the presence of NEB. Additionally, we investigated the molecular mechanisms of NEB in NO synthesis through activation of endothelial nitric oxide synthase and β_3 adrenoceptors.

2. Methods

The study was approved by the Research Ethics Committee of the Faculty of Medicine of Ribeirão Preto, Brazil (CAAE 37738620.0.0000.5440 approved on 19 October 2020 FMRP–USP) following the principles of the Helsinki Declaration, and all subjects gave written informed consent. Diagnosis criteria of PE were defined by the American College of Obstetricians and Gynecologists [5]. For each pool, 10 plasma samples from healthy pregnant (HP) women and 10 samples from pregnant women with PE were selected.

Human umbilical vein endothelial cells (HUVEC) (EA.hy 926) were cultured until reaching 80–90% of confluence. HUVEC were incubated in supplemented DMEM with 10% (v/v) plasma from PE or HP and 10 µM of NEB, 100 µM of L-name (NOS inhibitor), 3 µM of L-748,337 (β_3 antagonist), for 4 h. Cells were used until the 10th passage.

We used the DAF-FM compound for intracellular NO quantification (5 µM) (Invitrogen, Thermo Fisher Scientific, Carlsbad, CA, USA). The fluorescence signal was measured (excitation 495 nm, emission 535 nm) in a multifunctional plate reader (Synergy 4, BioTek, Winooski, VT, USA).

Replicates of 5 per group combined with treatments were performed in each experiment. When we compare three or more groups, we used One-Way ANOVA followed by the Holm-Sidak test. Statistical analyses were performed using GraphPad Prism 6.0 (GraphPad Software, San Diego, CA, USA) and for all tests, we considered a p-value ≤ 0.05 (two-tailed) significant.

3. Results

The general clinical parameters of women in the HP and PE groups whose plasma samples were collected and pooled for the in vitro studies are shown in Table 1.

Table 1. Clinical parameters of HP and PE pregnant women enrolled in the study.

Parameters	HP (n = 10)	PE (n = 10)
Maternal age (years)	24 ± 1	26 ± 1
SBP (mmHg)	110 ± 3	137 ± 6 *
DBP (mmHg)	71 ± 3	85 ± 4 **
GA at sampling (weeks)	36.25 ± 0.4	32.75 ± 1.7
BMI before pregnancy (Kg/m^2)	25 ± 1	30 ± 1 **
GA at delivery (weeks)	40 ± 1	32 ± 1 ***
Newborn weight (g)	3175 ± 166	2053 ± 358 *

Values are the means ± S.D. or percentage. HP, Healthy Pregnant; PE, preeclampsia; BMI, body mass index; SBP, systolic blood pressure; DBP, diastolic blood pressure; GA, gestational age. Values are mean ± standard error of the mean. Parametric variables were compared by Student t-test and non-parametric by Mann-Whitney test. * $p < 0.05$, ** $p < 0.01$, *** $p < 0.001$ vs healthy pregnant.

No differences were identified in the maternal age parameter between the groups. Systolic blood pressure (SBP) and diastolic blood pressure (DBP) is increased in the PE group when compared to the HP group (both $p < 0.001$). Furthermore, the index of body mass (BMI) before pregnancy is also increased in the PE group when compared with the HP group ($p < 0.001$), the gestational age (GA) at delivery, and the weight of the newborn were lower in PE compared to the HP group ($p < 0.001$ and $p < 0.05$, respectively). In addition, no differences were identified in the GA at sampling parameter between the groups. The PE samples were composed of 40% late-onset and 60% early-onset, according to GA at sampling (data not shown).

Endothelial cells incubated with HP plasma induce higher NO levels when compared with the PE group ($p < 0.05$) (Figure 1). In this group, NEB treatment increased NO levels ($p < 0.0001$) through eNOS activation, as we can identify in the group incubated with L-name (NOS inhibitor) ($p < 0.0001$) (Figure 1). To verify if NEB's action was mediated by the β_3 adrenergic receptor it was used the β_3 antagonist and we found that it reduces NO levels when compared to cultures without antagonist ($p < 0.0001$) (Figure 1).

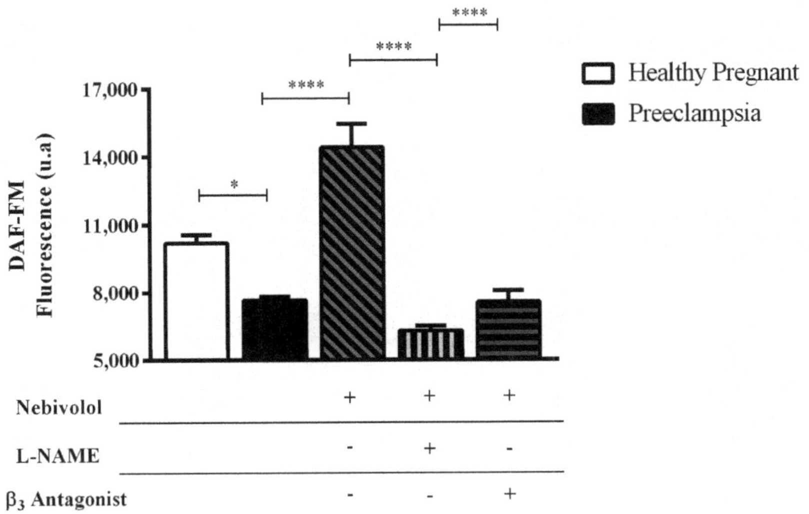

Figure 1. NO fluorescence intensity measured by DAF-FM. HUVECS were incubated with 10% (*v*/*v*) plasma from healthy pregnant (HP, white column) or preeclampsia (PE, black column) (n = 10 per group) and treatments for 4 h. Values are means ± S.E.M. Comparisons between groups were realized, assed by One-Way ANOVA followed by Holm-Sidak test. * ($p < 0.05$) HP vs PE; **** ($p < 0.0001$) PE vs PE + Nebivolol; PE + Nebivolol vs PE + L-NAME + Nebivolol; PE + Nebivolol vs PE + β_3 antagonist + Nebivolol.

4. Conclusions

Our results suggest that NEB acts through eNOS activation and β_3 adrenergic receptors in the endothelial cells following exposure to plasma from preeclamptic patients. The manuscript data lead us to further investigate other mechanisms involving the NO pathway and how nebivolol acts on them in a PE context. We believe that a further understanding of these mechanisms using our strategy model can offer, in the future, alternative management of PE and, consequently, reduce the repercussions of the disease in patients.

Author Contributions: Conceptualization, T.O.B.-P., P.R.N., M.B.M., A.L.V.d.R. and V.C.S.; Data curation, T.O.B.-P., P.R.N., M.B.M., A.L.V.d.R. and V.C.S.; Formal Analysis, T.O.B.-P.; Investigation, T.O.B.-P. and A.L.V.d.R.; Methodology, T.O.B.-P., P.R.N., M.B.M. and A.L.V.d.R.; Visualization, T.O.B.-P.; Funding acquisition, Project administration, Resources, Supervision, Validation, V.C.S.; Writing—original draft, T.O.B.-P.; Writing—review & editing, T.O.B.-P., P.R.N., M.B.M., A.L.V.d.R. and V.C.S. All authors have read and agreed to the published version of the manuscript.

Funding: This work was supported by the São Paulo Research Foundation (FAPESP-Brazil) [Grant Number: #2019/07230-8, #2020/14610-9, #2021/00141-0, #2021/01945-5], the Coordination for the Improvement of Higher Education Personnel (CAPES-Brazil) [Grant Number: #88887.647796/2021-00] and Brazilian National Council for Scientific and Technological Development (Cnpq-Brazil) [Grant Number: #308504/2021-6 #133901/2021.1].

Institutional Review Board Statement: The study was conducted according to the guidelines of the Declaration of Helsinki, and approved by the Research Ethics Committee of the Faculty of Medicine of Ribeirão Preto, Brazil (CAAE 37738620.0.0000.5440 approved on 19 October 2020 FMRP–USP).

Informed Consent Statement: Informed consent was obtained from all subjects involved in the study.

Data Availability Statement: Not applicable.

Acknowledgments: We thank Fabíola Taufic Monica Iglesias (UNICAMP) for the availability of reagents to perform the in vitro studies.

Conflicts of Interest: The authors declare no conflict of interest.

References

1. Van Nueten, L.; Taylor, F.R.; Robertson, J.I.S. Nebivolol vs atenolol and placebo in essential hypertension: A double-blind randomised trial. *J. Hum. Hypertens.* **1998**, *12*, 135–140. [CrossRef]
2. Dessy, C.; Saliez, J.; Ghisdal, P.; Daneau, G.; Lobysheva, I.I.; Frérart, F.; Belge, C.; Jnaoui, K.; Noirhomme, P.; Feron, O.; et al. Endothelial β_3-adrenoreceptors mediate nitric oxide-dependent vasorelaxation of coronary microvessels in response to the third-generation β-blocker nebivolol. *Circulation* **2005**, *112*, 1198–1205. [CrossRef]
3. Sandrim, V.C.; Dias, M.C.; de Carvalho Bovolato, A.L.; Tanus-Santos, J.E.; Deffune, E.; Cavalli, R.C. Plasma from pre-eclamptic patients induces the expression of the anti-angiogenic miR-195-5p in endothelial cells. *J. Cell. Mol. Med.* **2016**, *20*, 1198–1200. [CrossRef] [PubMed]
4. Viana-Mattioli, S.; Cinegaglia, N.; Bertozzi-Matheus, M.; Bueno-Pereira, T.O.; Dias-Caldeira, M.; Cavalli, R.C.; Sandrim, V.C. SIRT1-dependent effects of resveratrol and grape juice in an in vitro model of preeclampsia. *Biomed. Pharmacother.* **2020**, *131*, 110659. [CrossRef] [PubMed]
5. ACOG. ACOG Practice Bulletin Clinical Management Guidelines for Obstetrician Gynecologists. Gestational Hypertension and Preeclampsia. *Obstet. Gynecol.* **2020**, *135*, e237–e260. [CrossRef]

cells MDPI

Article

Low Dose of IL-2 Normalizes Hypertension and Mitochondrial Function in the RUPP Rat Model of Placental Ischemia

Evangeline Deer [1], Lorena M. Amaral [1], Nathan Campbell [1], Sarah Fitzgerald [1], Owen Herrock [1], Tarek Ibrahim [1] and Babbette LaMarca [1,2,3,*]

1 Department of Pharmacology and Toxicology, University of Mississippi Medical Center, Jackson, MS 39216, USA; edeer@umc.edu (E.D.); lmamaral@umc.edu (L.M.A.); ncampbell@umc.edu (N.C.); sjfitzgerald@umc.edu (S.F.); oherrock@umc.edu (O.H.); ibrahimt@uthscsa.edu (T.I.)

2 Department of Obstetrics and Gynecology, University of Mississippi Medical Center, Jackson, MS 39216, USA

3 Departments of Pharmacology, Physiology, and Obstetrics and Gynecology, Center for Excellence in Cardiovascular and Renal Research, University of Mississippi Medical Center, Jackson, MS 39216, USA

* Correspondence: bblamarca@umc.edu

Abstract: IL-2 is a cytokine released from CD4+T cells with dual actions and can either potentiate the inflammatory response or quell a chronic inflammatory response depending on its circulating concentration. IL-2 is elevated in many chronic inflammatory conditions and is increased during preeclampsia (PE). PE is characterized by new-onset hypertension during pregnancy and organ dysfunction and increasing evidence indicates that proinflammatory cytokines cause hypertension and mitochondrial (mt) dysfunction during pregnancy. The reduced uterine perfusion pressure (RUPP) model of placental ischemia is a rat model of PE that we commonly use in our laboratory and we have previously shown that low doses of recombinant IL-2 can decrease blood pressure in RUPP rats. The objective of this study was to determine the effects of a low dose of recombinant IL-2 on multi-organ mt dysfunction in the RUPP rat model of PE. We tested our hypothesis by infusing recombinant IL-2 (0.05 ng/mL) into RUPP rats on GD14 and examined mean arterial pressure (MAP), renal, placental and endothelial cell mt function compared to control RUPP. MAP was elevated in RUPP rats ($n = 6$) compared to controls ($n = 5$) (122 ± 5 vs. 102 ± 3 mmHg, $p < 0.05$), but was reduced by administration of LD recombinant IL-2 (107 ± 1 vs. 122 ± 5 mmHg, $n = 9$, $p < 0.05$). Renal, placental and endothelial mt ROS were significantly increased in RUPP rats compared to RUPP+ IL-2 and controls. Placental and renal respiration rates were reduced in RUPP rats compared to control rats but were normalized with IL-2 administration to RUPPs. These data indicate that low-dose IL-2 normalized multi-organ mt function and hypertension in response to placental ischemia.

Keywords: hypertension; preeclampsia; IL-2; inflammation; oxidative stress; placental ischemia

Citation: Deer, E.; Amaral, L.M.; Campbell, N.; Fitzgerald, S.; Herrock, O.; Ibrahim, T.; LaMarca, B. Low Dose of IL-2 Normalizes Hypertension and Mitochondrial Function in the RUPP Rat Model of Placental Ischemia. *Cells* **2021**, *10*, 2797. https://doi.org/10.3390/cells10102797

Academic Editors: Susanta K. Behura and Cord Brakebusch

Received: 15 July 2021
Accepted: 12 October 2021
Published: 19 October 2021

1. Introduction

Preeclampsia (PE) is a pregnancy associated disorder affecting 5–7% of pregnancies worldwide and is a well-known cause of maternal, fetal, neonatal morbidity and mortality [1]. PE is defined as new-onset hypertension and end-organ dysfunction during pregnancy occurring after the 20th week of gestation and is associated with chronic immune activation, proteinuria, fetal growth restriction and maternal endothelial dysfunction [2]. The only treatment for PE is delivery of the fetus, which is oftentimes pre-term. Therefore, additional investigation into the pathophysiological mechanisms that lead to the development of PE is necessary in order to develop potential therapies.

A normal pregnancy evolves with tightly controlled immune responses, whereas pregnancies diagnosed as PE exhibit a heightened pro-inflammatory immune response [3]. The complex immune response in PE has been associated with inflammatory immune cells and cytokines, which leads to the production of reactive oxygen species (ROS), increased expression of endothelin-1 (ET-1), sFlt-1 and autoantibodies to the angiotensin II type 1 receptor (AT1-AA) [4,5]. Importantly, in PE, reduced uterine perfusion may cause placental ischemia, a phenomenon that has been well demonstrated in the reduced uterine perfusion pressure (RUPP) rat model of preeclampsia [2,5–7]. Reactive oxygen species (ROS) are highly reactive free radicals that damage DNA, RNA, and protein, leading to cellular dysfunction and death. Oxidative stress is created during the imbalance between ROS and antioxidant defense of the cell [6,8]. In normal pregnancies, there is an increase in ROS compared to the non-pregnant state; however, ROS production is excessive in pathological states such as preeclampsia [6,9]. Vaka [6] examined mitochondrial (mt) dysfunction and ROS in the RUPP rat model of PE and found that mt dysfunction contributed to the hypertension. Although we know the importance of renal and placental mt dysfunction in hypertension in the RUPP model of PE, other avenues to lower excessive ROS or improve mt function need to be examined.

Many clinical studies have demonstrated that low-dose IL-2 ranging from 0.3×10^6 to 3.0×10^6 IU improved chronic inflammatory states and outcomes in patients with type 1 diabetes, ischemic heart disease, autoimmune liver disease, and lupus [10–13]. We have recently shown that low-dose IL-2, specifically (LD = 0.01 IU: 0.05 ng/mL), attenuated circulating and placental NK cells, normalized T regulatory cells, and lowered sFlt-1 and renal preproendothelin and blood pressure in the RUPP rats [14]. However, we do not know the effect of IL-2 on renal or placental mt dysfunction as a mechanism to improve hypertension. Therefore, we repeated our study and infused low-dose IL-2 (0.01 IU) into the RUPP rats and evaluated its effect on blood pressure and multi-organ mt function.

2. Materials and Methods

Timed-pregnant 12-week-old female Sprague Dawley (SD) rats (>240 g) were purchased from Envigo (Indianapolis, IN, USA) housed in an enclosed temperature-controlled room (75 °F) consisting of a 12:12 h light/dark cycle and free access to standard chow and water. All experiments were in compliance with the guidelines of the University of Mississippi Medical Center, and the animals were handled with care based on the approved protocol #1435 (12/1/2020) and published principles in the National Institutes of Health Guide for the Care of Animals and the Institutional Animal Care and Use Committee (IACUC).

Rats were divided into three groups consisting of normal pregnant rats (NP, n = 5), reduced uterine perfusion pressure rats (RUPP, n = 6), and RUPP rats treated with a low-dose (LD) treatment of recombinant IL-2 (RUPP + LD IL-2, n = 9, 0.05 ng/mL) (Recombinant IL-2, R and D Systems, Minneapolis, MN, USA). On day 14, the RUPP surgery was performed [15]. This surgical procedure is a model of preeclampsia in the rat and mirrors the pathophysiology of PE in women [6,15]. Surgical clips were placed just above the iliac bifurcation on the abdominal aorta and on ovarian arteries on the left and right side to reduce blood flow by approximately 40%. One group of pregnant RUPP rats received a low dose of recombinant IL-2 (0.05 ng/mL) infused intraperitoneally by a mini-osmotic pump (Alzet; Model 2002; Cupertino, CA, USA) inserted on day 14 of pregnancy. We have previously published that infusion of this dose of recombinant IL-2 into normal pregnant rats had no effect on blood pressure [14]. On day 18, all groups were inserted with indwelling carotid catheters [15]. Following the RUPP procedure, analgesics were provided to the rats, and included 5 mg/kg carprofen administered via subcutaneous injection and once daily for 2–3 days following RUPP surgical procedure, and 0.25% bupivacaine hydrochloride administered topically after carotid catheter insertion. On GD 19, blood pressure was measured with a pressure transducer (Cobe II tranducer CDX Sema, Aurora, CO, USA) and recorded continuously for one hour after a 30 min stabilization period as previously described [15], and is the average of several of the readings over a one-hour period. On gestation day 19, mean arterial pressure (MAP), fetal and placental weights were measured, and blood, placentas, and kidneys were collected for analysis of mitochondrial function.

2.1. Isolation of Mitochondria

Renal or placenta mitochondria were isolated from rats using differential centrifugation method [6,16]. Concisely, fresh tissues were rinsed and processed using a dounce homogenizer. The homogenate was centrifuged at 4000 rpm for 3 min at 4 °C. The supernatant was centrifuged at 10,000 rpm for 10 min at 4 °C, and the pellet was collected and suspended in 1 mL of Mito I buffer (250 mM sucrose, 10 mM HEPES, 1 mM EGTA 0.1% BSA, pH 7.2) and centrifuged at 10,000 rpm for 10 min at 4 °C. The collected pellet was suspended in 1 mL of Mito II (250 mM sucrose, 10 mM HEPES, 0.1% BSA, pH 7.2) and centrifuged at 10,000 rpm for 10 min at 4 °C. The final pellet was collected and suspended in 200 µL of Mito II buffer and used for respiration and ROS experiments.

2.2. Mitochondrial Respiration

Respiration in isolated mitochondria was measured using an Oxygraph 2K. The basal, state 2, state 3, state 4, and uncoupled respiration rates were measured using glutamate/malate, ADP, oligomycin, and FCCP (carbonyl cyanide-4-[trifluoromethoxy] phenylhydrazone), respectively [6]. Non-mitochondrial respiration was recorded with the use of Rotenone and antimycin A. The data collected were analyzed and expressed as pmol of oxygen consumed per second per milligram of mitochondrial protein.

2.3. Mitochondrial ROS

Mitochondrial hydrogen peroxide (H_2O_2) production in placental and renal mitochondria was determined by using amplex red assay [6,17]. Mitochondria (0.4 mg/mL) were incubated in a 96 well plate containing respiration buffer, superoxide dismutase (40 U/mL), horseradish peroxidase (4 U/mL), and succinate (10 mM). Amplex red (10 µM) was added to the wells last to start the reaction. The final volume of the wells used in the microplate was 200 µL. The real-time production of H_2O_2 was measured at 555/581 nm excitation/emission using a plate reader for 30 min at 25 °C. Sample controls (blanks without mitochondrial protein or amplex red) were included in the assay.

2.4. Endothelial Mitochondrial ROS

Mitochondrial-specific reactive oxygen species were measured using MitoSOX red, a fluorogenic dye that specifically targets the mitochondria in live cells. HUVECs (ATCC), passage 4, were grown to 70% confluency in 6 well culture plates in HUVEC complete growth medium [Medium 199-DMEM (50:50), 10% FBS, and 1% antimycotic/antibiotic] in a humidified atmosphere of 5% CO_2 at 37 °C. Cells were serum starved for 4 h prior to incubation with HUVEC complete growth media and 10% of individual serum from NP ($n = 5$), RUPP ($n = 4$), or RUPP + LD IL-2 ($n = 7$) sera overnight. Each experiment for individual rats were performed in duplicate and averaged together/animal. The data were then averaged for each group.

Media with serum was rinsed off and cells were incubated with MitoSOX red (5 µM) for 30 min at 37 °C. Antimycin A (100 µM) was utilized as a positive control. Serum free medium was added after washing the cells twice with DPBS and the cells were incubated for an additional 4 h. Cells were collected and analyzed in the FL2 channel of Gallios flow cytometer (Beckman Coulter, Brea, CA, USA).

2.5. IL-2 Cytokine Profile

Blood samples from NP, RUPP, and RUPP + LD IL-2 were collected and centrifuged at 825× *g* for 10 min at 4 °C. Serum samples were separated from the clot. The supernatant was obtain, aliquoted, and stored at −80 °C. IL-2 cytokine levels in serum samples from NP, RUPP, and RUPP + LD IL-2 were measured using a Bio-Plex immunoassay according to the manufacturer's instructions. Data were acquired using the BIO-PlexTM 200 system (Bio-Rad Laboratories, Hercules, CA, USA).

2.6. *Statistical Analysis*

All statistical analyses were performed with GraphPad Prism 7.02 software (GraphPad Software, San Diego, CA, USA). Results were reported as means ± SEM. Comparison of groups were assessed by one-way ANOVA with Bonferroni multiple comparisons test as post hoc analysis. Results were considered as statistically significant when $p < 0.05$.

3. Results

3.1. *IL-2 Significantly Lowered Blood Pressure in RUPP Rats*

Mean arterial pressure (MAP) was elevated in RUPP rats (n = 6) compared to NP controls (n = 5) (122 ± 5 vs. 102 ± 3 mmHg, $p < 0.05$), but was reduced by administration of LD IL-2 in RUPP rats (107 ± 1 vs. 122 ± 5 mmHg, n = 9, $p < 0.05$) (Figure 1). Placental weights were reduced in both RUPP rats (0.53 ± 0.03 g, $p < 0.05$) and RUPP + LD IL-2 rats (0.50 ± 0.02 g, $p < 0.05$) compared to NP controls (0.66 ± 0.04 g) (Table 1). Fetal weights were reduced in RUPP rats (1.99 ± 0.07 g, $p < 0.05$) and RUPP + IL-2 rats (1.95 ± 0.08 g, $p < 0.05$) compared to NP controls (2.27 ± 0.05 g) (Table 1). Percent reabsorptions and survivability were reduced in RUPP rats compared to NP controls and RUPP + LD IL-2 rats (Table 1).

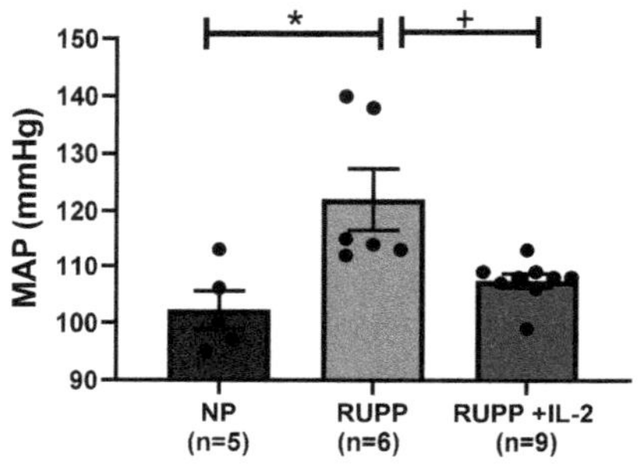

Figure 1. Mean arterial pressure was elevated in RUPP rats (n = 6) compared to NP rats (n = 5), but was normalized in RUPP rats administered a lose dose of IL-2 (n = 9). Results were reported as means ± SEM and considered statistically significant when $p < 0.05$. (* $p < 0.05$ vs. NP control; + $p < 0.05$ vs. RUPP).

Table 1. Placental Weights, Fetal Weights, Percent of Total Reabsorptions, and Percent Survival.

Animal Group	Placental Weight (g)	Fetal Weight (g)	% Reabsorptions	% Survived
NP	0.66 ± 0.04	2.27 ± 0.05	0 ± 0	100 ± 0
RUPP	0.53 ± 0.03 *	1.99 ± 0.07 *	19.6 ± 4 *	80.4 ± 4 *
RUPP + LD IL-2	0.50 ± 0.02 $^{+}$	1.95 ± 0.08 $^{+}$	0.43 ± 0.1 $^{\#}$	99.5 ± 0.1 $^{\#}$

* $p < 0.05$ RUPP vs. NP control; $^{+}$ $p < 0.05$ RUPP + LD IL-2 vs. NP control; $^{\#}$ $p < 0.05$ RUPP + LD IL-2 vs. RUPP.

3.2. *IL-2 Significantly Improved Multi-Organ mt Function in RUPP Rats*

Placental mitochondrial ROS, as measured by production of H_2O_2, was significantly elevated in RUPP rats (144.6 ± 14.18% gated, $p < 0.05$, n = 5) compared to NP controls (100 ± 12.34% gated, n = 5), but was normalized in RUPP + LD IL-2 rats (108.7 ± 7.38% gated, $p < 0.05$, n = 9) (Figure 2A). Renal mitochondrial ROS increased in real-time production of H_2O_2 in RUPP rats (127.1 ± 2.81% gated, $p < 0.05$, n = 5) in comparison to NP controls (100 ± 5% gated, n = 5), but was significantly reduced in RUPP + LD IL-2 rats (63.26 ± 3.57% gated, $p < 0.05$, n = 9), Figure 2B.

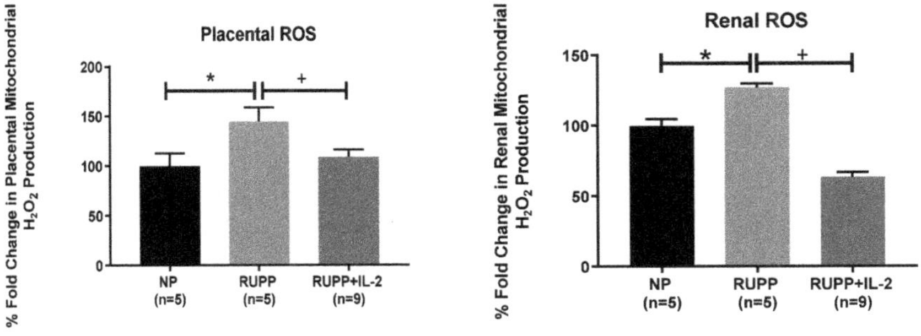

(A) (B)

Figure 2. There were increases in the (**A**) placental and (**B**) renal production of mitochondrial ROS in RUPP rats ($n = 5$) compared to NP rats ($n = 5$). Administration of a low dose IL-2 in RUPP rats ($n = 9$) normalized the production of mitochondrial ROS in the placenta and kidney. Results were reported as means ± SEM and considered statistically significant when $p < 0.05$. (* $p < 0.05$ vs. NP control; + $p < 0.05$ vs. RUPP).

Placental mitochondrial state 3 respiration, which is indicative of ATP produced from the addition of ADP, significantly decreased in RUPP rats ($n = 6$) (24.83 ± 15.53 pmol of O_2/s/mg, $p < 0.05$) compared to NP controls ($n = 5$) (132.9 ± 6.64 pmol of O_2/s/mg) (Figure 3A). State 3 placental respiration increased significantly in RUPP rats + LD IL-2 ($n = 6$) (157.3 ± 48.56 pmol of O_2/s/mg, $p < 0.05$) compared to RUPP rats, Figure 3A. Placental mitochondrial uncoupled respiration, which is indicative of electron transport chain function, was reduced significantly in RUPP rats ($n = 6$) (14.96 ± 3.89 pmol of O_2/s/mg, $p < 0.05$) compared to NP controls ($n = 5$) (91.02 ± 15.73 pmol of O_2/s/mg) (Figure 3B) but was significantly improved in RUPP + LD IL-2 rats ($n = 6$) (118.1 ± 35.42 pmol of O_2/s/mg, $p < 0.05$) compared to RUPP rats, Figure 3B.

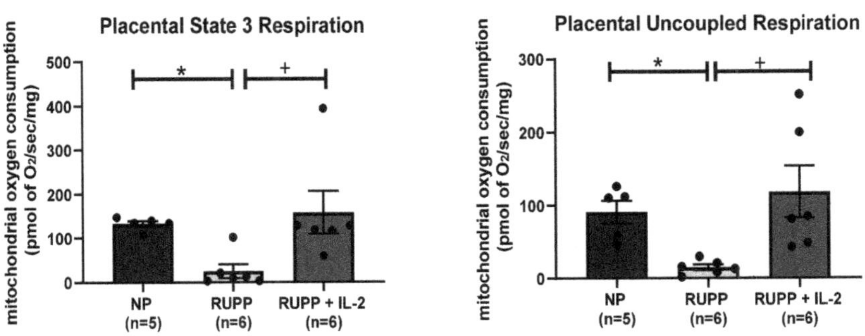

(A) (B)

Figure 3. There reductions in (**A**) state 3 and (**B**) uncoupled placental mitochondrial respiration in RUPP rats ($n = 6$) compared to RUPP + LD IL-2 ($n = 6$) and NP rats ($n = 5$). Results were reported as means ± SEM and considered statistically significant when $p < 0.05$. (* $p < 0.05$ vs. NP control; + $p < 0.05$ vs. RUPP).

Renal mitochondrial state 3 respiration, which is indicative of ATP produced from the addition of ADP, was significantly reduced in RUPP rats ($n = 5$) (138.4 ± 48.21 pmol of O_2/s/mg, $p < 0.05$) compared to NP controls ($n = 5$) (958 ± 200.6 pmol of O_2/s/mg) (Figure 4A), but was normalized in RUPP + LD IL-2 rats ($n = 6$) (904 ± 288 pmol of O_2/s/mg, $p < 0.05$), Figure 4A. Renal mitochondrial uncoupled respiration, which is indicative of electron transport chain function, decreased significantly in RUPP rats ($n = 5$) (68.1 ± 9.29 pmol of O_2/s/mg, $p < 0.05$) compared to NP controls ($n = 5$)

(476 ± 95.3 pmol of O_2/s/mg) (Figure 4B), but was increased in RUPP + LD IL-2 rats (n = 6) (824 ± 255.3 pmol of O_2/s/mg, $p < 0.05$) compared to RUPP, Figure 4B.

Placental RCR (state 3/state 4) significantly decreased in RUPP rats (n = 6) (0.79 ± 0.21 RCR, $p < 0.05$) compared to NP controls (n = 5) (1.77 ± 0.25 RCR) (Figure 5A), but was normalized in RUPP + LD IL-2 rats (n = 6) (1.77 ± 0.25 RCR, $p < 0.05$), Figure 5A. Although there was a decrease in renal RCR in RUPP rats (1.23 ± 0.37 RCR, n = 5) compared to NP controls (2.07 ± 0.35 RCR, n = 5), it was not significant nor was renal RCR improved in RUPP + LD IL-2 rats (1.30 ± 0.13 RCR, n = 5), Figure 5B.

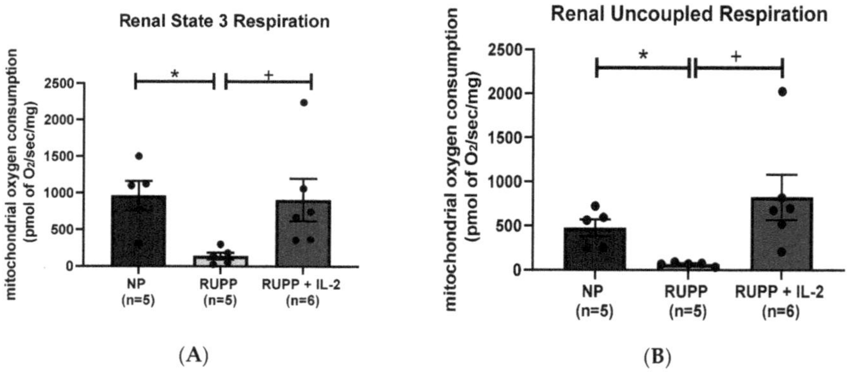

Figure 4. (**A**) State 3 and (**B**) uncoupled renal mitochondrial respiration were reduced in RUPP rats (n = 5) compared to NP rats (n = 5), but were normalized in RUPP + LD IL-2 rats (n = 6). Results were reported as means ± SEM and considered statistically significant when $p < 0.05$. (* $p < 0.05$ vs. NP control; + $p < 0.05$ vs. RUPP).

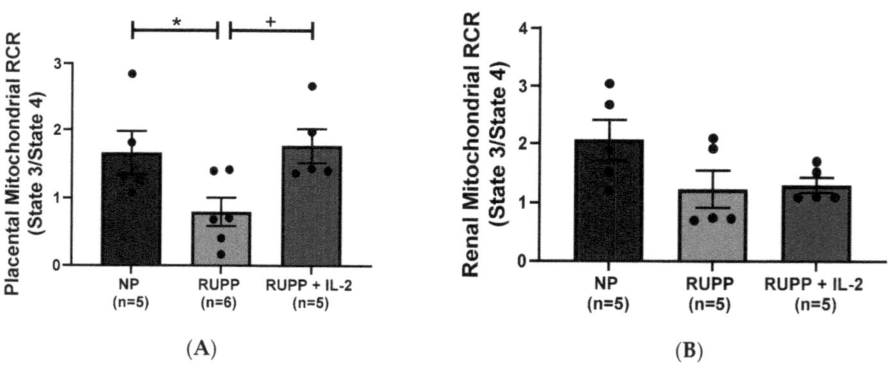

Figure 5. (**A**). Placental RCR (state 3/state 4) was reduced in RUPP (n = 6) compared to both NP rats (n = 5) and RUPP + LD IL-2 rats (n = 6). (**B**). Renal RCR (state3/state 4) was reduced in RUPP rats (n = 5) compared to NP rats (n = 5).There was no significant difference in RCR demonstrated in RUPP + LD IL-2 rats (n = 5) compared to RUPP rats. Results were reported as means ± SEM and considered statistically significant when $p < 0.05$. (* $p < 0.05$ vs. NP control; + $p < 0.05$ vs. RUPP).

Mt ROS significantly increased in endothelial cells, HUVECS, exposed to media containing sera from RUPP (n = 4) (6.38 ± 1.81% gated, $p < 0.05$) compared to NP controls (n = 5) (1.86 ± 0.6% gated) (Figure 6), but was normalized in RUPP + IL-2 rats (n = 7) (2.69 ± 0.53% gated, $p < 0.05$) compared to RUPP, Figure 6.

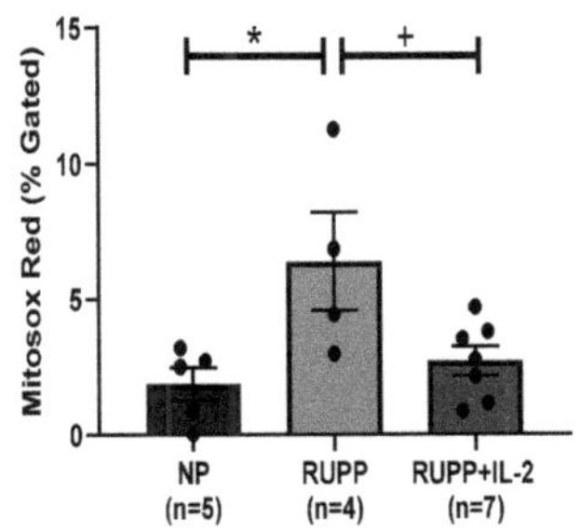

Figure 6. HUVECS incubated with RUPP sera (n = 4) exhibited an increase in endothelial mt dysfunction as demonstrated by an increase in mtROS compared to NP sera (n = 5), but co-incubation with RUPP + LD IL-2 sera (n = 7) normalized the production of mtROS. Results were reported as means ± SEM and considered statistically significant when $p < 0.05$. (* $p < 0.05$ vs. NP control; + $p < 0.05$ vs. RUPP).

3.3. Serum IL-2 Levels Are Increased in PE

As shown in Figure 7, IL-2 levels were higher RUPP + LD IL-2 (68 ± 16 pg/mL, $p < 0.05$) than RUPP control rats (5 ± 3.5 pg/mL, n = 4).

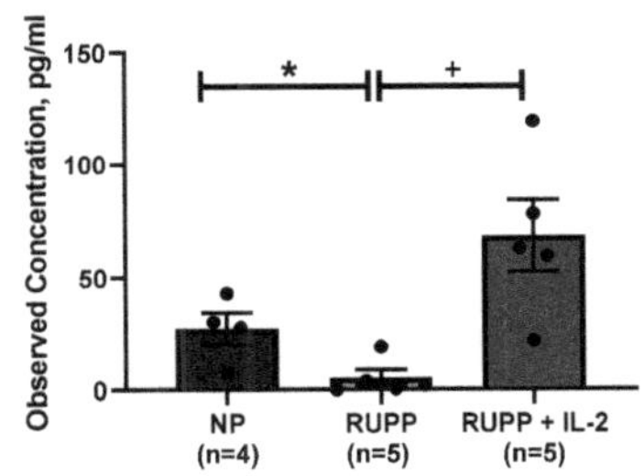

Figure 7. Comparison of IL-2 cytokine levels demonstrated that RUPP sera (n = 5) was lower compared to RUPP rats treated with a low dose of recombinant IL-2 (n = 5) and NP controls sera (n = 4). Results were reported as means ± SEM and considered statistically significant when $p < 0.05$. (* $p < 0.05$ vs. NP control; + $p < 0.05$ vs. RUPP).

4. Discussion

One hallmark of PE is multi-organ dysfunction, which can include a combination of renal, hepatic, neural, cardiac, placental or endothelial dysfunction. Appropriate cellular processes at the level of the mitochondria such as maintaining electron transfer, cellular respiration, and oxygen utilization are important for tissue homeostasis and function. In this study, we investigated the effects of recombinant IL-2 supplementation on PE characteristics such as hypertension and placental and renal mt dysfunction/ROS in RUPP rats. In addition, we measured endothelial mt dysfunction from cells exposed to circulating factors in sera from control normal pregnant rats, RUPP rats and RUPP rats + IL-2. The RUPP rat model of placental ischemia is a well characterized and well known model of preeclampsia that mimics the physiological features of humans, including hypertension, immune system abnormalities, systemic and renal vasoconstriction, and oxidative stress in the mother, and intrauterine growth restriction found in the offspring. Therefore, the RUPP model of placental ischemia has been shown by our lab and others as a useful model in studying the effects of placental ischemia. Although a limitation of our study is that we had a small number of animals in our experimental groups, the overall purpose of our study was to understand more about the effect of IL-2 and its role in inflammation in response to placental ischemia. The results of our study demonstrated that IL-2 normalized mean arterial pressure and significantly reduced the mt ROS in the kidney and placenta of

RUPP rats. Moreover, IL-2 reduced mt ROS from HUVECs exposed to sera from RUPP rats, indicating that IL-2 was able suppress circulating factors produced in response to placental ischemia that stimulate endothelial dysfunction. Importantly, renal and placental respiration were reduced in RUPP rats compared to normal pregnant rats. Low-dose recombinant IL-2 was able to normalize both renal and placental respiration, thus indicating that low-dose IL-2 was able to improve organ function in response to placental ischemia.

A compilation of research indicates that endothelial and mt dysfunction is readily observed in preeclamptic placentas [6,18–27]. Previously, McCarthy [28] showed vascular mtROS and decreased respiration in HUVECs exposed to sera from PE patients compared to HUVECS cultured with sera from normal pregnant women, thus indicating the importance of the release of soluble factors in the circulation to cause cellular mt dysfunction. Notably, we demonstrated that the blockade of circulating AT1-AA from human PE sera was able to attenuate mtROS in HUVECS cultured with PE sera with and without AT1-AA blockade [23]. In addition to the AT1-AA, our lab has recently investigated the importance of mt oxidative stress in PE pathology, and has linked reduced vascular endothelial mt respiration and mtROS with the presence of CD4+ T cells stimulated in response to placental ischemia [23–29]. In a previous study from our groups, we showed that LD recombinant IL-2 improved T regs cell in RUPP rats. In corroboration with our previous studies, this study demonstrated that infusion of low-dose IL-2 was able to normalize impaired mitochondrial function in tissues and vascular endothelium, thereby presenting a novel role for IL-2, possibly via improving T reg cells, in the pathology of preeclampsia. Importantly, PE is caused by multiple factors, and we investigate the role of various factors to contribute to the pathology of the disease and to possibly eliminate gaps in the literature. Although this study investigated IL-2, there is still more to understand about preeclampsia and the other factors that contribute to this disease. In addition, multiple factors can induce placental mitochondrial ROS production, and for example, we have previously shown that natural killer cells [4] and CD4+ T cells [29] cause mitochondrial dysfunction in RUPP rats and endothelial cells incubated with RUPP serum exhibit mitochondrial ROS [6]. HUVECS treated with serum from PE women and incubated with MitoSOX Red indicated that serum from PE women contained circulating factors which contribute to mitochondrial dysfunction and an increase in mt ROS in cultured human vascular endothelial cells, thereby demonstrating that an increase in oxidative stress contributes to endothelial dysfunction [6,23]. Furthermore, PE is associated with chronic immune activation, thereby leading to an increase in inflammatory cytokine production. This imbalance leads to chronic inflammation that is characterized by increases in pro-inflammatory cytokines and oxidative stress, (ROS), (endothelin (ET-1), and agonistic autoantibodies to the angiotensin II (Ang II), type 1 receptor (AT1-AA). These and other factors may influence mitochondrial activity and its contribution to maintenance of normal blood pressure during pregnancy. Discovering strategies that could potentially target mitochondrial stress via reducing mt oxidative stress and improving mitochondrial function is a finding that will greatly benefit maternal and fetal outcomes.

We recently showed a low-dose regimen of IL-2 consisting of three regimens between 0.01 and 0.05 IU into the RUPP rats significantly increased T Regs and decreased NK cells and hypertension during pregnancy [14]. We chose the lowest dose IL-2 from the previous study we performed because it had least detrimental effects on pup weight and survivability while still lowering the blood pressure and other factors associated with placental ischemia of pregnancy, such as sFlt-1 and renal endothelin-1. Moreover, this dose of IL-2 normalized circulating T regs in our previous study. Considering the importance of IL-2 in NK cell and T cells maturation, transformation and activity, we utilized this dose for our current study.

Regulatory T cells are necessary to maintain an immune steady state and to prevent autoimmune diseases. IL-2 has been touted for its unique ability for T cell expansion, function, and survival. When disrupted, IL-2-dependent balance of Treg and T effector cells causes autoimmunity and chronic inflammation. Recent studies have indicated that treatment with a low-dose IL-2 induces immune tolerance resulting in the suppression of an unwarranted immune responses and suggests that it may be a possible treatment of certain autoimmune disorders [30]. Because Treg cells cannot make their own IL-2, they depend on IL-2 produced by activated CD4+ T cells, therefore, linking T effector and T reg cell populations for immune homeostasis. In numerous autoimmune diseases, there is a decrease in the numbers and function of Treg cells [31–34] that is restored by exogenous low doses of IL-2 in mice [35,36] and humans [37–39]. Clinical trials of low-dose IL-2 in patients have had selective effects on T regs in healthy individuals, patients with hepatitis C virus-induced vasculitis, type 1 diabetes, and systemic lupus erythematosus [39–42]. Although some studies have reported that IL-2 did not alter or increase blood pressure in rats, it appears that the differences in the dose and frequency of IL-2 injection may be responsible for the difference in the lack of antihypertensive effect.

Previous studies have shown that IL-2 attenuated progression of hypertension in Dahl S rats, which was accompanied by improvements in renal dysfunction and cardiac hypertrophy [43,44]. In addition, treatment with IL-2 immune complex of hypertensive mice was shown to increase T regs and reduce aortic stiffening [44]. Moreover, a low dose of IL-2 administered in mice was able to prevent type 1 diabetes mellitus and improve the numbers of Tregs via their programming dependence on IL-2 [45]. Importantly, a high dose of IL-2 could produce lethal toxicity [46,47] and lead to the destruction of cells making high-dose IL-2 efficacious in treating metastatic cancer due to increasing the activity of natural killer cells toward the tumors. Therefore, the dose of IL-2 appears to be a determining factor in the imbalance between immune tolerance and destructive autoimmunity and is important to be established safely before moving forward as a treatment for any disease.

In order to successful maintain pregnancy, cytokines are necessary for maintaining a fetotolerant environment. Although IL-2, TNF-α, and IFN-γ are characteristic of T Helper 1-type immunity and induce a cytotoxic and an inflammatory reaction, IL-2 has been shown to improve inflammation during pregnancy without causing any detrimental effects [48]. In a normal pregnancy, IL-2 levels are decreased to concentrations necessary for the development, proliferation, and survival of T regs [49,50]. When IL-2 levels are high, pregnant women have a higher susceptibility for spontaneous abortion, preterm delivery, IUGR, and the development of PE [51,52]. High doses of IL-2 cause capillary damage, renal and liver damage, and hypotension [10–12]. Yet, higher doses of IL-2 have been coupled with additional drugs as a chemotherapeutic agent [10–12,53]. At high doses, NK cell proliferation and cytolytic activity are stimulated, which is an important component of some metastatic cancer therapies [53]. Furthermore, a high-dose treatment of IL-2 has been shown to be beneficial in treating metastatic cancer because of an increased activity of natural killer cells towards tumors. Therefore, high-dose treatments have been more common and utilized for longer periods of treatment compared to those utilizing lower doses of IL-2. However, these current studies show that in late pregnancy, low-dose IL-2 may help to lessen inflammation in response to placental ischemia which in turn lowers hypertensive molecules such as sFlt-1, ET-1 and mt ROS.

Cytolytic natural killer cells were also significantly decreased with LD IL-2 infusion in to RUPP and normal pregnant rats [14]. Because the immune cell profile was so very different between RUPP and RUPP+IL-2 in our previous study and because we have shown the importance of both T cells and NK cells to cause mitochondrial dysfunction, we examined the effects of IL-2 on mt function in RUPP rats. Mitochondria are an important source of ROS production, and the superoxide that is not able escape the mitochondria is reduced to hydrogen peroxide. Therefore, the highly reactive free radicals (ROS) damage the cellular contents and result in cellular dysfunction and cell death. Increased cell death within a tissue contributes to an overall dysfunction of that tissue and organ and therefore damage to mitochondrial function correlates with dysfunctional organ systems. HUVECS supplemented with serum from preeclamptic and normotensive women have shown impaired tube like structure formation and normal regular tube-like structure, respectively [9]. Furthermore, an in vitro model of HUVECS treated with PE serum demonstrates that

NADPH oxidase activity is increased and thus is important in O_2 formation [54–56]. We have shown that circulating factors such as the AT1-AA and TNF-α contribute to vascular endothelial cell mt dysfunction. In this study, we show that LD IL-2 improves not only renal and placental mt function in RUPP rats but that circulating factors stimulated by placental ischemia in the RUPP treated with LD IL-2 are decreased to the extent that RUPP sera no longer stimulates endothelial cell mt dysfunction and ROS.

5. Conclusions

In conclusion, supplementation of IL-2 significantly decreased the blood pressure in RUPP rats and lowered both placental and renal mt dysfunction/ROS and endothelial mt dysfunction/ROS. Moreover, in the current study, infusing a low dose of IL-2 mitigated the decrease in fetal reabsorptions, and thereby increased the percent survivability of the fetus in contrast to untreated RUPP rats. Although the previous study by Cunningham et al. [14] showed adverse fetal effects of LD IL-2, this study did not, which could be due to different operator's skills, yet both studies did demonstrate an increase in reabsorptions compared to NP controls. Therefore, coupled with our previous study, these results demonstrate that IL-2 normalized mt function in RUPP rats, which is associated with lower blood pressure and improved fetal survivability, thereby indicating that a potential therapeutic target for PE could be a carefully planned regimen of LD IL-2.

Author Contributions: Conceptualization, E.D. and B.L.; data curation, E.D., L.M.A.,N.C., S.F., O.H., T.I. and B.L.; funding acquisition, E.D., L.M.A. and B.L.; methodology, E.D., L.M.A., N.C., S.F., O.H., T.I. and B.L.; project administration, E.D., L.M.A. and B.L.; resources, B.L.; supervision, E.D., L.M.A. and B.L.; validation, E.D., L.M.A. and B.L.; visualization, E.D., L.M.A. and B.L.; writing—original draft, E.D. and B.L.; writing—review and editing, E.D. and B.L. All authors have read and agreed to the published version of the manuscript.

Funding: This study was supported by NIH grants RO1HD067541-10 (B.L.) and P20GM121334 (L.M.A., B.L.), American heart association (A.H.A.) early career award 19CDA34670055 (L.M.A.), and the T32 Trainee Grant T32-HL105324 (E.D.).

Institutional Review Board Statement: This study was conducted according to the guidelines of the University of Mississippi Medical Center, and the animals were handled according to the guiding principles published in the National Institutes of Health Guide for the Care of Animals and the Institutional Animal Care and Use Committee.

Informed Consent Statement: Not applicable.

Data Availability Statement: All data relevant to the study are included in the article.

Conflicts of Interest: The authors declare no conflict of interest.

References

1. Herse, F.; LaMarca, B. Angiotensin II Type 1 Receptor Autoantibody (AT1-AA)-Mediated Pregnancy Hypertension. *Am. J. Reprod. Immunol.* **2013**, *69*, 413–418. [CrossRef]
2. Cornelius, D.C.; Cottrell, J.; Amaral, L.M.; Lamarca, B. Inflammatory mediators: A causal link to hypertension during preeclampsia. *Br. J. Pharmacol.* **2019**, *176*, 1914–1921. [CrossRef] [PubMed]
3. Ma, Y.; Ye, Y.; Zhang, J.; Ruan, C.-C.; Gao, P.-J. Immune imbalance is associated with the development of preeclampsia. *Medicine* **2019**, *98*, e15080. [CrossRef] [PubMed]
4. Cunningham, M.W., Jr.; Vaka, V.R.; McMaster, K.; Ibrahim, T.; Cornelius, D.C.; Amaral, L.; Campbell, N.; Wallukat, G.; McDuffy, S.; Usry, N.; et al. Renal natural killer cell activation and mitochondrial oxidative stress; new mechanisms in AT1-AA mediated hypertensive pregnancy. *Pregnancy Hypertens.* **2019**, *15*, 72–77. [CrossRef] [PubMed]
5. LaMarca, B.; Parrish, M.R.; Wallace, K. Agonistic Autoantibodies to the Angiotensin II Type I Receptor Cause Pathophysiologic Characteristics of Preeclampsia. *Gend. Med.* **2012**, *9*, 139–146. [CrossRef] [PubMed]
6. Vaka, V.R.; McMaster, K.M.; Cunningham, M.W., Jr.; Ibrahim, T.; Hazlewood, R.; Usry, N.; Cornelius, D.C.; Amaral, L.M.; LaMarca, B. Role of mitochondrial dysfunction and reactive oxygen species in mediating hypertension in the reduced uterine perfusion pressure rat model of preeclampsia. *Hypertension* **2018**, *72*, 703–711. [CrossRef] [PubMed]

7. Novotny, S.R.; Wallace, K.; Heath, J.; Moseley, J.; Dhillon, P.; Weimer, A.; Wallukat, G.; Herse, F.; Wenzel, K.; Martin, J.N., Jr.; et al. Activating autoantibodies to the angiotensin II type I receptor play an important role in mediating hypertension in response to adoptive transfer of CD4+ T lymphocytes from placental ischemic rats. *Am. J. Physiol.-Regul. Integr. Comp. Physiol.* **2012**, *302*, R1197–R1201. [CrossRef]
8. Birben, E.; Sahiner, U.M.; Sackesen, C.; Erzurum, S.; Kalayci, O. Oxidative stress and antioxidant defense. *World Allergy Organ. J.* **2012**, *5*, 9–19. [CrossRef]
9. Sánchez-Aranguren, L.C.; Prada, C.E.; Riaño-Medina, C.E.; Lopez, M. Endothelial dysfunction and preeclampsia: Role of oxidative stress. *Front. Physiol.* **2014**, *5*, 372. [CrossRef]
10. Jeffery, H.C.; Jeffery, L.E.; Lutz, P.; Corrigan, M.; Webb, G.J.; Hirschfield, G.; Adams, D.; Oo, Y.H. Low-dose interleukin-2 promotes STAT-5 phosphorylation, Treg survival and CTLA-4-dependent function in autoimmune liver diseases. *Clin. Exp. Immunol.* **2017**, *188*, 394–411. [CrossRef]
11. Zhao, T.X.; Kostapanos, M.; Griffiths, C.; Arbon, E.L.; Hubsch, A.; Kaloyirou, F.; Helmy, J.; Hoole, S.P.; Rudd, J.H.F.; Wood, G.; et al. Low-dose interleukin-2 in patients with stable ischaemic heart disease and acute coronary syndromes (LILACS): Protocol and study rationale for a randomised, double-blind, placebo-controlled, phase I/II clinical trial. *BMJ Open* **2018**, *8*, e022452. [CrossRef] [PubMed]
12. Seelig, E.; Howlett, J.; Porter, L.; Truman, L.; Heywood, J.; Kennet, J.; Arbon, E.L.; Anselmiova, K.; Walker, N.M.; Atkar, R.; et al. The DILfrequency study is an adaptive trial to identify optimal IL-2 dosing in patients with type 1 diabetes. *JCI Insight* **2018**, *3*, e99306. [CrossRef] [PubMed]
13. Fishman, M.; Dutcher, J.P.; Clark, J.I.; Alva, A.; Miletello, G.P.; Curti, B.; Agarwal, N.; Hauke, R.; Mahoney, K.M.; Moon, H.; et al. Overall survival by clinical risk category for high dose interleukin-2 (HD IL-2) treated patients with metastatic renal cell cancer (mRCC): Data from the PROCLAIM(SM) registry. *J. Immunother. Cancer* **2019**, *7*, 84. [CrossRef] [PubMed]
14. Cunningham, M.W.; Amaral, L.M.; Campbell, N.E.; Cornelius, D.C.; Ibrahim, T.; Vaka, V.R.; LaMarca, B. Investigation of interleukin-2-mediated changes in blood pressure, fetal growth restriction, and innate immune activation in normal pregnant rats and in a preclinical rat model of preeclampsia. *Biol. Sex Differ.* **2021**, *12*, 1–10. [CrossRef] [PubMed]
15. Vaka, V.R.; Cunningham, M.W.; Deer, E.; Franks, M.; Ibrahim, T.; Amaral, L.M.; Usry, N.; Cornelius, D.C.; Dechend, R.; Wallukat, G.; et al. Blockade of endogenous angiotensin II type I receptor agonistic autoantibody activity improves mitochondrial reactive oxygen species and hypertension in a rat model of preeclampsia. *Am. J. Physiol. Integr. Comp. Physiol.* **2020**, *318*, R256–R262. [CrossRef] [PubMed]
16. Rebelato, H.J.; Esquisatto, M.A.M.; Moraes, C.; Amaral, M.E.C.; Catisti, R. Gestational protein restriction induces alterations in placental morphology and mitochondrial function in rats during late pregnancy. *J. Mol. Histol.* **2013**, *44*, 629–637. [CrossRef] [PubMed]
17. LaMarca, B.B.D.; Cockrell, K.; Sullivan, E.; Bennett, W.; Granger, J.P. Role of Endothelin in Mediating Tumor Necrosis Factor-Induced Hypertension in Pregnant Rats. *Hypertension* **2005**, *46*, 82–86. [CrossRef]
18. Beyramzadeh, M.; Dikmen, Z.G.; Erturk, N.K.; Tuncer, Z.S.; Akbiyik, F. Placental respiratory chain complex activities in high risk pregnancies. *J. Matern. Neonatal Med.* **2017**, *30*, 2911–2917. [CrossRef]
19. Illsinger, S.; Janzen, N.; Sander, S.; Schmidt, K.-H.; Bednarczyk, J.; Mallunat, L.; Bode, J.; Hagebölling, F.; Hoy, L.; Lücke, T.; et al. Preeclampsia and HELLP Syndrome: Impaired Mitochondrial Function in Umbilical Endothelial Cells. *Reprod. Sci.* **2010**, *17*, 219–226. [CrossRef]
20. Mando', C.; Marino, M.; Miriam, F.; De Palma, C.; Borelli, M.; Trabattoni, D.; Stampalija, T.; Ferrazzi, E.; Clementi, E.; Cetin, I. OS048 Mitochondrial content and function in placental cells and tissuesof preeclampsia and IUGR. *Pregnancy Hypertens. Int. J. Women's Cardiovasc. Health* **2012**, *2*, 203. [CrossRef]
21. Matsubara, S.; Minakami, H.; Sato, I.; Saito, T. Decrease in cytochrome c oxidase activity detected cytochemically in the placental trophoblast of patients with pre-eclampsia. *Placenta* **1997**, *18*, 255–259. [CrossRef]
22. Rodríguez, J.C.M.; Mendoza, P.Y.; Bernal, M.P.; Acosta, J.L.A. Oxidative stress level and placental histological changes during preeclampsia. *Ginecol. Obstet. Mex.* **2008**, *76*, 319–326.
23. Deer, E.; Vaka, V.R.; McMaster, K.M.; Wallace, K.; Cornelius, D.C.; Amaral, L.M.; Cunningham, M.W.; LaMarca, B. Vascular endothelial mitochondrial oxidative stress in response to preeclampsia: A role for angiotension II type 1 autoantibodies. *Am. J. Obstet. Gynecol. MFM* **2021**, *3*, 100275. [CrossRef] [PubMed]
24. Shibata, E.; Nanri, H.; Ejima, K.; Araki, M.; Fukuda, J.; Yoshimura, K.; Toki, N.; Ikeda, M.; Kashimura, M. Enhancement of Mitochondrial Oxidative Stress and Up-regulation of Antioxidant Protein Peroxiredoxin III/SP-22 in the Mitochondria of Human Pre-eclamptic Placentae. *Placenta* **2003**, *24*, 698–705. [CrossRef]
25. Shi, Z.; Long, W.; Zhao, C.; Guo, X.; Shen, R.; Ding, H. Comparative Proteomics Analysis Suggests that Placental Mitochondria are Involved in the Development of Pre-Eclampsia. *PLoS ONE* **2013**, *8*, e64351. [CrossRef]
26. Shanklin, D.R.; Sibai, B.M. Ultrastructural aspects of preeclampsia: II. Mitochondrial changes. *Am. J. Obstet. Gynecol.* **1990**, *163*, 943–953. [CrossRef]
27. He, L.; Wang, Z.; Sun, Y. Reduced amount of cytochrome c oxidase subunit I messenger RNA in placentas from pregnancies complicated by preeclampsia. *Acta Obstet. Gynecol. Scand.* **2004**, *83*, 144–148. [CrossRef]
28. McCarthy, C.M.; Kenny, L.C. Therapeutically targeting mitochondrial redox signalling alleviates endothelial dysfunction in preeclampsia. *Sci. Rep.* **2016**, *6*, 32683. [CrossRef]

29. Deer, E.; Reeve, K.E.; Amaral, L.M.; Vaka, V.R.; Franks, M.; Campbell, N.; Fitzgerald, S.; Herrock, O.T.; Ibrahim, T.; Cornelius, D.C.; et al. CD4+ T cells cause renal and placental mitochondrial oxidative stress as mechanisms of hypertension in response to placental ischemia. *Am. J. Physiol. Physiol.* **2021**, *320*, F47–F54. [CrossRef]
30. Ye, C.; Brand, D.; Zheng, S.G. Targeting IL-2: An unexpected effect in treating immunological diseases. *Signal Transduct. Target. Ther.* **2018**, *3*, 2. [CrossRef]
31. Szmyrka-Kaczmarek, M.; Kosmaczewska, A.; Ciszak, L.; Szteblich, A.; Wiland, P. Peripheral blood Th17/Treg imbalance in patients with low-active systemic lupus erythematosus. *Postęp. Hig. Med. Dośw.* **2014**, *68*, 893–898. [CrossRef]
32. Goldstein, J.D.; Pérol, L.; Zaragoza, B.; Baeyens, A.; Marodon, G.; Piaggio, E. Role of Cytokines in Thymus- Versus Peripherally Derived-Regulatory T Cell Differentiation and Function. *Front. Immunol.* **2013**, *4*, 155. [CrossRef]
33. Brusko, T.M.; Wasserfall, C.; Clare-Salzler, M.J.; Schatz, D.A.; Atkinson, M.A. Functional Defects and the Influence of Age on the Frequency of CD4+CD25+ T-Cells in Type 1 Diabetes. *Diabetes* **2005**, *54*, 1407–1414. [CrossRef]
34. Xing, Q.; Wang, B.; Su, H.; Cui, J.; Li, J. Elevated Th17 cells are accompanied by FoxP3+ Treg cells decrease in patients with lupus nephritis. *Rheumatol. Int.* **2012**, *32*, 949–958. [CrossRef]
35. Sgouroudis, E.; Kornete, M.; Piccirillo, C.A. IL-2 production by dendritic cells promotes Foxp3+regulatory T-cell expansion in autoimmune-resistant NOD congenic mice. *Autoimmunity* **2011**, *44*, 406–414. [CrossRef]
36. Zhang, H.; Chua, K.S.; Guimond, M.; Kapoor, V.; Brown, M.V.; A Fleisher, T.; Long, L.M.; Bernstein, D.; Hill, B.J.; Douek, D.C.; et al. Lymphopenia and interleukin-2 therapy alter homeostasis of CD4+CD25+ regulatory T cells. *Nat. Med.* **2005**, *11*, 1238–1243. [CrossRef]
37. Matsuoka, K.I.; Koreth, J.; Kim, H.T.; Bascug, G.; McDonough, S.; Kawano, Y.; Murase, K.; Cutler, C.; Ho, V.T.; Alyea, E.P.; et al. Low-dose interleukin-2 therapy restores regulatory T cell homeostasis in patients with chronic graft-versus-host disease. *Sci. Transl. Med.* **2013**, *5*, 179ra43. [CrossRef]
38. Koreth, J.; Matsuoka, K.-I.; Kim, H.T.; McDonough, S.M.; Bindra, B.; Alyea, E.P.; Armand, P.; Cutler, C.; Ho, V.T.; Treister, N.S.; et al. Interleukin-2 and Regulatory T Cells in Graft-versus-Host Disease. *N. Engl. J. Med.* **2011**, *365*, 2055–2066. [CrossRef] [PubMed]
39. Saadoun, D.; Rosenzwajg, M.; Joly, F.; Six, A.; Carrat, F.; Thibault, V.; Sene, D.; Cacoub, P.; Klatzmann, D. Regulatory T-Cell Responses to Low-Dose Interleukin-2 in HCV-Induced Vasculitis. *N. Engl. J. Med.* **2011**, *365*, 2067–2077. [CrossRef]
40. Ito, S.; Bollard, C.M.; Carlsten, M.; Melenhorst, J.J.; Biancotto, A.; Wang, E.; Chen, J.; Kotliarov, Y.; Cheung, F.; Xie, Z.; et al. Ultra-low Dose Interleukin-2 Promotes Immune-modulating Function of Regulatory T Cells and Natural Killer Cells in Healthy Volunteers. *Mol. Ther.* **2014**, *22*, 1388–1395. [CrossRef] [PubMed]
41. Hartemann, A.; Bensimon, G.; Payan, C.A.; Jacqueminet, S.; Bourron, O.; Nicolas, N.; Fonfrede, M.; Rosenzwajg, M.; Pharm, C.B.; Klatzmann, D. Low-dose interleukin 2 in patients with type 1 diabetes: A phase 1/2 randomised, double-blind, placebo-controlled trial. *Lancet Diabetes Endocrinol.* **2013**, *1*, 295–305. [CrossRef]
42. Von Spee-Mayer, C.; Siegert, E.; Abdirama, D.; Rose, A.; Klaus, A.; Alexander, T.; Enghard, P.; Sawitzk, B.; Hiepe, F.; Radbruch, A.; et al. Low-dose interleukin-2 selectively corrects regulatory T cell defects in patients with systemic lupus erythematosus. *Ann. Rheum. Dis.* **2016**, *75*, 1407–1415. [CrossRef] [PubMed]
43. Ishimitsu, T.; Uehara, Y.; Numabe, A.; Tsukada, H.; Ogawa, Y.; Yagi, S. Antihypertensive effect of interleukin-2 in salt-sensitive Dahl rats. *Hypertension* **1994**, *23*, 68–73. [CrossRef] [PubMed]
44. Majeed, B.; Tawinwung, S.; Eberson, L.S.; Secomb, T.W.; Larmonier, N.; Larson, D.F. Interleukin-2/Anti-Interleukin-2 Immune Complex Expands Regulatory T Cells and Reduces Angiotensin II-Induced Aortic Stiffening. *Int. J. Hypertens.* **2014**, *2014*, 1–12. [CrossRef]
45. Dwyer, C.; Ward, N.C.; Pugliese, A.; Malek, T.R. Promoting Immune Regulation in Type 1 Diabetes Using Low-Dose Interleukin-2. *Curr. Diabetes Rep.* **2016**, *16*, 1–10. [CrossRef]
46. Bluestone, J.A.; Herold, K.; Eisenbarth, G. Genetics, pathogenesis and clinical interventions in type 1 diabetes. *Nat. Cell Biol.* **2010**, *464*, 1293–1300. [CrossRef]
47. Van Belle, T.L.; Coppieters, K.T.; Von Herrath, M.G. Type 1 diabetes: Etiology, immunology, and therapeutic strategies. *Physiol. Rev.* **2011**, *91*, 79–118. [CrossRef]
48. Jørgensen, N.; Persson, G.; Hviid, T.V.F. The Tolerogenic Function of Regulatory T Cells in Pregnancy and Cancer. *Front. Immunol.* **2019**, *10*, 911. [CrossRef]
49. Shields, C.A.; McCalmon, M.; Ibrahim, T.; White, D.L.; Williams, J.M.; Lamarca, B.D.; Cornelius, D.C. Placental ischemia-stimulated T-helper 17 cells induce preeclampsia-associated cytolytic natural killer cells during pregnancy. *Am. J. Physiol. Integr. Comp. Physiol.* **2018**, *315*, R336–R343. [CrossRef]
50. Marzi, M.; Vigano, A.; Trabattoni, D.; Villa, M.L.; Salvaggio, A.; Clerici, M. Characterization of type 1 and type 2 cytokine production profile in physiologic and pathologic human pregnancy. *Clin. Exp. Immunol.* **1996**, *106*, 127–133. [CrossRef]
51. Calleja-Agius, J.; Brincat, M.P. Recurrent miscarriages: What is the role of cytokines? *Gynecol. Endocrinol.* **2008**, *24*, 663–668. [CrossRef]
52. Raghupathy, R. Cytokines as Key Players in the Pathophysiology of Preeclampsia. *Med. Princ. Pract.* **2013**, *22*, 8–19. [CrossRef]
53. Cornish, G.; Sinclair, L.V.; Cantrell, D. Differential regulation of T-cell growth by IL-2 and IL-15. *Blood* **2006**, *108*, 600–608. [CrossRef] [PubMed]

54. Maynard, S.E.; Min, J.-Y.; Merchan, J.; Lim, K.-H.; Li, J.; Mondal, S.; Libermann, T.A.; Morgan, J.P.; Sellke, F.W.; Stillman, I.E.; et al. Excess placental soluble fms-like tyrosine kinase 1 (sFlt1) may contribute to endothelial dysfunction, hypertension, and proteinuria in preeclampsia. *J. Clin. Investig.* **2003**, *111*, 649–658. [CrossRef] [PubMed]
55. Matsubara, K.; Matsubara, Y.; Hyodo, S.; Katayama, T.; Ito, M. Role of nitric oxide and reactive oxygen species in the pathogenesis of preeclampsia. *J. Obstet. Gynaecol. Res.* **2010**, *36*, 239–247. [CrossRef] [PubMed]
56. Cui, X.L.; Brockman, D.; Campos, B.; Myatt, L. Expression of NADPH oxidase isoform 1 (Nox1) in human placenta: Involvement in preeclampsia. *Placenta* **2006**, *27*, 422–431. [CrossRef] [PubMed]

cells MDPI

Article

Peripartum Investigation of Red Blood Cell Properties in Women Diagnosed with Early-Onset Preeclampsia

Beata Csiszar [1,2,*], Gergely Galos [2,3], Simone Funke [4], Dora Kinga Kevey [4], Matyas Meggyes [2,5], Laszlo Szereday [2,5], Peter Kenyeres [2,3], Kalman Toth [2,3] and Barbara Sandor [2,3]

1 Department of Anaesthesiology and Intensive Therapy, Medical School, University of Pécs, H-7624 Pécs, Hungary
2 Szentágothai Research Centre, University of Pécs, H-7624 Pécs, Hungary; galosgergely@gmail.com (G.G.); meggyes.matyas@pte.hu (M.M.); szereday.laszlo@pte.hu (L.S.); kenyeres.peter@pte.hu (P.K.); toth.kalman@pte.hu (K.T.); sandor.barbara@pte.hu (B.S.)
3 1st Department of Medicine, Medical School, University of Pécs, H-7624 Pécs, Hungary
4 Department of Obstetrics and Gynaecology, Medical School, University of Pécs, H-7624 Pécs, Hungary; funke.simone@pte.hu (S.F.); keveydorka@gmail.com (D.K.K.)
5 Department of Medical Microbiology and Immunology, Medical School, University of Pécs, H-7624 Pécs, Hungary
* Correspondence: csiszar.beata@pte.hu

Citation: Csiszar, B.; Galos, G.; Funke, S.; Kevey, D.K.; Meggyes, M.; Szereday, L.; Kenyeres, P.; Toth, K.; Sandor, B. Peripartum Investigation of Red Blood Cell Properties in Women Diagnosed with Early-Onset Preeclampsia. *Cells* **2021**, *10*, 2714. https://doi.org/10.3390/cells10102714

Academic Editors: Junie P. Warrington, Ana T. Palei, Mark W. Cunningham and Lorena M. Amaral

Received: 16 August 2021
Accepted: 5 October 2021
Published: 10 October 2021

Publisher's Note: MDPI stays neutral with regard to jurisdictional claims in published maps and institutional affiliations.

Abstract: We investigated peripartum maternal red blood cell (RBC) properties in early-onset preeclampsia (PE). Repeated blood samples were taken prospectively for hemorheological measurements at PE diagnosis (n = 13) or during 26–34 weeks of gestation in healthy pregnancies (n = 24), then at delivery, and 72 h postpartum. RBC aggregation was characterized by M index (infrared light transmission between the aggregated RBCs in stasis) and aggregation index (AI—laser backscattering from the RBC aggregates). We observed significantly elevated RBC aggregation (M index = 9.8 vs. 8.5; AI = 72.9% vs. 67.5%; $p < 0.001$) and reduced RBC deformability in PE ($p < 0.05$). A positive linear relationship was observed between AI and gestational age at birth in PE by regression analysis ($R^2 = 0.554$; $p = 0.006$). ROC analysis of AI showed an AUC of 0.84 (0.68–0.99) ($p = 0.001$) for PE and indicated a cutoff of 69.4% (sensitivity = 83.3%; specificity = 62.5%), while M values showed an AUC of 0.75 (0.58–0.92) ($p = 0.019$) and indicated a cutoff of 8.39 (sensitivity = 90.9% and specificity = 50%). The predicted probabilities from the combination of AI and M variables showed increased AUC = 0.90 (0.79–1.00) ($p < 0.001$). Our results established impaired microcirculation in early-onset PE manifesting as deteriorated maternal RBC properties. The longer the pathologic pregnancy persists, the more pronounced the maternal erythrocyte aggregation. AI and M index could help in the prognostication of early-onset PE, but further investigations are warranted to confirm the prognostic role before the onset of symptoms.

Keywords: early-onset preeclampsia; hemorheology; red blood cell aggregation; red blood cell deformability; erythrocyte

1. Introduction

Hypertensive disorders affect 5–10% of pregnancies worldwide and are still one of the leading causes of maternal and perinatal morbidity and mortality [1,2]. Preeclampsia (PE) is defined as new-onset hypertension accompanied by proteinuria and/or maternal organ and/or uteroplacental dysfunction manifesting at or after 20 weeks of gestation [3]. Early-onset PE develops before 34 weeks of gestation and has higher risks of maternal morbidity, perinatal death, and severe neonatal morbidity [4]. Short-term maternal consequences could be life-threatening, and women with a history of PE are more prone to suffering from cardiovascular diseases in later life [5,6]. The consequences also seriously affect the fetus since the placenta is not able to ensure adequate perfusion.

The pathophysiological explanation of early-onset PE is based on abnormal placentation, but the exact processes are still equivocal. It is suggested that impaired cytotrophoblast invasion into spiral arteries in the first trimester results in abnormal vascular remodeling with resistant vessels and elevated pressure [7]. The subsequent placental hypoxia leads to the release of pro-inflammatory factors, which might contribute to a pathological systemic endothelial response, characterized by increased capillary permeability, microvascular thrombosis, and sustained vascular hypertension, and consequently a manifestation of clinical signs and symptoms [8].

Red blood cells (RBCs) are the most abundant cells in the bloodstream affected by the deteriorated microcirculatory environment in PE, which leads to remarkable alterations in the erythrocyte morphology and the cell membrane [9]. The science of hemorheology investigates blood flow conditions, the physical properties of blood elements, including cellular and plasmatic components (i.e., RBC aggregation and deformability). Erythrocyte deformability is described as the ability of RBCs to adapt by deformation in response to mechanical forces to be able to cross over narrow capillaries. RBC aggregation means the rouleaux formation of RBCs under low flow conditions. Elevated RBC aggregation and decreased deformability can adversely affect tissue perfusion. The complex mechanisms of erythrocyte aggregation and deformability are discussed in more detail in the Supplementary Material (Supplementary Material for 1. Introduction section).

Altered RBC aggregation and deformability can be considered as the cause or the result of this complex circulatory disorder. Endothelial damage leads to microthrombus formation and erythrocyte fragmentation or hemolysis due to vascular mechanical forces in PE [9]. The blood viscosity at low shear conditions such as the placental intervillous space is mainly determined by RBC aggregation, and increased values potentially deteriorate placental microcirculation [10]. Enhanced RBC aggregation may have an additional increasing impact on peripheral vascular resistance, and the endothelial dysfunction may further increase blood pressure, thereby provoking a vicious cycle [11].

According to previous studies, maternal erythrocyte aggregation and deformability might deteriorate in PE leading to impaired blood flow at the uteroplacental cross-over [12]. However, limited information is available about the early-onset form.

2. Aim

Our research group intended to evaluate the peripartum maternal erythrocyte aggregation and membrane deformability alterations by repeated sampling in early-onset PE and to compare these with physiological changes in healthy pregnancies.

A more precise understanding of the peripartum pathologic alterations of maternal RBCs would help to identify new therapeutic perspectives targeting the restoration of the erythrocyte function and improvement of oxygen-carrying capacity. In addition, we sought to identify aggregation and deformability parameters possessing the strongest ability to diagnose early-onset PE. The determination of these factors may help to identify which parameter changes should be given special attention even before the onset of symptoms.

3. Materials and Methods

3.1. Population

Thirteen non-smoking women diagnosed with early-onset PE based on the International Society for the Study of Hypertension in Pregnancy criteria and admitted to the Department of Obstetrics and Gynecology, University of Pécs, were involved in this prospective, case-control study [3].

The control group consisted of 24 healthy, non-smoking, age- and gestation-matched women. Before the enrolment, a physician excluded any relevant comorbidities (chronic hypertension, diabetes mellitus, cardiovascular, hematologic, gynecologic disorders). In the first phase, we recruited 34 women who were evaluated to be healthy and expected to have a pregnancy without complications. Finally, 10 women were excluded from the control group due to peripartum complications (Figure 1).

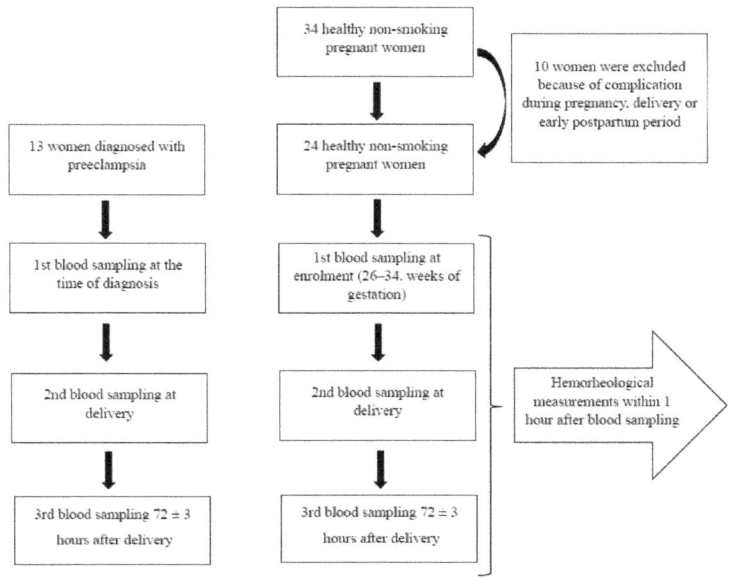

Figure 1. Process of recruitment and sample collection of women diagnosed with PE and healthy pregnant women.

In both groups, exclusion criteria were twin pregnancy, intrauterine developmental abnormality of the fetus, intrauterine infection, severe maternal anemia, participation in another study, or lack of signed informed consent.

All women gave their written informed consent. The study protocol was approved by the Regional and Local Research Ethics Committee at the Medical School, University of Pécs (Reference number: 6942-PTE 2018). The study protocol conforms to the ethical guidelines of the 1975 Declaration of Helsinki.

3.2. *Sample and Data Collection*

Data collection included maternal anamnestic information, comorbidities, symptoms, gestational age at birth. We recorded laboratory parameters at diagnosis and within 72 h after delivery and maternal physical parameters (height, weight, body mass index-BMI, heart rate, blood pressure) at enrolment. Neonatal physical parameters (birth weight, length, head circumference, shoulder width) and the Appearance, Pulse, Grimace, Activity, and Respiration (Apgar) score determined immediately after the birth (Apgar 1) and 5 min later (Apgar 5) were also recorded.

Among patients, the first blood sample was drawn at diagnosis. Among controls, the first sample was collected at enrolment (26–34 weeks of gestation), later discussed as "initial" values. In both groups, blood samples were drawn within the 1st hour after delivery and 72 $\pm$ 3 h later. Every time, 2 $\times$ 6 mL of peripheral blood from antecubital veins were collected into EDTA-Vacutainer tubes. The hemorheological measurements were performed within one hour after sampling in the Hemorheological Laboratory of the University of Pécs under standardized conditions. Figure 1 summarizes the process of recruitment and sample collection.

3.3. *RBC Aggregation*

RBC aggregation was determined with two different methods. The Myrenne aggregometer (model MA-1, Myrenne GmbH, Roetgen, Germany) measures infrared light transmission through the plasma gaps between the RBC aggregates between a transparent plate and a cone. The system rotates the injected 30 μL of blood for 10 s at high shear stress and disperses all pre-existing cell aggregates, then it instantly stops (M mode in stasis) or continues with reduced shear stress (M1 mode in low shear), which stimulates the aggregation and increases the light transmission. The aggregation is determined by the quantity of light transmission. The two dimensionless indices (M, M1) increase with enhanced erythrocyte aggregation [13]. More details about the measurement and the instrument can be found in the Supplementary Material (Supplementary Material for 3. Materials and Methods section; Supplementary Figure S1).

The Laser-assisted Optical Rotational Cell Analyzer (LORCA, R&R Mechatronics, Hoorn, Netherlands) determines the erythrocyte aggregation by detecting laser backscattering from the RBC aggregates. Briefly, a 1 mL blood sample is injected in the gap between the outer, rotating cylinder, and the inner, static cylinder of the instrument, and RBCs are disaggregated at a high shear rate until the motor rapidly stops. The intensity of reflected light is measured and plotted as a function of time. The aggregation index (AI) is calculated during the first 10 s of the measurement [14].

3.4. RBC Deformability

The erythrocyte deformability at different shear stresses was determined by LORCA as well. Briefly, a 20 μL blood sample is diluted in a viscous medium (polyvinylpyrrolidone) and injected between the cylinders of the instrument. A laser-diode is projected through the fluid, the light diffracts on the RBCs resulting in a diffraction pattern on a diaphragm, which will be analyzed. Increasing shear stresses elongate the RBCs, and the diffraction pattern changes from a circular to an elliptical shape. We could express RBC deformability as elongation index (EI) given at each shear stress [15]. More details about the aggregation and deformability measurements by LORCA can be found in the Supplementary Material (Supplementary Material for 3. Materials and Methods Section; Supplementary Figure S2).

3.5. Statistical Analysis

Statistical analysis was evaluated by IBM SPSS Statistics® 27.0. Continuous variables are reported as mean ± standard deviation or medians and interquartile ranges, while categorical variables are presented as frequencies and percentages. After testing the normality by the Kolmogorov–Smirnov test, Mann–Whitney U-test or Student T-test and chi-square test were used to compare data between groups. Bivariate correlation analysis was performed calculating Spearman's correlation coefficient (rho). Multiple regression analysis with various models including aggregation and deformability variables and maternal factors considering the principle of multicollinearity was performed to reveal which factors could predict gestational age at delivery. The diagnostic power of variables was assessed using the area under the curve (AUC) of the receiver operating characteristic (ROC) curve. The predicted probabilities were calculated from the combination of initial AI and M variables produced by binary logistic regression analysis. $p \leq 0.05$ was considered statistically significant.

4. Results

4.1. Population Characteristics

Table 1 presents basic demographic and clinical data of the groups. The mean values of the maximum measured systolic and diastolic blood pressure in the PE group were 180 ± 18 mmHg and 112 ± 13 mmHg. Women with PE had significantly higher rate of preterm deliveries and infants with low birth weight, length, head circumference, shoulder width, intrauterine growth restriction (IUGR) and lower values of Apgar 1 and 5 as expected. In the PE group, all patients received antihypertensive medications (e.g., methyldopa), antenatal corticosteroid prophylaxis, and magnesium sulfate. The women in the control group did not take any medications regularly.

Table 1. Maternal and neonatal demographic and clinical data.

	Baseline Characteristics	Preeclampsia	Control	p-Value
Maternal	Age (years)	29.08 ± 2.13	30.42 ± 1.39	0.589
	Gestation age at first sampling (weeks)	29.69 ± 0.67	28.71 ± 0.48	0.240
	Systolic blood pressure at admission (mmHg)	160 (146–175)	120 (119–130)	<0.001
	Diastolic blood pressure at admission (mmHg)	100 (90–110)	80 (70–80)	<0.001
	Heart rate at admission (/min)	82 (80–95)	82 (77–88)	0.320
	Body height (m)	1.63 ± 0.06	1.67 ± 0.07	0.094
	Body weight (kg)	78 (70–87)	80 (70–86)	0.824
	BMI (kg/m^2)	30.1 (26.8–33.5)	28.2 (25.7–30.2)	0.227
	Change in body weight (kg)	10 ± 7	13 ± 4	0.132
	Mode of delivery: cesarean section (n; %)	13; 100%	9; 37.5%	<0.001
	Length of hospital stay (day)	8 (6–15)	4 (4–5)	<0.001
Neonatal	Gestational age at birth (weeks)	30.23 ± 0.86	39.06 ± 0.28	<0.001
	Birth weight (gram)	1355.83 ± 157.69	3420.00 ± 89.04	<0.001
	Birth length (cm)	36.67 ± 1.64	49.76 ± 0.80	<0.001
	Head circumference (cm)	27.81 ± 1.01	34.24 ± 0.35	<0.001
	Shoulder width (cm)	28.09 ± 1.30	36.88 ± 0.57 cm	<0.001
	Apgar 1	7.0 (6.0–8.0)	9.0 (9.0–9.0)	<0.001
	Apgar 5	9.0 (8.0–9.0)	10.0 (10.0–10.0)	<0.001
	IUGR (n; %)	7; 53.8%	0	<0.001

The results were expressed as the mean value ± standard deviation of the mean or median and interquartile range. BMI: body mass index; IUGR: intrauterine growth restriction.

4.2. Routinely Measured RBC Laboratory Parameters

We found significantly elevated RBC count (4.29 ± 0.12 vs. 3.91 ± 0.05 T/L; p = 0.009), hematocrit (36.3% ± 2.7% vs. 34.3% ± 1.9%; p = 0.019), and hemoglobin (12.658 ± 0.260 vs. 11.791 ± 0.138 g/dL; p = 0.003), and significantly lower mean corpuscular volume (MCV) (84.71 ± 2.85 vs. 87.86 ± 4.56 fL; p = 0.037) in PE diagnosis compared to controls.

4.3. RBC Aggregation

RBC aggregation measured with two different methods was significantly elevated in PE compared to controls. M indices were higher in PE at diagnosis (9.8 ± 0.4 vs. 8.5 ± 0.2; p = 0.007) and at delivery (10.7 ± 0.8 vs. 8.0 ± 0.4; p = 0.002) compared to those in healthy pregnancies. In controls, the M value increased until 72 h after birth compared to values at enrolment or delivery. This alteration was not observed in PE, where M values were continuously high (Figure 2a). We did not find a significant difference concerning M1 values.

The AI values were significantly higher at the initial measurement in PE than among controls (72.9% ± 3.5% vs. 67.5% ± 3.9%; p < 0.001) (Figure 2b). Investigating the values in control women showed that RBC aggregation significantly increased from the first blood sampling at enrolment to the time of delivery (AI: 67.5% ± 0.8% vs. 71.1% ± 1.0% p = 0.003), while this elevation was not visible in PE.

To exclude the potentially disturbing impact of the mode of delivery, additional statistical analyses were performed involving all preeclamptic patients (n = 13; 100%) and nine women (37.5%) in the control group who had cesarean section. In addition, aggregation parameters were evaluated within the control group according to the mode of delivery. Our subgroup analyses showed no significant effect of the mode of delivery on the RBC aggregation results in the postpartum period (Supplementary Material for 4. Results Section; Supplementary Figures S3–S5).

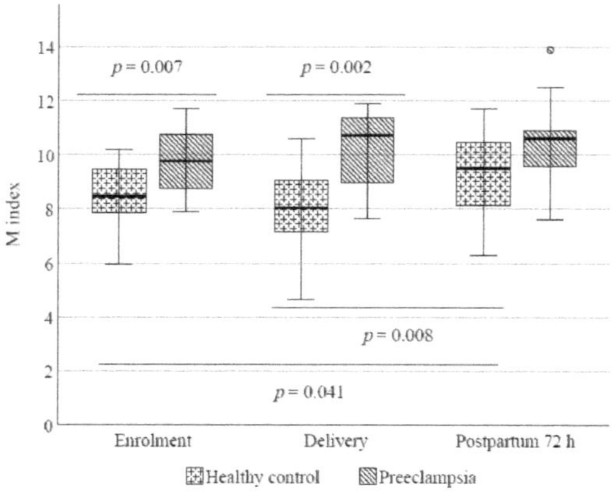

(a)

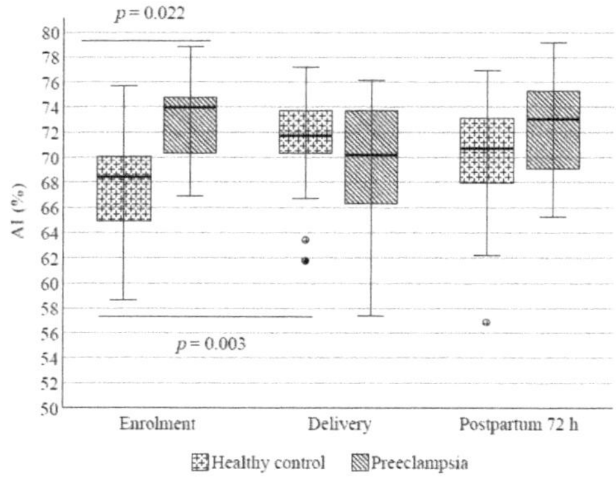

(b)

Figure 2. (**a**,**b**) M index and AI values reflecting RBC aggregation in PE and control group at the three investigated time points. Box plot diagrams show the median with interquartile ranges and the minimum and maximum values.

A positive significant correlation was observed between initial AI measured at diagnosis and the weeks of gestation at delivery in PE ($R^2 = 0.554$; $p = 0.006$) (Figure 3). A multiple regression analysis was run to predict gestational age at delivery from AI and EI at 5.33 shear stress and maternal systolic blood pressure at admission. These variables were statistically significant predicting gestational age at delivery, $p < 0.029$, $R^2 = 0.657$. Out of the three variables, only AI added statistical significance to the prediction of gestational age at delivery, $p < 0.010$.

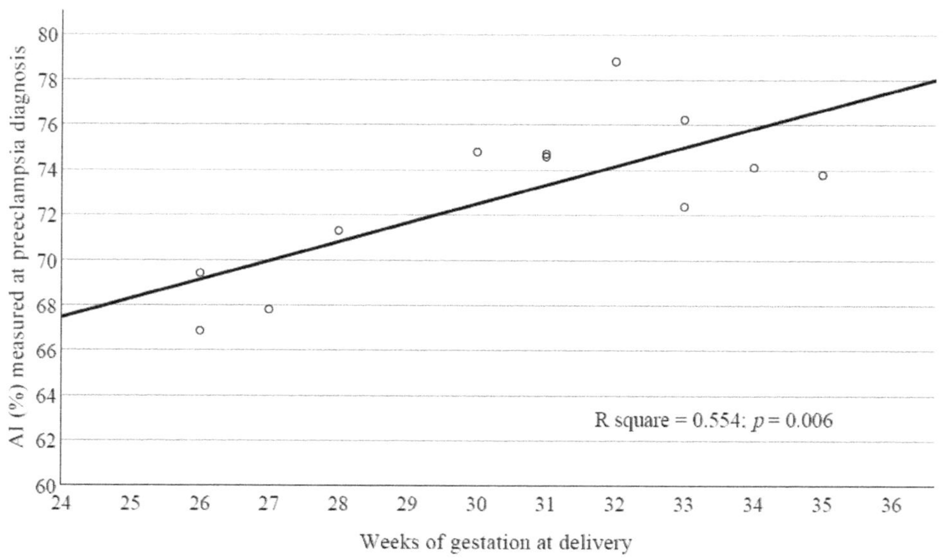

Figure 3. Linear regression analysis of initial AI measured at diagnosis of PE and weeks of gestation at birth.

4.4. RBC Deformability

We found statistically significant deterioration of erythrocyte deformability under medium shear stresses (9.49, 5.33, and 3 Pa) at PE diagnosis compared to controls (Figure 4).

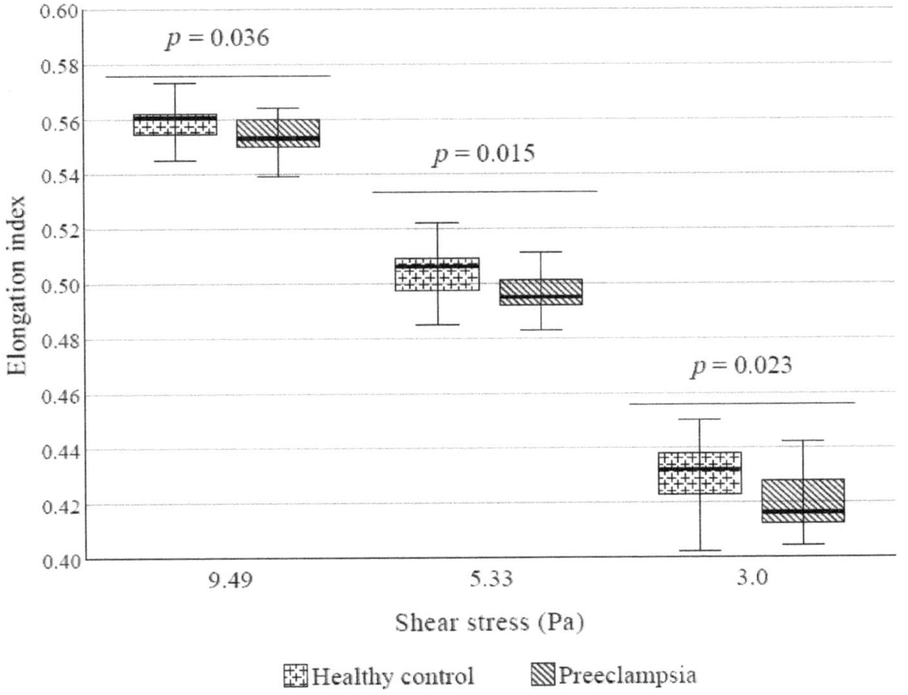

Figure 4. Deformability (EI values) at medium shear stresses in preeclampsia and control group. Box plot diagrams show the median with interquartile ranges and the minimum and maximum values.

Analyzing changes within groups, we found that the RBC deformability improved until postpartum 72 h in PE compared to the values at diagnosis or during delivery (Supplementary Table S1a). Our results did not show any elevation in EI values in controls. Considerably decreased EI values were found at high shear stress during delivery, but after 72 h, the deformability returned to similar values as measured at enrolment (Supplementary Table S1b).

4.5. Indicators of RBC Aggregation for Preeclampsia Diagnosis

ROC analysis was carried out with maternal aggregation parameters in the first investigated time point to test the diagnostic power for PE. The analysis of initial AI indicated a cutoff point of 69.4% for PE with an AUC of 0.837 [0.684–0.990] ($p = 0.001$) (sensitivity = 83.3%; specificity = 62.5%). ROC analysis of initial M values showed an AUC of 0.750 [0.576–0.924] ($p = 0.019$) and indicated a cutoff of 8.39 (sensitivity = 90.9%; specificity = 50%). The predicted probabilities from the combination of initial AI and M variables produced by binary logistic regression analysis showed slightly increased AUC with 0.900 ([0.789–1.000] $p < 0.001$) comparing the ones of AI or M value per se as classifiers for PE (Figure 5).

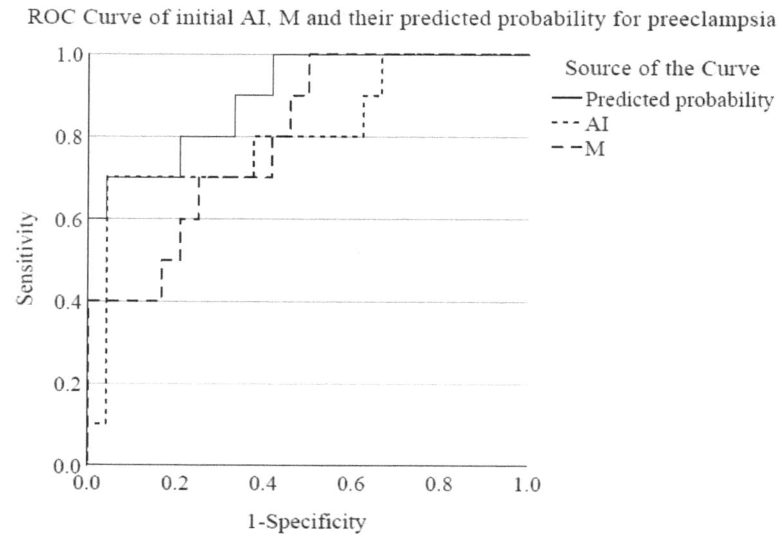

Figure 5. ROC curve for preeclampsia comparing initial AI and M values per se measured at diagnosis or enrolment and their combination expressed as a predicted probability.

5. Discussion

Our research findings are intended to emphasize and confirm the role of maternal hemorheological alterations in pathophysiological processes affecting microcirculation in earlyonset PE. Our investigations focused mainly on peripartum alterations of RBC pro-perties including their aggregation and deformability. The complex pathophysiological background of early-onset PE including inflammatory, genetic, immunological, hemodynamic changes, oxidative stress, and disturbed trophoblastic invasion finally leads to vascular endothelial damage, vasospasm, and hypertension. These pathophysiological mechanisms alter the biophysical properties of RBCs including their aggregation and deformation abilities. Deterioration of these factors (i.e., elevated RBC aggregation and reduced deformability) leads to increased viscosity, which may hinder the proper tissue perfusion in the intervillous space of the placenta, and hence the erythrocytes are not able to ensure adequate oxygen delivery to the fetus. The damaged erythrocytes may enhance the progression of the endothelial dysfunction and maternal circulatory disorders.

5.1. Routinely Measured RBC Laboratory Parameters

During normal pregnancy, total blood volume, plasma volume, and RBC mass are extensively increasing. Additionally, the plasma volume is elevating proportionally more than the RBC mass, resulting in lower hemoglobin concentrations from physiological hemodilution [16]. However, we observed slightly elevated RBC count and hemoglobin values in early-onset PE, which are in line with previous findings [17]. Initial MCV was slightly decreased in our PE group compared to that in controls, contrary to previous investigations, according to whom significantly increased MCV occurred in PE [17,18]. Importantly, in these studies, the mean gestational age was higher in the PE and control groups as well than in our study. According to a recent publication analyzing the osmotic and mechanical stability of erythrocytes, it is suggested that erythrocytes with lower volume and lower hemoglobin content are osmotically more stable. They assumed that lower MCV values could be a mechanism of compensatory mechanical selection, which is beneficial in PE [19]. However, the desired range of MCV throughout pregnancies complicated by PE remains unclear.

5.2. RBC Aggregation

According to our results, elevated initial maternal AI and M values proved to be the most promising indicators of PE with high sensitivity and acceptable specificity. Moreover, we gained more favorable AUC by their combination compared to that in their individual analysis. The elevated RBC aggregation at PE diagnosis compared to that in controls refers to impaired microcirculation in early-onset PE, which is also supported by the positive relationship between RBC aggregation at PE diagnosis and gestational age at delivery. This association suggests that the longer the pathologic pregnancy persists, the worse the AI values are, reflecting enhanced maternal RBC aggregation. We did not observe a significant relationship in healthy pregnancies concerning their AI value and their gestational age at enrolment, contrary to PE. These observations suggest that in normal pregnancy, the gestational age alone did not influence AI values, this association was specific to PE. The results of multiple regression analysis also supported the significant relationship of the AI values and the gestational weeks at delivery. It should be noted that initial AI values in both groups were independent of hemoglobin values, so they did not affect our measurement results in either group. Analyzing the changes within the groups, we can conclude that RBC aggregation increased within the first 72 postpartum hours in healthy pregnancies. Contrary, continuously elevated RBC aggregation was observed in PE, and we did not find a significant difference between the three investigated time points.

The cesarean section rate was significantly lower in the healthy control group, as expected. The hypothesis was raised that the difference in the mode of delivery may influence the results measured at delivery and in the postpartum period, since different physiological processes occur according to the mode of delivery regardless of the underlying obstetric disease. In contrast, our subgroup analyses showed that RBC aggregation values were not significantly affected by this variable. After excluding mothers who gave birth in a natural way, similar differences were revealed between the control and preeclamptic groups.

The RBC aggregation activity depends on the plasma protein levels and intrinsic properties of the erythrocytes such as alterations affecting the cell membrane [20]. In PE, changes of cellular factors are suggested to be more responsible for the pronounced aggregation propensity [10]. Increased RBC aggregation can be attributed to the reduced sialic acid content of the cell membrane that weakens repulsive forces [21] and conformational changes of the membrane enhancing erythrocyte aggregation [22]. A most recent study also confirmed decreased erythrocytic membrane anionic charge due to the reduced sialic acid content of the erythrocyte cell membrane in women with PE/eclampsia, which may be responsible for the aggregation of erythrocytes in these conditions [23].

Although we did not measure blood coagulation substances in the present study, the increased RBC aggregation in healthy mothers in the postpartum period compared to values before delivery is thought to be mainly due to increasing plasma protein (e.g., fibrinogen, immunoglobulin M, and alpha 2-macroglobulin) levels, since these are known to enhance the aggregating activity of RBCs [24]. During normal pregnancy, the RBC aggregation increases despite hemodilution, and this observation has been previously mainly explained via the nearly linear increase of fibrinogen synthesis in the liver [24]. In the third trimester of normal pregnancy, the physiological reserve of blood coagulation substances is accumulating to maintain a hypercoagulability state, probably enhancing erythrocyte aggregation as well, such changes are more pronounced after childbirth to prevent postpartum hemorrhage. With the progression of PE, the locally enforced coagulation state leads to the consumption of the coagulation substances, resulting in an overall low coagulation state. Compared with healthy pregnant women, PE patients had decreased fibrinogen values and a relatively hypocoagulable state was described [25]. Although literature data show elevated fibrinogen levels in healthy pregnancies, in our PE group, the RBC aggregation was still higher throughout the study period since aggregation is influenced not only by the external (plasmatic components such as fibrinogen) but also by the intrinsic properties of RBCs, which are likely to play a greater role in PE [10]. It is also conceivable that the longer time interval between enrolment and delivery in the control group (Table 1) made the physiological increase of the RBC aggregation values possible.

5.3. RBC Deformability

Initial EI values at medium shear stresses were decreased in PE reflecting altered RBC deformability. The reason for reduced RBC deformability in PE is suggested to be chronic inflammation and hypoxia, which leads to an increased concentration of free radicals inducing changes in RBC membrane properties and subsequent increase of intracellular Ca^{2+} [26]. In addition, an increased Ca^{2+} pump activity leads to adenosine triphosphate (ATP) depletion in RBCs resulting in poor deformability [27].

The relatively rapid improvement of deformability in the postpartum period in PE may presumably occur due to the termination of gravidity, which has maintained the disease. It can, therefore, be hypothesized that RBC deformability may be a sensitive and early marker of the normalization of maternal microcirculation after delivery. The deformation ability of erythrocytes depends on complex biophysical properties: the internal cell viscosity, membrane viscoelasticity (e.g., spectrin–F-actin network; phospholipid bilayer; pump activity, ATP pool, intracellular Ca^{2+} concentration), surface–volume ratio, and erythrocyte morphology. The consequence of increased internal cell viscosity or membrane rigidity or elevated surface–volume ratio is reduced elongation ability [28]. External factors may provoke rapid change in a number of these characteristics leading to alteration in the deformability. In the study of Schauf et al., reduced RBC deformability was described in pregnancies complicated by PE and/or IUGR. They reported that deformability values returned within 5 days after delivery to the RBC deformability observed in early pregnancy, and these mentioned changes were already visible on the first and second postpartum days [29]. These results are in line with our findings. They did not discuss the reason for deformability improvement in the postpartum period; however, they suggested that intracellular Ca^{2+} concentration plays a key role in the regulation of RBC mechanical properties. An increase in intracellular Ca^{2+} concentration leads to a reduction of RBC deformability. It was shown that intracellular Ca^{2+} concentration is elevated in PE, which might explain the reduction of RBC deformability [30]. It was previously described that the diminution in the activity of the RBC Ca^{2+}-ATPase enzyme in PE is related to an enhanced level of lipid peroxidation of the plasma membranes [31]. The placenta is thought to be the central organ that regulates this condition and the major source of reactive oxygen species during pregnancy, which deteriorate the erythrocyte membrane's properties [32]. At the same time, when delivery occurs, the maintaining cause of the disease is also removed. Consequently, it can be suggested that a relatively rapid change may happen in erythrocyte membrane or cytoskeletal properties (e.g., due to the improved pump activity or phospholipid bilayer status). However, these are only assumptions, and no precise explanations have been found in the literature to date.

5.4. Previous Data Concerning RBC Properties

Recent studies on RBC deformability and aggregation in PE seem to be in conflict. Some authors found decreased erythrocyte deformability in PE [12,17,33] and uteroplacental hypoperfusion with subsequent RBC membrane damage has also been described [27]. Others did not observe any significant deformability alteration in PE [34,35]. Tranquilli et al. reported increased erythrocyte aggregation in PE [22], while other authors did not observe significant deterioration [36,37]. In line with our observations, Heilmann et al. found enhanced RBC aggregation and reduced deformability by high shear stress in patients with severe PE, suggesting that hemorheological parameters play an important role in the microcirculation of the intervillous space of the placenta [17].

Most of the studies dealing with hemorheological alterations in PE were published more than 2–3 decades ago. As in the case of PE definitions, the measurement methods and interpretation of results may have changed since then; therefore, it is necessary to re-evaluate these results. Moreover, the diagnostic criteria for the early-onset type have been broadened and clarified [3]. The distinction between the early- and late-onset forms is increasingly recognized since they have a different pathophysiological background and clinical features, thus they should be investigated completely separately. Early-onset PE is suggested to mainly originate from defective placentation, whilst the late-onset form may develop due to maternal genetic predisposition to cardiovascular and metabolic disease [38]. Therefore, it would be desirable to treat the two types separately from a hemorheological point of view as well.

Regarding RBC aggregation, previous investigations mostly applied a Myrenne aggregometer per se [12,17], while only a few reported results were measured by LORCA [36]. To the best of our knowledge, there are no data in the literature that examine these two measurement methods simultaneously in early-onset PE as we have done. It can be seen from our report that the two types of measurement methods gave similar results in terms of RBC aggregation, moreover, their combination would even raise the diagnostic power.

Although in this study we intended to evaluate exclusively the erythrocyte properties, it is also desirable to mention the role of the platelets in PE, as they are the other most important cellular elements of blood. In addition, platelet aggregation activity is a relevant hemorheological parameter as well. However, we would also like to note that platelet characteristics do not influence the RBC aggregation and deformability measurement results. A recent systematic review suggests higher platelet activation and lower platelet aggregation or no difference [39]. However, results concerning platelet aggregation in PE are controversial. Platelet activation, aggregation, and adhesion markers are also essential components of the pathogenesis of PE, which is also supported by the efficacy of aspirin prophylaxis. However, further investigations are required to prove the predictive value in PE due to the conflicting results.

5.5. Screening

In the past decade, extensive efforts have been made to develop an efficient screening algorithm to identify high-risk patients [40]. Currently, the best screening model in the first trimester combines multiple examinations containing maternal risk factors, comorbid conditions, physical parameters, mean arterial pressure, placental growth factor level, and uterine artery pulsatility index. Women identified to be at high risk should receive 150 mg acetylsalicylic acid prophylaxis daily commencing at $11\text{–}14^{+6}$ weeks of gestation until either 36 weeks, when delivery occurs, or when the disease is diagnosed [41].

Determination of factors and specific biomarkers used in the current screening algorithm is difficult to perform, time-consuming, and contains costly measurements requiring the involvement of qualified professionals. In contrast, we investigated a cheap method with easy implementation and indicators that are quantitative, objective, and can be blinded to other clinical characteristics. Excluding the purchase of the instruments, hemorheological test costs are minimal and can be easily performed without special training. The above-detailed measurements require a quite small amount of blood as a sample and provide quickly obtainable results. Therefore, further investigations are suggested to reveal whether RBC aggregation and deformability parameters, especially AI and M values really possess the potential for susceptibility or risk biomarkers in the early stage of PE before the onset of symptoms. Moreover, their inclusion in the screening algorithm should be considered.

5.6. Strengths and Limitations

The strength of our study is the prospective, case-control design with repeated blood sampling in the peripartum period to evaluate the kinetics and changes of RBC properties. Using two methods together (Myrenne, measuring the increase of light transmission through plasma gaps between RBC aggregates, and LORCA, detecting laser backscattering from the RBC aggregates) provides more reliable information on RBC aggregation properties. The low total number of enrolled patients and its single-center nature with local treatment strategies and guidelines could limit the generalizability of our findings.

6. Conclusions

The most remarkable findings of our research are the potential diagnostic power of elevated AI and M index reflecting enhanced RBC aggregation in early-onset PE. A positive linear relationship was observed between initial AI and gestational age at birth in PE. Reduced initial EI values were found in PE, reflecting impaired erythrocyte deformability at diagnosis, the values of which were improved within three days after delivery. RBC properties could help in the prognostication of early-onset PE, but further investigations are warranted to confirm the prognostic role before the onset of symptoms.

Supplementary Materials: The following are available online at https://www.mdpi.com/article/10.3390/cells10102714/s1. Supplementary material for 1. Introduction section; Supplementary material for 3. Materials and methods section: Supplementary Figure S1. and Supplementary Figure S2; Supplementary material for 4. Results section: Supplementary Figure S3. and Supplementary Figure S4. and Supplementary Figure S5. and Supplementary Table S1a. Elongation index on different shear stresses in PE group; Supplementary Table S1b. Elongation index on different shear stresses in healthy pregnant women.

Author Contributions: Conceptualization, B.C., S.F., M.M., L.S. and B.S.; data curation, B.C., G.G., S.F. and D.K.K.; funding acquisition, M.M., L.S. and K.T.; methodology, B.C., G.G. and D.K.K.; project administration, B.C. and B.S.; resources, L.S. and K.T.; supervision, P.K., K.T. and B.S.; validation, P.K.; visualization, B.C.; writing—original draft, B.C. and B.S.; writing—review and editing, M.M., L.S. and K.T. All authors have read and agreed to the published version of the manuscript.

Funding: Project no. TKP2020-IKA-08 has been implemented with the support provided from the National Research, Development, and Innovation Fund of Hungary, financed under the 2020-4.1.1-TKP2020 funding scheme. This research was funded by grants from the National Research, Development and Innovation Office (NKFIH K119529). This study was supported by grants from the University of Pecs Medical School (KA-2021-20), the Janos Bolyai Research Scholarship of the Hungarian Academy of Sciences to M.M. and the ÚNKP-20-5 New National Excellence Program of the Ministry for Innovation and Technology from the source of the National Research, Development, and Innovation Fund.

Institutional Review Board Statement: The study was conducted according to the guidelines of the Declaration of Helsinki and approved by the Local Ethics Committee of University of Pécs (Reference number: 6942-PTE 2018).

Informed Consent Statement: Written informed consent was obtained from all subjects involved in the study.

Data Availability Statement: All data relevant to the study are included in the article or uploaded as Supplementary Materials.

Acknowledgments: The authors would like to thank the dedicated work of the colleagues of the 1st Department of Medicine and Hemorheological Laboratory, the Department of Obstetrics and Gynecology, and the Department of Medical Microbiology and Immunology of the University of Pécs, Medical School. We would like to express our special thanks to the women who volunteered to participate in our study.

Conflicts of Interest: The authors declare no conflict of interest. The funders had no role in the design of the study; in the collection, analyses, or interpretation of data; in the writing of the manuscript; or in the decision to publish the results.

References

1. Villar, J.; Carroli, G.; Wojdyla, D.; Abalos, E.; Giordano, D.; Baaqeel, H.; Farnot, U.; Bergsjø, P.; Bakketeig, L.; Lumbiganon, P.; et al. Preeclampsia, gestational hypertension and intrauterine growth restriction, related or independent conditions? *Am. J. Obstet. Gynecol.* **2006**, *194*, 921–931. [CrossRef] [PubMed]
2. Say, L.; Chou, D.; Gemmill, A.; Tunçalp, Ö.; Moller, A.-B.; Daniels, J.; Gülmezoglu, A.M.; Temmerman, M.; Alkema, L. Global causes of maternal death: A WHO systematic analysis. *Lancet Glob. Health* **2014**, *2*, e323–e333. [CrossRef]
3. Brown, M.A.; Magee, L.A.; Kenny, L.C.; Karumanchi, S.A.; McCarthy, F.; Saito, S.; Hall, D.R.; Warren, C.E.; Adoyi, G.; Ishaku, S. The hypertensive disorders of pregnancy: ISSHP classification, diagnosis & management recommendations for international practice. *Pregnancy Hypertens.* **2018**, *13*, 291–310. [PubMed]

4. Lisonkova, S.; Sabr, Y.; Mayer, C.; Young, C.; Skoll, A.; Joseph, K. Maternal Morbidity Associated With Early-Onset and Late-Onset Preeclampsia. *Obstet. Gynecol.* **2014**, *124*, 771–781. [CrossRef] [PubMed]
5. Vahedi, F.A.; Gholizadeh, L.; Heydari, M. Hypertensive Disorders of Pregnancy and Risk of Future Cardiovascular Disease in Women. *Nurs. Women's Health* **2020**, *24*, 91–100. [CrossRef]
6. McDonald, S.D.; Malinowski, A.; Zhou, Q.; Yusuf, S.; Devereaux, P. Cardiovascular sequelae of preeclampsia/eclampsia: A systematic review and meta-analyses. *Am. Heart J.* **2008**, *156*, 918–930. [CrossRef]
7. Staff, A.C.; Fjeldstad, H.E.; Fosheim, I.K.; Moe, K.; Turowski, G.; Johnsen, G.M.; Alnaes-Katjavivi, P.; Sugulle, M. Failure of physiological transformation and spiral artery atherosis: Their roles in preeclampsia. *Am. J. Obstet. Gynecol.* 2020. [CrossRef] [PubMed]
8. Powe, C.E.; Levine, R.J.; Karumanchi, S.A. Preeclampsia, a disease of the maternal endothelium: The role of antiangiogenic factors and implications for later cardiovascular disease. *Circulation* **2011**, *123*, 2856–2869. [CrossRef]
9. Hernández Hernández, J.D.; Villaseñor, O.R.; Del Rio Alvarado, J.; Lucach, R.O.; Zárate, A.; Saucedo, R.; Hernández-Valencia, M. Morphological changes of red blood cells in peripheral blood smear of patients with pregnancy-related hypertensive disorders. *Arch. Med. Res.* **2015**, *46*, 479–483. [CrossRef]
10. Gamzu, R.; Barshtein, G.; Tsipis, F.; Lessing, J.B.; Berliner, A.S.; Kupferminc, M.J.; Eldor, A.; Yedgar, S. Pregnancy-induced hypertension is associated with elevation of aggregability of red blood cells. *Clin. Hemorheol. Microcirc.* **2002**, *27*, 163–169.
11. Baskurt, O.K.; Meiselman, H.J. Erythrocyte aggregation: Basic aspects and clinical importance. *Clin. Hemorheol. Microcirc.* **2013**, *53*, 23–37. [CrossRef]
12. Soliman, A.A.; Csorba, R.; Yilmaz, A.; Tsikoaras, P.; von Tempelhoff, G.F. Rheologic results and their correlation to hemostatic changes in patients with moderate and severe preeclampsia: An observational cross-sectional study. *Clin. Hemorheol. Microcirc.* **2015**, *59*, 1–15. [CrossRef] [PubMed]
13. Vayá, A.; Falcó, C.; Fernández, P.; Contreras, T.; Valls, M.; Aznar, J. Erythrocyte aggregation determined with the Myrenne aggregometer at two modes (M0, M1) and at two times (5 and 10 sec). *Clin. Hemorheol. Microcirc.* **2003**, *29*, 119–127. [PubMed]
14. Hardeman, M.R.; Dobbe, J.G.; Ince, C. The Laser-assisted Optical Rotational Cell Analyzer (LORCA) as red blood cell aggregometer. *Clin. Hemorheol. Microcirc.* **2001**, *25*, 1–11. [PubMed]
15. Rabai, M.; Detterich, J.A.; Wenby, R.B.; Hernandez, T.M.; Toth, K.; Meiselman, H.J.; Wood, J.C. Deformability analysis of sickle blood using ektacytometry. *Biorheology* **2014**, *51*, 159–170. [CrossRef]
16. Chesley, L.C. Plasma and red cell volumes during pregnancy. *Am. J. Obstet. Gynecol.* **1972**, *112*, 440–450. [CrossRef]
17. Heilmann, L.; Rath, W.; Pollow, K. Hemorheological changes in women with severe preeclampsia. *Clin. Hemorheol. Microcirc.* **2004**, *31*, 49–58. [PubMed]
18. Elgari, M.M.; Khabour, O.F.; Alhag, S.M. Correlations between changes in hematological indices of mothers with preeclampsia and umbilical cord blood of newborns. *Clin. Exp. Hypertens.* **2019**, *41*, 58–61. [CrossRef]
19. de Freitas, M.A.R.; da Costa, A.V.; Medeiros, L.A.; Cunha, L.M.; Coutinho Filho, U.; Garrote Filho, M.D.S.; Diniz, A.L.D.; Penha-Silva, N. The role of the erythrocyte in the outcome of pregnancy with preeclampsia. *PLoS ONE* **2019**, *14*, e0212763. [CrossRef]
20. Krüger-Genge, A.; Sternitzky, R.; Pindur, G.; Rampling, M.; Franke, R.P.; Jung, F. Erythrocyte aggregation in relation to plasma proteins and lipids. *J. Cell. Biotechnol.* **2019**, *5*, 65–70. [CrossRef]
21. Tranquilli, A.L.; Curzi, C.M.; Moretti, N.; Nanetti, L.; Mazzanti, L. 362 Reduction of sialic acid and membrane fluidity in erythrocyte from preeclamptic patients as the ground for increased aggregation. *Am. J. Obstetr. Gynecol.* **2001**, *185*, S180. [CrossRef]
22. Tranquilli, A.L.; Gioele Garzetti, G.; De Tommaso, G.; Boemi, M.; Lucino, E.; Fumelli, P.; Romanini, C. Nifedipine treatment in preeclampsia reverts the increased erythrocyte aggregation to normal. *Am. J. Obstetr. Gynecol.* **1992**, *167*, 942–945. [CrossRef]
23. Sen, P.; Ghosh, D.; Sarkar, C. Erythrocytic membrane anionic charge, sialic acid content, and their correlations with urinary glycosaminoglycans in preeclampsia and eclampsia. *Scand. J. Clin. Lab. Investig.* **2020**, *80*, 343–347. [CrossRef] [PubMed]
24. Huisman, A.; Aarnoudse, J.G.; Krans, M.; Huisjes, H.J.; Fidler, V.; Zijlstra, W.G. Red cell aggregation during normal pregnancy. *Br. J. Haematol.* **1988**, *68*, 121–124. [CrossRef]
25. Xu, C.; Li, Y.; Zhang, W.; Wang, Q. Analysis of perinatal coagulation function in preeclampsia. *Medicine* **2021**, *100*, e26482. [CrossRef] [PubMed]
26. Friederichs, E.; Meiselman, H.J. Effects of calcium permeabilization on RBC rheologic behavior. *Biorheology* **1994**, *31*, 207–215. [CrossRef] [PubMed]
27. Schauf, B.; Aydeniz, B.; Pijnenborg, R.; Wallwiener, D. Connection between the deformability of the erythrocytes and intraerythrocyte ATP in pregnancy. *Geburtshilfe Und Frauenheilkunde* **2002**, *62*, 477–480. [CrossRef]
28. Mohandas, N.; Clark, M.R.; Jacobs, M.S.; Shohet, S.B. Analysis of factors regulating erythrocyte deformability. *J. Clin. Investig.* **1980**, *66*, 563–573. [CrossRef] [PubMed]
29. Schauf, B.; Lang, U.; Stute, P.; Schneider, S.; Dietz, K.; Aydeniz, B.; Wallwiener, D. Reduced red blood cell deformability, an indicator for high fetal or maternal risk, is found in preeclampsia and IUGR. *Hypertens Pregnancy* **2002**, *21*, 147–160. [CrossRef]
30. Ray, J.; Vasishta, K.; Kaur, S.; Majumdar, S.; Sawhney, H. Calcium metabolism in pre-eclampsia. *Int. J. Gynaecol. Obstet.* **1999**, *66*, 245–250. [CrossRef]

31. Oviedo, N.J.; Benaim, G.; Cervino, V.; Proverbio, T.; Proverbio, F.; Marín, R. The plasma membrane Ca2+-ATPase protein from red blood cells is not modified in preeclampsia. *Biochim. Biophys. Acta* **2006**, *1762*, 381–385. [CrossRef] [PubMed]
32. Srour, M.A.; Bilto, Y.Y.; Juma, M.; Irhimeh, M.R. Exposure of human erythrocytes to oxygen radicals causes loss of deformability, increased osmotic fragility, lipid peroxidation and protein degradation. *Clin. Hemorheol. Microcirc.* **2000**, *23*, 13–21. [PubMed]
33. Gresele, P.; Guerciolini, R.; Nenci, G.G. Erythrocyte deformability changes in normal pregnancy and pre-eclampsia. *Br. J. Haematol.* **1982**, *52*, 340–342. [CrossRef]
34. Inglis, T.C.; Stuart, J.; George, A.J.; Davies, A.J. Haemostatic and rheological changes in normal pregnancy and pre-eclampsia. *Br. J. Haematol.* **1982**, *50*, 461–465. [CrossRef] [PubMed]
35. Lang, G.D.; Lowe, G.D.; Walker, J.J.; Forbes, C.D.; Prentice, C.R.; Calder, A.A. Blood rheology in pre-eclampsia and intrauterine growth retardation: Effects of blood pressure reduction with labetalol. *Br. J. Obstet. Gynaecol.* **1984**, *91*, 438–443. [CrossRef]
36. Pepple, D.J.; Hardeman, M.R.; Mullings, A.M.; Reid, H.L. Erythrocyte deformability and erythrocyte aggregation in preeclampsia. *Clin. Hemorheol. Microcirc.* **2001**, *24*, 43–48. [PubMed]
37. Ozanne, P.; Miller, F.C.; Meiselman, H.J. RBC aggregation in diabetic and pre-eclamptic pregnancy. *Clin. Hemorheol. Microcirc.* **1983**, *3*, 145–154. [CrossRef]
38. Burton, G.J.; Redman, C.W.; Roberts, J.M.; Moffett, A. Pre-eclampsia: Pathophysiology and clinical implications. *BMJ* **2019**, *366*, l2381. [CrossRef]
39. Jakobsen, C.; Larsen, J.B.; Fuglsang, J.; Hvas, A.M. Platelet function in preeclampsia—A systematic review and meta-analysis. *Platelets* **2019**, *30*, 549–562. [CrossRef]
40. Rolnik, D.L.; Wright, D.; Poon, L.C.; O'Gorman, N.; Syngelaki, A.; de Paco Matallana, C.; Akolekar, R.; Cicero, S.; Janga, D.; Singh, M.; et al. Aspirin versus Placebo in Pregnancies at High Risk for Preterm Preeclampsia. *N. Engl. J. Med.* **2017**, *377*, 613–622. [CrossRef]
41. Poon, L.C.; Shennan, A.; Hyett, J.A.; Kapur, A.; Hadar, E.; Divakar, H.; McAuliffe, F.; da Silva Costa, F.; von Dadelszen, P.; McIntyre, H.D.; et al. The International Federation of Gynecology and Obstetrics (FIGO) initiative on pre-eclampsia: A pragmatic guide for first-trimester screening and prevention. *Int. J. Gynaecol. Obstet.* **2019**, *145* (Suppl. 1), 1–33. [CrossRef] [PubMed]

cells

MDPI

Study Protocol

PROVE—Pre-Eclampsia Obstetric Adverse Events: Establishment of a Biobank and Database for Pre-Eclampsia

Lina Bergman [1,2,3,*], Karl Bergman [4], Eduard Langenegger [3], Ashley Moodley [3], Stephanie Griffith-Richards [5], Johan Wikström [6], David Hall [3], Lloyd Joubert [7], Philip Herbst [8], Sonja Schell [3], Teelkien van Veen [8], Michael Belfort [9], Stephen Y. C. Tong [10,11], Susan Walker [10,11], Roxanne Hastie [10,11] and Catherine Cluver [3,10]

1 Department of Women's and Children's Health, Uppsala University, 75309 Uppsala, Sweden
2 Department of Obstetrics and Gynaecology, Institute of Clinical Sciences, Sahlgrenska Academy, University of Gothenburg, 41650 Gothenburg, Sweden
3 Department of Obstetrics and Gynaecology, Faculty of Medicine and Health Sciences, Stellenbosch University, Cape Town 7505, South Africa; langen@sun.ac.za (E.L.); ashleymoodley69@gmail.com (A.M.); drh@sun.ac.za (D.H.); sonjaschell@sun.ac.za (S.S.); cathycluver@hotmail.com (C.C.)
4 Department of Molecular and Clinical Medicine, Institution of Medicine, Sahlgrenska Academy, University of Gothenburg, 41650 Gothenburg, Sweden; karl.bergman@gu.se
5 Division of Radiodiagnosis, Department of Medical Imaging and Clinical Oncology, Faculty of Medicine and Health Sciences, Stellenbosch University, Cape Town 7505, South Africa; drsteph21@hotmail.com
6 Department of Surgical Sciences, Radiology, Uppsala University, 75309 Uppsala, Sweden; johan.wikstrom@radiol.uu.se
7 Division of Cardiology, Department of Medicine, Medicine and Health Sciences, Stellenbosch University, Cape Town 7505, South Africa; lloydjoubert@gmail.com
8 Department of Obstetrics and Gynaecology, University Medical Center Groningen, 9713 Groningen, The Netherlands; teelkien@gmail.com (T.v.V.); herbst@hotmail.com (P.H.)
9 Department of Obstetrics and Gynaecology, Baylor College of Medicine, Houston, TX 77004, USA; belfort@bcm.edu
10 Translational Obstetrics Group, Department of Obstetrics and Gynaecology, University of Melbourne, Parkville, VIC 3010, Australia; stong@unimelb.edu.au (S.Y.C.T.); spwalker@unimelb.edu.au (S.W.); hastie.r@unimelb.edu.au (R.H.)
11 Mercy Perinatal, Mercy Hospital for Women, Heidelberg, VIC 3010, Australia
* Correspondence: lina.bergman.2@gu.se

Citation: Bergman, L.; Bergman, K.; Langenegger, E.; Moodley, A.; Griffith-Richards, S.; Wikström, J.; Hall, D.; Joubert, L.; Herbst, P.; Schell, S.; et al. PROVE—Pre-Eclampsia Obstetric Adverse Events: Establishment of a Biobank and Database for Pre-Eclampsia. *Cells* **2021**, *10*, 959. https://doi.org/10.3390/cells10040959

Academic Editors: Ana T. Palei, Junie P. Warrington, Mark W. Cunningham and Lorena M. Amaral

Received: 19 March 2021
Accepted: 16 April 2021
Published: 20 April 2021

Abstract: Pre-eclampsia is a leading cause of maternal and perinatal morbidity and mortality. The burden of disease lies mainly in low-middle income countries. The aim of this project is to establish a pre-eclampsia biobank in South Africa to facilitate research in the field of pre-eclampsia with a focus on phenotyping severe disease.The approach of our biobank is to collect biological specimens, detailed clinical data, tests, and biophysical examinations, including magnetic resonance imaging (MRI) of the brain, MRI of the heart, transcranial Doppler, echocardiography, and cognitive function tests.Women diagnosed with pre-eclampsia and normotensive controls are enrolled in the biobank at admission to Tygerberg University Hospital (Cape Town, South Africa). Biological samples and clinical data are collected at inclusion/delivery and during the hospital stay. Special investigations as per above are performed in a subset of women. After two months, women are followed up by telephonic interviews. This project aims to establish a biobank and database for severe organ complications of pre-eclampsia in a low-middle income country where the incidence of pre-eclampsia with organ complications is high. The study integrates different methods to investigate pre-eclampsia, focusing on improved understanding of pathophysiology, prediction of organ complications, and potentially future drug evaluation and discovery.

Keywords: pre-eclampsia; eclampsia; pulmonary oedema; biobank; database

1. Introduction

Pre-eclampsia is one of the most serious complications of pregnancy and affects 3–8% of pregnancies worldwide [1]. It is a leading cause of maternal and perinatal morbidity and mortality [2]. Globally, pre-eclampsia is responsible for more than 60,000 maternal deaths annually, and in South Africa, hypertensive disorders of pregnancy account for 14% of all maternal deaths [3].

The pathogenesis of pre-eclampsia is not yet fully elucidated. It likely involves maternal, foetal, and placental factors, resulting in relative placental under-perfusion, hypoxia, and ischaemia [1]. The placenta then releases anti-angiogenic factors causing endothelial dysfunction resulting in a multi-system disorder [4]. Hypertension is the hallmark feature and other systems including renal, cardiovascular, respiratory, haematological, hepatic, cerebrovascular, and the fetoplacental unit can be involved [5]. In its most severe form, it can cause seizures (eclampsia), cerebrovascular events, pulmonary oedema, cardiac failure, acute kidney injury, haemolysis, elevated liver enzymes, low platelet (HELLP) syndrome, disseminated intravascular coagulation, sub-capsular liver haematoma or rupture, and even death [6–10].

Pre-eclampsia is a disease that is only observed in humans. No animal models of pre-eclampsia have the ability to exactly mimic the true pre-eclampsia state due to differences in placentation among mammals [11]. It is imperative that critical laboratory observations are made on human tissues.

Biobanks enable research for which clinical information and analyses from biological samples are efficiently used and combined in more than one project. Biobanks can also be used in collaboration with international biobanks to achieve sufficient power and generalisability in rare outcomes.

Low-and-middle-income countries (LMIC) have a higher incidence of pre-eclampsia, more complications, and bear the major burden of global maternal and perinatal deaths [6]. For these reasons, it makes sense to establish a biobank in an area with a high incidence of pre-eclampsia. Researchers from Tygerberg Hospital, Stellenbosch University, have been investigating pre-eclampsia for many years [12–16]. Tygerberg hospital is a referral center servicing a population of over 2 million people. The Obstetrics Department only manages complicated pregnancies (such as those affected by severe, or preterm pre-eclampsia) and delivers approximately 8000 women per year. The high-risk referral service is supported by maternal-foetal medicine sub-specialists, critical care specialists, and neonatologists. The occurrence of pre-eclampsia and its complications is high in South Africa which is expected for a low-and-middle-income country. Unfortunately, the exact incidence is not known. The academic centre at Tygerberg Hospital is, therefore, an ideal service to embed a pre-eclampsia biobank.

The research team at Tygerberg Hospital also has established collaborations with leading academic centres in Australia, Sweden, The Netherlands, the United Kingdom, and the United States of America. Such collaborations enable linkage of a unit that has an extremely high incidence of pre-eclampsia, and in particular, pre-eclampsia with severe features, with other research leaders in the field of pre-eclampsia. These collaborations can enable us to make important discoveries about why different organs are affected in pre-eclampsia, findings that may improve the clinical care of women.

2. Aims and Objectives

To establish a pre-eclampsia database and biobank at Tygerberg Hospital, Stellenbosch University and to facilitate research in pre-eclampsia, detailed data are collected according to an international consensus of core predictors and outcomes in pre-eclampsia research [17]. The aims of this approach are to enable the discovery of possible predictors and diagnostic tools for organ complications in pre-eclampsia and to develop the understanding of the underlying pathophysiology further with a focus on different organ complications. We collect biological specimens including plasma, saliva, cerebrospinal fluid, urine, placenta, and umbilical cord blood (mixed venous–arterial). In subgroups, we perform specialised examinations to assess organ dysfunction, reveal underlying pathophysiological pathways, and aim to relate the findings to predictors and biomarkers. These examinations include cerebral magnetic resonance imaging (MRI), cardiac MRI, transcranial Doppler, echocardiography, and cognitive function testing.

3. Materials and Methods

The study is registered at the International Standard Randomised Controlled Trial Number (ISCRTN) with trial registration number ISRCTN10623443.

3.1. Study Design and Population

We include women who are admitted to Tygerberg Hospital for the birth of their baby. Cases (women with pre-eclampsia) are included before or after delivery depending on presentation. Controls (women with normotensive pregnancies) are recruited when admitted for delivery or at a visit to the outpatient clinic. For the controls, samples are collected at inclusion and delivery. Cases are seen daily and clinical information and samples are collected serially until discharge. All participants are contacted telephonically in the postpartum period. Data are collected prospectively and from the medical charts on maternal health and neonatal outcomes (Figure 1).

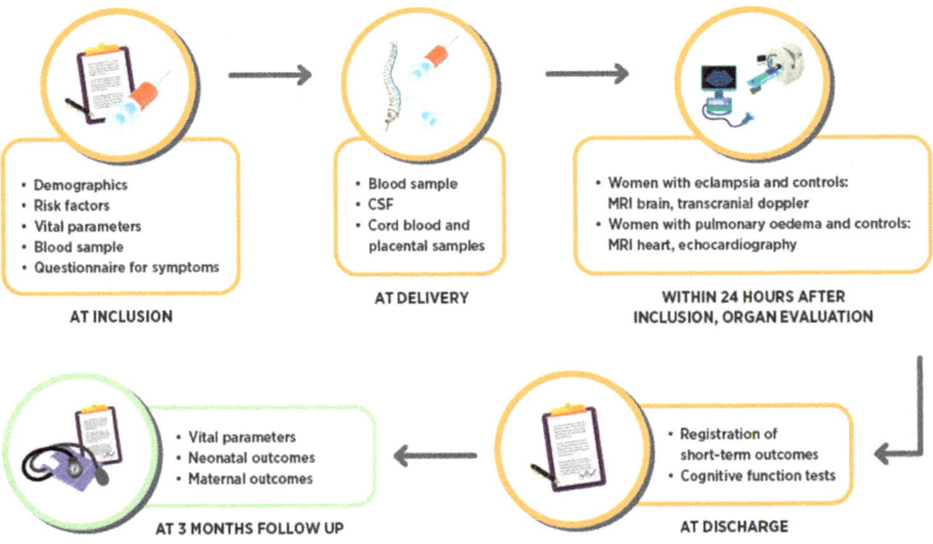

Figure 1. Flow chart of the study.

Inclusion and Exclusion Criteria

All women with pre-eclampsia who are admitted to Tygerberg hospital are eligible for inclusion. Our controls are women who deliver at Tygerberg without the hypertensive disease. Pre-eclampsia is defined according to the International Society for the Study of Hypertension in Pregnancy (ISSHP) [18]. Both women with pre-eclampsia without severe features and pre-eclampsia with end-organ complications are eligible for inclusion. End-organ complications are defined according to the ISSHP classification system and include haemolysis, elevated liver enzymes, and low platelets (HELLP) syndrome, eclampsia, pulmonary oedema, renal impairment, and acute severe hypertension [18]. All parameters are included in the online database described below, and women can thus be classified according to different classification systems. All participants must be competent to provide informed consent before enrolment. If they are minors, they need to read and sign the assent form, and a parent or guardian must sign consent. All individual studies originating from the biobank require a unique ethical approval.

3.2. Collection of Clinical Information

Clinical information is collected on datasheets and entered into an online secure password-protected Research Electronic Data Capture (REDCap) database. The data collection includes information on the participant's age, gravidity, parity, medical history, current pregnancy information, and outcome of the pregnancy. Only the principal investigators have access to patient identifying information and each participant is allocated a study number. The variables collected in the database correspond to the recommended core demographics, predictors, and outcomes listed in *"Strategy for standardization for pre-eclampsia research design"* by the international group Co-Lab for harmonisation of pre-eclampsia research (Table 1) [17]. The women are also interviewed at inclusion with a short questionnaire regarding signs and symptoms specific for pre-eclampsia (Table 1).

Table 1. Clinical information collected in the database RedCap.

Variables at Inclusion	Variables During Hospital Stay	Variables at Discharge and at 3 Months Follow Up
Demographics	**Delivery details**	**Discharge**
Age (c)	Gestation at delivery (c)	Date of discharge (N/A)
Date of birth (N/A)	Indication for delivery (n)	Number of days in hospital (c)
Hospital folder number (N/A)	Mode of delivery (n)	Adverse maternal outcomes (n)
Address (N/A)	Place and supervision of delivery (n)	*Eclampsia*
Contact number (N/A)	Medication used before delivery (n)	*Recurrent eclampsia*
Race (n)	Maternal plasma sample taken (b)	*Severe hypertension*
Marital status (n)	Cerebrospinal fluid sample taken (b)	*Stroke*
Total years of full-time education (o)		*PRES syndrome*
Current job situation (n)	**Neonatal outcomes**	*Cortical blindness*
Current living situation (n)	Liveborn (b)	*Severe renal impairment*
Number of people cohabitating (o)	Sex (b)	*Dialysis*
Diagnosis at inclusion (n)	Intubation at birth (b)	*Pulmonary edema*
Singleton pregnancy (b)	Birth weight (c)	*Inotropic support*
Plasma sample taken at inclusion (b)	APGAR score at 5 min (o)	*Signs of bleeding or DIC*
	Cord blood taken (b)	
Medical history	Neonatal outcome (n)	
Gravidity (o)		
Parity (o)	**Placental outcomes**	
Number of miscarriages (o)	Placental sample taken (b)	
Number of previous terminations (o)	Date for placental sample (N/A)	
Outcome of previous viable pregnancies (n)	Placental weight (c)	*HELLP syndrome*
Previous pre-eclampsia, number of times (o)		*Liver enzymes > 500 IU/L*
Previous pre-eclampsia before 34 weeks (b)	**Daily hospital forms**	*Liver rupture or hematoma*
New paternity (b)	Days since admitted (n)	*Admitted to ICU*
Fertility treatment (n)	Highest systolic blood pressure (c)	*Maternal death*
HIV status (b)	Highest diastolic blood pressure (c)	*PPH*
If HIV positive, viral load (c)	Lowest oxygen saturation (c)	*Laparotomy for any reason apart from caesarean section*
Tobacco use (n)	Highest respiratory rate (c)	*Glasgow Coma Score < 13*
Alcohol use (n)	Intubated (b)	*Myocardial infarction*
Methamphetamine use (b)	CPAP (b)	*Intubation*
Diabetes (n)	Antihypertensive medication (n)	*Sepsis*
Previous cardiovascular disease (n)	Magnesium sulphate (b)	*Coma*
Chronic hypertension(b)	Nitroglycerin/Tridil (b)	*Venous thromboembolism*
If chronic hypertensive, number of medications (o)	Diuretic (b)	*Intrauterine fetal death*
Anaemia (n)	Special investigations performed (n)	*Perinatal or infant mortality*
Neurological disease (n)	Seen by other specialities (n)	*Abruptio placenta*
Respiratory disease (n)	Haemoglobin (c)	*Mild to moderate pre-eclampsia without complications*
Renal disease (n)	Platelets (c)	*None (pregnant control)*
Inflammatory bowel disease (b)	Urea (c)	
Autoimmune disease (o)	Creatinine (c)	**Cognitive function testing**
Depression (b)	Arterial pH (c)	MoCa (scanned)
Heredity for pre-eclampsia (o)	Arterial pO2 (c)	CFQ (c)
First degree relative with hypertensive disorders (b)	Arterial pCO2 (c)	
Partner's heredity for pre-eclampsia (b)	Arterial lactate (c)	**Long-term outcome 3-12 months**
Pre-pregnancy weight (c)	Magnesium (c)	Date (N/A)
Pre-pregnancy height (c)	Blood taken for freezing (b)	Neonatal death within 6 weeks after expected due date (b)
Antenatal care (n)		
Gestation at first presentation for antenatal care (c)	**Transcranial doppler**	Neonatal death at any point after discharge (b)
Systolic blood pressure at booking (c)	Date and time for examination (N/A)	Maternal systolic blood pressure (c)
Diastolic blood pressure at booking (c)	Oxygen saturation when start of measurement (c)	Maternal diastolic blood pressure (c)
Proteinuria at booking (o)	Pulse at start of examination (c)	Is mother on blood pressure treatment (b)
On hypertensive treatment at booking (b)	Systolic blood pressure at start of examination (c)	Final hypertensive diagnosis (n)
Aspirin use in this pregnancy (b)	Diastolic blood pressure at start of examination (85)	Any persisting neurological symptoms (b)
Gestation aspirin was started (c)	Oxygen (n)	Glasgow Outcome Scale (o)
Calcium use in this pregnancy (b)	End-tidal CO_2 (c)	
Gestation calcium was started (c)	Magnesium sulphate at examination (b)	
Estimated due date (N/A)	Magnesium sulphate before examination (b)	
Estimated due date calculated by (o)	If yes, how many hours since finished (o)	
	Nitroglycerin/Tridil at examination (b)	
	Antihypertensive medication at examination (o)	

Table 1. *Cont.*

Variables at Inclusion	Variables During Hospital Stay	Variables at Discharge and at 3 Months Follow Up
	GCS score at examination (c)	
	Neurological deficits at start of examination (text)	
	Depth left side (c)	
	Systolic velocity left side (c)	
	Mean velocity left side (c)	
	Diastolic velocity left side (c)	
	Cerebral perfusion pressure left side (c)	
	Autoregulatory index left side (c)	
	Depth right side (c)	
Symptoms before inclusion	Systolic velocity right side (c)	
Edema (n)	Mean velocity right side (c)	
Visual disturbances (n)	Diastolic velocity right side (c)	
Time of onset for visual disturbances (n)	Cerebral perfusion pressure right side (c)	
Headache (n)	Autoregulatory index right side (c)	
Headache onset (n)		
Epigastric/abdominal pain (n)	**Brain MRI**	
Tightness in the chest (b)	MRI performed (b)	
Shortness of breath (b)	Date of MRI (N/A)	
Focal neurological deficits (n)	*MR data saved in the electronic radiology system for later entry*	
Tendon reflexes (n)		
Nausea (b)	**Echocardiography and MRI heart**	
Vomit (b)	Diagnosis of pulmonary edema in relation to delivery (n)	
Confusion (b)	Date of echocardiography (N/A)	
Twitching or jerking arms or legs (b)	Date of cardiac MRI (N/A)	
People around her noticing that she was absent in mind (b)	Inclusion criteria (o)	
Difficulty concentrating (b)	If pulmonary oedema, how was diagnosis set (o)	
Speech affected (b)	Blood sample (b)	
Hearing affected (b)	LV diameter in diastole (c)	
Mood changes (n)	LV end-systolic diameter (c)	
Anxiety (b)	Interventricular septal thickness in diastole (c)	
Feel like end of the world was coming (b)	Posterior wall thickness in diastole (c)	
Dizziness (b)	LV mass (c)	
Weakness/paralysis (b)	LV mass index (c)	
Other symptoms not listed (free text)	Relative wall thickness (c)	
Highest blood pressure before inclusion /fit (c)	Left ventricular ejection fraction (by Simpson's biplane) (c)	
Highest diastolic blood pressure before inclusion /fit (c)	Fractional shortening (c)	
Blood pressure recorded before/after fit (n)	Tissue doppler SA wave velocity (c)	
Date of convulsion if applicable (N/A)	Myocardial performance index (c)	
Fitted postpartum (b)	Global longitudinal strain (c)	
Proteinuria (o)	Regional wall motion abnormalities (c)	
Lowest platelet count (c)	E wave height (c)	
Highest AST (c)	A wave height (c)	
Lowest haemoglobin (c)	E:A ratio (c)	
Highest creatinine (c)	E wave deceleration time (c)	
Blood pressure medication before the event/inclusion (n)	Isovolumetric relaxation time (c)	
Magnesium sulphate before the event/inclusion (b)	E' septal (c)	
If eclamptic fit, where did it occur (n)	E' lateral (c)	
	E'average (c)	
	E:E'septal (c)	
	E:E'lateral (c)	
	E:E'average (c)	
	Pulmonary vein pulse wave doppler (c)	
	Left atrial diameter (c)	
	Left atrial area (c)	
	Left atrial volume (c)	
	Left atrial volume indexed to BSA (c)	

APGAR; appearance, pulse, grimace, activity, respiration, AST; aspartate transaminase, CFQ; Cognitive Failure Questionnaire, CPAP; continuous positive airway pressure, GCS; Glasgow Coma Scale, HELLP; haemolysis, elevated liver enzymes, low platelets, HIV; human immune deficiency virus, ICU; intensive care unit, LV; left ventricle, MoCA; Montreal Cognitive Assessment, MRI; magnetic resonance imaging, PPH; postpartum haemorrhage, PRES; posterior reversible encephalopathy syndrome. Nominal (n); binary (b); ordinal (o); continuous (c).

3.3. *Sample Collection and Storage*

Samples are collected by research midwives. After a blood sample is drawn, it is centrifuged, aliquoted, and frozen within two hours of collection. The same procedure is performed for cerebrospinal fluid and umbilical cord blood. Saliva and urine are frozen directly and are not centrifuged. Placental samples are taken at four sites on the maternal and foetal surfaces of the placenta. They are stored in RNAlater for at least 48 h and then frozen. All samples are stored in a minus 70 °C freezer which has an electronic temperature monitoring system. All samples are labelled only with the study number, sample number, and date. Samples collected for the biobank are presented in Table 2.

Table 2. Biological samples collected for the PROVE biobank.

Type of Sample	Sampling	Volume	Tubes, Processing
Blood	At inclusion, at delivery, during hospital stay	Maximum of 12 mL/sample	EDTA plasma, spinned, 1 mL aliquots
Placenta: foetal surface	At delivery	RNAlater, 4 × 1 cm^3	Removal of RNA later, 2 mL aliquots
Placenta: foetal surface	At delivery	Frozen sections 1 cm^3	2 mL aliquots
Placenta: maternal surface	At delivery	RNAlater, 4 × 1 cm^3	Removal of RNA later, 2 mL aliquots
Placenta: maternal surface	At delivery	Frozen sections 1 cm^3	2 mL aliquots
Cord Blood (mixed venous–arterial)	At delivery	Maximum 12 mL	EDTA plasma, spinned, 1 mL aliquots
Urine	At inclusion	Maximum 12 mL	Sterile tube, spinned, 1 mL aliquots
Cerebrospinal fluid	At delivery	Maximum 2 mL	Sterile tube, spinned, 1 mL aliquots
Saliva	At inclusion	Stored in a 3 cc cup	2 mL aliquots

EDTA; ethylenediaminetetraacetic acid, RNA; ribonucleic acid.

3.4. *Questionnaires and Biophysical Examinations*

3.4.1. Cognitive Function Testing

Cognitive function is assessed subjectively and objectively as close to discharge as possible. The subjective function is assessed using the Cognitive Failure Questionnaire (Supplemental Figure S1). Cognitive function is also assessed objectively using the Montreal Cognitive Assessment (MoCA) (Supplemental Figure S2). A detailed description of the cognitive function tests can be found in the Supplementary Methods Section.

3.4.2. Brain MRI

For women with eclampsia and women with pre-eclampsia with severe features without eclampsia as a control group, we perform MRI examinations on entry into the study. We use a 1.5 Tesla scanner, used in clinical practice, at the Department of Radiology at Tygerberg University Hospital. The MRI protocol includes sequences for evaluation of brain morphology including infarcts, oedema, and haemorrhages, as well as assessment of arterial spasm and cerebral perfusion. Imaging sequence information can be found in the Supplementary Methods Section.

3.4.3. Transcranial Doppler

Transcranial Doppler examination is performed in a subgroup of women to evaluate the cerebral perfusion pressure and dynamic cerebral autoregulation on entry into the study and for some women also before discharge from the hospital and in addition before and after vasoactive medication. Both women with normotensive pregnancies, pre-eclampsia without severe features, and pre-eclampsia with severe features are eligible for transcranial Doppler examination. This is performed by locating the middle cerebral artery bilaterally by the Doppler technique and simultaneously recording continuous blood pressure and end-tidal CO_2. A detailed description of the methodology can be found in the Supplementary Methods Section.

3.4.4. Echocardiography and Cardiac MRI

A subgroup of cases (women with pulmonary oedema) and controls (women with pre-eclampsia without pulmonary oedema and women with normotensive pregnancies) have echocardiograms and cardiac MRI on entry into the study. Investigations are performed by two cardiologists according to a predefined protocol that can be found in the Supplementary Methods Section.

3.5. Statistics

Results from echocardiography, MRI, and cerebral Doppler will be calculated by two independent interpreters blinded to groups and entered manually into the database. MR data will be analysed in collaboration with Uppsala University by neuroradiologists, blinded to groups. Blood samples will be analysed for placental, cardiac-, renal-, neurological- and endothelial biomarkers, using standardised platforms when available and for the remaining analyses through manual enzyme-linked immunosorbent assay (ELISA) analyses in duplicates. Inter- and intra-assay coefficients of variation will be aimed at below 10%.

Demographics will be presented as medians or means as appropriate by distribution. Comparison between groups will be analysed by Student's *t*-test or Mann–Whitney u-test with means or medians and confidence intervals or interquartile range, as appropriate according to the distribution of the variables. When comparing multiple groups, the Kruskal–Wallis test or one-way ANOVA will be used as appropriate according to the distribution of the variables. Correlations will be analysed by Pearson's r or Spearman's rho, as appropriate by the distribution of the variable. Regression analyses, unadjusted and adjusted, will be performed to adjust for known confounding variables. A total of 10 cases per variable at the lowest will be considered appropriate to avoid overfitting of the model. All statistical analyses will be performed in SPSS or R.

Power Calculations

Tygerberg University hospital has approximately 8000 high-risk deliveries yearly. Pre-eclampsia affects a large proportion of these deliveries but the exact numbers are not known.

Prospective power calculations for some investigations will not be possible since no reference values or results exist. In addition, the biobank is designed to be open for future research on pre-eclampsia where the research question to date might not be known. Some examples of power calculations for planned analyses are given below.

Cerebral blood flow regulation: To detect a difference in dynamic cerebral autoregulation index of 1, 2 with a standard deviation (SD) of 1.5, 25 women are required in each group [19]. Cerebral biomarkers: To detect a difference of 4 pg/mL between cases and controls for NfL with an SD of 5, 25 women are required per group [20].

In order to extend the comparisons to correlations and sub-analyses, the initial sample size for PROVE was set to 100 women with eclampsia, 50 women with pulmonary oedema, and 50 women in each control group before the first round of analyses are initiated.

3.6. Patient and Public Involvement

No input from patients has been solicited in the creation of the database and biobank.

3.7. Ethics and Dissemination

Due to the frequency, morbidity, and mortality associated with pre-eclampsia, there is a great scientific and social value to undergird this research. Blanket and broad consent have been avoided, and only research linked to the topic of pre-eclampsia will be conducted. The risk of adverse events is very low. Peripheral blood collection is carried out by an experienced midwife/doctor at the time of routine blood collection. Cerebrospinal fluid is collected from a discarded sample from the anaesthetist during the administration of spinal anaesthesia. MRI imaging in selected participants with eclampsia or cerebral signs has fewer side effects than conventional imaging. Performing transcranial Doppler measurements holds no risks. Participant information databases will be routinely backed up to prevent loss due to technical issues, and samples within the biobank will be secured via routine laboratory safety nets (e.g., the availability of backup freezer space, documentation of usage, etc.).

Blood samples are collected at the same time as routine blood testing, and thus, no additional discomfort is inflicted. Other samples are collected from tissue and fluid that are usually discarded.

No publications will come directly from the biobank. The biobank will provide a source of material for further studies that will each need individual ethics approval.

4. Collection to Date

From April 2018 until March 2020, 244 women have been included in the biobank. The biobank is an ongoing project that will continuously enrol women with pre-eclampsia and controls with yearly updates to the local Health Research Ethics Committee. In Table 3, the background data of these women are presented according to the type of organ complication at inclusion. The biobank recently restarted inclusions after the COVID-19 pandemic in March 2021 and will continue to include women according to the research questions and sample size described above.

Table 3. Background characteristics of the women included in the biobank until March 2020, by complication at inclusion.

	Neurology *	Pulmonary Oedema	HELLP/Renal Impairment	Pre-Eclampsia **	Normotensive
n	86	43	23	54	38
AT BASELINE					
Maternal age (years)	22.8 (6.1)	28.8 (6.9)	29.2 (6.8)	26.1 (5.8)	29.1 (6.3)
Nulliparous n (%)	60 (70)	18 (42)	7 (30)	28 (52)	9 (24)
BMI (kg/m^2)	26.4 (8.1)	32.2 (9.2)	29.8 (6.4)	29.1 (7.3)	27.6 (6.8)
Missing ***	16	5	7	3	4
HIV n (%)	10 (12)	11 (26)	6 (26)	8 (15)	7 (18)
Chronic hypertension n (%)	7 (8)	2 (5)	5 (22)	9 (17)	0 (0)
Gestation at first presentation for antenatal care (weeks)	17.4 (8)	15.3 (7.6)	18.0 (6.9)	16.0 (7.4)	16.6 (7.7)
AFTER INCLUSION					
Gestation at delivery (weeks)	33.4 (4.3)	31.8 (5.0)	30.6 (5.0)	34.0 (4.2)	36.1 (3.9)
Mode of delivery n (%)					
Vaginal	25 (29)	11 (26)	8 (35)	15 (28)	8 (21)
Planned CS	1 (1)	2 (5)	0 (0)	5 (9)	24 (63)
Emergency CS	60 (70)	30 (70)	15 (65)	34 (63)	6 (16)
Birthweight (grams)	2093 (906)	1748 (969)	1315 (588)	2078 (961)	2761 (877)
OUTCOMES n (%)					
Severe hypertension	35 (41)	29 (67)	16 (70)	14 (26)	0 (0)
Eclampsia	82 (95)	0 (0)	0 (0)	0 (0)	0 (0)
Recurrent eclampsia	26 (30)	0 (0)	0 (0)	0 (0)	0 (0)
GCS < 13	22 (26)	0 (0)	0 (0)	0 (0)	0 (0)
Stroke	3 (4)	0 (0)	0 (0)	0 (0)	0 (0)
Blindness	1 (1)	0 (0)	0 (0)	0 (0)	0 (0)
Pulmonary oedema	5 (6)	43 (100)	3 (13)	0 (0)	0 (0)
HELLP	19 (22)	7 (16)	22 (96)	1 (2)	0 (0)
Renal impairment	16 (19)	5 (12)	9 (39)	1 (2)	0 (0)
Admitted to ICU	10 (12)	6 (14)	1 (4)	0 (0)	0 (0)
Postpartum haemorrhage	9 (11)	3 (7)	2 (13)	3 (5)	2 (4)
Intubation	15 (17)	6 (14)	1 (4)	0 (0)	0 (0)
Intrauterine foetal death	12 (14)	8 (19)	7 (30)	5 (9)	0 (0)
Venous thromboembolism	2 (2)	1 (2)	0 (0)	0 (0)	0 (0)
Abruptio placentae	5 (6)	2 (5)	2 (9)	2 (4)	1 (3)
GOS at two months					
Mild or no disability	63 (96)	33 (100)	10 (100)	42 (100)	26 (100)
Severe disability	1 (1)	0 (0)	0 (0)	0 (0)	0 (0)
Death	2 (3)	0 (0)	0 (0)	0 (0)	0 (0)
Missing	20	10	13	12	12

* Eclampsia, stroke, or blindness. ** Pre-eclampsia with or without acute severe hypertension, without organ complications. *** Missing values for BMI due to missing antenatal charts or antenatal charts without registration of BMI.; BMI, body mass index; GCS, Glasgow Coma Scale, GOS, Glasgow Outcome Scale, HELLP, haemolysis, elevated liver enzymes, and low platelets; HIV, human immunodeficiency virus; ICU, intensive care unit; continuous variables are presented as means with standard deviations. Categorical variables are presented as numbers with percentages.

5. Discussion

This project aims to establish a biobank and database for severe organ complications of pre-eclampsia. This is only feasible in LMIC where the incidence of pre-eclampsia and its complications are high. To our knowledge, the project is the first biobank of its kind in an LMIC, integrating different methods and engaging clinical and preclinical researchers to investigate the organ complications of pre-eclampsia. The project has a unique focus on improving the understanding of the pathophysiology, the possibility for prediction of organ complications, and over time, the hope for drug evaluation and drug discovery for pre-eclampsia with severe features.

This particular field of research is hampered by small retrospective studies that lack sufficient validation or power [21]. Organ complications of pre-eclampsia are complex and difficult to investigate due to their low incidence in high-income countries, their sudden onset, and the restriction often imposed by research methods in pregnant women. By using a database that adheres to recommended variables and biosamples in pre-eclampsia research, the results from this study can be merged with results from other studies. This will enable the investigation of infrequent rare outcomes by increasing sample size, this providing valuable scientific knowledge, and support to the women who are affected by this serious condition [17]. PROVE holds biological samples and data from special investigations such as MRI and cerebral blood flow measurements as well as detailed demographic data. Taken together, this will enable us to examine the disease phenotype deeply and achieve an improved understanding of the pathophysiology behind the various debilitating organ complications.

Another known biobank of pregnancy complication in an LMIC is PREPARE [22]. PREPARE collects biospecimens and clinical data from women with pre-eclampsia in Brazil and also uses the same recommended variables as PROVE does from the CoLab initiative [17]. The aim of PREPARE is to include a large sample size of women in order to validate existing biomarkers and diagnostic tools in an LMIC setting, whereas in PROVE, we have a smaller sample size, but we also perform special investigations, focusing on underlying pathophysiological mechanisms. These biobanks are complementary to each other and also enable the merging of data and biomarker results. Other examples of existing biobanks (mostly in HIC) are the Baby Biobank [23], Peribank [24], and IMPACT study [25] that focus on the prediction of disease, genetics, and microbiome.

Studies assessing neurological complications of pre-eclampsia are mostly small retrospective case-control and preclinical studies investigating cerebral blood flow alterations, cerebral biomarkers, MRI findings, and clinical signs and symptoms. Results indicate altered dynamic cerebral autoregulation and increased peripheral concentrations of cerebral biomarkers in women with pre-eclampsia, compared to normotensive women [19,20,26]. However, these results are from smaller populations and have not been investigated in women with clinical manifestations of cerebral complications such as eclampsia. In PROVE, we add knowledge to this field by adding similar investigations in a group of women with severe disease. Again, this should improve the understanding of the pathophysiology underlying cerebral oedema, eclampsia, and intracerebral haemorrhage in pre-eclampsia. Cerebral biomarkers and cognitive function deficits in previous reports have also mainly been based on women with less severe disease [20,26,27]. Results from PROVE will confirm or dismiss the role of cerebral biomarkers and cognitive function in relation to the severity of disease in a dose–response fashion. Clinical signs and symptoms are poorly investigated and in a recent systematic review by our group, we showed that clinical signs and symptoms currently used such as headache and visual disturbances are generally poor predictors of eclampsia, and included studies were mainly composed of case–control studies [28]. Our study collects signs and symptoms in a predefined protocol, enabling the discovery of new promising predictors of eclampsia, as opposed to predictors already collected in clinical practice. These data will be combined with data collected in a similar fashion in other countries such as Pakistan, Solomon Islands, and Sweden. MRI findings are predominantly retrospective studies in which the prevalence of cerebral oedema in eclampsia varies from

60 to 100%, and MRI has been performed mostly for by clinical indications. The prevalence of cerebral oedema in other severe forms of pre-eclampsia is still unknown [29,30]. Thus, there is a need for larger, studies incorporating prospectively collected information with predefined variables and systematically conducted investigations of the cerebral function in women with pre-eclampsia, with and without neurological complications.

Pulmonary oedema is characterised by fluid retention in the alveoli causing swelling of the lungs. Fluid is normally kept within the capillaries due to healthy endothelial cells and a balance between the opposing capillary hydrostatic pressure and colloid oncotic pressure. If this balance is deranged or the endothelial layer in the alveoli is disrupted, pulmonary oedema can follow [31]. Pulmonary oedema secondary to pre-eclampsia can theoretically be caused by any of these pathways, explained by increased hydrostatic pressures due to increased afterload and left ventricular diastolic dysfunction in combination with capillary leak and interstitial oedema [32–34]. There is a paucity of data regarding the pathophysiological pathways in pulmonary oedema and how to individualise treatment depending on the underlying causes such as heart failure treatment versus lowering of blood pressure and fluid restriction.

The diagnosis of pulmonary oedema in pre-eclampsia is based on clinical and radiological features, and echocardiography is not a prerequisite for diagnosis [31]. Thus, the contribution of the heart in pre-eclampsia complicated by pulmonary oedema remains largely unknown. There is, therefore, a need to evaluate diagnostic cardiac biomarkers and/or echocardiography in the management of women with pre-eclampsia and pulmonary oedema regarding the involvement of the heart to improve short- and long-term prognosis. Data from PROVE will contribute to this gap in knowledge by a deeper characterisation of cardiac tissue abnormalities, cardiac function, and cardiac biomarkers in women with pulmonary edema versus women with pre-eclampsia that is not complicated by pulmonary oedema and normotensive pregnancies, respectively.

In summary, PROVE aims to improve knowledge of the pathophysiology behind the development of severe organ complications of pre-eclampsia, enabling early identification and treatment and in addition, over time, identification of new drug targets. PROVE is open to all researchers after application to the research team and ethical approval. More information can be found at www.preeclampsiaresearch.com (accessed on 17 April 2021). Through the high-quality standardised collection of data and biosamples in accordance with the requirements of CoLab [17], PROVE will contribute to data- and biosample sharing in infrequent outcomes, enabling analyses of larger datasets. This is urgently needed in order to contribute to the sustainable development goals and improve women's health.

Supplementary Materials: The following are available online at https://www.mdpi.com/article/10.3390/cells10040959/s1, Figure S1: The Cognitive Failures Questionnaire, Figure S2: The Montreal Cognitive Assessment (MoCa) test, Supplementary Methods.

Funding: This study is supported by STINT; The Swedish Foundation of International Cooperation in Research and Higher Education, Pre-eclampsia Foundation, Märta Lundqvist's foundation, Ester Åsberg's foundation, Jane and Dan Olsson's foundation, Center for Clinical research Dalarna, the Swedish Society of Medicine, the Swedish Research Council and the Mercy Health Foundation.

Institutional Review Board Statement: This study has ethical approval (protocol number N18/03/034, Federal Wide assurance number 00001372, Institutional Review Board number IRB0005239) received on 28 February 2018 and updated yearly with the last update in May 2020.

Informed Consent Statement: Informed consent was obtained from all subjects involved in the study. All participating women are given oral and written information and sign an informed consent before participation in the study.

Data Availability Statement: The datasets used and/or analysed during the current study are available from the corresponding author on reasonable request.

Acknowledgments: We would like to thank Zukiswa Magogotye for valuable assistance in recruiting the women to the study.

Conflicts of Interest: The authors declare no conflict of interest.

References

1. Mol, B.W.J.; Roberts, C.T.; Thangaratinam, S.; Magee, L.A.; de Groot, C.J.M.; Hofmeyr, G.J. Preeclampsia. *Lancet* **2016**, *387*, 999–1011. [CrossRef]
2. Duley, L. The Global Impact of Pre-eclampsia and Eclampsia. *Semin. Perinatol.* **2009**, *33*, 130–137. [CrossRef] [PubMed]
3. Moodley, J. Maternal deaths due to hypertensive disorders of pregnancy. Saving Mothers 2014–2016. In *7th Triennial Report on Confidential Enquiries into Maternal Deaths in South Africa*; Department of Health: Pretoria, South Africa, 2018.
4. Levine, R.J.; Maynard, S.E.; Qian, C.; Lim, K.-H.; England, L.J.; Yu, K.F.; Schisterman, E.F.; Thadhani, R.; Sachs, B.P.; Epstein, F.H.; et al. Circulating Angiogenic Factors and the Risk of Preeclampsia. *N. Engl. J. Med.* **2004**, *350*, 672–683. [CrossRef] [PubMed]
5. Young, B.C.; Levine, R.J.; Karumanchi, S.A. Pathogenesis of Preeclampsia. *Annu. Rev. Pathol. Mech. Dis.* **2010**, *5*, 173–192. [CrossRef]
6. Ghulmiyyah, L.; Sibai, B. Maternal Mortality from Preeclampsia/Eclampsia. *Semin. Perinatol.* **2012**, *36*, 56–59. [CrossRef]
7. Mackay, A.P.; Berg, C.J.; Atrash, H.K. Pregnancy-Related Mortality from Preeclampsia and Eclampsia. *Obstet. Gynecol.* **2001**, *97*, 533–538. [CrossRef]
8. Melchiorre, K.; Sutherland, G.R.; Watt-Coote, I.; Liberati, M.; Thilaganathan, B. Severe Myocardial Impairment and Chamber Dysfunction in Preterm Preeclampsia. *Hypertens. Pregnancy* **2010**, *31*, 454–471. [CrossRef]
9. Zhang, J.; Meikle, S.; Trumble, A. Severe Maternal Morbidity Associated with Hypertensive Disorders in Pregnancy in the United States. *Hypertens. Pregnancy* **2003**, *22*, 203–212. [CrossRef]
10. Hall, D.R.; Grové, D.; Carstens, E. Early pre-eclampsia: What proportion of women qualify for expectant management and if not, why not? *Eur. J. Obstet. Gynecol. Reprod. Biol.* **2006**, *128*, 169–174. [CrossRef]
11. Sunderland, N.; Hennessy, A.; Makris, A. Animal Models of Pre-eclampsia. *Am. J. Reprod. Immunol.* **2010**, *65*, 533–541. [CrossRef]
12. Hall, D.R.; Odendaal, H.J.; Steyn, D.W.; Grove, D.B. Expectant management of early onset, severe pre-eclampsia: Maternal outcome. *BJOG Int. J. Obstet. Gynaecol.* **2000**, *107*, 1252–1257. [CrossRef]
13. Steyn, D.W.; Odendaal, H.J. Randomised controlled trial of ketanserin and aspirin in prevention of pre-eclampsia. *Lancet* **1997**, *350*, 1267–1271. [CrossRef]
14. Odendaal, H.J.; Pattinson, R.C.; Bam, R.; Grove, D.; Kotze, T.J. Aggressive or expectant management for patients with severe preeclampsia between 28–34 weeks' gestation: A randomized controlled trial. *Obstet. Gynecol.* **1990**, *76*, 1070–1075.
15. Cluver, C.A.; Hannan, N.J.; van Papendorp, E.; Hiscock, R.; Beard, S.; Mol, B.W.; Theron, G.B.; Hall, D.R.; Decloedt, E.H.; Stander, M.; et al. Esomeprazole to treat women with preterm preeclampsia: A randomized placebo controlled trial. *Am. J. Obstet. Gynecol.* **2018**, *219*, 388.e1–388.e17. [CrossRef]
16. Hall, D.R. Understanding expectant management of pre-eclampsia: Review. *Obstet. Gynaecol. Forum* **2016**, *26*, 22–27.
17. Myatt, L.; Redman, C.W.; Staff, A.C.; Hansson, S.; Wilson, M.L.; Laivuori, H.; Poston, L.; Roberts, J.M. Strategy for Standardization of Preeclampsia Research Study Design. *Hypertension* **2014**, *63*, 1293–1301. [CrossRef]
18. Brown, M.A.; Magee, L.A.; Kenny, L.C.; Karumanchi, S.A.; McCarthy, F.P.; Saito, S.; Hall, D.R.; Warren, C.E.; Adoyi, G.; Ishaku, S. The hypertensive disorders of pregnancy: ISSHP classification, diagnosis & management recommendations for international practice. *Pregnancy Hypertens.* **2018**, *13*, 291–310.
19. Van Veen, T.R.; Panerai, R.B.; Haeri, S.; Griffioen, A.C.; Zeeman, G.G.; Belfort, M.A. Cerebral autoregulation in normal pregnancy and preeclampsia. *Obstet Gynecol.* **2013**, *122*, 1064–1069. [CrossRef]
20. Bergman, L.; Zetterberg, H.; Kaihola, H.; Hagberg, H.; Blennow, K.; Akerud, H. Blood-based cerebral biomarkers in preeclampsia: Plasma concentrations of NfL, tau, S100B and NSE during pregnancy in women who later develop preeclampsia—A nested case control study. *PLoS ONE* **2018**, *13*, e0196025. [CrossRef]
21. Bergman, L.; Torres-Vergara, P.; Penny, J.; Wikström, J.; Nelander, M.; Leon, J.; Tolcher, M.; Roberts, J.M.; Wikström, A.-K.; Escudero, C. Investigating Maternal Brain Alterations in Preeclampsia: The Need for a Multidisciplinary Effort. *Curr. Hypertens. Rep.* **2019**, *21*, 72. [CrossRef]
22. De Oliveira, L.; Dias, M.A.B.; Jeyabalan, A.; Payne, B.; Redman, C.W.; Magee, L.; Poston, L.; Chappell, L.; Seed, P.; von Dadelszen, P.; et al. Creating biobanks in low and middle-income countries to improve knowledge—The PREPARE initiative. *Pregnancy Hypertens.* **2018**, *13*, 62–64. [CrossRef]
23. Leon, L.J.; Solanky, N.; Stalman, S.E.; Demetriou, C.; Abu-Amero, S.; Stanier, P.; Regan, L.; Moore, G.E. A new biological and clinical resource for research into pregnancy complications: The Baby Bio Bank. *Placenta* **2016**, *46*, 31–37. [CrossRef]
24. Antony, K.M.; Hemarajata, P.; Chen, J.; Morris, J.; Cook, C.; Masalas, D.; Gedminas, M.; Brown, A.; Versalovic, J.; Aagaard, K. Generation and validation of a universal perinatal database and biospecimen repository: PeriBank. *J. Perinatol.* **2016**, *36*, 921–929. [CrossRef]
25. Bergman, L.; Sandström, A.; Jacobsson, B.; Hansson, S.; Lindgren, P.; Larsson, A.; Imberg, H.; Conner, P.; Kublickas, M.; Carlsson, Y.; et al. Study for Improving Maternal Pregnancy and Child ouTcomes (IMPACT): A study protocol for a Swedish prospective multicentre cohort study. *BMJ Open* **2020**, *10*, e033851. [CrossRef]
26. Evers, K.S.; Atkinson, A.; Barro, C.; Fisch, U.; Pfister, M.; Huhn, E.A.; Lapaire, O.; Kuhle, J.; Wellmann, S. Neurofilament as Neuronal Injury Blood Marker in Preeclampsia. *Hypertension* **2018**, *71*, 1178–1184. [CrossRef]

27. Elharram, M.; Dayan, N.; Kaur, A.; Landry, T.; Pilote, L. Long-Term Cognitive Impairment After Preeclampsia: A Systematic Review and Meta-analysis. *Obstet Gynecol.* **2018**, *132*, 355–364. [CrossRef]
28. Hastie, R.; Brownfoot, F.C.; Cluver, C.A.; Walker, S.P.; Hesselman, S.; Tong, S.; Bergman, L. The predictive value of the signs and symptoms preceding eclampsia: A systematic review. *Pregnancy Hypertens.* **2019**, *17*, S15. [CrossRef]
29. Zeeman, G.G.; Cunningham, F.G. Posterior reversible encephalopathy syndrome in 46 of 47 patients with eclampsia. *Am. J. Obstet. Gynecol.* **2014**, *210*, 378–379. [CrossRef] [PubMed]
30. Camara-Lemarroy, C.R.; Escobedo-Zúñiga, N.; Villarreal-Garza, E.; García-Valadez, E.; Góngora-Rivera, F.; Villarreal-Velázquez, H.J. Posterior reversible leukoencephalopathy syndrome (PRES) associated with severe eclampsia: Clinical and biochemical features. *Pregnancy Hypertens.* **2017**, *7*, 44–49. [CrossRef] [PubMed]
31. Zlatnik, M.G. Pulmonary edema: Etiology and treatment. *Semin. Perinatol.* **1997**, *21*, 298–306. [CrossRef]
32. Anthony, J.; Damasceno, A.; Ojjii, D. Hypertensive disorders of pregnancy: What the physician needs to know. *Cardiovasc. J. Afr.* **2016**, *27*, 104–110. [CrossRef] [PubMed]
33. Guirguis, G.F.; Aziz, M.M.; Liang, C.B.; Williams, S.F.; Apuzzio, J.J.; Bilinski, R.; Mornan, A.J.; Shah, L.P. Is preeclampsia an independent predictor of diastolic dysfunction? A retrospective cohort study. *Pregnancy Hypertens.* **2015**, *5*, 359–361. [CrossRef] [PubMed]
34. Gandhi, S.; Sun, N.; Park, A.L.; Hladunewich, M.; Silversides, C.K.; Ray, J.G. The Pulmonary Edema Preeclampsia Evaluation (PEPE) Study. *J. Obstet. Gynaecol. Can.* **2014**, *36*, 1065–1070. [CrossRef]

cells

MDPI

Review

Tocotrienol in Pre-Eclampsia Prevention: A Mechanistic Analysis in Relation to the Pathophysiological Framework

Zaleha Abdullah Mahdy [1,*], Kok-Yong Chin [2,*], Nik Lah Nik-Ahmad-Zuky [3], Aida Kalok [1] and Rahana Abdul Rahman [1]

[1] Department of Obstetrics and Gynaecology, Universiti Kebangsaan Malaysia Medical Centre, Cheras, Kuala Lumpur 56000, Malaysia; aidahani.mohdkalok@ppukm.ukm.edu.my (A.K.); drrahana@ukm.edu.my (R.A.R.)

[2] Department of Pharmacology, Faculty of Medicine, Universiti Kebangsaan Malaysia, Cheras, Kuala Lumpur 56000, Malaysia

[3] Department of Obstetrics and Gynaecology, School of Medical Sciences, Universiti Sains Malaysia, Kubang Kerian 16150, Malaysia; nikzuky@usm.my

* Correspondence: zaleha@ppukm.ukm.edu.my (Z.A.M.); chinkokyong@ppukm.ukm.edu.my (K.-Y.C.)

Abstract: The pathophysiology of pre-eclampsia involves two major pathways, namely systemic oxidative stress and subsequent generalised inflammatory response, which eventually culminates in endothelial cell injury and the syndrome of pre-eclampsia with multi-organ dysfunction. Aspirin has been used to reduce the risk of pre-eclampsia, but it only possesses anti-inflammatory properties without any antioxidant effect. Hence, it can only partially alleviate the problem. Tocotrienols are a unique form of vitamin E with strong antioxidant and anti-inflammatory properties that can be exploited as a preventive agent for pre-eclampsia. Many preclinical models showed that tocotrienol can also prevent hypertension and ischaemic/reperfusion injury, which are the two main features in pre-eclampsia. This review explores the mechanism of action of tocotrienol in relation to the pathophysiology of pre-eclampsia. In conclusion, the study provides sufficient justification for the establishment of a large clinical trial to thoroughly assess the capability of tocotrienol in preventing pre-eclampsia.

Keywords: anti-inflammation; antioxidant; hypertension; ischaemia; vitamin E

Citation: Mahdy, Z.A.; Chin, K.-Y.; Nik-Ahmad-Zuky, N.L.; Kalok, A.; Abdul Rahman, R. Tocotrienol in Pre-Eclampsia Prevention: A Mechanistic Analysis in Relation to the Pathophysiological Framework. *Cells* **2022**, *11*, 614. https://doi.org/10.3390/cells11040614

Academic Editors: Lorena M. Amaral and Alexander E. Kalyuzhny

Received: 15 December 2021
Accepted: 9 February 2022
Published: 10 February 2022

1. Introduction

Pre-eclampsia is a clinical syndrome characterised by raised blood pressure and multi-organ dysfunction during the second half of pregnancy, labour or the puerperium in a previously normotensive woman [1]. It is the leading cause of maternal and perinatal mortality and morbidity [2]. The global incidence of pre-eclampsia has increased from 16.30 million in 1990 to 18.08 million in 2019 (a 17.80% increase) despite global improvement in maternal health [3]. Chronic diseases, such as diabetes mellitus, obesity, chronic hypertension, renal disease, and connective tissue disorders, are risk factors of pre-eclampsia [4].

Pre-eclampsia tends to recur, but a window of opportunity for its screening and prevention exists. The current management of pre-eclampsia involves screening (by risk factors, imaging, or biomarkers), counselling, continuous and meticulous monitoring and pharmacological control of blood pressure before and after delivery, as well as the prophylactic use of low-dose aspirin in high-risk women, magnesium sulphate to prevent eclampsia, and betamethasone to promote fetal lung maturity in pregnancies <34 weeks [5]. Mapping the pathophysiology of a disease as accurately as possible is necessary to justify the use of the preventive measure and institute prevention by pharmacoprophylaxis. Over the years, many candidate drugs have been clinically tested, the most established being aspirin and calcium [6]. Low-dose aspirin (81 mg/day) has been recommended by the US Preventive Task Force as the preventive drug, because it reduces pre-eclampsia risk and has multiple benefits on maternal and neonatal outcomes [7]. However, a recent meta-analysis revealed that aspirin at doses lower than 150 mg/day could not remarkably prevent preterm pre-eclampsia [8]. In addition, low-dose aspirin may be contraindicated in patients with hypersensitivity to non-steroidal anti-inflammatory drugs and gastrointestinal and genitourinary bleeding tendencies [9]. Aspirin might also cross the placenta to the foetus and pose a theoretical risk of foetal intracranial haemorrhage despite the lack of epidemiological evidence [10]. For calcium supplementation, the risk reduction is only about 20%, and is higher amongst populations with low calcium intake [11]. Recent studies have demonstrated the potential of biologics, such as recombinant vascular endothelial growth factor (VEGF) and placental growth factor, in managing pre-eclampsia, but they are still in the development phase [12]. Currently, the pursuit of an alternative agent for pre-eclampsia prevention is still ongoing.

Oxidative stress has been associated with pre-eclampsia in many observational studies [13,14]. Thus, many researchers have advocated for the potential role of antioxidants in preventing pre-eclampsia [15]. The antioxidants that have been actively investigated include coenzyme Q10 [16], melatonin [17,18], and curcumin [19]. However, meta-analyses have reported that antioxidants are ineffective in pre-eclampsia prevention, and the effects on foetal outcomes are heterogeneous [20,21]. Results of trials involving vitamin E, particularly α-tocopherol, have been hugely disappointing [22]. However, the effects of another form of vitamin E, tocotrienol, in pre-eclampsia prevention have been less explored. Only one study examined the effects of tocotrienol-rich fraction (TRF) derived from palm oil with promising outcomes [23]. In this study, TRF reduced the risk of pre-eclampsia by 64% [23]. Before embarking on expanding such clinical trials further, scrutinising the currently accepted pathophysiology of pre-eclampsia and the potential targets of tocotrienol in general would be prudent. Therefore, through this review, we set out to critically evaluate the scientific merits of tocotrienol as a preventive agent of pre-eclampsia.

2. Current Concept on the Pathophysiology of Pre-Eclampsia

The pathophysiology of pre-eclampsia has evolved progressively over the years, and the most recent concept embraces oxidative stress and the inflammatory process [24]. The pathogenesis of pre-eclampsia involves maternal and placental pathways. The source of insult originates from maternal biochemical anomalies, due to chronic diseases or placental developmental anomalies in the form of restricted spiral artery invasion by foetal trophoblasts [25]. Maternal origins of insult may arise from pre-existing diseases, such as diabetes mellitus, chronic hypertension, and obesity, whereas placental anomalies may be the result of immunological incompatibility between maternal and paternal human leukocyte antigens. Maternal metabolic diseases like ischaemia/reperfusion of the placental bed releases biochemical signals into the maternal circulation. These signals converge at the endothelium and form the final common pathway that results in pre-eclampsia, which is marked by endothelial dysfunction, raised blood pressure, and potential multi-organ complications. The humoral signals that produce pre-eclampsia and endothelial dysfunction consist of oxidative and inflammatory agents [12,25].

In the early stage of pregnancy, oxygen concentration within the trophoblast is maintained at a relatively low level to avoid teratogenic effects on the foetus. The hypoxic condition also up-regulates hypoxia-inducible factor 1 alpha, which, in turn, increases endothelial nitric oxide (NO) synthase to produce NO radicals that facilitate trophoblast proliferation and invasion. Towards the end of the first trimester, uteroplacental circulation is developed to allow a higher concentration of oxygen to reach the embryo to facilitate cell differentiation and foetal development. Oxidative stress develops easily at the syncytiotrophoblast, which is a special epithelium necessary for the transport of solute to the foetus and the production of hormones. It has a membrane rich in unsaturated fatty acid, but low in antioxidant enzymes; thus, it is susceptible to oxidative damage (summarised in [26]).

Defective placental implantation onto the uterine wall impairs uterine spiral arteriole remodelling by extravillous trophoblasts, which subsequently reduces placental perfusion.

The resultant cycles of placental ischaemia and reperfusion produce oxidative stress [27]. Notably, the hypoxia/reperfusion cycles activate xanthine oxidase, nicotinamide adenine dinucleotide phosphate oxidase, and electron transport chain (complexes I and III) in the mitochondria to generate oxygen radicals. These radicals are reduced to hydrogen peroxide by manganese, copper, and zinc superoxide dismutase in the intermembranous space of the mitochondria, and subsequently reduced to water by glutathione peroxidase or catalase [24]. Maternal metabolic derangements could result in impaired antioxidant defence and higher oxidants, which allow these radical oxygen species to cause cellular and tissue damage. Oxidative stress can also activate the nuclear factor kappa-B (NF-κB) pathway, which mediates the inflammation process [24]. Damaged tissues release cell debris antiangiogenic factors, such as soluble fms-like tyrosine kinase-1 (sFlt-1), soluble endoglin, and cytokines [27]. sFlt-1 binds to VEGF and reduces its bioavailability to maternal endothelial cells, which results in the reduction of NO production and triggers vasoconstriction [28]. It also sensitises human umbilical artery endothelial cells to the action of pro-inflammatory cytokines [29]. The hypoxic condition up-regulates p53 expression (indicative of pro-apoptotic signalling) and down-regulates Bcl-2 expression (indicative of anti-apoptotic signalling), which further escalates the apoptosis of syncytiotrophoblasts [30,31]. Oxidative stress also activates Wnt signalling and matrix metalloproteinases (MMPs), and thus promotes the invasiveness of trophoblasts and increases the risk of pre-eclampsia [32].

Uterine immune cells are a major regulator of inflammation. During placentation, various immune cells, such as macrophages, natural killer cells, T-cells, B-cells, and dendritic cells, guide the proper invasion of trophoblasts [33]. In pre-eclampsia, chronic placental and peripheral inflammation, marked by increased tumour necrosis factor-α (TNFα) and interleukin (IL)-6 and decreased IL-10 and IL-4, are observed [34]. In particular, TNFα triggers endothelial dysfunction by reducing NO production, which activates the NF-κB pathway, propagates inflammation, increases endothelin-1 (ET-1), and causes vasoconstriction [35]. TNFα, IL-6, and IL-17 also activate B-cells to secrete agonistic autoantibody against angiotensin II type-1 (AT-1) receptor, which, upon binding, causes the release of proinflammatory cytokines as well as ET-1 and sFlt-1 in the vasculature, and subsequently causes vasoconstriction [36]. Overall, the pathophysiology of pre-eclampsia is summarised in Figure 1.

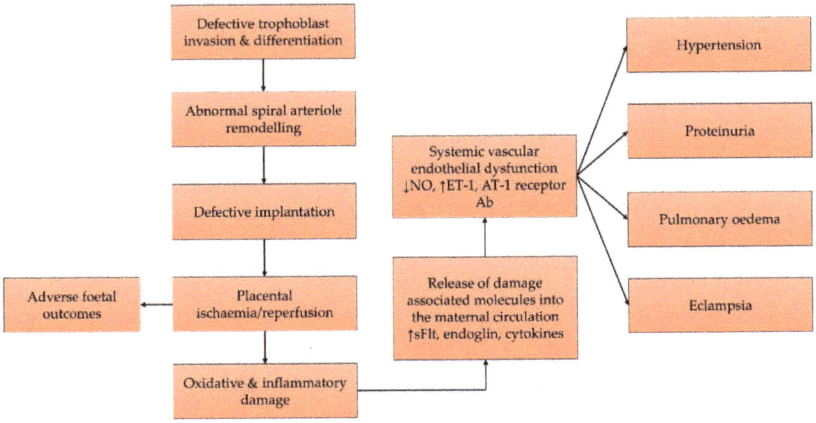

Figure 1. Pre-eclampsia begins with defective placental implantation, which leads to recurrent ischaemia and reperfusion cycles that result in oxidative and inflammatory damages to the tissue. Damage-associated molecules cause systemic endothelial dysfunction and various features of pre-eclampsia. Abbreviations: Ab, antibody; AT-1, angiotensin-1; ET-1, endothelin-1; NO, nitric oxide; sFlt-1, soluble fms-like tyrosine kinase 1.

3. Vitamin E: Tocopherol and Tocotrienol

Although α-tocopherol was the first form of vitamin E to be recognised, natural vitamin E has eight lipophilic forms, including four tocopherols (α, β, γ and δ) and four tocotrienols (α, β, γ and δ) [37]. Tocopherols are saturated forms of vitamin E, whereas tocotrienols are the unsaturated forms [38]. α-Tocopherol is more ubiquitous in nature, whereas tocotrienol can be found in oil palm kernel, annatto bean, and rice bran. Tocotrienol isomers are usually found in mixtures of varying compositions in natural sources [39]. For instance, palm-derived TRF consists of a mixture of ≈75% tocotrienol (α, β, γ and δ) and ≈25% α-tocopherol, whereas annatto tocotrienol consists of ≈90% δ-tocotrienol and ≈10% γ-tocotrienol [40]. The effect of α-tocopherol on the treatment efficacy of tocotrienol mixtures is a subject of debate [41], because α-tocopherol competes with other vitamin E isomers to bind to a-tocopherol transfer protein (ATTP) in the liver before being released into the circulation [42]. α-Tocopherol preferentially binds with ATTP; thus, it has a higher bioavailability than other vitamin E isomers; the lower bioavailability might hinder the biological activities of the isomers. However, a comparative study did not find any remarkable difference in the skeletal and metabolic activities of palm TRF (containing α-tocopherol) and annatto tocotrienol (lacking α-tocopherol) in rats [43,44].

Vitamin E has been actively investigated for its effects in managing pre-eclampsia, because it could tackle several pathophysiological pathways of the disease, theoretically. Firstly, vitamin E is a well-established antioxidant [45,46] that can suppress the oxidative damage caused by repeated placental ischaemia/reperfusion. Secondly, vitamin E and its metabolites are anti-inflammatory agents [47–49] that could suppress systemic inflammation caused by the release of damage-associated molecules. Thirdly, vitamin E supplementation may regulate blood pressure, but the evidence is debatable [50].

Until recently, most studies have focused on α-tocopherol, which is the predominant form of vitamin E in body tissues. A previous Cochrane review demonstrated that vitamin E (α-tocopherol) combined with other supplements had no effect on pre-eclampsia [51]. This result was because most women have sufficient vitamin E from their diet, and a short-term supplementation is unlikely to have an effect on pregnant women. In addition, most comparative studies showed that α-tocopherol is less efficacious than tocotrienols in managing chronic diseases [40,52,53]. These limitations justify the shift of attention from α-tocopherol to tocotrienol in pre-eclampsia management.

4. Antioxidant Effects of Tocotrienol

Tocopherols and tocotrienols share similar structures, such as an electron-rich chromanol ring and a long carbon tail. The carbon tail of tocopherols is saturated, whereas the tail of tocotrienols contains three double bonds (Figure 2) [54]. The structural difference between tocopherols and tocotrienols dictates their antioxidant potentials on a lipid membrane. Compared with α-tocopherol, α-tocotrienol has a stronger ability to disorganise lipid bilayers and is distributed more homogeneously within the lipid bilayers. α-Tocotrienol is also closer to the surface of the membrane, which facilitates its interaction with free radicals and the redox recycling process [46,55].

Nuclear factor erythroid 2-related factor 2 (NRF2) is a master regulator of cellular antioxidants, as it governs the expression of various antioxidants in response to oxidative stress [56]. Palm TRF up-regulates NRF2 expression, which increases the nuclear translocation of NRF2 in mouse liver [57]. Moreover, δ-tocotrienol can stabilise the NRF2 activation induced by 5-fluorouracil in human oral keratinocytes, which can cause a sustained increase in the expression of heme oxygenase-1 and NAD(P)H:quinone oxidoreductase-1 [58]. Another study on the antioxidant effects of δ-tocotrienol osteoblasts found that the activation of the NRF2 and PI3K/AKT pathways is necessary for the protection of δ-tocotrienol against oxidative stress [59].

The antioxidant effects of tocotrienol, through direct electron scavenging activities or the activation of the NRF2 system, are translated in vivo. Tocotrienol isomers or mixtures exert antioxidant effects in various disease models, such as osteoporosis [60–62], gastric ulcers [63], cardiovascular diseases [64,65], and metabolic syndrome [44]. Interestingly, a recent clinical trial demonstrated the more pronounced antioxidant benefits of TRF amongst women than men [66]. A recent meta-analysis of human clinical trials showed that tocotrienol supplementation can improve the redox status of patients, as evidenced by the markedly reduced levels of lipid peroxidation products [67]. Thus, the antioxidant effects of tocotrienol can be exploited to prevent the oxidative stress caused by placental ischaemia/reperfusion injury in pre-eclampsia.

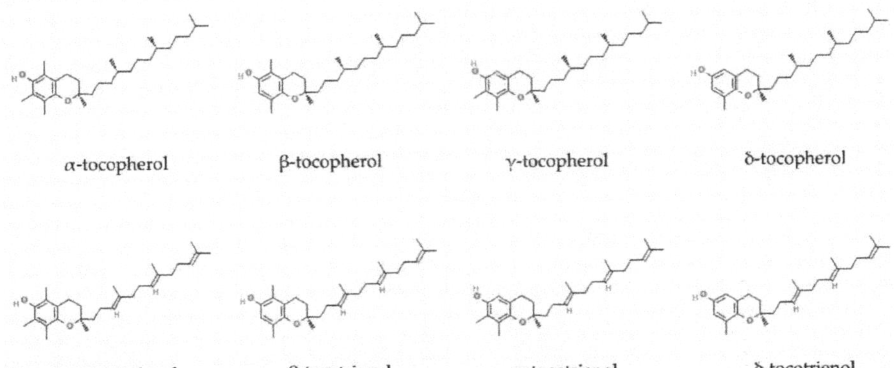

α-tocopherol β-tocopherol γ-tocopherol δ-tocopherol

α-tocotrienol β-tocotrienol γ-tocotrienol δ-tocotrienol

Figure 2. Molecular structures of tocopherols and tocotrienols. They consist of a chromanol ring and a long carbon tail. Tocopherols and tocotrienols have saturated and unsaturated carbon tails, respectively. Each isomer is distinct from each other by the position of the methyl side chain on the chromanol ring. The structures were obtained from PubChem.

5. Anti-Inflammatory Effects of Tocotrienol

The NF-κB pathway is an important regulatory pathway for inflammation [68]. Previous studies showed that γ- and δ-tocotrienols inhibit the TNFα-induced activation of NF-κB by increasing intracellular dihydroceramides (a type of sphingolipid), cellular stress, and A20 [69,70]. A20 is a negative regulator of the NF-κB pathway by acting as a deubiquitinase [71]. γ-Tocotrienol suppresses the phosphorylation and degradation of TNF-induced NF-κB inhibitor-α (IκBα) by preventing IκBα kinase activation and thus, abolishes the phosphorylation and nuclear translocation of p65 and the transcription of NF-κB-dependent reporter genes [72]. This mechanism, together with the suppression of CCAAT/enhancer-binding protein beta, explains the suppressive effects of γ-tocotrienol on the expression of IL-6 and granulocyte colony-stimulating factor in primary bone marrow-derived macrophages, including chemokine response in the pathogenesis of asthma [70].

NF-κB is also indispensable for the expression of cyclooxygenase, which is responsible for converting arachidonic acid to prostaglandin precursors [73]. δ-Tocotrienol selectively decreases the expression of cyclo-oxygenase (COX)-2 and 5-lipoxygenase, but not COX-1, in lipopolysaccharide-induced microglial cells [74]. This decrease is accompanied by a reduction in the expression of inducible NO synthase and IL-1β [74]. The anticancer activities of γ-tocotrienol are also attributed to its ability to suppress the expression of COX-2, MMP-2, and MMP-9 proteins [75]. A comparison of the anti-inflammatory effects of tocotrienol isomers and RAW264.7 macrophages showed that only α-tocotrienol can remarkably suppress TNFα expression [76]. Besides, all tocotrienol isomers can reduce the expression of IL-6 and COX-2, but γ-tocotrienol is less efficacious in decreasing prostaglandin E_2 release. The COX-down-regulating activities of tocotrienol isomers are not shared by α-tocopherol [76]. Another study on the anti-asthma potential of vitamin E reported that γ-tocotrienol is more effective than α-, γ-, and δ-tocopherols in suppressing IL-13-stimulated eotaxin-3 production in human lung epithelial A549 cells [77]. These studies highlighted that the anti-inflammatory effects of each tocotrienol isomer against each cytokine could be different, and a mixture with the full spectrum of tocotrienols, such as palm TRF, could be more effective in suppressing inflammation.

The anti-inflammatory effects of tocotrienol mixtures can be explored with the aim of suppressing the inflammation caused by maternal metabolic conditions or inflammation in pre-eclampsia. Besides, animal studies also suggested that tocotrienols could prevent aspirin-induced gastric lesions by limiting lipid peroxidation [78]. Tocopherols and tocotrienols have gastroprotective effects against oxidative stress, but only tocotrienols can block stress-induced changes in gastric acidity and gastrin level [79]. Palm TRF increases the level of the constituently expressed COX-1 enzyme and prostaglandin E_2 to exert its gastric protective effects [80]. γ-Tocopherol prevents, whereas α-tocopherol worsens, aspirin-induced gastric injury [81]. Considering these studies, tocotrienols could mitigate the gastrointestinal bleeding side effects of aspirin. Nonetheless, a recent meta-analysis of human clinical trials showed that the effects of tocotrienol supplementation on circulating inflammatory mediators were not consistent [67].

6. Role of Tocotrienol in Regulating Blood Pressure

Hypertension is one of the main features of pre-eclampsia; hence, the effects of tocotrienol in regulating blood pressure deserve a closer look. Numerous preclinical studies reported the blood pressure-regulating mechanism of tocotrienol. Newaz et al. [82] supplemented male spontaneously hypertensive rats (SHR) with γ-tocotrienol (15, 30 and 150 mg/kg diet) for 3 months and found that γ-tocotrienol reduced systolic blood pressure at all doses. This beneficial effect was accompanied by a reduction in circulating malondialdehyde and nitrite levels and an increase in NO synthase activity in the blood vessels [82]. Muharis et al. [83] showed that the exposure of the aortic ring of SHR to palm TRF enhanced acetylcholine, but not sodium nitroprusside-induced relaxation [83]. α-Tocopherol has similar effects as TRF. However, this study did not pinpoint any specific mechanism of tocotrienol.

In diet-induced metabolic syndrome models, palm TRF and annatto tocotrienol supplementation (60 and 100 mg/kg body weight, respectively) for 8 weeks decreased the blood pressure and other metabolic derangements in rats, such as hyperglycemia and dyslipidemia. These changes are associated with the reduction of circulating IL-6 and IL-1α, but not TNFα [44,84]. Other researchers proposed that palm TRF could exert antihypertensive effects at a lower dose (60 mg/kg) and a shorter period (4 weeks) in rats fed with high-fat diet [85]. The effectiveness of individual vitamin E isomers (α-tocopherol, α-, γ-, and δ-tocotrienols; each at 85 mg/kg) against metabolic syndrome induced by high-fat high-carbohydrate diet in rats was also examined. Only γ- and δ-tocotrienols can normalise the blood pressure of rats with metabolic syndrome [86].

Hyperhomocysteinemia is a classical risk factor of hypertension that can lead to endothelial cell damage and reduced blood vessel flexibility [87]. Palm TRF supplementation (60 and 150 mg/kg diet) for 5 weeks in rats fed with high-methionine diet decreased the levels liver, heart, and aortic lipid peroxidation products, increased heart glutathione peroxidase and circulating NO, but did not change the systolic pressure of the animals [88–90]. The ineffectiveness of tocotrienol in normalising blood pressure in hyperhomocysteinemia could be due to the more extensive damage on endothelial tissues, but further studies are needed.

The effects of self-emulsified palm TRF on blood pressure and arterial stiffness were also tested in a randomised controlled trial. Rasool et al. reported that self-emulsified palm TRF (50, 100, and 200 mg) supplementation in healthy men for 2 months did not modify their blood pressure and lipid profile, but remarkably reduced pulse wave velocity and augmentation index. The lack of effects of palm TRF on these subjects is expected because they were not hypertensive [91]. By contrast, Aminuddin et al. demonstrated that palm TRF at 100 mg daily from 12–16 gestational weeks to delivery substantially reduced the risk of pregnancy-induced hypertension (odds ratio [OR]: 0.254; 95% confidence interval [CI]: 0.07–0.93) and pre-eclampsia (OR: 0.030, 95% CI: 0.001–0.65) amongst pregnant women [23]. However, the CI values for both effects were very large, which implies that the effect might not be consistent for all patients.

Based on these studies, tocotrienol might be able to prevent hypertension by normalising the circulating NO level and reducing inflammation and oxidative stress. By extension, these protective effects suggest that tocotrienol has a protective role in suppressing hypertension caused by pre-eclampsia, because the principal mechanisms of blood pressure regulation in various health conditions are similar. However, most of the hypotensive studies were conducted in rats, which have a different metabolic rate and life span compared with humans [92]. Most of the studies summarised were performed in adult rats. Cheng et al. [85] demonstrated that at least 4 weeks of treatment is necessary for TRF to reverse hypertension. Approximately 10.5 days in rats are equivalent to 1 year in humans at adulthood [92], which means that humans need to consume tocotrienol for at least 2.67 years to effectively reverse hypertension. Given this limitation, a direct translation of the effects of tocotrienol on blood pressure from existing animal models may be problematic because it is not tested in a pre-eclampsia model. Preliminary data from Aminuddin et al. [23] suggested that tocotrienol could be used as a preventive agent rather than a reversal agent, based on animal studies. More comprehensive human clinical trials with larger sample size are required to prove the effects of TRF in preventing hypertension amongst women with pre-eclampsia.

7. Role of Tocotrienol in Protecting against Ischaemic Injury

Placental ischaemic injury is a major feature of pre-eclampsia. Thus, the protective effects of tocotrienol against ischaemic injury would be of interest. An investigation of the human placentome revealed that γ-tocopherol is preferentially deposited at the basal plate but not in the human umbilical cord, whereas β-tocopherol is found at both sites. The levels of β- and γ-tocopherols are higher than that of α-tocopherol. Moreover, β- and γ-tocotrienols are found in minute amounts in the placentome [93]. These observations showed that vitamin E isomers (tocotrienols and tocopherols) are deposited at the placentome and probably have a functional role as parts of the non-enzymatic antioxidant defence system.

Although the effects of tocotrienol on placental ischaemia have not been examined, they have been investigated in other organs, particularly in the brain and heart. Palm TRF (100 mg twice daily) supplementation in mongrel canines for 10 weeks before transient middle cerebral artery occlusion was able to attenuate the ischaemic lesion and retained white matter fibre tract connectivity by improving the circulation at the stroke area. Palm TRF also stimulated the expression of arteriogenic markers and suppressed MMP-2 activity at the stroke area [94]. Mice supplemented with commercialised palm TRF (Tocovid, 200 mg/kg/day for 1 month) and subjected to transient middle cerebral artery occlusion showed reduced brain lesion, oxidative damage, and autophagy/apoptosis markers, as well as increased NRF2 pathway activation and improved function [95]. Mishima et al. differentiated the individual protective effects of vitamin E isomers and reported that α-tocotrienol, α-tocopherol, and γ-tocopherol are more effective than γ-tocotrienol, δ-tocopherol, and δ-tocotrienol in reducing brain infarct lesions after middle cerebral artery occlusion in mice [96]. This finding was further confirmed by another study, which reported that α-tocotrienol protected against ischaemic cerebral injury via the up-regulation of multidrug resistance-associated protein 1 and the down-regulation of its modulator miR-199a-5p [97].

For ischaemic heart disease, the supplementation of male rats with red palm oil (7 g/kg diet for 6 weeks) rich in vitamin E (including tocotrienol), B-carotenes, and various fatty acids preserved cardiac output, cGMP production, and total polyunsaturated fatty acid content of the rodent heart subjected to ischaemic/reperfusion injury [98]. cGMP production is triggered by NO [99]; thus, red palm oil could increase NO production by NO synthase, but this mechanism was not examined in the study. Das et al. [100] compared the protective effects of α-, β-, γ-, and δ-tocotrienol isomers (20 mol/kg/day for 4 weeks, after 4 weeks on 2% cholesterol diet) on excised proline heart using an ischaemic/reperfusion model. α and γ-Tocotrienols suppress the expression of genes modulating ET-1, MMP-2, MMP9, and thyroid hormone-responsive protein Spot14, and increased TGFβ [100].

Although direct evidence is lacking, the examples aforementioned highlight several targets of placental ischaemia/reperfusion injury that are modulated by TRF. These targets include NRF2, which modulates cellular antioxidant response [101]; NO/cGMP pathway, which plays a major role in vasoconstriction [99]; and MMPs, which are important for vascular and uterine remodelling [99]. TRF could suppress MMP production by extravillous trophoblasts and decidual cells at the early stage of pregnancy, which contradicts its other beneficial effects in preventing pre-eclampsia. This aspect should be investigated in a valid preclinical model of pre-eclampsia.

8. Safety Profile of Palm TRF amongst Pregnant Women

Other important aspects concerning the use of palm TRF amongst pregnant women are bleeding risk and reproductive toxicity. Vitamin E supplementation (400 IU tocopherol for 8 years) has been associated with increased risk of haemorrhagic stroke amongst male physicians in the Physicians' Health Study II [102], but this effect was not shown in the Women's Health Study (600 IU tocopherol for 10 years) [103]. Real-world safety data for tocotrienol is lacking. A subacute toxicity study showed that palm TRF above 500 mg/kg body weight can prolong bleeding and clotting time in female mice [104]. Based on the body surface conversion formula [105], this dose is equivalent to 2400 mg for a human adult weighing 60 kg; hence, the daily dose used in the clinical trial on pre-eclampsia (100 mg) was much lower. With regard to the concurrent use of tocotrienol and other anticoagulants, a recently published randomised trial reported that the concurrent use of aspirin, clopidogrel, and palm TRF could reduce the incidence of aspirin resistance [106]. Only one study on the female reproductive toxicity of palm TRF was reported. Supplementation of palm TRF (200, 500, and 1000 mg/kg) 2 weeks before pregnancy and continuing throughout the pregnancy (for about 21 days) did not result in a remarkable change in pup number, body weight, and body length [107]. However, no further assessment was done on the dams or pups.

A recent study highlighted that α-tocotrienol could reduce the viability, growth, migration, epithelial–mesenchymal transition, invasion, and angiogenesis of trophoblasts in vitro. These effects were mediated by the down-regulation of miR-429 and the up-regulation of zinc finger E-box-binding homeobox (ZEB). The same study revealed that placenta sampled from pre-eclamptic women expressed higher miR-429 and lower ZEB levels [108]. These findings imply that δ-tocotrienol could be toxic to trophoblasts. Given these observations, we postulated that a natural mixture with lower δ-tocotrienol could be safer for embryos. Compared with annatto tocotrienol with ≈90% δ-tocotrienol in the mixture, palm TRF usually contains less than 20% of δ-tocotrienol, which makes palm TRF a safer supplementation [40]. However, this speculation requires further validation from preclinical studies.

9. Future Research

Only one clinical trial has looked at a tocotrienol-predominant preparation in preventing pre-eclampsia [23]. In this randomised double-blind placebo-controlled trial involving 299 primigravidae without any other risk factors for pre-eclampsia, the risk of pre-eclampsia was remarkably reduced by 83%. Vitamin E was prescribed in the form of TRF, at a dose of 100 mg daily. Repeating this trial on a population with additional risk factors for pre-eclampsia, such as a history of pre-eclampsia and current diabetes mellitus, chronic hypertension, or other medical disorders that increase the risk of developing pre-eclampsia, would be of interest.

As prophylactic agents against pre-eclampsia, the combined use of aspirin and tocotrienols should be considered and investigated, as tocotrienols seem to be protective against aspirin-induced gastric lesions and, therefore, add a potential benefit besides affording antioxidant activity. Whether tocotrienols will enhance the anti-inflammatory property of aspirin is another pertinent research question. Embarking on an in vitro or animal study may be helpful to scrutinise this hypothesis before conducting a clinical trial combining tocotrienols and aspirin for pre-eclampsia prevention. Which preparation of vitamin E should be used in this endeavour is another question to ponder. Perhaps TRF, with its unique combination of tocotrienols and tocopherol, can be considered, given the abundance of supporting evidence of the benefit of TRF from studies on non-communicable diseases [39].

10. Conclusions

The pathophysiology of pre-eclampsia involves oxidative stress and inflammation activation. Anti-inflammatory agents, such as aspirin, only act upon a part of the disease mechanism, which results in a partial remedy in the preventive process. Tocotrienols are forms of vitamin E that possess potent antioxidant and anti-inflammatory characteristics; hence, they are potentially more appropriate preventive agents to combat pre-eclampsia, compared with tocopherol. Tocotrienols might also prevent hypertension and ischaemia/reperfusion injury in the placenta (Figure 3). Notably, the effects of tocotrienol on hypertension are contentious, because a long duration of treatment is required. Most of the mechanisms proposed are theoretical, because they are not proven yet in experimental models of pre-eclampsia. This limitation necessitates more fundamental studies on this topic. To date, only a single-centre small-scale clinical trial investigated the effects of TRF on pregnant mothers. Therefore, a large randomised clinical trial on a population of pregnant women at high risk of pre-eclampsia is warranted to shed more light on this matter.

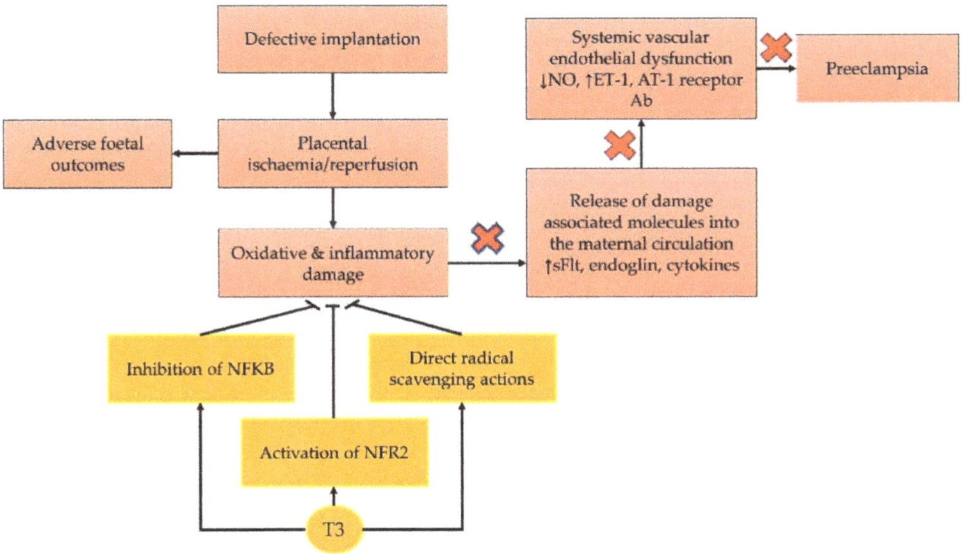

Figure 3. Theoretical framework of how tocotrienol prevents pre-eclampsia. Tocotrienol exerts anti-inflammatory and antioxidant actions, which prevent the oxidative and inflammatory damages caused by ischaemia. The same properties also prevent hypertension in pre-eclampsia. Abbreviation: Ab, antibody; AT-1, angiotensin-1; ET-1, endothelin-1; NF-κB, nuclear factor kappa-B; NO, nitric oxide; NRF2; nuclear factor erythroid 2-related factor 2; T3, tocotrienol; sFlt-1, soluble fms-like tyrosine kinase 1.

Author Contributions: Conceptualisation, Z.A.M., A.K. and R.A.R.; writing—original draft preparation, Z.A.M. and K.-Y.C.; writing—review and editing, Z.A.M., N.L.N.-A.-Z., A.K. and R.A.R.; visualisation, K.-Y.C.; supervision, Z.A.M.; project administration, Z.A.M. and K.-Y.C. All authors have read and agreed to the published version of the manuscript.

Funding: The authors are supported by Universiti Kebangsaan Malaysia via grant number FF-2019-371/1.

Institutional Review Board Statement: Not applicable.

Informed Consent Statement: Not applicable.

Conflicts of Interest: The authors declare no conflict of interest.

References

1. Brown, M.A.; Magee, L.A.; Kenny, L.C.; Karumanchi, S.A.; McCarthy, F.P.; Saito, S.; Hall, D.R.; Warren, C.E.; Adoyi, G.; Ishaku, S.; et al. Hypertensive Disorders of Pregnancy: ISSHP Classification, Diagnosis, and Management Recommendations for International Practice. *Hypertension* **2018**, *72*, 24–43. [CrossRef] [PubMed]
2. Say, L.; Chou, D.; Gemmill, A.; Tuncalp, O.; Moller, A.B.; Daniels, J.; Gulmezoglu, A.M.; Temmerman, M.; Alkema, L. Global causes of maternal death: A WHO systematic analysis. *Lancet Glob. Health* **2014**, *2*, e323–e333. [CrossRef]
3. Wang, W.; Xie, X.; Yuan, T.; Wang, Y.; Zhao, F.; Zhou, Z.; Zhang, H. Epidemiological trends of maternal hypertensive disorders of pregnancy at the global, regional, and national levels: A population-based study. *BMC Pregnancy Childbirth* **2021**, *21*, 364. [CrossRef] [PubMed]
4. Bartsch, E.; Medcalf, K.E.; Park, A.L.; Ray, J.G.; High Risk of Pre-eclampsia Identification Group. Clinical risk factors for pre-eclampsia determined in early pregnancy: Systematic review and meta-analysis of large cohort studies. *BMJ* **2016**, *353*, i1753. [CrossRef]
5. Rana, S.; Lemoine, E.; Granger, J.P.; Karumanchi, S.A. Preeclampsia: Pathophysiology, Challenges, and Perspectives. *Circ. Res.* **2019**, *124*, 1094–1112. [CrossRef] [PubMed]
6. Ronsmans, C.; Campbell, O. Quantifying the fall in mortality associated with interventions related to hypertensive diseases of pregnancy. *BMC Public Health* **2011**, *11* (Suppl. 3), S8. [CrossRef] [PubMed]
7. U. S. Preventive Services Task Force; Davidson, K.W.; Barry, M.J.; Mangione, C.M.; Cabana, M.; Caughey, A.B.; Davis, E.M.; Donahue, K.E.; Doubeni, C.A.; Kubik, M.; et al. Aspirin Use to Prevent Preeclampsia and Related Morbidity and Mortality: US Preventive Services Task Force Recommendation Statement. *JAMA* **2021**, *326*, 1186–1191. [CrossRef] [PubMed]
8. Van Doorn, R.; Mukhtarova, N.; Flyke, I.P.; Lasarev, M.; Kim, K.; Hennekens, C.H.; Hoppe, K.K. Dose of aspirin to prevent preterm preeclampsia in women with moderate or high-risk factors: A systematic review and meta-analysis. *PLoS ONE* **2021**, *16*, e0247782. [CrossRef]
9. American College of Obstetricians and Gynecologists ACOG Committee Opinion No. 743: Low-Dose Aspirin Use During Pregnancy. *Obstet. Gynecol.* **2018**, *132*, e44–e52. [CrossRef] [PubMed]
10. Atallah, A.; Lecarpentier, E.; Goffinet, F.; Doret-Dion, M.; Gaucherand, P.; Tsatsaris, V. Aspirin for Prevention of Preeclampsia. *Drugs* **2017**, *77*, 1819–1831. [CrossRef]
11. Hofmeyr, G.J.; Lawrie, T.A.; Atallah, A.N.; Torloni, M.R. Calcium supplementation during pregnancy for preventing hypertensive disorders and related problems. *Cochrane Database Syst. Rev.* **2018**, *10*, CD001059. [CrossRef]
12. Phipps, E.A.; Thadhani, R.; Benzing, T.; Karumanchi, S.A. Pre-eclampsia: Pathogenesis, novel diagnostics and therapies. *Nat. Rev. Nephrol.* **2019**, *15*, 275–289. [CrossRef] [PubMed]
13. Cohen, J.M.; Kramer, M.S.; Platt, R.W.; Basso, O.; Evans, R.W.; Kahn, S.R. The association between maternal antioxidant levels in midpregnancy and preeclampsia. *Am. J. Obstet. Gynecol.* **2015**, *213*, 695.e1–695.e13. [CrossRef]
14. Taravati, A.; Tohidi, F. Comprehensive analysis of oxidative stress markers and antioxidants status in preeclampsia. *Taiwan J. Obstet. Gynecol.* **2018**, *57*, 779–790. [CrossRef] [PubMed]
15. Raijmakers, M.T.; Dechend, R.; Poston, L. Oxidative stress and preeclampsia: Rationale for antioxidant clinical trials. *Hypertension* **2004**, *44*, 374–380. [CrossRef]
16. Marin, R.; Chiarello, D.I.; Abad, C.; Rojas, D.; Toledo, F.; Sobrevia, L. Oxidative stress and mitochondrial dysfunction in early-onset and late-onset preeclampsia. *Biochim. Biophys. Acta Mol. Basis Dis.* **2020**, *1866*, 165961. [CrossRef] [PubMed]
17. Hobson, S.R.; Gurusinghe, S.; Lim, R.; Alers, N.O.; Miller, S.L.; Kingdom, J.C.; Wallace, E.M. Melatonin improves endothelial function in vitro and prolongs pregnancy in women with early-onset preeclampsia. *J. Pineal Res.* **2018**, *65*, e12508. [CrossRef]
18. Chuffa, L.G.A.; Lupi, L.A.; Cucielo, M.S.; Silveira, H.S.; Reiter, R.J.; Seiva, F.R.F. Melatonin Promotes Uterine and Placental Health: Potential Molecular Mechanisms. *Int. J. Mol. Sci.* **2019**, *21*, 300. [CrossRef]
19. Filardi, T.; Varì, R.; Ferretti, E.; Zicari, A.; Morano, S.; Santangelo, C. Curcumin: Could This Compound Be Useful in Pregnancy and Pregnancy-Related Complications? *Nutrients* **2020**, *12*, 3179. [CrossRef]
20. Salles, A.M.; Galvao, T.F.; Silva, M.T.; Motta, L.C.; Pereira, M.G. Antioxidants for preventing preeclampsia: A systematic review. *Sci. World J.* **2012**, *2012*, 243476. [CrossRef]

21. Tenorio, M.B.; Ferreira, R.C.; Moura, F.A.; Bueno, N.B.; Goulart, M.O.F.; Oliveira, A.C.M. Oral antioxidant therapy for prevention and treatment of preeclampsia: Meta-analysis of randomized controlled trials. *Nutr. Metab. Cardiovasc. Dis.* **2018**, *28*, 865–876. [CrossRef] [PubMed]
22. Conde-Agudelo, A.; Romero, R.; Kusanovic, J.P.; Hassan, S.S. Supplementation with vitamins C and E during pregnancy for the prevention of preeclampsia and other adverse maternal and perinatal outcomes: A systematic review and metaanalysis. *Am. J. Obstet. Gynecol.* **2011**, *204*, 503.e1–503.e12. [CrossRef]
23. Aminuddin, N.A.; Sutan, R.; Mahdy, Z.A. Role of Palm Oil Vitamin E in Preventing Pre-eclampsia: A Secondary Analysis of a Randomized Clinical Trial Following ISSHP Reclassification. *Front. Med.* **2020**, *7*, 596405. [CrossRef]
24. Tenorio, M.B.; Ferreira, R.C.; Moura, F.A.; Bueno, N.B.; de Oliveira, A.C.M.; Goulart, M.O.F. Cross-Talk between Oxidative Stress and Inflammation in Preeclampsia. *Oxidative Med. Cell. Longev.* **2019**, *2019*, 8238727. [CrossRef] [PubMed]
25. Phipps, E.; Prasanna, D.; Brima, W.; Jim, B. Preeclampsia: Updates in Pathogenesis, Definitions, and Guidelines. *Clin. J. Am. Soc. Nephrol.* **2016**, *11*, 1102–1113. [CrossRef]
26. Chiarello, D.I.; Abad, C.; Rojas, D.; Toledo, F.; Vazquez, C.M.; Mate, A.; Sobrevia, L.; Marin, R. Oxidative stress: Normal pregnancy versus preeclampsia. *Biochim. Biophys. Acta Mol. Basis Dis.* **2020**, *1866*, 165354. [CrossRef] [PubMed]
27. Aouache, R.; Biquard, L.; Vaiman, D.; Miralles, F. Oxidative Stress in Preeclampsia and Placental Diseases. *Int. J. Mol. Sci.* **2018**, *19*, 1496. [CrossRef] [PubMed]
28. Jena, M.K.; Sharma, N.R.; Petitt, M.; Maulik, D.; Nayak, N.R. Pathogenesis of Preeclampsia and Therapeutic Approaches Targeting the Placenta. *Biomolecules* **2020**, *10*, 953. [CrossRef]
29. Cindrova-Davies, T.; Sanders, D.A.; Burton, G.J.; Charnock-Jones, D.S. Soluble FLT1 sensitizes endothelial cells to inflammatory cytokines by antagonizing VEGF receptor-mediated signalling. *Cardiovasc. Res.* **2011**, *89*, 671–679. [CrossRef]
30. Chen, B.; Longtine, M.S.; Sadovsky, Y.; Nelson, D.M. Hypoxia downregulates p53 but induces apoptosis and enhances expression of BAD in cultures of human syncytiotrophoblasts. *Am. J. Physiol. Cell Physiol.* **2010**, *299*, C968–C976. [CrossRef]
31. Ishioka, S.; Ezaka, Y.; Umemura, K.; Hayashi, T.; Endo, T.; Saito, T. Proteomic analysis of mechanisms of hypoxia-induced apoptosis in trophoblastic cells. *Int. J. Med. Sci.* **2006**, *4*, 36–44. [CrossRef]
32. Zhuang, B.; Luo, X.; Rao, H.; Li, Q.; Shan, N.; Liu, X.; Qi, H. Oxidative stress-induced C/EBPbeta inhibits beta-catenin signaling molecule involving in the pathology of preeclampsia. *Placenta* **2015**, *36*, 839–846. [CrossRef]
33. Yang, F.; Zheng, Q.; Jin, L. Dynamic Function and Composition Changes of Immune Cells During Normal and Pathological Pregnancy at the Maternal-Fetal Interface. *Front. Immunol.* **2019**, *10*, 2317. [CrossRef] [PubMed]
34. Aggarwal, R.; Jain, A.K.; Mittal, P.; Kohli, M.; Jawanjal, P.; Rath, G. Association of pro- and anti-inflammatory cytokines in preeclampsia. *J. Clin. Lab. Anal.* **2019**, *33*, e22834. [CrossRef]
35. Zhang, H.; Park, Y.; Wu, J.; Chen, X.; Lee, S.; Yang, J.; Dellsperger, K.C.; Zhang, C. Role of TNF-alpha in vascular dysfunction. *Clin. Sci.* **2009**, *116*, 219–230. [CrossRef] [PubMed]
36. Xia, Y.; Kellems, R.E. Angiotensin receptor agonistic autoantibodies and hypertension: Preeclampsia and beyond. *Circ. Res.* **2013**, *113*, 78–87. [CrossRef] [PubMed]
37. Jiang, Q. Natural forms of vitamin E: Metabolism, antioxidant, and anti-inflammatory activities and their role in disease prevention and therapy. *Free Radic. Biol. Med.* **2014**, *72*, 76–90. [CrossRef] [PubMed]
38. Aggarwal, B.B.; Sundaram, C.; Prasad, S.; Kannappan, R. Tocotrienols, the vitamin E of the 21st century: Its potential against cancer and other chronic diseases. *Biochem. Pharm.* **2010**, *80*, 1613–1631. [CrossRef]
39. Wong, S.K.; Kamisah, Y.; Mohamed, N.; Muhammad, N.; Masbah, N.; Fahami, N.A.M.; Mohamed, I.N.; Shuid, A.N.; Saad, Q.M.; Abdullah, A.; et al. Potential Role of Tocotrienols on Non-Communicable Diseases: A Review of Current Evidence. *Nutrients* **2020**, *12*, 259. [CrossRef] [PubMed]
40. Chin, K.Y.; Ima-Nirwana, S. The biological effects of tocotrienol on bone: A review on evidence from rodent models. *Drug Des. Dev. Ther.* **2015**, *9*, 2049–2061. [CrossRef]
41. Chin, K.Y.; Ima-Nirwana, S. The effects of alpha-tocopherol on bone: A double-edged sword? *Nutrients* **2014**, *6*, 1424–1441. [CrossRef]
42. Hosomi, A.; Arita, M.; Sato, Y.; Kiyose, C.; Ueda, T.; Igarashi, O.; Arai, H.; Inoue, K. Affinity for alpha-tocopherol transfer protein as a determinant of the biological activities of vitamin E analogs. *FEBS Lett.* **1997**, *409*, 105–108. [CrossRef]
43. Wong, S.K.; Chin, K.-Y.; Ima-Nirwana, S. The Effects of Tocotrienol on Bone Peptides in a Rat Model of Osteoporosis Induced by Metabolic Syndrome: The Possible Communication between Bone Cells. *Int. J. Environ. Res. Public Health* **2019**, *16*, 3313. [CrossRef] [PubMed]
44. Wong, S.K.; Chin, K.-Y.; Suhaimi, F.H.; Ahmad, F.; Ima-Nirwana, S. The effects of palm tocotrienol on metabolic syndrome and bone loss in male rats induced by high-carbohydrate high-fat diet. *J. Funct. Foods* **2018**, *44*, 246–254. [CrossRef]
45. Hidalgo, M.; Rodriguez, V.; Kreindl, C.; Porras, O. Biological Redox Impact of Tocopherol Isomers Is Mediated by Fast Cytosolic Calcium Increases in Living Caco-2 Cells. *Antioxidants* **2020**, *9*, 155. [CrossRef]
46. Serbinova, E.; Kagan, V.; Han, D.; Packer, L. Free radical recycling and intramembrane mobility in the antioxidant properties of alpha-tocopherol and alpha-tocotrienol. *Free Radic. Biol. Med.* **1991**, *10*, 263–275. [CrossRef]
47. Jiang, Q. Metabolism of natural forms of vitamin E and biological actions of vitamin E metabolites. *Free Radic. Biol. Med.* **2021**, *179*, 375–387. [CrossRef]
48. Pein, H.; Ville, A.; Pace, S.; Temml, V.; Garscha, U.; Raasch, M.; Alsabil, K.; Viault, G.; Dinh, C.P.; Guilet, D.; et al. Endogenous metabolites of vitamin E limit inflammation by targeting 5-lipoxygenase. *Nat. Commun.* **2018**, *9*, 3834. [CrossRef] [PubMed]

49. Qureshi, A.A.; Reis, J.C.; Papasian, C.J.; Morrison, D.C.; Qureshi, N. Tocotrienols inhibit lipopolysaccharide-induced pro-inflammatory cytokines in macrophages of female mice. *Lipids Health Dis.* **2010**, *9*, 143. [CrossRef]
50. Emami, M.R.; Safabakhsh, M.; Alizadeh, S.; Asbaghi, O.; Khosroshahi, M.Z. Effect of vitamin E supplementation on blood pressure: A systematic review and meta-analysis. *J. Hum. Hypertens.* **2019**, *33*, 499–507. [CrossRef]
51. Rumbold, A.; Ota, E.; Hori, H.; Miyazaki, C.; Crowther, C.A. Vitamin E supplementation in pregnancy. *Cochrane Database Syst. Rev.* **2015**, *18*, CD004069. [CrossRef]
52. Pang, K.L.; Chin, K.Y. The Role of Tocotrienol in Protecting Against Metabolic Diseases. *Molecules* **2019**, *24*, 923. [CrossRef] [PubMed]
53. Wong, S.K.; Chin, K.Y.; Suhaimi, F.H.; Ahmad, F.; Ima-Nirwana, S. Vitamin E As a Potential Interventional Treatment for Metabolic Syndrome: Evidence from Animal and Human Studies. *Front. Pharm.* **2017**, *8*, 444. [CrossRef] [PubMed]
54. Niki, E.; Abe, K. Chapter 1: Vitamin E: Structure, Properties and Functions. In *Vitamin E: Chemistry and Nutritional Benefits*; Niki, E., Ed.; Royal Society of Chemistry: London, UK, 2019; pp. 1–11. [CrossRef]
55. Suzuki, Y.J.; Tsuchiya, M.; Wassall, S.R.; Choo, Y.M.; Govil, G.; Kagan, V.E.; Packer, L. Structural and dynamic membrane properties of alpha-tocopherol and alpha-tocotrienol: Implication to the molecular mechanism of their antioxidant potency. *Biochemistry* **1993**, *32*, 10692–10699. [CrossRef]
56. Bellezza, I.; Giambanco, I.; Minelli, A.; Donato, R. Nrf2-Keap1 signaling in oxidative and reductive stress. *Biochim. Biophys. Acta Mol. Cell Res.* **2018**, *1865*, 721–733. [CrossRef]
57. Atia, A.; Alrawaiq, N.S.; Abdullah, A. Tocotrienols Activate Nrf2 Nuclear Translocation and Increase the Antioxidant- Related Hepatoprotective Mechanism in Mice Liver. *Curr. Pharm. Biotechnol.* **2021**, *22*, 1085–1098. [CrossRef]
58. Takano, H.; Momota, Y.; Kani, K.; Aota, K.; Yamamura, Y.; Yamanoi, T.; Azuma, M. gamma-Tocotrienol prevents 5-FU-induced reactive oxygen species production in human oral keratinocytes through the stabilization of 5-FU-induced activation of Nrf2. *Int. J. Oncol.* **2015**, *46*, 1453–1460. [CrossRef]
59. Casati, L.; Pagani, F.; Limonta, P.; Vanetti, C.; Stancari, G.; Sibilia, V. Beneficial effects of delta-tocotrienol against oxidative stress in osteoblastic cells: Studies on the mechanisms of action. *Eur. J. Nutr.* **2020**, *59*, 1975–1987. [CrossRef] [PubMed]
60. Chin, K.Y.; Ima-Nirwana, S. Vitamin E as an Antiosteoporotic Agent via Receptor Activator of Nuclear Factor Kappa-B Ligand Signaling Disruption: Current Evidence and Other Potential Research Areas. *Evid. Based Complement. Altern. Med.* **2012**, *2012*, 747020. [CrossRef]
61. Chin, K.Y.; Mo, H.; Soelaiman, I.N. A review of the possible mechanisms of action of tocotrienol—A potential antiosteoporotic agent. *Curr. Drug Targets* **2013**, *14*, 1533–1541. [CrossRef]
62. Mohamad, N.V.; Ima-Nirwana, S.; Chin, K.Y. Therapeutic potential of annatto tocotrienol with self-emulsifying drug delivery system in a rat model of postmenopausal bone loss. *Biomed. Pharm.* **2021**, *137*, 111368. [CrossRef] [PubMed]
63. Nur Azlina, M.F.; Kamisah, Y.; Chua, K.H.; Ibrahim, I.A.; Qodriyah, H.M. Preventive Effects of Tocotrienol on Stress-Induced Gastric Mucosal Lesions and Its Relation to Oxidative and Inflammatory Biomarkers. *PLoS ONE* **2015**, *10*, e0139348. [CrossRef] [PubMed]
64. Nafeeza, M.I.; Norzana, A.G.; Jalaluddin, H.L.; Gapor, M.T. The effects of a tocotrienol-rich fraction on experimentally induced atherosclerosis in the aorta of rabbits. *Malays. J. Pathol.* **2001**, *23*, 17–25.
65. Newaz, M.A.; Nawal, N.N. Effect of gamma-tocotrienol on blood pressure, lipid peroxidation and total antioxidant status in spontaneously hypertensive rats (SHR). *Clin. Exp. Hypertens.* **1999**, *21*, 1297–1313. [CrossRef]
66. Goon, J.A.; Nor Azman, N.H.E.; Abdul Ghani, S.M.; Hamid, Z.; Wan Ngah, W.Z. Comparing palm oil tocotrienol rich fraction with alpha-tocopherol supplementation on oxidative stress in healthy older adults. *Clin. Nutr. ESPEN* **2017**, *21*, 1–12. [CrossRef]
67. Khor, B.H.; Tiong, H.C.; Tan, S.C.; Wong, S.K.; Chin, K.Y.; Karupaiah, T.; Ima-Nirwana, S.; Abdul Gafor, A.H. Effects of tocotrienols supplementation on markers of inflammation and oxidative stress: A systematic review and meta-analysis of randomized controlled trials. *PLoS ONE* **2021**, *16*, e0255205. [CrossRef]
68. Mussbacher, M.; Salzmann, M.; Brostjan, C.; Hoesel, B.; Schoergenhofer, C.; Datler, H.; Hohensinner, P.; Basilio, J.; Petzelbauer, P.; Assinger, A.; et al. Cell Type-Specific Roles of NF-kappaB Linking Inflammation and Thrombosis. *Front. Immunol.* **2019**, *10*, 85. [CrossRef]
69. Yang, C.; Jiang, Q. Vitamin E delta-tocotrienol inhibits TNF-alpha-stimulated NF-kappaB activation by up-regulation of anti-inflammatory A20 via modulation of sphingolipid including elevation of intracellular dihydroceramides. *J. Nutr. Biochem.* **2019**, *64*, 101–109. [CrossRef]
70. Wang, Y.; Jiang, Q. gamma-Tocotrienol inhibits lipopolysaccharide-induced interlukin-6 and granulocyte colony-stimulating factor by suppressing C/EBPbeta and NF-kappaB in macrophages. *J. Nutr. Biochem.* **2013**, *24*, 1146–1152. [CrossRef]
71. Shembade, N.; Harhaj, E.W. Regulation of NF-kappaB signaling by the A20 deubiquitinase. *Cell. Mol. Immunol.* **2012**, *9*, 123–130. [CrossRef]
72. Ahn, K.S.; Sethi, G.; Krishnan, K.; Aggarwal, B.B. Gamma-tocotrienol inhibits nuclear factor-kappaB signaling pathway through inhibition of receptor-interacting protein and TAK1 leading to suppression of antiapoptotic gene products and potentiation of apoptosis. *J. Biol. Chem.* **2007**, *282*, 809–820. [CrossRef] [PubMed]
73. Newton, R.; Kuitert, L.M.; Bergmann, M.; Adcock, I.M.; Barnes, P.J. Evidence for involvement of NF-kappaB in the transcriptional control of COX-2 gene expression by IL-1beta. *Biochem. Biophys. Res. Commun.* **1997**, *237*, 28–32. [CrossRef] [PubMed]
74. Tan, S.W.; Israf Ali, D.A.B.; Khaza'ai, H.; Wong, J.W.; Vidyadaran, S. Cellular uptake and anti-inflammatory effects of palm oil-derived delta (delta)-tocotrienol in microglia. *Cell. Immunol.* **2020**, *357*, 104200. [CrossRef] [PubMed]

75. Zhang, Y.H.; Ma, K.; Liu, J.R.; Wang, H.X.; Tian, W.X.; Tu, Y.H.; Sun, W.G. gamma-tocotrienol inhibits the invasion and migration of human gastric cancer cells through downregulation of cyclooxygenase-2 expression. *Oncol. Rep.* **2018**, *40*, 999–1007. [CrossRef]
76. Yam, M.L.; Abdul Hafid, S.R.; Cheng, H.M.; Nesaretnam, K. Tocotrienols suppress proinflammatory markers and cyclooxygenase-2 expression in RAW264.7 macrophages. *Lipids* **2009**, *44*, 787–797. [CrossRef]
77. Wang, Y.; Moreland, M.; Wagner, J.G.; Ames, B.N.; Illek, B.; Peden, D.B.; Jiang, Q. Vitamin E forms inhibit IL-13/STAT6-induced eotaxin-3 secretion by up-regulation of PAR4, an endogenous inhibitor of atypical PKC in human lung epithelial cells. *J. Nutr. Biochem.* **2012**, *23*, 602–608. [CrossRef]
78. Nafeeza, M.I.; Fauzee, A.M.; Kamsiah, J.; Gapor, M.T. Comparative effects of a tocotrienol-rich fraction and tocopherol in aspirin-induced gastric lesions in rats. *Asia Pac. J. Clin. Nutr.* **2002**, *11*, 309–313. [CrossRef]
79. Azlina, M.F.; Nafeeza, M.I.; Khalid, B.A. A comparison between tocopherol and tocotrienol effects on gastric parameters in rats exposed to stress. *Asia Pac. J. Clin. Nutr.* **2005**, *14*, 358–365.
80. Nur Azlina, M.F.; Kamisah, Y.; Chua, K.H.; Qodriyah, H.M. Tocotrienol Attenuates Stress-Induced Gastric Lesions via Activation of Prostaglandin and Upregulation of COX-1 mRNA. *Evid. Based Complement. Altern. Med.* **2013**, *2013*, 804796. [CrossRef]
81. Jiang, Q.; Moreland, M.; Ames, B.N.; Yin, X. A combination of aspirin and gamma-tocopherol is superior to that of aspirin and alpha-tocopherol in anti-inflammatory action and attenuation of aspirin-induced adverse effects. *J. Nutr. Biochem.* **2009**, *20*, 894–900. [CrossRef]
82. Newaz, M.A.; Yousefipour, Z.; Nawal, N.; Adeeb, N. Nitric oxide synthase activity in blood vessels of spontaneously hypertensive rats: Antioxidant protection by gamma-tocotrienol. *J. Physiol. Pharmacol.* **2003**, *54*, 319–327. [PubMed]
83. Muharis, S.P.; Top, A.G.; Murugan, D.; Mustafa, M.R. Palm oil tocotrienol fractions restore endothelium dependent relaxation in aortic rings of streptozotocin-induced diabetic and spontaneously hypertensive rats. *Nutr. Res.* **2010**, *30*, 209–216. [CrossRef] [PubMed]
84. Wong, S.K.; Chin, K.Y.; Suhaimi, F.H.; Ahmad, F.; Ima-Nirwana, S. Exploring the potential of tocotrienol from Bixa orellana as a single agent targeting metabolic syndrome and bone loss. *Bone* **2018**, *116*, 8–21. [CrossRef] [PubMed]
85. Cheng, H.S.; Ton, S.H.; Tan, J.B.L.; Abdul Kadir, K. The Ameliorative Effects of a Tocotrienol-Rich Fraction on the AGE-RAGE Axis and Hypertension in High-Fat-Diet-Fed Rats with Metabolic Syndrome. *Nutrients* **2017**, *9*, 984. [CrossRef] [PubMed]
86. Wong, W.Y.; Ward, L.C.; Fong, C.W.; Yap, W.N.; Brown, L. Anti-inflammatory gamma- and delta-tocotrienols improve cardiovascular, liver and metabolic function in diet-induced obese rats. *Eur. J. Nutr.* **2017**, *56*, 133–150. [CrossRef] [PubMed]
87. Ganguly, P.; Alam, S.F. Role of homocysteine in the development of cardiovascular disease. *Nutr. J.* **2015**, *14*, 6. [CrossRef] [PubMed]
88. Kamisah, Y.; Norsidah, K.Z.; Azizi, A.; Faizah, O.; Nonan, M.R.; Asmadi, A.Y. Palm tocotrienol-rich fraction inhibits methionine-induced cystathionine beta-synthase in rat liver. *J. Physiol. Biochem.* **2015**, *71*, 659–667. [CrossRef]
89. Norsidah, K.Z.; Asmadi, A.Y.; Azizi, A.; Faizah, O.; Kamisah, Y. Palm tocotrienol-rich fraction improves vascular proatherosclerotic changes in hyperhomocysteinemic rats. *Evid. Based Complement. Altern. Med.* **2013**, *2013*, 976967. [CrossRef]
90. Norsidah, K.Z.; Asmadi, A.Y.; Azizi, A.; Faizah, O.; Kamisah, Y. Palm tocotrienol-rich fraction reduced plasma homocysteine and heart oxidative stress in rats fed with a high-methionine diet. *J. Physiol. Biochem.* **2013**, *69*, 441–449. [CrossRef]
91. Rasool, A.H.; Rahman, A.R.; Yuen, K.H.; Wong, A.R. Arterial compliance and vitamin E blood levels with a self emulsifying preparation of tocotrienol rich vitamin E. *Arch. Pharmacal Res.* **2008**, *31*, 1212–1217. [CrossRef]
92. Agoston, D.V. How to Translate Time? The Temporal Aspect of Human and Rodent Biology. *Front. Neurol.* **2017**, *8*, 92. [CrossRef]
93. Li, Z.M.; Benker, B.; Bao, Q.; Henkelmann, B.; Corsten, C.; Michalke, B.; Pauluschke-Frohlich, J.; Flisikowski, K.; Schramm, K.W.; De Angelis, M. Placental distribution of endogenous and exogenous substances: A pilot study utilizing cryo-sampled specimen off delivery room. *Placenta* **2020**, *100*, 45–53. [CrossRef]
94. Rink, C.; Christoforidis, G.; Khanna, S.; Peterson, L.; Patel, Y.; Khanna, S.; Abduljalil, A.; Irfanoglu, O.; Machiraju, R.; Bergdall, V.K.; et al. Tocotrienol vitamin E protects against preclinical canine ischemic stroke by inducing arteriogenesis. *J. Cereb. Blood Flow Metab.* **2011**, *31*, 2218–2230. [CrossRef]
95. Shang, J.; Yan, H.; Jiao, Y.; Ohta, Y.; Liu, X.; Li, X.; Morihara, R.; Nakano, Y.; Fukui, Y.; Shi, X.; et al. Therapeutic Effects of Pretreatment with Tocovid on Oxidative Stress in Postischemic Mice Brain. *J. Stroke Cerebrovasc. Dis.* **2018**, *27*, 2096–2105. [CrossRef] [PubMed]
96. Mishima, K.; Tanaka, T.; Pu, F.; Egashira, N.; Iwasaki, K.; Hidaka, R.; Matsunaga, K.; Takata, J.; Karube, Y.; Fujiwara, M. Vitamin E isoforms alpha-tocotrienol and gamma-tocopherol prevent cerebral infarction in mice. *Neurosci. Lett.* **2003**, *337*, 56–60. [CrossRef]
97. Park, H.A.; Kubicki, N.; Gnyawali, S.; Chan, Y.C.; Roy, S.; Khanna, S.; Sen, C.K. Natural vitamin E alpha-tocotrienol protects against ischemic stroke by induction of multidrug resistance-associated protein 1. *Stroke* **2011**, *42*, 2308–2314. [CrossRef] [PubMed]
98. Esterhuyse, A.J.; Toit, E.D.; Rooyen, J.V. Dietary red palm oil supplementation protects against the consequences of global ischemia in the isolated perfused rat heart. *Asia Pac. J. Clin. Nutr.* **2005**, *14*, 340–347. [PubMed]
99. Russwurm, M.; Russwurm, C.; Koesling, D.; Mergia, E. NO/cGMP: The past, the present, and the future. *Methods Mol. Biol.* **2013**, *1020*, 1–16. [CrossRef]
100. Das, S.; Mukherjee, S.; Lekli, I.; Gurusamy, N.; Bardhan, J.; Raychoudhury, U.; Chakravarty, R.; Banerji, S.; Knowlton, A.A.; Das, D.K. Tocotrienols confer resistance to ischemia in hypercholesterolemic hearts: Insight with genomics. *Mol. Cell. Biochem.* **2012**, *360*, 35–45. [CrossRef]
101. da Costa, R.M.; Rodrigues, D.; Pereira, C.A.; Silva, J.F.; Alves, J.V.; Lobato, N.S.; Tostes, R.C. Nrf2 as a Potential Mediator of Cardiovascular Risk in Metabolic Diseases. *Front. Pharm.* **2019**, *10*, 382. [CrossRef]

102. Sesso, H.D.; Buring, J.E.; Christen, W.G.; Kurth, T.; Belanger, C.; MacFadyen, J.; Bubes, V.; Manson, J.E.; Glynn, R.J.; Gaziano, J.M. Vitamins E and C in the prevention of cardiovascular disease in men: The Physicians' Health Study II randomized controlled trial. *JAMA* **2008**, *300*, 2123–2133. [CrossRef] [PubMed]
103. Lee, I.M.; Cook, N.R.; Gaziano, J.M.; Gordon, D.; Ridker, P.M.; Manson, J.E.; Hennekens, C.H.; Buring, J.E. Vitamin E in the primary prevention of cardiovascular disease and cancer: The Women's Health Study: A randomized controlled trial. *JAMA* **2005**, *294*, 56–65. [CrossRef] [PubMed]
104. Ima-Nirwana, S.; Nurshazwani, Y.; Nazrun, A.S.; Norliza, M.; Norazlina, M. Subacute and Subchronic Toxicity Studies of Palm Vitamin E in Mice. *J. Pharmacol. Toxicol.* **2011**, *6*, 166–173. [CrossRef]
105. Nair, A.B.; Jacob, S. A simple practice guide for dose conversion between animals and human. *J. Basic Clin. Pharm.* **2016**, *7*, 27–31. [CrossRef] [PubMed]
106. Slivka, A.; Rink, C.; Paoletto, D.; Sen, C.K. Platelet function in stroke/transient ischemic attack patients treated with tocotrienol. *FASEB J.* **2020**, *34*, 11838–11843. [CrossRef]
107. Norazlina, M.; Norzalyana, N.; Ima-Nirwana, S. The toxic effects of palm vitamin e on the reproductive system of female mice. *J. Toxicol. Res.* **2013**, *3*, 25–28.
108. Shi, M.; Chen, X.; Li, H.; Zheng, L. Delta-tocotrienol suppresses the migration and angiogenesis of trophoblasts in preeclampsia and promotes their apoptosis via miR-429/ ZEB1 axis. *Bioengineered* **2021**, *12*, 1861–1873. [CrossRef]

cells MDPI

Article

Glucose Transporter 9 (GLUT9) Plays an Important Role in the Placental Uric Acid Transport System

Benjamin P. Lüscher [1,2], Christiane Albrecht [3,4], Bruno Stieger [4,5], Daniel V. Surbek [1,2] and Marc U. Baumann [1,2,*]

1 Department of Obstetrics and Gynecology, University Hospital of Bern, University of Bern, CH-3010 Bern, Switzerland; l_beni@hotmail.com (B.P.L.); daniel.surbek@insel.ch (D.V.S.)
2 Department of Biomedical Research, University Hospital of Bern, University of Bern, CH-3010 Bern, Switzerland
3 Institute of Biochemistry and Molecular Medicine, University of Bern, CH-3012 Bern, Switzerland; christiane.albrecht@ibmm.unibe.ch
4 Swiss National Center of Competence in Research, NCCR TransCure, University of Bern, CH-3012 Bern, Switzerland; bruno.stieger@uzh.ch
5 Department of Clinical Pharmacology and Toxicology, University Hospital of Zürich, University of Zurich, CH-8091 Zurich, Switzerland
* Correspondence: marc.baumann@insel.ch; Tel.: +41-(0)31-632-10-10; Fax: +41-31-632-12-27

Abstract: Background: Hyperuricemia is a common laboratory finding in pregnant women compromised by preeclampsia. A growing body of evidence suggests that uric acid is involved in the pathogenesis of preeclampsia. Glucose transporter 9 (GLUT9) is a high-capacity uric acid transporter. The aim of this study was to investigate the placental uric acid transport system, and to identify the (sub-) cellular localization of GLUT9. Methods: Specific antibodies against GLUT9a and GLUT9b isoforms were raised, and human villous (placental) tissue was immunohistochemically stained. A systemic GLUT9 knockout (G9KO) mouse model was used to assess the placental uric acid transport capacity by measurements of uric acid serum levels in the fetal and maternal circulation. Results: GLUT9a and GLUT9b co-localized with the villous (apical) membrane, but not with the basal membrane, of the syncytiotrophoblast. Fetal and maternal uric acid serum levels were closely correlated. G9KO fetuses showed substantially higher uric acid serum concentrations than their mothers. Conclusions: These findings demonstrate that the placenta efficiently maintains uric acid homeostasis, and that GLUT9 plays a key role in the placental uric acid transport system, at least in this murine model. Further studies investigating the role of the placental uric acid transport system in preeclampsia are eagerly needed.

Keywords: glucose transporter 9; GLUT9; uric acid; preeclampsia

Citation: Lüscher, B.P.; Albrecht, C.; Stieger, B.; Surbek, D.V.; Baumann, M.U. Glucose Transporter 9 (GLUT9) Plays an Important Role in the Placental Uric Acid Transport System. *Cells* **2022**, *11*, 633. https://doi.org/10.3390/cells11040633

Academic Editors: Mark W. Cunningham, Junie P. Warrington, Ana T. Palei and Lorena M. Amaral

Received: 1 January 2022
Accepted: 8 February 2022
Published: 11 February 2022

1. Introduction

Preeclampsia, characterized by hypertension and proteinuria during pregnancy, contributes substantially to perinatal morbidity of both the mother and her child. There is a growing body of evidence that indicates that uric acid plays a role in the pathogenesis of preeclampsia. [1,2] Uric acid is the final metabolic product of purine metabolism in humans and great apes. The plasma uric acid concentration in humans is higher (180–420 µmol/L) than in other mammalian species (30–120 µmol/L) [3], which is due to the mutational silencing of the liver enzyme uricase. Uric acid levels are regulated within relatively tight limits. Although it is a potent antioxidant, its limited solubility leads to metabolic disease such as diabetes, gout, kidney stone disease and hypertension, already with small increases above normal levels [4–9]. Plasma urate levels decrease in the first trimester of pregnancy by at least 25%, return to normal levels in the second trimester and then increase towards the end of pregnancy [10]. In pregnancies complicated by preeclampsia, elevated uric acid serum levels are commonly observed [1]. An elevated uric acid serum concentration during early pregnancy is considered to be an independent risk factor of gestational hypertension and preeclampsia [11]. Of note, several studies have shown a correlation between the severity of the disease, fetal outcome and uric acid serum levels [12–15].

The exact mechanisms of the transplacental uric acid transport system remain unclear. Glucose transporter 9 (GLUT9, also known as SLC2A9) is a member of the glucose transporter family. Its sequence is similar to other members of the GLUT family. However, the glucose transport activity of GLUT9 is low, although GLUT9 was shown to be a high-capacity urate transporter [16]. There are two splice variants, GLUT9a and GLUT9b, which differ in the N-terminal domain [17,18]. GLUT9a consists of 12 exons with a length of 540 amino acids, while GLUT9b consists of 13 exons with a total length of 512 amino acids.

The aim of our study was to investigate the placental uric acid transport system using immunohistochemical staining to localize GLUT9a and GLUT9b, and to analyze the transplacental transport capacity in vivo using a murine knockout model. Hence, we used a systemic GLUT9 knockout (G9KO) mouse model. We hypothesized that homozygous knockout fetuses lacking GLUT9 in the liver, as well as in their placentae, would develop hyperuricemia.

2. Materials and Methods

2.1. Placental Tissue

Placentae from normal pregnancies were collected following elective caesarean sections. The exclusion criteria were fetal anomalies, intrauterine growth restriction, diabetes, (pregnancy-induced) hypertension, anemia, infectious disease, drug use or other medical or obstetric complications. Written informed consent was obtained using a protocol approved by the ethical committee of the Canton of Berne, CH-3010 Bern, Switzerland IRB#178/03, approved on 18 November 2013.

2.2. Antibodies

For generating polyclonal antibodies against GLUT9a and GLUT9b, peptides coupled C-terminally to keyhole limpet hemocyanin (KLH) were obtained from PolyPeptide Laboratoire France, Strasbourg, France. The following peptides were used: GLUT9a: MARKQNRNSKELGLVC; GLUT9b: MKLSKKDRGEDEESDC. Rabbit antisera were generated using a protocol described previously [19]. The animal experiments were approved by the local authority supervising animal studies.

2.3. Immunofluorescence

Human embryonic kidney cells (HEK-293) were cultured in minimal essential medium (Invitrogen) containing 10% heat-inactivated fetal bovine serum albumin (FBS), 50 units/mL penicillin and 50 μL/mL streptomycin. The cells were plated on 30 mm dishes and transfected with a total amount of 1 μg of DNA/dish and 3 μL lipofectamin 2000/dish (Invitrogen). The cells were fixed on glass cover slips in ice-cold methanol (−20 °C) for 20 s. Following washing with phosphate-buffered saline (PBS) (Inselspital, CH-3010 Bern, Switzerland), the cells were incubated in the presence of the primary antibodies, i.e., rabbit anti-hGLUT9a or −9b antibodies (1:500 diluted in PBS + 0.5% Bovine serum albumin (BSA; Sigma, Switzerland)), for 30 min at 4 °C. After washing three times with ice-cold PBS + 0.5% (*w*/*v*) BSA (PBS/BSA), the cells were incubated with the secondary antibody Alexa fluor 488 goat anti-rabbit IgG (H + L) A11008 (Cross-Adsorbed Secondary Antibody, Alexa Fluor 488 from Thermo Fisher Scientific, catalog # A-11008, RRID AB_143165 Invitrogen, Life Technologies Europe B.V., CH-6300 Zug, Switzerland), for 40 min at 4 °C (1:300 in PBS/BSA). After washing three times with ice-cold PBS/BSA, the slides were analyzed using a LEICA DM6000B microscope.

2.4. Immunohistochemistry

The placentae were cut and washed in cold PBS. Villous tissue was dissected into 5–10 mm fragments for histological examination. Villous tissue was fixed in a formaldehyde solution (4% (*v*/*v*); Merck, Whitehouse Station, NJ, USA) for 2–4 h at room temperature (RT), followed by 4 °C for a total time of 24–48 h. Fixed placentae were embedded in paraffin and sectioned into 3 μm coronal slices. After deparaffinization of the slides, the target was retrieved in a Tris-EDTA buffer (10 mM Tris Base, 1 mM EDTA solution, 0.05% (*w*/*v*) Tween 20, pH 9.0) via heat treatment in a pressure cooker for 15 min. The slides were washed in PBS and Tween 20 0.1% and blocked with goat serum 10% (*v*/*v*) and BSA 1% (*w*/*v*) in PBS. Human GLUT9 isoforms were detected with rabbit anti-hGLUT9a and −9b antisera (see above) and the Dako Cytomation EnVision System-HRP (DAKO, Glostrup, Denmark). The slides were washed in PBS and Tween 20 0.1% (2 × 5 min) and incubated with the endogenous peroxidase block solution for 15 min at room temperature. A peroxidase-labeled polymer was applied to the slides for 30 min at RT, followed by 3 washes in PBS (5 min each) and the addition of 3,3′-diaminobenzidine in chromogen solution in buffer substrate for 10–30 min, according to the manufacturer's instructions. Slides were rinsed in H_2O, counterstained with hematoxylin (Sigma-Aldrich, St. Louis, MO, USA) for 2 min, then rinsed with tap water for 1 min, dehydrated in a series of ethanol baths (70%, 95%, 100%, *v*/*v*) and xylene and mounted with Eukitt (Sigma-Aldrich, St. Louis, MO, USA). A negative control performed in the absence of the primary antibody revealed no detectable signal (Supplementary Materials: Figure S2).

2.5. G9KO Mice

G9KO mice were a generous gift from Bernard Thorens (Center for Integrative Genomics, University of Lausanne, Lausanne, Switzerland) and have been characterized previously [20]. Mice were kept in colonies of 5 mice per cage with 12 h day/night cycles and free access to food and water. Female mice were prepared for mating by placing the animals in a cage where male animals had been kept previously, 3 days ahead of mating. For mating, 2 female animals were placed together with 1 male animal. On the next day at 7:30 AM, the animals were separated, and females were checked for plaque formation. Plaque detection was defined as being on day 0.5 of gestation. The mating of heterozygous GLUT9+/− animals resulted in a typical Mendelian off-spring ratio of 25:50:25 for (wild type):(heterozygous):(knockout).

Heterozygous animals were crossbred to obtain wild-type (WT), which also served as controls, and knockout (KO) pups within the same litter. Maternal mice were fed with either standard chow or standard chow +1 g/kg inosine after mating. At day 18.5 after mating, maternal mice were sacrificed, and uric acid serum levels were measured. Fetuses from these animals were isolated and blood was collected via decapitation for uric acid measurements. Plasma uric acid was analyzed using a Roche/Hitachi 902 robot system (Roche); no dilution of fetal sera was needed.

2.6. Statistical Analysis

The investigators performing the statistical analysis were blinded. Data are expressed as the mean ± SEM. Graph prism software was used for statistical analysis. Differences between means were tested using Student's *t*-test or by linear regression analysis where appropriate. A *p*-value of <0.05 was considered to be significant.

3. Results

3.1. Localization of GLUT9 Isoforms in Human Placenta

Villous (placental) tissue was embedded in paraffin and histological sections were prepared. Immunohistochemical staining against GLUT9b revealed a distinct signal co-localizing with the microvillous (apical) membrane of syncytiotrophoblasts. An analogous staining procedure using rabbit anti-GLUT9a antibodies also demonstrated the presence of hGLUT9a in microvillous membranes. Neither GLUT9a- nor GLUT9b was detectable in the basolateral membrane of syncytiotrophoblasts (Figure 1).

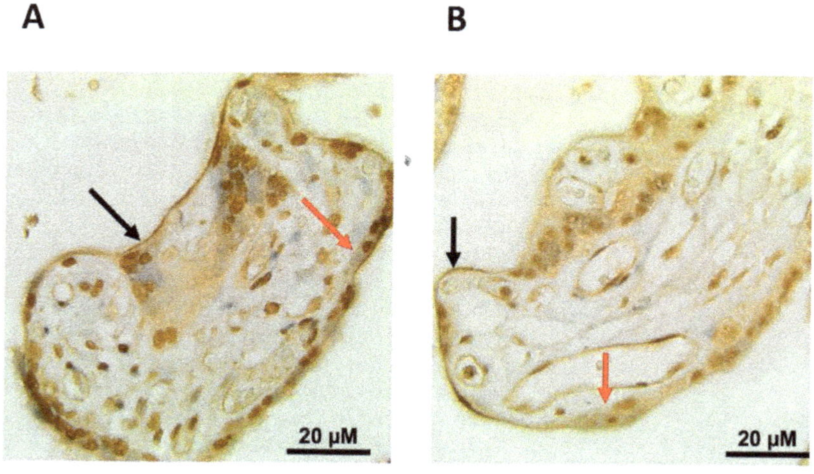

Figure 1. Immunohistological stainings of placental GLUT9a and GLUT9b. The black arrow marks the microvillus (apical) membrane, and the red arrow depicts the basal membrane of the syncytiotrophoblast in placental villi. Scale bar represents 20 μm. (**A**) Staining of a term placenta using anti-GLUT9a antibodies. (**B**) Staining of a term placenta using anti-GLUT9b antibodies. Both antibodies co-localize with the villous (apical) membrane of the syncytiotrophoblast, while no immunohistological signal is detectable in the basal membrane of the syncytiotrophoblast.

3.2. Transplacental Uric Acid Transport

To investigate the transplacental uric acid transport system, different models covering a wide range of maternal and fetal uric acid serum concentrations were used. In WT fetuses and in their mothers, uric acid levels were not different (n = 7, 27.43 ± 2.52 versus 28.25 ± 2.98 μmol/L, $p = 0.70$, NS, paired Student's t-test). The serum levels in the fetal and maternal circulation ranging from 22 to 38 and 24 to 37 μmol/L, respectively, correlated very closely (n = 7, $p > 0.001$, $r^2 = 0.95$, Figure 2A). To increase the severity of fetal hyperuricemia, maternal mice were challenged by a chow diet and inosine supplementation. Similarly, uric acid levels of WT fetuses and their mothers were not different (n = 7, 47.14 ± 6.60 versus 48.14 ± 4.85 μmol/L, $p = 0.74$, NS, paired Student's t-test). The serum levels in the fetal and maternal circulation ranging from 29 to 65 and 32 to 63 μmol/L, respectively, correlated very closely (n = 7, $p > 0.01$, $r^2 = 0.85$, Figure 2B). In contrast, under a normal chow diet, G9KO fetuses showed 4.95-fold higher uric acid serum levels than their mothers (n = 7, 142 ± 6.12 versus 29 ± 2.19 μmol/L, $p < 0.001$, paired Student's t-test). Fetal serum levels correlated with those found in the maternal circulation (n = 7, ranging from 123 to 169 and 24 to 37 μmol/L, respectively, $p < 0.005$, NS, $r^2 = 0.86$, Figure 3A). When challenged with inosine supplementation, WT fetuses showed 1.72-fold higher uric acid serum concentrations than fetuses from mothers under a normal chow diet (n = 7, 47.12 ± 6.60 versus 27.43 ± 2.52, $p < 0.05$, paired Student's t-test). Serum levels in the fetal and maternal circulation ranging from 28 to 65 and 32 to 63 mmol/L, respectively, correlated closely ($p < 0.01$, $r^2 = 0.85$, Figure 3B). Upon inosine supplementation, G9KO fetuses showed 4.10-fold higher uric acid serum levels than their mothers (n = 7, 192.88 ± 12.44 versus 47 ± 4.20 μmol/L, $p < 0.005$, paired Student's t-test). Fetal serum levels did not correlate with those found in the maternal circulation (n = 7, ranging from 155 to 259 and 32 to 63 μmol/L, respectively, $p = 0.12$, NS, $r^2 = 0.35$, Figure 3B).

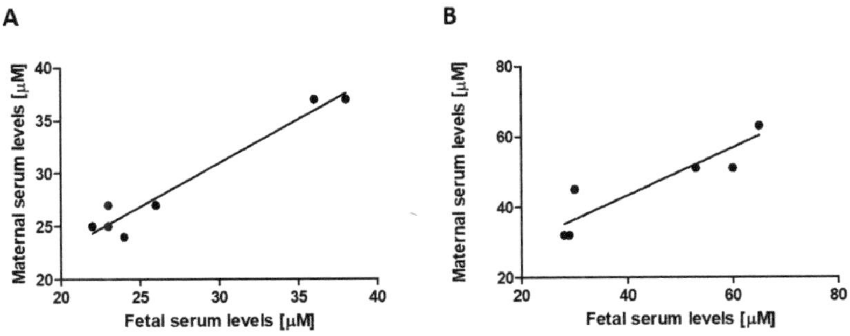

Figure 2. Correlation between maternal and (wild-type) fetal uric acid levels: (**A**) normal diet, and (**B**) diet with inosine supplementation.

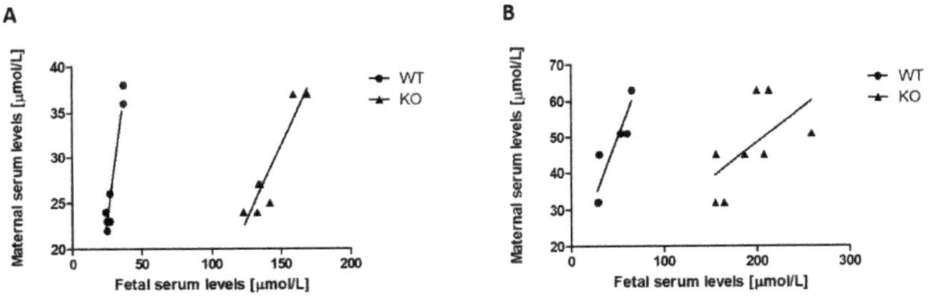

Figure 3. Correlation between maternal and (G9KO) fetal uric acid levels: (**A**) normal diet, and (**B**) diet with inosine supplementation.

4. Discussion

Hyperuricemia is a common laboratory finding in pregnant women compromised by preeclampsia. This circumstance is usually considered secondary to altered maternal kidney function. Hyperuricemia in preeclampsia, however, may also be explained by inflammatory and ischemic processes which subsequently increase uric acid production. A growing body of studies suggests that uric acid not only is a bystander but also plays an important role in the pathogenesis of preeclampsia [1,2]. Despite its important antioxidant function, uric acid has deleterious effects at higher concentrations or in a hypoxic environment. Several studies showed that elevated uric acid serum levels correlate with adverse maternal and perinatal outcomes [12–15]. The fetoplacental unit continuously produces uric acid which cannot be further metabolized in humans due to the mutational silencing of the liver enzyme uricase. To prevent its accumulation and its deleterious sequelae, uric acid has to be transported across the placenta into the maternal circulation, which allows renal secretion. In the proximal tubule of the kidney, GLUT9 transports uric acid across the basolateral membrane into the blood, as part of the reabsorption process, and thus it plays an important role in the homeostasis of serum uric acid levels in humans.

Knowledge regarding the placental uric acid transport system is scarce. In this study, we aimed to localize GLUT9 at different sites in the placenta. Immunohistochemical analysis and Western blotting initially failed due to the poor quality of commercially available antibodies against GLUT9. Thus, we raised specific antibodies against GLUT9a and GLUT9b in rabbits using N-terminal peptides of the GLUT9a and GLUT9b proteins. Immunohistochemical staining against placental GLUT9a and GLUT9b revealed distinct signals co-localizing with the microvillus (apical) membrane of syncytiotrophoblasts. Neither GLUT9a nor GLUT9b was detectable in the basal membrane of syncytiotrophoblasts. The simultaneous presence of GLUT9a and GLUT9b in the microvillus (apical) membrane of syncytiotrophoblasts may facilitate uric acid transportation from the fetoplacental unit into the maternal circulation, which is crucial for fetal well-being and survival. Our group has previously demonstrated that, in contrast to GLUT9b, GLUT9a is regulated by iodine [21]. The question of whether the localization of placental GLUT9a, i.e., on the microvillus membrane, is important for the regulation of transplacental uric acid transport capacity remains to be solved. Interestingly, we found both GLUT9 isoforms to be exclusively expressed in the microvillus (apical) membrane facing the maternal side, but not in basolateral membranes, of syncytiotrophoblasts. This is in contrast to the localization patterns of other cell types such as renal tubular cells, where GLUT9a and -9b are expressed in opposite plasma membrane domains. In tubular kidney cells, GLUT9a is expressed on the basal side, whereas GLUT9b was found to be expressed exclusively on apical membranes [22]. The distinct expression pattern in renal tubular cells ensures that GLUT9b reabsorbs uric acid from the urine, and that GLUT9a transports it across the basal membrane back into the circulation, thus maintaining uric acid homeostasis. Both GLUT9a and GLUT9b isoforms are high-capacity uric acid transporters. They are differentially expressed along the nephron but share similar transport capacities [16]. Since the transport characteristics of GLUT9a- and GLUT9b-mediated uric acid transport are congruent, the presence or absence of these isoforms in subcellular localization has an impact on transport capacity. Further, the presence of GLUT9a endorses the potential to regulate uric acid transport via iodine. Whether these regulatory mechanisms play a role in renal tubular cell re-uptake or in transplacental uric acid transport remains to be elucidated. The syncytiotrophoblast shows a specific expression pattern which, in turn, enables a unique transport pathway. The organic anion transporter-4 (OAT4) is another important placental uric acid transporter, which was shown to be expressed in the basal membranes of the syncytiotrophoblast. However, unlike GLUT9, OAT4 is unidirectional and only transports uric acid into the cell [23]. Our results suggest that uric acid is transported exclusively in the feto–maternal (but not in the materno–fetal) direction, which may help to protect the fetus from deleterious effects due to uric acid accumulation in the placenta.

To evaluate the placental potential to warrant uric acid homeostasis, we analyzed maternal and fetal uric acid serum levels in mice. We demonstrated that uric acid serum levels in the fetal and in the maternal circulation correlated very closely, even when uric acid levels were raised to supra-physiological levels by a diet with inosine supplementation. This observation highlights the ability of the placenta to efficiently maintain uric acid homeostasis between the maternal and fetal compartments using its uric acid transport system, in order to transport uric acid from the fetoplacental unit into the maternal circulation where it can be finally excreted by the kidneys and the intestines.

To assess the role of GLUT9 in the placental uric acid transport system, we used a murine G9KO model. In wild-type fetal mice, uric acid is transported by GLUT9 (which is expressed on hepatic cells) into these cells and, in turn, is further metabolized into allantoin by the liver enzyme uricase. Another pathway to prevent uric acid accumulation in these animals is transplacental uric acid transportation via GLUT9. Homozygous G9KO fetuses also lacking liver GLUT9, which prevents access to hepatic uricase, will develop hyperuricemia. Interestingly, these animals show exceedingly high uric acid serum levels. This finding indicates that the lack of placental GLUT9 is responsible for the inability to maintain placental uric acid homeostasis, and that GLUT9 is—at least in our animal model—the only relevant placental uric acid transporter. Our study has some limitations. First, due to the relatively small number of animals investigated, similar experiments with a larger sample size would corroborate our results. Second, data obtained from animal experiments cannot necessarily be extrapolated to human diseases.

Pro-inflammatory stimuli such as elevated uric acid levels during early pregnancy may facilitate the development of preeclampsia [24]. Further studies investigating the role of the placental uric acid transport system in preeclampsia are eagerly needed. These novel insights into the placental and renal uric acid transport systems may enable the development of therapeutic and preventive strategies for hyperuricemia-related diseases such as preeclampsia.

Supplementary Materials: The following supporting information can be downloaded at: https://www.mdpi.com/article/10.3390/cells11040633/s1, Figure S1: Test of GLUT9-specific antibodies using transfected HEK-293 cells.; Figure S2: Negative control of GLUT9-specific antibodies with term placentae.

Author Contributions: Conceptualization, B.P.L., D.V.S. and M.U.B.; methodology, D.V.S. and M.U.B.; software, B.P.L. and M.U.B.; validation, M.U.B.; formal analysis, B.P.L., M.U.B., C.A. and B.S.; investigation, B.P.L., C.A. and B.S.; resources, M.U.B. and D.V.S.; data curation, B.P.L. and M.U.B.; writing—original draft preparation, B.P.L., M.U.B. and D.V.S.; writing—review and editing, B.P.L. and M.U.B.; visualization, B.P.L. and M.U.B.; supervision, D.V.S. and M.U.B.; project administration, M.U.B.; funding acquisition, D.V.S. and M.U.B. All authors have read and agreed to the published version of the manuscript.

Funding: This study was financially supported by the Swiss National Science Foundation via the National Center of Competence in Research (NCCR) TransCure, University of Bern, CH-3012 Bern, Switzerland (Grant No. 35900 and Grant No. 51NF40-185544).

Institutional Review Board Statement: This study was conducted in accordance with the Declaration of Helsinki and approved by the Institutional Review Board (or Ethics Committee) of the Canton of Berne, Switzerland IRB#178/03, approved on 18 November 2013.

Informed Consent Statement: Informed consent was obtained from all subjects involved in the study.

Data Availability Statement: All data can be provided by the authors on request.

Acknowledgments: We would like to thank Bernard Thorens (Center for Integrative Genomics, University of Lausanne, Lausanne, Switzerland) who provided G9KO mice as a generous gift.

Conflicts of Interest: The funders had no role in the design of the study; in the collection, analyses or interpretation of data; in the writing of the manuscript; or in the decision to publish the results.

References

1. Bainbridge, S.A.; Roberts, J.M. Uric acid as a pathogenic factor in preeclampsia. *Placenta* **2008**, *29*, 67–72. [CrossRef]
2. Martin, A.C.; Brown, M.A. Could uric acid have a pathogenic role in pre-eclampsia? *Nat. Rev. Nephrol.* **2010**, *6*, 744–748. [CrossRef]
3. Hediger, M.A.; Johnson, R.J.; Miyazaki, H.; Endou, H. Molecular physiology of urate transport. *Physiology* **2005**, *20*, 125–133. [CrossRef]
4. Bo, S.; Cavallo-Perin, P.; Gentile, L.; Repetti, E.; Pagano, G. Hypouricemia and hyperuricemia in type 2 diabetes: Two different phenotypes. *Eur. J. Clin. Investig.* **2001**, *31*, 318–321. [CrossRef]
5. Feig, D.I.; Kang, D.H.; Nakagawa, T.; Mazzali, M.; Johnson, R.J. Uric acid and hypertension. *Curr. Hypertens. Rep.* **2006**, *8*, 111–115. [CrossRef]
6. Hjortnaes, J.; Algra, A.; Olijhoek, J.; Huisman, M.; Jacobs, J.; van der Graaf, Y.; Visseren, F. Serum uric acid levels and risk for vascular diseases in patients with metabolic syndrome. *J. Rheumatol.* **2007**, *34*, 1882–1887.
7. Onat, A.; Uyarel, H.; Hergenç, G.; Karabulut, A.; Albayrak, S.; Sarı, I.; Yazıcı, M.; Keleş, I. Serum uric acid is a determinant of metabolic syndrome in a population-based study. *Am. J. Hypertens.* **2006**, *19*, 1055–1062. [CrossRef]
8. Panoulas, V.F.; Milionis, H.J.; Douglas, K.M.; Nightingale, P.; Kita, M.D.; Klocke, R.; Elisaf, M.S.; Kitas, G.D. Association of serum uric acid with cardiovascular disease in rheumatoid arthritis. *Rheumatology* **2007**, *46*, 1466–1470. [CrossRef]
9. Rathmann, W.; Haastert, B.; Icks, A.; Giani, G.; Roseman, J.M. Ten-year change in serum uric acid and its relation to changes in other metabolic risk factors in young black and white adults: The CARDIA study. *Eur. J. Epidemiol.* **2007**, *22*, 439–445. [CrossRef]
10. Powers, R.W.; Bodnar, L.M.; Ness, R.B.; Cooper, K.M.; Gallaher, M.J.; Frank, M.P.; Daftary, A.R.; Roberts, J.M. Uric acid concentrations in early pregnancy among preeclamptic women with gestational hyperuricemia at delivery. *Am. J. Obstet. Gynecol.* **2006**, *194*, 160.e1–160.e8. [CrossRef]
11. Chen, Y.; Ou, W.; Lin, D.; Lin, M.; Huang, X.; Ni, S.; Chen, S.; Yong, J.; O'Gara, M.C.; Tan, X.; et al. Increased Uric Acid, Gamma-Glutamyl Transpeptidase and Alkaline Phosphatase in Early-Pregnancy Associated With the Development of Gestational Hypertension and Preeclampsia. *Front. Cardiovasc. Med.* **2021**, *8*, 756140. [CrossRef]
12. Voto, L.S.; Illia, R.; Darbon-Grosso, H.A.; Imaz, F.U.; Margulies, M. Uric acid levels: A useful index of the severity of preeclampsia and perinatal prognosis. *J. Perinat. Med.* **1988**, *16*, 123–126. [CrossRef]
13. Roberts, J.M.; Bodnar, L.M.; Lain, K.Y.; Hubel, C.A.; Markovic, N.; Ness, R.B.; Powers, R.W. Uric acid is as important as proteinuria in identifying fetal risk in women with gestational hypertension. *Hypertension* **2005**, *46*, 1263–1269. [CrossRef]
14. Wu, Y.; Xiong, X.; Fraser, W.D.; Luo, Z.C. Association of uric acid with progression to preeclampsia and development of adverse conditions in gestational hypertensive pregnancies. *Am. J. Hypertens.* **2012**, *25*, 711–717. [CrossRef]
15. Livingston, J.R.; Payne, B.; Brown, M.; Roberts, J.M.; Côté, A.M.; Magee, L.A.; von Dadelszen, P.; PIERS Study Group. Uric Acid as a predictor of adverse maternal and perinatal outcomes in women hospitalized with preeclampsia. *J. Obstet. Gynaecol. Can.* **2014**, *36*, 870–877. [CrossRef]

16. Bibert, S.; Hess, S.K.; Firsov, D.; Thorens, B.; Geering, K.; Horisberger, J.D.; Bonny, O. Mouse GLUT9: Evidences for a urate uniporter. *Am. J. Physiol.-Ren. Physiol.* **2009**, *297*, F612–F619. [CrossRef]
17. Augustin, R.; Carayannopoulos, M.O.; Dowd, L.O.; Phay, J.E.; Moley, J.F.; Moley, K.H. Identification and characterization of human glucose transporter-like protein-9 (GLUT9): Alternative splicing alters trafficking. *J. Biol. Chem.* **2004**, *279*, 16229–16236. [CrossRef]
18. Phay, J.E.; Hussain, H.B.; Moley, J.F. Cloning and expression analysis of a novel member of the facilitative glucose transporter family, SLC2A9 (GLUT9). *Genomics* **2000**, *66*, 217–220. [CrossRef]
19. Stieger, B.; Hagenbuch, B.; Landmann, L.; Höchli, M.; Schroeder, A.; Meier, P.J. In situ localization of the hepatocytic Na^+/Taurocholate cotransporting polypeptide in rat liver. *Gastroenterology* **1994**, *107*, 1781–1787. [CrossRef]
20. Preitner, F.; Bonny, O.; Laverrière, A.; Rotman, S.; Firsov, D.; da Costa, A.; Metref, S.; Thorens, B. Glut9 is a major regulator of urate homeostasis and its genetic inactivation induces hyperuricosuria and urate nephropathy. *Proc. Natl. Acad. Sci. USA* **2009**, *106*, 15501–15506. [CrossRef]
21. Lüscher, B.; Surbek, D.; Clemençon, B.; Huang, X.; Albrecht, C.; Marini, C.; Hediger, M.; Baumann, M.U. Different Pharmacological Properties of GLUT9a and GLUT9b: Potential Implications in Preeclampsia. *Cell. Physiol. Biochem.* **2019**, *53*, 508–517. [PubMed]
22. Kimura, T.; Takahashi, M.; Yan, K.; Sakurai, H. Expression of SLC2A9 isoforms in the kidney and their localization in polarized epithelial cells. *PLoS ONE* **2014**, *9*, e84996. [CrossRef]
23. Hagos, Y.; Stein, D.; Ugele, B.; Burckhardt, G.; Bahn, A. Human renal organic anion transporter 4 operates as an asymmetric urate transporter. *J. Am. Soc. Nephrol.* **2007**, *18*, 430–439. [CrossRef] [PubMed]
24. Staff, A.C.; Johnsen, G.M.; Dechend, R.; Redman, C.W.G. Preeclampsia and uteroplacental acute atherosis: Immune and inflammatory factors. *J. Reprod. Immunol.* **2014**, *101–102*, 120–126. [CrossRef] [PubMed]

MDPI
St. Alban-Anlage 66
4052 Basel
Switzerland
Tel. +41 61 683 77 34
Fax +41 61 302 89 18
www.mdpi.com

Cells Editorial Office
E-mail: cells@mdpi.com
www.mdpi.com/journal/cells

www.ingramcontent.com/pod-product-compliance
Lightning Source LLC
Chambersburg PA
CBHW061614220926
51850CB00005B/509
9783036564111